This book aims to give a comprehensive survey of the chemical and physical parameters governing proton conduction. It includes descriptions of the preparation, structures and properties of typical materials (glasses, crystals, ceramics, metals, organic and inorganic polymers) and of devices.

Chemistry of Solid State Materials

Proton conductors
Solids, membranes and gels – materials and devices

Chemistry of Solid State Materials

Series Editors
A. R. West, Dept of Chemistry, University of Aberdeen
H. Baxter, formerly at the Laboratory of the Government Chemist, London

Proton conductors

Solids, membranes and gels – materials and devices

Edited by
Philippe Colomban
Laboratoire de Physique de la Matière Condensée,
Centre National de la Recherche Scientifique,
Ecole Polytechnique, Palaiseau, France

CAMBRIDGE
UNIVERSITY PRESS

CAMBRIDGE UNIVERSITY PRESS
Cambridge, New York, Melbourne, Madrid, Cape Town, Singapore, São Paulo

Cambridge University Press
The Edinburgh Building, Cambridge CB2 8RU, UK

Published in the United States of America by Cambridge University Press, New York

www.cambridge.org
Information on this title: www.cambridge.org/9780521383172

First published 1992
This digitally printed version 2008

A catalogue record for this publication is available from the British Library

Library of Congress Cataloguing in Publication data

Proton conductors: solids, membranes, and gels: materials and
devices/edited by Philippe Colomban.
   p. cm.—(Chemistry of solid state materials)
Includes bibliographical references and index.
ISBN 0-521-38317-X
1. Energy band theory of metals.   2. Protons.   3. Electric
conductors.   4. Solid state physics.   5. Solid state chemistry.
I. Colomban, Philippe.   II. Series.
QC176.8.E4P768   1992
530.4'16—dc20 91-43012 CIP

ISBN 978-0-521-38317-2 hardback
ISBN 978-0-521-07890-0 paperback

Contents

Contents

Contents

Contents

Contents

Contents

Contents

Contents

Contents

Contents

Contributors

Prof. G. Alberti
Dipartimento di Chimica, Laboratorio di Chimica Inorganica, Universita di Perugia, via Elce di Sotto, 8, 06100 Perugia, Italy

Dr J. C. Badot
Laboratoire de Chimie de la Matière Condensée, UA 302 CNRS, ENSCP, 11 rue P. et M. Curie, 75005 Paris, France

Prof. S. V. Bhat
Department of Physics, Indian Institute of Science, Bangalore 560012, India

Dr O. Bohnke
Laboratoire d'Electrochimie des Solides, UA 436 CNRS, Université de Franche-Comté, 25030 Besançon, France

Dr M. Casciola
Dipartimento di Chimica, Laboratorio di Chimica Inorganica, Universita di Perugia, via Elce di Sotto, 8, 06100 Perugia, Italy

Dr A. M. Chippindale
Inorganic Chemistry Laboratory, University of Oxford, South Parks Road, Oxford OX1 3QR, UK

Prof. A. Clearfield
Department of Chemistry, Texas A&M University, College Station, Austin, Texas 77843, USA

Dr Ph. Colomban
Laboratoire de Physique de la Matière Condensée, UA1254 CNRS, Ecole Polytechnique, 91128 Palaiseau, France
and
ONERA, BP72, 92322 Chatillon, France

Dr J. Conard
CRSOI-CNRS, 1B rue de la Férollerie, 45045 Orléans Cedex, France

Contributors

Dr F. Devreux
Laboratoire de Physique de la Matière Condensée, UA1254 CNRS, Ecole Polytechnique, 91128 Palaiseau, France

Dr P. G. Dickens
Inorganic Chemistry Laboratory, University of Oxford, South Parks Road, Oxford OX1 3QR, UK

Prof. R. Frech
Department of Chemistry, University of Oklahoma, Norman, Oklahoma 73019, USA

Dr F. Freund
Department of Physics, San José State University, San José, CA 95192, USA

Dr H. Fuzellier
Laboratoire de Chimie Minérale Appliquée, Université de Nancy 1, BP 239, 54506 Vandoeuvre-les-Nancy, France

Dr C. Gowach
Laboratoire de Physico-Chimie des Systèmes Polyphasés-UA 330 CNRS, Route de Mende, BP 5051, 34033 Montpellier Cedex, France

Dr J. Guitton
Laboratoire d'Ionique et d'Electrochimie du Solide, INPG-ENSEEG, BP 75, 38402 St Martin d'Hères, France

Prof. H. Ikawa
Tokyo Institute of Technology, Faculty of Engineering, Department of Inorganic Materials, O-Okayama, Meguro-ku, Tokyo 152, Japan

Prof. H. Iwahara
Synthetic Crystal Research Laboratory, Faculty of Engineering, Nagoya University, Furo-Cho, Chikusa-Ku, Nagoya 464, Japan

Dr J.-P. Korb
Laboratoire de Physique de la Matière Condensée, UA1254 CNRS, Ecole Polytechnique, 91128 Palaiseau, France

Dr K.-D. Kreuer
Max-Planck Institut für Festkörperforschung, Heisenbergstrasse 1, 7000 Stuttgart 80, Germany
and
Conducta GmbH & Co, Dieselstrasse 24, D-7016 Gerlingen, Germany

Prof. E. Krogh Andersen
Department of Chemistry, Odense University, Campusvej 55, DK-5230 Odense M, Denmark

Contributors

Dr I. G. Krogh Andersen
Department of Chemistry, Odense University, Campusvej 55, DK-5230 Odense M, Denmark

Dr J. C. Lassègues
Laboratoire de Spectroscopie Moléculaire et Crystalline, Université de Bordeaux I, 351 Cours de la Libération, 33405 Talence, France

Prof. N. Miura
Kyushu University 39, Department of Materials Science and Technology, Graduate School of Engineering Sciences, Kasuga-shi, Fukuoka 816, Japan

Prof. P. S. Nicholson
Ceramic Engineering Research Group, Department of Materials Science and Engineering, McMaster University, 1280 Main Street West, Hamilton, Ontario, Canada L8S 4L7

Dr A. Novak
CNRS, Laboratoire de Spectrochimie IR et Raman, 2 rue Henry Dunant, 94320 Thiais, France

Dr M. Pham-Thi
Laboratoire Central de Recherche Thomson-CSF, Domaine de Corbeville, 91404 Orsay, France

Dr C. Poinsignon
Laboratoire d'Ionique et d'Electrochimie du Solide, INPG-ENSEEG, BP 75, 38402 St Martin d'Hères, France

Prof. A. Potier
Laboratoire des Acides Minéraux, Université de Montpellier, UA 79 CNRS, Place E. Bataillon, 34095 Montpellier, France

Dr G. Pourcelly
Laboratoire de Physico-Chimie des Systèmes Polyphasés-UA 330 CNRS, Route de Mende, BP 5051, 34033 Montpellier Cedex, France

Prof. J. Rozière
Laboratoire des Acides Minéraux, Université de Montpellier, UA 79 CNRS, Place E. Bataillon, 34095 Montpellier, France

Dr I. A. Ryzhkin
Institute of Solid State Physics, Academy of Science of the USSR, 142432, Chernogolovka, Moscow Distr., CIS

Dr J. Y. Sanchez
Laboratoire d'Ionique et d'Electrochimie du Solide, INPG-ENSEEG, BP 75, 38402 St Martin d'Hères, France

Contributors

Dr E. Skou
Department of Chemistry, Odense University, Campusvej 55, DK-5230 Odense M. Denmark

Dr R. C. T. Slade
Department of Chemistry, University of Exeter, Stocker Road, Exeter EX4 QD, UK

Dr I. Svare
Physics Department, University of Trondheim, Unit-NTH, Sem Saelandsvei 9, N 7034 Trondheim, Norway

Prof. J. O. Thomas
Institute of Chemistry, University of Uppsala, Box 531, 751-21 Uppsala, Sweden

Prof. N. Yamazoe
Kyushu University 39, Department of Materials Science and Technology, Graduate School of Engineering Sciences, Kasuga-shi, Fukuoka 816, Japan

Preface

Since the discovery of ice conductivity more than one hundred years ago in Japan, proton transport in solids has aroused considerable interest. Currently, proton conductors appear interesting because of protonic transport in biophysical processes, and – as with many other ionic conductors – since they can be used in numerous electrochemical devices such as batteries, fuel cells, chemical sensors, electrochromic displays and supercapacitors. Furthermore, energy systems based on hydrogen are a possible answer to prevent earth pollution. The number of materials, crystalline or amorphous, organic or inorganic, solids and gels, where proton transport is known to play an important role, has increased during the last few years. At the same time, a better understanding of proton transfer mechanism seems to have been reached. Recently, the debate on cold fusion has made most scientists aware of the proton peculiarity.

It is hardly an exaggeration to say that everything about hydrogen is unique. Mendeleev could not find a place for this element in the Periodic Table. The hydrogen ion 'H^+' is an ion without an electron, a bare proton and so, protons are usually solvated. The literature about proton transport is also unique and must be read with caution. For instance, early measurements on ice by Eigen, de Maeyer & Spatz showed a measurable ionic conductivity which was explained in proton defect terms, protons diffusing by quantum-mechanical tunnelling, and the carrier concentration could be adjusted by doping. In 1973, Von Hippel, Runch & Westphal showed that conductivity measurements were perturbed by surface and interfacial phenomena. More recently Petrenko and co-workers showed that the defect concentration close to the surface of ice crystals is higher than in the interior. This can be compared with oxide ceramics where a layer a few micrometres thick near the surface is always different from the bulk, because of a diffusion gradient. Controversy exists also in the cases of the fast proton conductors such as $H_3OUO_2PO_4 . 3H_2O$, $\alpha\ Zr(HPO_4)_2 . nH_2O$ and β''-aluminas and is linked to the prominent role of surfaces for protonic materials and the influence of water partial pressure on thermal stability. As recently as 1983 Ernsberger cast doubt on the existence of true fast protonic conduction. Nevertheless, most of the reviews on ionic diffusion do not mention protons.

This book aims to give a survey of the chemical and physical parameters concerning (fast) proton conduction, the preparation, structures and properties of typical materials, the dynamics of charge transport and some examples of devices using proton transport.

The distinction between 'conductors' and 'insulators' depends on the conductivity which can vary by more than ten orders of magnitude. The charge carriers can be either electrons (or holes) or ions (or vacancies). Ionic conductivity implies a diffusion of matter inside the solid (gel) and/or at its surface; and thus most ionic conductors are also very good ion-exchangers. Materials (oxides, nitrides, halides or polymers) can be divided into four groups according to their electrical properties.

(i) 'Insulators' with a residual conductivity lower than $10^{-10} \, \Omega^{-1}$ cm^{-1} where the electronic conduction is generally also of the same order of magnitude.

(ii) 'Ionic' conductors in which the presence of structural defects leads to an ionic conductivity between 10^{-9} and $10^{-6} \, \Omega^{-1} \, cm^{-1}$.

(iii) 'Superionic' conductors with a conductivity higher than $10^{-5} \, \Omega^{-1}$ cm^{-1} and usually between 10^{-4} and $10^{-2} \, \Omega^{-1} \, cm^{-1}$. The main difference between superionic and ionic conductors concerns the activation energy (E_a) of the ionic conductivity: for the former, which are also called fast ionic conductors or solid electrolytes, E_a is lower than 0.4 eV while values varying between 0.6 and 1 eV are usually observed for 'normal' ionic conductors. A low activation energy implies that good electrical properties are present far from the melting point. In the latter two types of materials, the electronic conductivity is usually low.

(iv) Finally, there are also mixed conductors exhibiting both good ionic and electronic (metallic or semi-conductor) conductivity. These materials are used as non-blocking electrodes.

Protonic conduction is a particular case of ionic conduction; however, the small dimension of the proton implies that there may be some similarities with electronic conduction. The existence of a bare proton is restricted to very special cases (e.g. in a plasma, solar wind, or synchrotron-ring). In condensed matter, however, the proton reacts strongly with the environment because of the absence of electron shells to shield the nucleus. In metals, the proton interacts with a conduction electron and proton transport is correlated with some elastic and inelastic displacements (polarization) of electrons and can be assisted by phonons. In electronic insulators and semi-conductors, in which the Debye length is larger than the interatomic distances, the proton penetrates the valence electron shell and directional (covalent) bonds occur: tunnelling can be expected, but fast proton diffusion is generally not possible. The proton can also interact with certain molecules (e.g. solvation by oxygen, water or ammonia) in the liquid state and multinuclear protonic species are formed, leading to 'classical' ions such as OH^-, H_3O^+ or NH_4^+. Two main mechanisms of proton transfer have been recognized: the 'vehicle mechanism', where the proton is supported by the solvated species, and the Grotthuss mechanism which involves a correlated jump of the bare proton between host molecules and correlated reorientation of the latter.

Much in the same way that the beginning of the investigation into fast ionic conduction in solids is usually ascribed to the discovery of Na^+ ion conductivity in Na β-alumina ceramics by Ford Motor Research Group, so research into fast proton conductors was initiated with the Du Pont patent on Nafion© perfluorocarbon sulphonic acid polymer ($\sigma_{300K} \approx 5 \times 10^{-2} \, \Omega^{-1} \, cm^{-1}$) in 1972, in connection with the Gemini fuel-cell program, and by Potier's group who showed proton conductivity in oxonium perchlorate in 1973. Large-scale studies with the possible applications to low-temperature fuel-cells (EEC Research Program initiated by J. Jensen) and microionic devices with the work on hydrogen uranyl phosphate, HUP (by the Leeds group) began after 1980. These investigations were followed by the discovery of high conductivity above 150 °C in stable acid sulphates by Russian groups and more recently by an increased interest in gels which are nearly always good proton conductors. Finally, earlier works and the role of mineralogists in the understanding of proton mobility in solids should be mentioned. For instance, the first notion of a 'quasi-liquid state' of ions in a solid (HUP) is due to Beintema in 1938 ('these ions may thus be considered as true vagabond ions') while Wilkins described many structures containing mobile water and oxonium ions in 1974.

Unlike metals or semi-conductors for which the concept of the surface is well-defined, the analogous concept in protonic materials corresponds to a much thicker 'layer' similar to that in ceramics: it is a zone where the atmospheric pressure in contact with the solid induces a high

Preface

concentration gradient and thus important diffusion phenomena which are accelerated at temperatures close to the melting or decomposition point. In proton conductors, it is usually not possible to have both a high conductivity and thermal stability. The latter is usually low because of the weakness of hydrogen bonds. The presence of defects in the surface layers facilitates 'absorption' of various molecular entities until a liquid phase is formed. The versatility of hydrogen bonds which can be 'stretched' from strong to weak ones helps to obtain a quasi-continuous passage from the solid to liquid state via intermediate states such as particles with covered surface or soaked porous materials.

Progress in the understanding of superionic conduction is due to the use of various advanced techniques (X-ray (neutron) diffuse scattering, Raman spectroscopy and a.c.-impedance spectroscopy) and – in the particular case of protons – neutron scattering, nuclear magnetic resonance, infrared spectroscopy and microwave dielectric relaxation appear to be the most powerful methods. A number of books about solid electrolytes published since 1976 hardly mention proton conductors and relatively few review papers, limited in scope, have appeared on this subject. Proton transfer across biological membranes has received considerable attention but is not considered here (see references for more details).

The present book contains many contributions covering very different aspects and is organized as follows.

The first part describes chemical and physical parameters necessary for fast proton conduction and proposes a classification of different kinds of proton conductors. The methods used to determine the protonic nature of the charge carriers are given. The importance of partial water pressure, the role of defects and surface phenomena are discussed.

The second part treats the chemistry, structures and electrical properties of typical materials, from hydrogen bronzes to polymers via ice, hydroxides, acid sulphates, layer hydrates, inorganic ion exchangers, gels, porous media and mixed inorganic–organic polymers. These materials are compared with liquid and molten salt conductors, intercalated graphites and metal hydrides and have been chosen in order to illustrate the different behaviour of the proton: it has 'electron-like' properties in some oxides and hydrides, ion-like behaviour in some other oxides or liquid-state behaviour such as encountered in solution covered particles or pores of a gel.

The third part discusses the approaches leading to an understanding

xxiv

of proton dynamics and the methods used to study the motions on a short-range (local), long-range and macroscopic scale are given.

The fourth part concerns interpretation of three main conductivity mechanisms: the 'electron-like' type, the proton jump in statically or dynamically disordered solids and the 'quasi-liquid' state in hydrated materials giving rise to either vehicle or 'Grotthuss' mechanism.

Finally, the last part deals with applications, in particular with high-current electrochemical systems for energy production or storage, (micro)-ionic components using insertion or blocking electrodes and the devices such as MnO_2, PbO_2 batteries where the role of protons has been neglected for a long time.

ONERA, January 1992   Ph. Colomban

References

Books on superionic conductors

(Reviews on proton conductors are indicated)

G. D. Mahan and W. L. Roth (eds), *Superionic Conductors* (Plenum Press, New York (1976)).

S. Geller (ed), *Solid Electrolytes* (Springer Verlag, Berlin (1977)).

P. Hagenmuller and W. Van Gool (eds), *Solid Electrolytes* (Academic Press, New York (1978)).

M. S. Whittingham and A. J. Jacobson (eds), *Intercalation Chemistry* (Academic Press, New York (1982)).

T. A. Wheat, A. Ahmad and A. K. Kuriakose (eds), *Progress in Solid Electrolytes* (CANMET (Energy, Mines and Resources), Ottawa (1983)) (P. J. Wiseman, *Particle Hydrates as Proton Conductors*).

T. Takahashi (ed), *High Conductivity Solid Ionic Conductors* (World Publishing Co. Singapore (1988) (F. W. Poulsen, *Proton Conduction in Solids*).

R. Laskar and S. Chandra (eds), *Recent Developments in Superionic Conductors* (Academic Press, New York (1989)) (S. Chandra, *Proton Conductors*).

Reviews on protonic conduction

J. Bruinink, Proton Migration in Solids, *J. Appl. Electrochem.* **2** (1972) 239–49.

L. Glasser, Proton Conduction and Injection in Solids, *Chem. Rev.* **75** (1975) 21–65.

J. Jensen, *Energy Storage* (Newnes–Butterworths Technical Publishers, London (1980)).

W. A. England, M. G. Cross, A. Hamnett, P. J. Wiseman and J. B. Goodenough, Fast Proton Conduction in Inorganic Ion Exchange Compounds, *Solid State Ionics*, **1** (1980) 231–49.

Preface

Ph. Colomban and A. Novak, Hydrogen containing β and β″ Aluminas, *Solid State Protonic Conductors* I. J. Jensen and M. Kleitz (eds) (Odense University Press (1982)) 153–201.

S. Chandra, Fast Proton Transport in Solids, *Mater. Sci. Forum* **1** (1984) 153–70.

K.-D. Kreuer, Fast Proton Conduction in Solids, *Proc. Electrochem. Soc.* (Manganese Dioxide Electrode Theory, Electrochem. Appl.) **85.5** (1985) 21–47.

C. I. Ratcliffe and D. E. Irish, The Nature of the Hydrated Proton. Part One: The Solid and Gaseous States, in *Water Science Review 2*, F. Franks (ed) (Cambridge University Press, Cambridge (1986)) 149–214.

Ph. Colomban and A. Novak, Proton Transfer and Superionic Conductivity in Solids and Gels, *J. Mol. Struct.* **177** (1988) 277–308.

K.-D. Kreuer, Fast Proton Transport in Solids, *J. Mol. Struct.* **177** (1988) 265–76.

A. Clearfield, Role of Ion Exchange in Solid State Chemistry, *Chem. Rev.* **88** (1988) 125–48.

Proton transfer in biological areas

J. F. Nagle and S. Tristram-Nagle, Hydrogen Bonded Chain Mechanism for Proton Conduction and Proton Pumping, *J. Membrane Biol.* **74** (1983) 1–14.

D. Hadzi, Proton Transfer in Biological Mechanisms, *J. Mol. Struct.* **177** (1988) 1–21.

L. Packe (ed), Protons and Membrane Functions, *Methods in Enzymology* **127** (1986), Academic Press.

Other references

Solid State Protonic Conductors for Fuel Cells and Monitors, I, II, III, (Odense University Press 1982, 1983, 1985) edited by J. Jensen, M. Kleitz, A. Potier and J. B. Goodenough, *IV*, edited by R. C. T. Slade, *Solid State Ionics* **35** (1989) and *V*, edited by G. Alberti, *Solid State Ionics* **46** (1991).

C. J. T. de Grotthuss, Mémoire sur la décomposition de l'eau et des corps qu'elle tient en dissolution à l'aide de l'électricité galvanique, *Ann. Chim.* **LVIII** (1806) 54–74.

W. E. Ayrton and J. Perry, Ice as an electrolyte, *Proc. Phys. Soc.* Vol. **2** (1877) 171–82.

J. Beintema, On the Composition and the Crystallography of Autunite and the Meta-Autunites, *Rec. Trav. Chim. Pays. Bas* **57** (1938) 155–75.

A. Potier and D. Rousselet, Conductivité électrique et diffusion du proton dans le perchlorate d'oxonium, *J. Chimie Physique* **70** (1973) 873–8.

M. Eigen, L. de Maeyer and H. C. Spatz, Kinetic Behavior of Protons and Deuterons in Ice Crystals, *Ber. Bunsenges. Physik. Chem.* **68** (1964) 19–29.

A. Von Hippel, A. H. Runch and W. B. Westphal, Ice Chemistry: Is Ice a Proton Semi-conductor?, in *Physics and Chemistry of Ice*, E. Whalley (ed) (Royal Society of Canada, Ottawa (1973)) 236–41.

Preface

R. W. T. Wilkins, A. Mateens and G. W. West, The Spectroscopic Study of Oxonium Ions in Minerals, *Ann. Miner.* **59** (1974) 811–19.

P. E. Childs, A. T. Howe and M. G. Shilton, Studies of Layered Uranium(VI) Compounds, *J. Solid State Chem.* **334** (1980) 341–6.

A. I. Baranov, N. M. Shchagina and L. A. Shuvalov, Superion Conductivity and Phase Transition in $CsHSO_4$ and $CsHSeO_4$ Crystals, *Sov. Pis'ma Zh. Eksp. Teor. Fiz.* **36** (1982) 381–4.

F. M. Ernsberger, The Nonconformist ion, *J. Am. Ceram. Soc.* **66** (1983) 747–50.

Yu. A. Ossipyan and V. F. Petrenko, The Physics of Ice, *Europhysics News* **19** (1988) 61–4.

Symbols

Chapter 3

V	vacancy
D	diffusion coefficient
C	concentration
e	charge
σ	conductivity
E	enthalpy

Chapter 4

C	concentration
D	diffusion coefficient
e	charge
σ	conductivity
E	enthalpy
f	correlation (or Haven) factor
B	mobility
σ_0	prefactor
σ_i	ionic conductivity
σ_e	electronic conductivity
t_i	ionic transport number

Chapter 20

A	pre-exponential factor
E_a	activation energy
k	Boltzmann constant
pKa	ionization constant of acids
pK	ionization constant of polymers
pK_0	ionization constant of polymers at $\alpha = 0$
T	absolute temperature
T_0	ideal glass transition temperature
T_g	glass transition temperature

Symbols

x	number of acid moles per polymer repeat unit
α	degree of ionization
σ	specific conductivity
ν	stretching vibrational frequency

Chapter 21

D_t	self-diffusion coefficient
D^+	abnormal proton self-diffusion coefficient
D_r	rotational diffusion constant
σ	specific conductivity
τ	residence time

Chapter 22

γ	nuclear gyromagnetic ratio
$\hbar$	$1/2\pi$ Planck's constant
I	nuclear spin
H_0	static field
H_1	amplitude of the r.f. field
ω	frequency of the r.f. field
$\tilde{\sigma}$	chemical shift tensor
$\tilde{J}$	indirect spin–spin coupling tensor
Q	electric quadrupole moment of the nucleus
V	electric field gradient
$\tilde{C}$	spin rotation tensor
T_1	spin–lattice relaxation time in the Zeeman field
$T_{1\rho}$	spin–lattice relaxation time in the rotating frame
T_2	spin–spin relaxation time
τ_C	correlation time
$\nu = 1/\tau_{CO}$	attempt frequency
E_a	potential barrier

Chapter 24

$I_{\rho\sigma}{}^{jk}$	intensity of scattered light polarized in the σ direction due to a transition from state j to state k induced by incident light polarized in the ρ direction
$\alpha_{\rho\sigma}{}^{jk}$	$\rho\sigma$ component of the Raman polarizability tensor originating in a transition from state j to state k
$\alpha_{\rho\sigma}{}^{m}$	$\rho\sigma$ component of the Raman polarizability tensor originating in a fundamental vibrational transition of normal mode m in the electronic ground state
P_σ	σ component of the electric moment operator
$\nu(X)$	vibrational frequency of species X
$M(X)$	molecular mass of species X

Symbols

$I(X)$	moment of inertia of species X
V_0	barrier height for motion from one site to an adjacent site
d	separation between adjacent sites
Γ	bandwidth (full width at half maximum intensity)
Γ_{vib}	vibrational contribution to the bandwidth
τ_c	correlation time for a vibrational mode coupled to a disordering process
ΔU	activation energy for a disordering process which is coupled to a vibrational mode
E_a	activation energy for thermally activated reorientational motion
D	diffusion coefficient

Chapter 25

σ^*	complex conductivity
σ'	real conductivity
σ''	imaginary conductivity
J	current density
E	electric field
ε_0	vacuum permittivity
ε'	real permittivity
ε''	imaginary permittivity
ε^*	complex permittivity
ε_∞	optical permittivity
ε_s	static permittivity
n^*	complex refractive index
n^B	Bose–Einstein occupation number
$tg\delta$	tangent loss
I	Raman scattered intensity
$\alpha(\omega)$	infrared absorption
J_0, J_1	Bessel functions
c	light velocity
α	Debye deviation parameter
τ	Debye relaxation time (collective)
τ'	Debye relaxation time (individual)
f_c	loss peak frequency
μ	dipole moment
C	Currie constant
k	Boltzmann constant
T	Curie–Weiss temperature

Chapter 26

B_0	magnetic field
D	self-diffusion coefficient
G	magnetic field gradient
M	magnetization

Symbols

n	$[H_3O^+]/[H_2O]$
R	resistance
T_1	longitudinal spin relaxation time
T_2	transversal spin relaxation time
γ	gyromagnetic ratio
σ	specific conductivity

Chapter 28

$R_z(t)$	normalized longitudinal magnetization time decay
$R_{xy}(t)$	normalized transverse magnetization time decay
$1/T_1$	longitudinal or spin–lattice relaxation rate
$1/T_2$	transverse or spin–spin relaxation rate
W_n	distribution of pore sizes
D	diffusion coefficient
D_f	fractal dimension
$A(G)/$	
$\quad A(0)$	attenuation of the spin–echo signal
G	intensity of the pulsed gradient field
k	permeability
Φ	tortuosity

Chapter 30

τ	relaxation time
τ_0	residential time
τ_1	time of flight
$P(\omega)$	Fourier transform of the autocorrelation function of particle velocities
$S_{inc}(Q,\omega)$	spectral density
Q	momentum transfer
$\sigma(\omega)$	frequency dependent conductivity
E_0	activation energy
η	viscosity
r_1	radius of mobile species
k	Boltzmann constant
e	charge
T_0	equilibrium glass transition temperature
R_τ	electrical relaxation time
m_i	effective mass of mobile species
V_{eff}	effective potential
γ	damping

Chapter 31

A	amplification factor $(D_{\Lambda H^+}/D_{H_2O})$
a	jump distance

Symbols

B	fraction of broken hydrogen bonds
D	self-diffusion coefficient
K	number of hydration spheres for pure water
K^+	number of hydration spheres surrounding the H_3O^+ ions
m	molarity
N	cluster radius for pure water
n	$[H_3O^+]/[H_2O]$
T_1	longitudinal spin relaxation time
T_2	transversal spin relaxation time
v_L	longitudinal sound velocity
Λ	equivalent conductance
σ	specific conductivity
$1/\tau_C$	cluster formation rate
$1/\tau_D$	diffusion rate
$1/\tau_P$	polarization rate of hydrogen bonds
k	Boltzmann constant
T	temperature

PART I · THE HYDROGEN BOND AND PROTONIC SPECIES

1 The hydrogen bond and chemical parameters favouring proton mobility in solids

ANTOINE POTIER

Proton conduction in solids via hydrogen bonding was suggested in 1950 by Ubbelohde & Rogers[1]. Later, the Grotthuss mechanism or translocation[2] was proposed to explain the conductivity of para-electric potassium dihydrogen phosphate, KDP[3], and claimed soon after also for $H_3O^+ClO_4^-$, oxonium perchlorate, OP[4]. Reviews that classified protonic superionic conductors, PSC, appeared from 1980 and were based successively on the ion-exchange properties of materials[5, 7], their structures[6, 7] and their conduction mechanisms[8-10].

In this chapter, the first part (Sections 1.1–1.3) deals with the conduction mechanisms, while the second part (Sections 1.4 and 1.5) points out the significant structural and chemical factors leading to the different conduction mechanisms. The hydrogen bond is a common feature and serves as Ariadne's thread*.

1.1 From ionic to protonic conduction

1.1.1 Ionic conduction

Ionic conductors may be divided[11] into three classes depending on their defect concentrations (i) dilute point defects, dpd ($\sim 10^{18}$ defects cm^{-3}), (ii) concentrated point defects, cpd, ($\sim 10^{18-20}$ cm^{-3}) and (iii) liquid-like or molten salt sublattice materials, mss, ($\sim 10^{22}$ cm^{-3}).

For dilute and concentrated point defect materials, examples are, respectively, NH_4ClO_4 ($\sigma_{250°C} = 10^{-9}$ Ω^{-1} cm^{-1}, $E = 1.4$ eV) and CeF_3

---

* The daughter of Minos and Pasiphae who gave Theseus the thread by which he escaped from the Labyrinth.

$(\sigma_{240°C} = 6 \times 10^{-4} \, \Omega^{-1} \, cm^{-1}, E = 0.26 \, eV)$. The conductivity occurs by an ion hopping mechanism.

In molten salt sublattice materials, practically all the ions in the sublattice are available for motion with an excess of available sites per cation as in e.g. Na β-alumina, $(\sigma_{25°C} = 1.4 \times 10^{-2} \, \Omega^{-1} \, cm^{-1}, E = 0.16 \, eV)$. This leads to a high degree of disorder of these cations. The site occupancies and the conductivity characteristics for some of these salts are given in reference 12, pp. 49 and 53. Conduction is favoured by a levelling of the energy profile along the conduction pathway[13].

Crystalline solid electrolytes have been subdivided[14] into soft ionic crystals such as β-PbF$_2$ and hard covalent crystals such as β-alumina. The conduction mechanism can be pictured as involving a 'liquid-like' charge carrier array moving in the vibrating potential energy profile set up by the immobile counterions.

1.1.2 From ionic to protonic conduction

Proton conduction might be expected, *a priori*, to occur either by a mechanism of lone proton migration or proton-carried migration (as an entity such as H_3O^+). The occurrence of these processes is now established and, in both cases, the previously proposed subdivisions (dpd, cpd and mss) are relevant.

Thus, lone proton migration (proton translocation or Grotthuss process) occurs in oxonium perchlorate (cpd or perhaps mss) and potassium hydrogen phosphate (mss).

Proton-carried migration occurs in oxonium β-alumina and hydrogen uranyl phosphate, HUP. Both are of the mss type and belong to the specific subclass of 'vehicle mechanisms' in which H_3O^+ is the mobile species.

1.2 The lone proton migration mechanism (translocation)

1.2.1 The simplest model for translocation and Bjerrum defects

The translocation mechanism is shown in Fig. 1.1a, b. It results from the coupling of (a) displacement of H^+ along a hydrogen bond and (b) transport of the H^+ ion from this hydrogen bond to the following one. Such a mechanism can occur only in the presence of L defects, as shown in the Bjerrum theory of ice conductivity (reference 15 and

Chapters 10 and 11). In this theory, the occurrence of doubly occupied sites (Doppelsetzung, D) and/or empty sites (Lehrstehle, L) in a translocation mechanism (Fig. 1.1b) is postulated.

The necessity of having Bjerrum defects present is illustrated for the network of water molecules in Fig. 1.1c. From (i) to (ii), two H^+ transfers occur simultaneously, in a chain sequence under the influence of an applied field V. In (ii) however, this has resulted in the creation of a reverse field, V' which induces the protons to hop back to their original position. Consequently, long-range H^+ conduction cannot occur.

Consider now, in Fig. 1.1c (iii), the effect of allowing the central water molecule in the network to rotate. This creates a pair of L and D defects in adjacent hydrogen bond positions and, on departure of H^+ from the vicinity of this L, D couple, the V' field is destroyed. Conduction can clearly continue, therefore, and the arrival of a new H^+ ion from the left restores the chain element in (iv) to its original position (i).

1.2.1.1 Translocation pathway models

Four types of translocation have been proposed[8,9]. These are as follows.

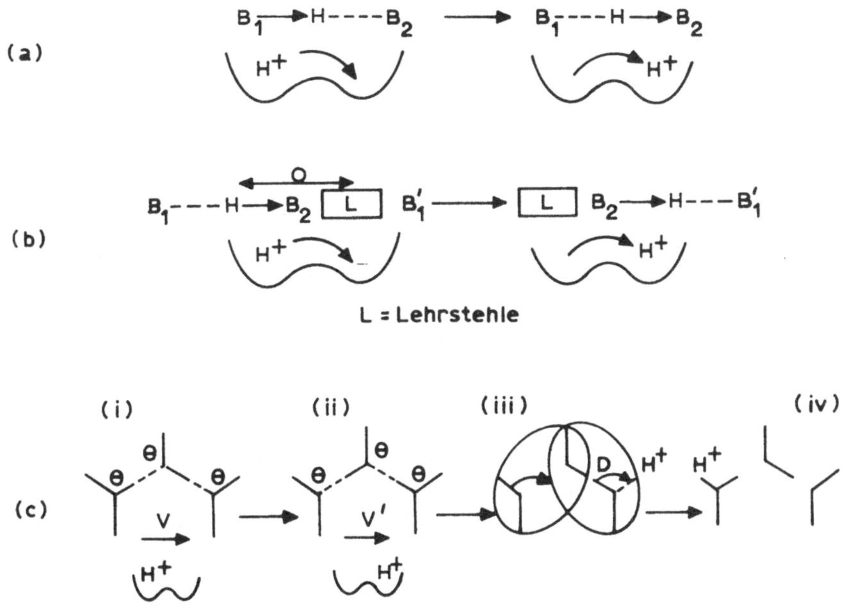

Fig. 1.1.

T_1: anion–cation translocation, e.g. in OP[4]. The T_1 (D/L) mechanism is detailed in Fig. 2 of reference 16.

T_2: anion chain translocation, e.g. in KDP. The T_2 mechanism for $H_2PO_4^-$ is given in Fig. 1 of reference 3.

T_3: the translocation of H^+ along a protonated water chain (identical to that for ice); it is detailed in reference 17 for HUP.

T_4: mixed translocation, for instance in $SnO_2.2H_2O$ where[5] a globular anion $(SnO_{2+x}H_x)^{x-}$ gives a proton to an aqueous chain thereby creating a protonated aqueous chain.

1.2.1.2 From the Bjerrum L, D defects to the molten proton sublattice

Our present state of knowledge does not really confirm or disprove the possible existence of Bjerrum defects, although the potential curve of the proton for L defect propagation in OP has been calculated[16] and appears quite satisfactory. The effect of L and D defects can also be regarded as that occurring from a statistical dynamic disorder of the protons. For example, in crystalline $CsDSO_4$, four H per unit cell are distributed over eight sites in (e) positions and 16 sites in (f) positions[18]. This disorder in the H sites and a high site/H^+ ratio may lead to a quasi-liquid proton sublattice.

1.2.2 A theory for translocation, length of the H bond and crystal field effects

The energy profile for H^+ transport in the translocation model, involving (a) displacement of H^+ along a hydrogen bond, with a transfer energy barrier E_{bar} and (b) transport of H^+ from this hydrogen bond to the next, with a bond breaking energy barrier E_{bond}, is plotted against the oxygen–oxygen distance, d_{o-o}, in Fig. 1.2 for the ion $H_5O_2^+$ in vacuum[19a, b]. With increasing d_{o-o}, E_{bond} decreases but E_{bar} increases, as in Fig. 1.2b. If we assume that the maximum values for these energy barriers are similar to the activation energy for protonic conduction, E_σ, which is usually less than ~ 0.9 eV (or ~ 20 kcal mol^{-1}), then values for d_{o-o} over the range A–B should permit protonic conduction. A refinement of this model in the crystalline state takes account of environmental effects and the energy E_{bond} is replaced by the energy barrier to reorientation of a water molecule, E_{reor}. Since E_{reor} is less than E_{bond}, dashed curve in Fig. 1.2a, the range of permitted values for d_{o-o} is extended to A–C. It can also be seen from Fig. 1.2a that there is an optimum value of d_{o-o} for translocation, corresponding to the cross-over of the two curves for E_{bar} and E_{reor}.

Consider now the acid–base couple, $HClO_4$–H_2O[16] shown in Fig. 1.3a. In vacuum, it has a potential energy curve with one minimum, corresponding to the isolated Lewis complex $H_2O \ldots HClO_4$. On immersing the complex in a Madelung potential, i.e. in a crystal lattice, a double minimum potential curve results. If allowance is then made for polarizability, or an Onsager reaction field, the ionic complex $H_3O^+HClO_4^-$ becomes preferentially stabilized.

The translocation mechanism in this system involves three steps, Fig. 1.3b. These include displacement of H within hydrogen bonds and rotation of the H_3O^+ ions. The effect of the crystal field on the H bond produces the conditions for translocation. This results from the hyperpolarizability of the H bond[19].

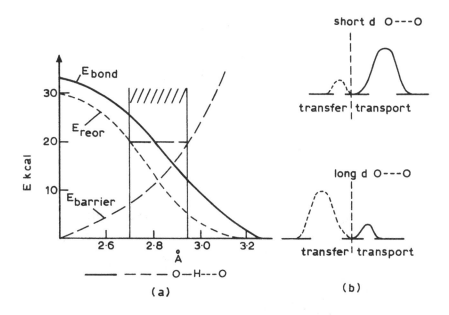

Fig. 1.2.

1.3 Proton-carrying mechanisms (the vehicle mechanism or V-mechanism)

1.3.1 Proton-carrying mechanism of the ionic conduction type

As seen previously (Section 1.1.2), the best known example is oxonium β-alumina where the proton carrier is H_2O. In β-alumina, the oxonium ions occupy only BR sites ($E_\sigma = 0.8$ eV; $\sigma_{25\,°C} = 10^{-11}\ \Omega^{-1}\ cm^{-1}$) while in β″-alumina, oxonium ions can occupy both prismatic and tetrahedral sites ($E_\sigma = 0.6$ eV; $\sigma_{25\,°C} = 2 \times 10^{-9}\ \Omega^{-1}\ cm^{-1}$). Conduction depends on

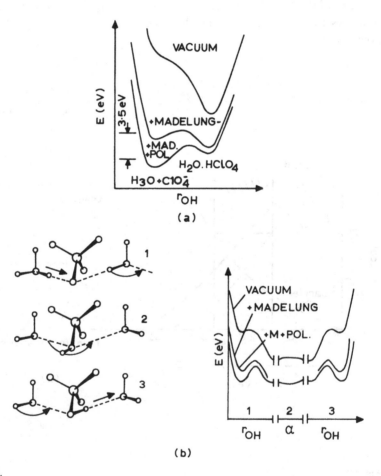

Fig. 1.3.

an excess of available sites. The mechanism can be close to either a cpd or an mss type.

1.3.2 *Proton-carrying mechanism for water-rich materials: the vehicle mechanism (V)*

In this case, the T_3 translocation (Section 1.2.1.1) is often accepted as the likely mechanism. In 1982, however, a new mechanism called the 'vehicle mechanism' was proposed[20]. At first sight, it resembles the previous proton-carrying process (Fig. 1.3a). It differs, however, in that the proton migrates in one direction as OH_3^+, NH_4^+, etc. bonded to a 'vehicle' such as H_2O, NH_3 etc. whereas the 'unladen' vehicles move in the opposite directon (see, for example, Fig. 1.5b). The explanation given here is that proposed by Rabenau. In this mechanism of cooperative motion, in which the counter-flow of protons and vehicles is essential, voids are created by thermal activation, as in free volume theory.

1.4 Structural effects

Proton-conducting materials have rather special structures[6,7], ion-exchange and acid-base properties[5,7]. It is necessary to take into account the presence of either cation vacancies (such as H_3O^+ vacancies) and voids or H^+ site vacancies.

The structural chemistry in these materials is essentially that of H bonds, that is to say the behaviour of different pairs of bases towards H^+, for example H_2O and ClO_4^- towards H^+ in

$$H_2O + H^+ClO_4^- \rightarrow H_3O^+ + ClO_4^- \text{ or } H_2O \cdot HClO_4$$

This leads not only to the partial or total internal ionization of the material[5] but also influences the conduction mechanism.

The crystal structures of proton conductors depend on (a) the anionic part of the material, (b) the cationic part and (c) the defects. Anions can be: (1) monomeric spheres such as ClO_4^-; (2) polymeric layers as in β-alumina; (3) polymeric channelled skeletons as in antimonic acid hydrates; and (4) drowned oxoanion clusters in water, as in heteropolyacid hydrates.

The cationic part of the material can be: (i) the proton in oxy-hydroxyanions, (ii) the oxonium ion in OP; and (iii) the hydrated oxonium ion in HUP. A crystal structure is therefore described by anion–cation combinations such as 1.ii for OP (see Fig. 11.1 p. 166).

Note that most of the proposed combinations are written as ionic formulae. It must be pointed out (Table 1.1), however, that, in acid hydrates, the acidic function is often not entirely ionized. This factor is taken into account as and when necessary.

Finally, the occurrence of different types of defects can serve as a guide in the identification of the H^+ migration mechanism. The defects can be H_3O^+ (or $H_5O_2^+$) defects that occur either as (a) thermally activated (Frenkel) excess sites, such as are present in ionic cpd materials, (b) voids in a quasi-liquid sublattice[20], (c) proton defects in the same sense as oxonium defects in a rigid sublattice or (d) proton voids occurring in a quasi-liquid proton sublattice.

1.4.1   Monomeric anions

The oxyhydroxy anion salts, KDP ($\sigma_{25\,°C} = 10^{-8}\,\Omega^{-1}\,cm^{-1}$; $E_\sigma = 0.55$ eV), CsHSO$_4$ ($\sigma_{150\,°C} = 10^{-2}\,\Omega^{-1}\,cm^{-1}$; $E_\sigma = 0.33$ eV) (see Chapter 11) and Cs$_3$H(SeO$_4$)$_2$ ($\sigma_{190\,°C} = 10^{-3}\,\Omega^{-1}\,cm^{-1}$; $E_\sigma = 0.36$ eV) have layered structures. Vacant H^+ defects have been identified[18]. The more numerous the defects, the flatter the potential along the conduction pathway. KDP is of the (c) type (T_2 mechanism) while the others are of the (d) type (quasi-liquid H^+ sublattice).

In the orthorhombic phase of OP ($\sigma_{25\,°C} = 3.5 \times 10^{-4}\,\Omega^{-1}\,cm^{-1}$; $E_\sigma = 0.36$ eV), rows of H_3O^+ ions are oriented along the three crystallographic axes. Excess oxonium sites occur only by thermal activation and the contribution of ionic cpd to conductivity is probably weak[4]. On the other hand, an excess of available proton sites[21,22] lowers considerably the barrier to H_3O^+ rotation. The reorientation energy, E_{reor} (see Fig. 1.2a), is much less than the activation energy for conduction, E_σ, which is essentially given by the energy barrier, E_{bar} to displacement within the hydrogen bond[16]. This is consistent with long bonds, d_{o-o}, in OP: $2.92 < d_{o-o} < 2.99$ Å.

HUP[23] (see p. 261) and UO$_2$(H$_2$PO$_4$)$_2$.3H$_2$O[24] have layered structures with H-bonded PO$_4{}^{3-}$ ions. The analysis[18] of H^+ defects in HUP leads to a low energy along the pathway so that a (T_3) conduction mechanism is possible. On the other hand, measurement of ^{18}O and H diffusivities have demonstrated a vehicle, V, mechanism by oxonium ions[25].

1.4.2 Layered polymeric anions (see Chapters 13 and 23)

The best example is H_3O^+ β-alumina ($\sigma_{25°C} = 10^{-10}\,\Omega^{-1}\,cm^{-1}$; $E_\sigma = 0.6\,eV$); it is a relatively poorly conducting material with H_3O^+ ions hopping into vacant sites; conduction is not a V process. This material is completely ionized owing to the poor basicity of the layers (Section 1.5.1.1). In the case of $(NH_4^+)(H_3O^+)$ β″-alumina single crystal[26], ($\sigma_{25°C} = 10^{-3}\,\Omega^{-1}\,cm^{-1}$; $E_\sigma = 0.3\,eV$), the data do not lead to an unambiguous choice between T_3 and V mechanisms, nor do H diffusion pulse field gradient, PFG or NMR measurements[27].

The best example of an oxonium hydrate salt is $H_3O^+(H_2O)_n$ β″-alumina ($\sigma_{25°C} = 6.5 \times 10^{-6}\,\Omega^{-1}\,cm^{-1}$; $E_\sigma = 0.15\,eV$). The low value of the activation energy is an argument for a V process and the low conductivity might be due to the small number of available sites for such a process, owing to saturation of the crystal. However, serious arguments for a T_3 translocation mechanism have been given[28,29].

1.4.3 Channelled polymeric anions

Polyantimonic acid hydrates can be written either as $Sb_2O_5 \cdot mH_2O$ or $HSbO_3 \cdot nH_2O$. Crystals with $n = 1, 1/2, 1/3, 1/4$ are known to possess interconnected channels. The activation energies are near $0.4\,eV$ while $\sigma_{25°C}$ decreases with n for all except the cubic $n = 1$, Fd3 phase. All reported studies indicate that conductivity occurs by translocation. For $n = 1$ the channel is populated by a continuous H bond network (Section 1.5.1.2) which is sometimes disrupted. The mechanism seems to result from the small diameter of the channels.

Ammoniated zeolites present the counter example (see Chapter 14). Large cavities are connected by short channels, the bottleneck effect of which governs the conductivity laws[30]. The log $\sigma/(1/T)$ curves are representative of a V process (Fig. 1 in reference 30). They do not obey an Arrhenius law.

1.4.4 Drowned oxoanion clusters

Dioxide dihydrates and heteropolyacid hydrates, HPA[31] (see Chapter 18), are representative examples. Following reference 5, in $SnO_2 \cdot 2H_2O$ ($\sigma_{25°C} = 10^{-4}\,\Omega^{-1}\,cm^{-1}$; $E_\sigma = 0.2\,eV$), 25 Å anion-clusters are surrounded by a bed of water forming a quasi-liquid phase (intercluster gap ∼6 Å). SnO_2 is partly hydroxylated (Section 1.5.1.1). A V process can be inferred from the proposed E_σ value[5] jointly with a cluster–water translocation that cannot be disregarded *a priori*.

HPA hydrates[31–33], $n = 6$, 14, 21, 29, give 12 Å clusters such as $[PW_{12}O_{40}]^{3-}$. They are very good PSC with e.g. for $n = 21$, $\sigma_{25°C} = 0.15\,\Omega^{-1}\,cm^{-1}$ and $E_\sigma = 0.15\,eV$. A vehicle mechanism has been proposed from these data[34,35]. In $HPA \cdot 21H_2O$, however, most of the water-cluster anion H bonds are long while the intrawater bonds are of medium length so that, with practically independent anions, the possibility of a T_3 process cannot be completely disregarded[32].

It can be concluded that the H^+ q.l. sublattice really exists as well as the oxonium q.l. sublattice. The experimental distinction between them remains subtle; even PFG measurements can have different interpretations and only the challenge of experimental facts[34] can shed some new light.

1.5 Chemical 'equilibrium' and the 'ionic defect': towards a chemical classification

1.5.1 The acidity–basicity effect, the 'ionic defect' and the H bond hyperpolarizability

1.5.1.1 Acidity in solid materials, surface acidity, condensation

Acidity–basicity effects must always be taken into account to explain the properties of oxygenated material. Thus, in strong acids, proton transfer to water is complete but this is not so for weaker acids and amphoteric materials.

Acidity–basicity effects in proton-conducting materials depend not only on the atomic and bonding properties of the central anionic species, but also on the properties of the aggregate surface and on the state of aggregation as can be observed in a discussion of the true compositions of weak and medium strength acidic materials such as tin dioxide, β''-alumina and antimonic acid hydrates (Table 1.1).

Taking as a measure of acidity the true oxonium concentration (mole fraction) in the cationic sublattice, the order of increasing acidity seems to be

$$SnO_2 < \beta''\text{-}Al_2O_3 < n = 1, n = 1/2 < n = 1/4(HSbO_3 \cdot nH_2O)$$

The behaviour of antimonic acid depends on the water/Sb_2O_5 ratio in $Sb_2O_5 \cdot mH_2O$ compounds. That of SnO_2 and β''-alumina is more difficult

Table 1.1.

Compounds	$N_{H_3O^+}$	Ref.
$[SnO_{1.9}(OH)_{0.7}]^{-0.5} (H_3O^+)_{0.5} (H_2O)_{0.9}$	0.36	(5)
$[Mg_{0.88}Al_{10.34}O_{17}]^{-1.5}(H_3O^+)_{1.5} (H_2O)_{1.5}$	0.5	(26)
and for the antimonic acid hydrates $HSbO_3 \cdot nH_2O$		
$n = 1/4$ $[Sb_4O_{9.8}(OH)_{1.2}]^{-0.8} (H_3O^+)_{0.8} (H_2O)_{0.2}$	0.8	(35)
$n = 1/2$ $[Sb_4O_{10.8}(OH)_{0.2}]^{-1.8} (H_3O^+)_{1.8} (H_2O)_{1.25}$	0.6	(35)
$n = 1$ $[Sb_4O_{10.4}(OH)_{1.6}]^{-2.4}(H_3O^+)_{2.4} (H_2O)_{1.6}$	0.6	(36, 37)

to understand: (i) though less acidic than the others, SnO_2 gives nevertheless an acid cationic phase, (ii) β''-alumina (Al_2O_3–MgO) appears to be more acidic than SnO_2 and this is implied in its method of preparation[38], although from elementary knowledge Mg and Al oxides are weaker acids than is tin hydroxide.

When condensation of an oxoacid occurs, this leads to a decrease in the surface *Lewis basicity* of the anionic part or an increase in the *Hammett acidity*, $-H_0$. This is shown as follows, for a selection of silica alumina catalysts[38].

	montmorillonite	kaolinite	$SiO_2 \cdot Al_2O_3$	SiO_2–MgO
$-H_0$	5.6 to 8.2	5.6 to 8.2	<8.2	-3 to -1.5

This effect is well-known from Gutmann's rules[39] and from quantum-chemical calculations[40].

It arises because: the more condensed a material is, the less charged is its surface. Effectively, a decrease of the superficial charge or, more specifically, of the charge/condensation ratio, corresponds to a decrease in basicity of the anion.

In the present example, the second ratio is taken as: (true anionic charge)/(number of oxygens in the formula) and leads to

	SnO_2	$n = 1$	$n = 1/2$	β''-alumina	$n = 1/4$
n_{e-}/n_0	0.42	0.20	0.16	0.09	0.07

This puts β''-alumina to the right of SnO_2 and the relative order of antimonic acid hydrates is preserved. The greater condensation in β''-alumina is the leading factor in this reversal of the expected acidity.

1.5.1.2 *Fundamentals of acid–base interactions and proton transfer*

As shown in Fig. 1.3a, the crystal field effect gives rise to a two-minima potential energy curve. The motion of H^+ from left to right is given for $AH/H_2O = 1$, (id = ionic defect (II)) by

$$A^- \ldots H^+\text{–}OH_2 \rightleftharpoons A\text{–}H \ldots OH_2; \quad K_{id} = \frac{|AH \cdot H_2O|}{|A^- \cdot H_3O^+|}$$
$$\text{(I)} \qquad\qquad \text{(II)}$$

Note: the energy difference, E_{min}, between the two minima, does not give an estimate for K_{id}, owing to the importance of the entropy effect.

The second crystal field effect leads to stabilizaton of the states (I) and (II) by reference to water and acid molecules *in vacuo* and therefore of the H bond strength E_b. The competition between states (I) and (II) arises from the relative basicities of A^- and H_2O towards H^+.

In Fig. 1.4a, H_2O is the reference base and, except for SnO_2, the materials considered are acid monohydrates (T_1). On the left are the conjugate anions of the acids, A^-. For A^- bases weaker than H_2O, the H bonds are longer and H_3O^+ is obtained. For A^- bases stronger than H_2O the H bond is also longer but the acid is 'not' ionized. This figure does not consider the competition between states (I) and (II) but concentrates on the more occupied site.

A case of interest is that of antimonic acid, one of whose forms is written[39,40] as $HSbO_3 \cdot H_2O$. Two sites, of types (I) and (II) respectively

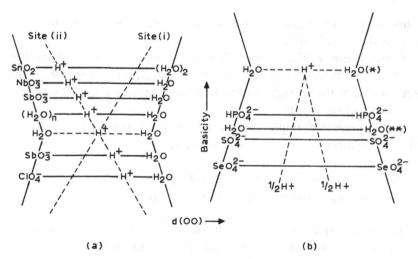

(a)   (b)

Fig. 1.4. H bond length and basicity competition. (In b, (*) 'in vacuum', (**) in chloroaurate.)

coexist and in the course of dehydration ($n < 1$) site (I) is the first to empty.

In Fig. 1.4b, symmetric BHB bonds are considered. The less basic is B, the weaker and longer is the H bond. Simultaneously, two minima appear and the energy barrier E_{br} increases (Fig. 1.3a, b). The polarization effect is due to the central anion atom P for $H_2PO_4^-$ and to the external field in the chloroaurate.

1.5.1.3 Competition between translocation and vehicular processes

Two types of alternative mechanisms can occur: between T_1 and T_3 translocation and between translocation and vehicular processes.

(i) Between (T_1) and (T_3) translocation. Consider, first, the competition between an A^- base and either a water molecule or a water oligomer towards a proton. Quantum calculations[41] show that the proton affinity of $(H_2O)_n$ increases with n, and therefore the $(H_2O)_nH^+$ cation is more free towards A^- when n increases (Fig. 1.4a). Consequently, proton transfer related to a process (from or to the anion), is less possible. In a crystal, $H_5O_2^+$ is more free than H_3O^+.

Consider now the effect of n in a T_3 translocation. The H bond distances in the oligomer increase with n and consequently the T_3 process is favoured (this is valid only for independent oligomers). Thus, T_3 translocation is favoured by a low value for the anion basicity and lengthening of the water chain.

(ii) Consider now the possibility of translocation along a $[H^+(H_2O)_n]_x$ chain. This could occur[36] for $HSbO_3 \cdot H_2O$ with $H_3O^+/H_2O = 1$. The individual chain is shown schematically in Fig. 1.5a with tri- and tetracoordinated oxygen. In this chain, each $H_2O(H_3O^+)$ element is polarized by all the others. The H bond is asymmetric but, owing to the chain statistics, two sites for $(\frac{1}{2})H$ occur. All of the conditions for a translocation step are present. Attack by H^+ on the right will give a T_3 chain to the left. This process will occur alone if parallel chains are isolated as in parallel channels, each of relatively low cross-section.

In the case of layered structures (Fig. 1.5b), the perturbation of the sublattice increases with increasing number of parallel chains. The occurrence of voids in a quasi-liquid sublattice is then possible and a V mechanism can compete with a T_3 one. Typical examples occur with zeolites.

The cases of tin oxide hydrate and HPA remain. The well-determined structure of HPA, with a long anion–cation H bond and medium length

The hydrogen bond and protonic species

H bonds in the spatial water aggregates, favour the pure V (mss) process. For tin dioxide, its amphoteric behaviour towards water must lead to medium or long anion–cation H bonds and a partition of protons between anions and the water bed. The longer the anion–water bed H bonds, the more difficult is the proton transfer, and the remaining motion is T_3 or, more probably, V.

1.5.2 Acidity–basicity effects in materials

The previous considerations can be used: (i) to obtain information on the true equilibrium in a crystal (such as that of $HSbO_3 \cdot H_2O$ in the Besse crystal[37]) and on the possibility of a translocation mechanism, and (ii) to displace an equilibrium with the aid of foreign atoms, for example to increase the acidity of an acid (as when going from $H_2W_{12}O_{40}{}^{6-}$ to $PW_{12}O_{40}{}^{3-}$). A preliminary observation is that all the materials considered have a central atom with its maximum oxidation number.

1.5.2.1 The influence of central or hetero-elements

These elements do not intervene directly in protonic conduction, but notably do modify the behaviour of materials. They can act either by modifying the basicity of the anion or by their dynamical effect on disorder.

(i) Coordinated central element.
(a) Monomeric anions. The increase in conductivity of

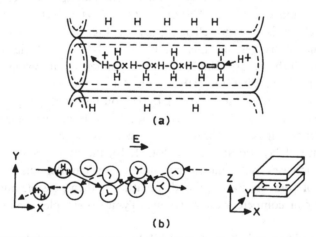

(a)

(b)

Fig. 1.5. $H_2O(H_3O^+)_n$ motion: (a) translation in isolated channels, (b) vehicle mechanism in a layer.

14

$Rb_3H(MO_4)_2$ on going from M = S to M = Se is correlated with the greater thermal agitation of Se^{42} and also the weaker electronegativity of Se (Fig. 1.4b) leading to a more difficult T_2 translocation process. In the cases of HUP and HUAs, their nearly identical conductivities[24] are related to the identical pK of the acids.

(b) Homopolyacids. For $HMO_3 \cdot H_2O$ (M = Sb, Nb, Ta) pyrochlore[43], the slight decrease in conductivity follows the order of acidities (Fig. 1.4a).

(c) Heteropolyacids $H_3PM_{12}O_{40} \cdot 29H_2O$ (M = Mo, W)[34,44]. A similar effect occurs to that in (b). The polymeric structure of these materials is related to the acidity.

(ii) Heteroelements.

(a) Monomeric anions. The heteroelement can be external, as in $M_3'H(MO_4)_2$ where the ammonium salts are more conducting than Rb salts probably owing to a greater degree of disorder[42], but a lot of examples, such as $M'HMO_4$ (M = S, Se; M' = K, Rb, Cs), show a decrease of σ from Cs to K occurring through an increase in the polarizing power of the M' cation; with Cs, and H bond skeleton is more nearly free.

For HUP, the basicity of HPO_4^{2-} is decreased owing to the doubly charged uranyl ion which is a strong acceptor (from the O of the phosphate) and thus releases the proton.

(b) Polymeric anions. The simplest examples are $H_3PM_{12}O_{40} \cdot nH_2O$ derived from the homopolyacid $H_4(H_2PM_{12}O_{40} \cdot nH_2O)$, where the heteroelement is central. The insertion of P in the cluster decreases the surface charge not only because of the P valence but perhaps also by the great acceptor power of P that enhances electron release from the surface.

The final case is that of zeolites, where, as seen before, the ratio Si/Al is the indicator for acidity.

This attempt to give a coherent classification of PSC is incomplete because the effects of phase changes with increasing temperature are not specifically analysed, although conductivity and structural changes can be interpreted simultaneously in terms of the effects of thermal motion.

1.6   References

1. A. R. Ubbelohde and S. E. Rogers, *Trans. Faraday. Soc.* **46** (1950) 1051–60.
2. C. T. J. Grotthuss, *Ann. Chem.* **LVIII** (1806) 54–74.

The hydrogen bond and protonic species

3. M. O'Keeffe and C. T. Perrino, *J. Phys. Chem. Solids* **28** (1967) 211–18.
4. D. Rousselet, Thèse (CNRS A.D. 2659) (1968) Montpellier; D. Rousselet and A. Potier, *J. Chimie Physique* **70** (1973) 873–8.
5. W. A. England, M. G. Cross, A. Hamnett, P. J. Wiseman and J. B. Goodenough, *Solid State Ionics* **1** (1980) 231–49.
6. J. Jensen in *Solid State Protonic Conductors*, I. J. Jensen and M. Kleitz (eds) (Odense University Press (1982)) 3–20.
7. J. Goodenough in *Solid State Protonic Conductors II*, J. Goodenough, J. Jensen and M. Kleitz (eds) (Odense University Press (1983)) 123–43.
8. A. Potier, *De la Microélectronique à la Microionique'* Ph. Colomban (ed) – (Ecole Polytechnique 3–4 mars 1983, Palaiseau) 149–74.
9. A. Potier, *Optica Pura y Applicada* **21** (1988) 7–18.
10. Ph. Colomban and A. Novak, *J. Mol. Struct.* **177** (1988) 277–308.
11. M. J. Rice and W. L. Roth, *J. Solid State Chem.* **4** (1972) 294.
12. S. Chandra, *Superionic Solids* (North-Holland (1981)) 54.
13. H. Schultz, *Ann. Rev. Mater. Sci.* **12** (1982) 351–76.
14. M. A. Rattner and A. Nitzan, *Solid State Ionics* **28–30** (1988) 3–33.
15. N. Bjerrum, *Kgl. Danske Videnskab Setskab Mat-fys*; *Science* **115** (1952) 385–91.
16. J. Angyan, M. Allavena, M. Picard, A. Potier and O. Tapia, *J. Chem. Phys.* **77** (1982) 9.
17. L. Bernard, A. N. Fitch, A. T. Howe, A. F. Wright and B. E. F. Fender, *J. Chem. Soc. (Chem. Comm.)* (1981) 784–6; T. K. Halstead, N. Boden, L. D. Clark and C. G. Clark, *J. Solid State Chem.* **47** (1933) 25–230; L. Bernard, A. Fitch, A. F. Wright, B. E. F. Fender and A. T. Howe, *Solid State Ionics* **5** (1981) 459–62.
18. A. I. Baranov, B. V. Merinov, A. V. Tregubchenko, V. P. Khiznichenko, L. A. Shuvalov and N. M. Shchagina, *Solid State Ionics*, **36** (1989) 279–82.
19. (a) R. Janoschek, E. G. Weidemann and G. Zundel, *J. Am. Chem. Soc.* **94** (1972) 2387–96; (b) S. Scheiner, *Int. J. Quantum Chem.* **7** (1980) 199–206.
20. K. D. Kreuer, A. Rabeneau and W. Weppner, *Angew. Chem. Int. Ed.* **21** (1982) 208–9.
21. (a) M. Pham-Thi, J. F. Herzog, M. H. Herzog-Cance, A. Potier and C. Poinsignon, *J. Mol. Struct.* **195** (1989) 293; (b) M. H. Herzog-Cance, M. Pham-Thi and A. Potier, *J. Mol. Struct.* **196** (1989) 291–305.
22. K. Czarniecki, S. A. Janik, J. M. Janik, G. Pytatz, M. Rachlewska and T. Waluga, *Physica B* **85** (1977) 291.
23. B. Morosin, *Acta Cryst.* **34** (1978) 372–34.
24. R. Mercier, M. Pham-Thi and Ph. Colomban, *Solid State Ionics* **15** (1985) 113–26.
25. K. D. Kreuer, A. Rabenau and R. Messer, *Appl. Phys. Chem. A* **32** (1983) 45–53.
26. J. O. Thomas and G. C. Farrington, *Acta Cryst.* **B29** (1983) 227–35.
27. Y. T. Tsai, S. Smoot, D. H. Whitmore, J. C. Tarczon and W. P. Halperin, *Solid State Ionics*, **9–10** (1983) 1033–40.

28. A. Teitsma, M. Sayer and S. L. Segel, *Mat. Res. Bull.* **15** (1980) 1611.
29. Ph. Colomban, G. Lucazeau, R. Mercier and A. Novak, *J. Chem. Phys.* **67** (1977) 5244.
30. K.-D. Kreuer, W. Weppner, and A. Rabenau, *Mat. Res. Bull.* **17** (1982) 501–9.
31. O. Nakamura and I. Ogino, *Mat. Res. Bull.* **17** (1982) 231–4.
32. C. J. Clark and D. Hall, *Acta Cryst.* **B32** (1976) 1545–7.
33. G. M. Brown, M. R. Noe-Spirlet, W. R. Busing and A. H. Levy, *Acta Cryst.* **B33** (1977) 1038–46; M.-R. Spirlet and W. R. Busing, *Acta Cryst.* **B34** (1977) 907–1030.
34. K. D. Kreuer, M. Hampele, K. Dolde and A. Rabenau, *Solid State Ionics* **28–30** (1988) 589–93; K. D. Kreuer, *J. Mol. Struct.* **177** (1983) 265–76.
35. (a) H. Arribart, Y. Piffard and C. Doremieux-Morin, *Solid State Ionics* **7** (1982) 91–9; (b) H. Arribart and Y. Piffard, *Solid State Comm.* **45** (1983) 571–5.
36. G. Doremieux-Morin, J. P. Fraissard, J. Besse and R. Chevalier, *Solid State Ionics* **17** (1985) 93–100.
37. R. G. Bell and M. T. Weller, *Solid State Ionics* **28–30** (1988) 601–6.
38. H. A. Benesi, *J. Am. Chem. Soc.* **78** (1956) 5490.
39. V. Gutmann in *The Donor Acceptor Approach to Molecular Interactions* (Plenum Press (1978)).
40. W. L. Mortier, J. Sauer, J. A. Lercher and H. Noller, *J. Phys. Chem.* **88** (1984) 905–17.
41. M. D. Newton and S. Ehrenson, *J. Am. Chem.* **93** (1971) 4981–3.
42. I. Brach, Thésis Montpellier (1986).
43. U. Chowdry, J. R. Baskley, A. D. English and A. W. Sleight, *Mat. Res. Bull.* **17** (1982) 917–33.
44. R. C. T. Slade, J. Barker, H. A. Presman and J. H. Strange, *Solid State Ionics* **28–30** (1988) 594–680.

2 Protonic species and their structures

DEBORAH J. JONES and JACQUES ROZIERE

2.1   General introduction

For solid protonic conductors the broad term 'protonic species' includes all those hydrogen-containing entities formed, perhaps only transiently during the conduction process, by ionic or molecular groups of sufficient basicity. This description includes the wide range of anionic protonated species in anhydrous protonic conductors: OH^-, protonated tetrahedral ions such as $HSeO_4^-$, HPO_4^{2-} and dimeric XO_4 groups such as $H(SeO_4)_2^{3-}$, in addition to the cationic proton hydrates $[H(H_2O)_n]^+$, including H_3O^+, $H_5O_2^+$, $H_7O_3^+$ etc., and combinations of hydrogen with nitrogen: ammonium NH_4^+, hydrazinium $N_2H_5^+$ and hydrazinium (2+) (dihydrazinium or hydrazonium), $N_2H_6^{2+}$. Structural and spectroscopic data on the anionic protonic species encountered in anhydrous materials are included in later chapters and will not be referred to further here, except in cases where the counterion is ammonium or a proton hydrate.

Detailed information on the proton hydrate series in the solid state is available, and the reader is referred to articles reviewing structural studies[1,2] and vibrational spectroscopic studies of proton hydrates[3,4], but also general overviews[5,6], the most recent being the comprehensive review of 1986[7].

The extensiveness of the proton hydrate series is not mirrored in the analogous nitrogen containing family. Of the isoelectronic pairs OH_3^+/NH_4^+ and $O_2H_5^+/N_2H_7^+$, in the solid state only the former is known. Comparison with $H_5O_2^+$ in crystals is thus limited either to theoretical calculations on $N_2H_7^+$ or to bulkier ions such as $[NH_4(NH_3)_4]^+$ involving N—H—N hydrogen bonds. In contrast, $N_2H_5^+$ and $N_2H_6^{2+}$ salts form a distinct class of proton conductors.

2.2   *Proton hydrates*

2.2.1   *Introduction*

The hydrogen ion H^+ cannot exist as a free species in condensed phases; its hydration has long fascinated chemists and physicists. Existence of the hydrated proton was first postulated to explain the catalytic effect of the proton in esterification[8] and later to rationalize the conduction of aqueous sulphuric acid solutions[9]. The concept of electrolytic dissociation and consequent conduction in aqueous solutions is a forerunner of the modern notion of the salts themselves as solid electrolytes in the absence of any solvating medium. The parallel is particularly clear for strong mineral acid hydrates where several acid/water compositions of ionic character exist, many of which are proton conducting, and in which proton hydrates H_3O^+ and $H_5O_2^+$ have been identified[10–16].

Of course, despite much physical and chemical evidence for the existence of H_3O^+ (and $H_5O_2^+$),[17] its identification in the solid state awaited the use of X-ray diffraction techniques and in particular, the observation that a quasi-identical X-ray powder photograph is given by solid perchloric acid monohydrate and ammonium perchlorate[18]. Detailed descriptions of the structure, geometry and bonding arrangements of protonated species are obtained from diffraction studies on the solid state especially, nowadays, by neutron diffraction, where identification of hydrogen relative to other atoms is favoured.

Theoretical calculations on the structure and vibrational spectra of $[H(H_2O)_n]^+$ typically refer to free, isolated entities; consequently, the results[19] can be compared only with those of gas phase experiments, although simulated electrostatic effects of a crystal environment have been included for an isolated H_3O^+ ion[20].

2.2.2   *Structural studies of proton hydrates*

In this section, a summary of the salient structural features of H_3O^+ and $H_5O_2^+$ is followed by a limited number of examples selected to show the structural environment of the proton hydrate. The literature up to 1986 has been thoroughly reviewed[7] and solid electrolyte materials reported since this date are highlighted where possible.

In the solid state, the oxonium ion is a discrete species in which all three hydrogen atoms are equivalent. As such, the term 'proton hydrate' is inappropriate, implying as it does a system in which one proton differs in its bonding arrangement from those in the water molecule[1]. In crystalline compounds, H_3O^+ is strongly hydrogen bonded to counteranions. The resulting arrangement distorts the symmetry to a greater or lesser extent, so that ideal C_{3v} symmetry is rarely observed.

In contrast, the diaquahydrogen ion, $H_5O_2^+$, is characterized by a short O—H—O hydrogen bond of length generally 2.4–2.5 Å, which links the two water molecules, although shorter examples are known[21]. Depending on local symmetry and crystal environment, the proton may be centred in the hydrogen bond or located closer to one of the water molecules. In three examples where neutron diffraction techniques have been applied to compounds in which no symmetry constraints operate on the short O—O bond, the hydrogen atoms were all slightly off-centred by 0.01–0.09 Å in hydrogen bonds of length 2.44–2.45 Å[22]. The conformation of the $H_5O_2^+$ ion varies from twisted to a regular *trans* form.

Although high coordination of water around a proton is unlikely, hydration of H_3O^+ and $H_5O_2^+$ must be considered. Complete use of the hydrogen bond-donating capacity leads to triaquaoxonium ($H_9O_4^+$, Eigen type[23]) and to tetraaqua(diaquahydrogen) ions, $H_{13}O_6^+$. In fact, these ions rarely exist as isolated species, tending to interlink to form more extensively hydrogen bonded networks such as chains, rings and cages. One notable exception is the discrete $H_{13}O_6^+$ ion in $[(C_9H_{18})_3(NH)_2Cl]Cl \cdot HCl \cdot 6H_2O$, the stability of which is considered to be due to the presence of the bulky anion[24].

Whereas H_3O^+ and $H_5O_2^+$ were first observed and systematically studied in strong inorganic acid hydrates, the past two decades have witnessed an immense broadening of the range of counteranion types that have been studied as well as extension to higher degrees of hydration. The progressive evolution towards construction of hydrogen bonded sublattices is particularly marked in the series of successive hydrates of inorganic *acids* such as $HClO_4$[10] or HBF_4[16]. Consider the 1 and 5.5 hydrates of *perchloric acid*. In the former, the discrete oxonium ion is embedded in an environment of hydrogen bonded acceptor centres of the polyatomic anion[11]; in the higher hydrate it is the anionic species which is encapsulated in a cage-like cationic structure[16], reminiscent of the inverse arrangement of H_3O^+ complexed by polycyclic crown ethers. It appears that the relative amount and overall dimensions of the anionic component are controlling factors in the distinction between isolated (discrete) and non-isolated (chains, rings etc.) proton hydrate species, although the specific anion also influences the nature of the cationic complex formed.

Protonic species and their structures

This leads us to suggest the following classification of proton hydrates based on the way in which the various protonic species are involved in the overall structural arrangement:

- discrete H_3O^+, $H_5O_2^+$ encapsulated in a macrocyclic ether or enclosed in a three-dimensional framework;
- discrete H_3O^+, $H_5O_2^+$ in a two-dimensional anionic framework or networked $H_3O^+/H_5O_2^+/H_2O$ parallel to a two-dimensional anionic framework;
- chains and rings of $H_3O^+/H_5O_2^+/H_2O$ and isolated anions;
- discrete anions encapsulated in a cationic clathrate structure.

In the first of these categories, the encapsulated oxonium ion is not expected to be labile, and a conduction path is difficult to envisage when the cavity of each polycyclic ring is occupied. Attention is given to this topic, nevertheless, by virtue of its novelty in the provision of a new class of oxonium salts. Complexation of the oxonium ion from mineral acid solutions by addition of a macrocyclic crown ether can be considered as an extension of the earliest work in this field. Infrared spectroscopy provided the first evidence for complexed oxonium ions from $HClO_4$ and $HPtF_6$ solutions containing dicyclohexyl-18-crown-6[25](18-crown-6 = $C_{12}H_{24}O_6$). Complexed H_3O^+ salts of 18-crown-6 with a variety of anions have since been reported[26]. Single crystal X-ray diffraction studies on substituted and non-substituted 18-crown-6 salts[27–31] indicate that the oxonium ion donates three hydrogen bonds to alternate oxygen atoms of the polyether ring (Fig. 2.1) with an average bond length[27–30] of 2.67(6) Å. However, hydrogen bonding to the second set of O atoms is also possible – the average O—H--O distance[27–29] is 2.82(6) Å – serving to further stabilize the penetration of H_3O^+ in the ring. Non-cyclic polyethers or oligomeric glymes are usually considered inferior ligands to the cyclic ethers, but a recent report describes $[CH_3O(CH_2CH_2O)CH_3 . H_3O]$ $[Mo(O)Br_4(H_2O)]$ in which the oxonium–glyme hydrogen bond is significantly shorter than those described above, with an average length 2.57 Å. This is presumably helped by the non-rigid conformation of the glyme molecule[32]. The oxonium ion can be occluded in strikingly similar cavities to those in cyclic ethers, by the hollows formed in two- or three-dimensional anionic macrostructures, as in layered $H_3O[(Al_3(H_2PO_4)_6(HPO_4)_2] . 4H_2O$[33].

The geometry of the oxonium ion in a crown, or cage-like surroundings in general represents a compromise between the dual tendencies of adopting a regular ionic arrangement with high coordination number and the formation of linear hydrogen bonds. It would seem that a crystal environment incorporating multiple hydrogen bond acceptor sites offers the possibility of reorientation of oxonium or diaquahydrogen ions. Such orientational disorder of protonic species must favour proton transfer between various hydrogen bond acceptor sites, and hence protonic conduction properties. For example, in the high-temperature phase of perchloric acid monohydrate, orientational disorder of H_3O^+ within the three-dimensional anionic sublattice described by the ClO_4^- ions facilitates cation–anion proton transfer[11a]. Similarly, disordered $H_5O_2^+$ ions are held in a framework of $[W_{12}PO_{40}]^{3-}$ ions in dodecatungstophosphoric acid-6 hydrate[34].

Reduced dimensionality of the anionic framework to a layered 'macroanionic' structure constitutes the second class in the scheme outlined above. The definition of planes within the structure describes a preferential conduction pathway and protonic conduction is expected to be anisotropic. Examples of discrete $H_5O_2^+$ and H_3O^+ ions hydrogen bonded only to the anionic framework are plentiful[35]. When hydration increases, the number of degrees of freedom provided by two-dimensionality allows hydrogen bonded proton hydrates to develop as chains[36] or as a distinct two-dimensional lattice extending in directions parallel to the

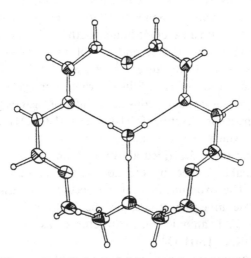

Fig. 2.1. Encapsulated H_3O^+ in the 18-crown-6 macrocycle (reproduced with permission from reference 30).

macroanionic layers. An extended network of disordered $H_3O^+/H_5O_2^+/$ H_2O exists in, for example, the torbernite minerals[37].

When macroanionic substructure occurs, the formation of rings or chains involving proton hydrate species is favoured. For instance, the $H_7O_3^+$ ion is often encountered as a chain of $H_5O_2^+$ and H_2O species[38] or as helical chains spiralling with anions[39]. More recently, $H_5O_2^+/H_2O$ chains and $H_7O_3^+$ ions have been observed in $H_2SO_4.6.5H_2O$[40]. Extension of interlinking to form isolated proton hydrate rings is also observed[41], with an excellent example provided by the cyclic $H_{14}O_6^{2+}$ ion[42]. Finally, chains and rings can be joined to form a cation-dominated framework of pronounced two-dimensional character, with interspersed anions between the layers[14,43] of the type, for instance, recently determined crystallographically in metastable $H_2SO_4.8H_2O$[40].

The final class consists of a proton hydrate network which is complete in three dimensions and encloses a globular anion in its void(s). As early as 1955, '$HPF_6.6H_2O$' was reported to adopt a cubo-octahedral cage of hydrogen bonded water molecules, the centre of which was occupied by the guest PF_6^- ion[44]. The cage was protonic by implication. Proton and ^{19}F NMR data were consistent with this description only if the host lattice stoichiometry were modified to $H_3O^+.HF.H_2O$[45]; this structure is also adopted by the arsenic and antimony congeners[46,47] and is shown in Fig. 2.2. Recent work has shown the presence of cationic host lattices in a range of *superacid hydrates*[16,48]. For example the polyhedral clathrate hydrates $HPF_6.7.67H_2O$, $HBF_4.5.75H_2O$, and $HClO_4.5.5H_2O$ all possess the type I (cubic 12 Å) crystal structure. In the phosphorus compound, guest–host hydrogen bond interactions of the type O—H--F are moderately weak with bond lengths *c.* 2.9 Å. To maintain a three-dimensional, fully hydrogen bonded host lattice, one H_2O is replaced by a (statistically disordered) HF molecule for each transferable proton, whereas in the HBF_4 and $HClO_4$ hydrates, creation of vacancies probably occurs[16].

Two underlying features have emerged: all examples of polyhedral cationic hydrates so far reported are (i) salts of *monobasic* acids and (ii) highly hydrated. For example, polymeric ribbons of alternating MF_6^- and H_3O^+ ions have been observed in oxonium salts of superacids[49] and recent results on $H_2SiF_6.6$ and $9.5H_2O$[50] and their titanium analogues[51], simply show the presence of $H_5O_2^+$.

As these are low melting compounds, the field of cationic clathrates could be developed only by the careful control and systematic use of appropriate low-temperature crystallization techniques. These have allowed the variety of known stable hydrates in the phase diagrams of acid/water systems to be extended[16,48]. Such developments indicate that acid hydrates and oxonium salts are likely to remain a rich area of study in years to come.

2.3 Nitrogenous protonic species

2.3.1 Introduction

Neither the existence nor the geometry of the ammonium ion NH_4^+ or its solvates, or the hydrazinium ions $N_2H_5^+$ and $N_2H_6^{2+}$, have ever been in doubt, nor has any debate taken place comparable to that on the identity and bonding arrangements of proton hydrates. For example, the ammonium ion has always been considered to be a monovalent species, in the same way as alkali metal cations, and its ionic radius (1.48 Å) equals that of Rb^+. Such acceptance comes in part from the fact that in aqueous solutions of ammonium salts the ion is little perturbed and remains identifiable[52]. These experimental observations are related to the proton transfer properties of H_3O^+ and NH_4^+ or $N_2H_5^+$ and so, intrinsically, to the nature of the hydrogen bond formed to acceptor species and in particular the shape of the proton potential energy function. In the case of more weakly hydrogen bonded nitrogen-containing protonic species,

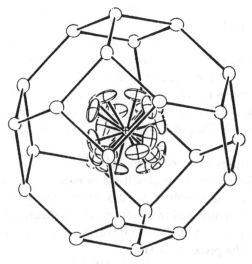

Fig. 2.2. Environment of encapsulated disordered PF_6^- in a $[4^66^8]$ void in $HPF_6 . 5H_2O . HF$ (reproduced with permission from reference 47); hydrogen positions not shown.

the N—H--O potential energy profile is highly asymmetric, and proton transfer properties are unfavoured compared with those of the corresponding oxonium salt in aqueous solution.

2.3.2   The ammonium ion

One of the consequences of the acid–base properties of ions H_3O^+ and NH_4^+ is of importance in the proton conductivity of solid oxonium and ammonium containing compounds. Consider the two schematic mechanisms of proton conduction developed up to now: a charge transfer process from a protonic species to an anion or neutral molecule (Grotthuss mechanism), and the so-called 'vehicle mechanism', in which the proton migrates with H_2O or NH_3 as a 'vehicle–proton complex'[53]. In cases where proton transfer is involved, the conductivity of solid ammonium salts will always be less than that of the corresponding oxonium compound. Ammonium perchlorate illustrates this point: its conductivity[54] at $250\,°C$ is c. $10^{-9}\,\Omega^{-1}\,cm^{-1}$; that of H_3OClO_4[55] at $25\,°C$ is $10^{-4}\,\Omega^{-1}$ cm^{-1}.

When other mechanisms of proton conduction are possible, as in ion exchange materials, conduction properties are reasonably well conserved on replacement of H_3O^+ by NH_4^+ in the structure[56-58]. Nevertheless, the conductivity in such cases is generally greater for the oxonium salt than for its ammonium analogue, and it is assumed that some degree of proton transfer still occurs. More than one species may be involved in diffusion in the vehicular process. For example, in $(NH_4)_4Fe(CN)_6.1.5H_2O$[59] and $(NH_4)_{10}W_{12}O_{41}.5H_2O$[60], diffusing NH_4^+ and H_2O were identified in the former, and NH_4^+ and H_3O^+ (from electrolysis of water of crystallization) in the latter.

The flexibility of hydrogen bonds contributes to the high degree of polymorphism exhibited by ammonium salts, many of which show disorder – either static or dynamic – and to temperature dependent physical properties such as ferroelectricity – paraelectricity–superionicity. In paraelectric NH_4HSeO_4 and the trigonal, superionic phase of $(NH_4)_3H(SeO_4)_2$, selenate groups are linked by hydrogen bonds into chains in the former[61], and in the latter, into three-dimensional networks[62]. Disorder in these compounds is pronounced and includes orientational disorder of the ammonium ion. It is of interest to relate these observations to the possibility of two concomitant mechanisms of proton conduction, for the acid selenate anions and NH_4^+.

The hydrogen bond and protonic species

It may be expected that because of its large ionic radius NH_4^+ should have a coordination number similar to those of K^+, Rb^+ or Cs^+, namely between 6 and 12. Its appreciable tendency for hydrogen bond formation, however, can act to reduce its coordination to hydrogen bond acceptors to 4 or 5[63].

The environment of the ammonium ion plays an important rôle in defining possible NH_4^+ reorientations. This aspect has been systematically studied by infrared spectroscopy at varying temperature, and by using isotopically substituted ammonium as a probe[64]. The situation in the ammonium halides has been extensively described[65]. Here, apart from the fluoride where the environment of NH_4^+ is quasi-tetrahedral[66], phase transitions lead to additional effects. In NH_4I for instance, three phases exist: I > 259 K; II 231–259 K; and III < 231 K. In phases III and II, each NH_4^+ is coordinated by eight iodide ions, whereas in the high temperature phase of NaCl type structure, the ammonium environment is octahedral[67], and free rotation about one N—H—I axis has been suggested[68]. Another example in which NH_4^+ can adopt a tetrahedral symmetry is the high temperature phase of $(NH_4)_2SiF_6$[69], in which NH_4^+ lies at the centre of a cuboctahedron of 12 F^- ions providing the possibility of threefold disorder.

Trigonal ammonium environments are found in the low temperature polymorph of $(NH_4)_2SiF_6$[70] and $(NH_4)_2Pb(SO_4)_2$[71], where NH_4^+ occupies sites of symmetry C_{3v}. The existence of several competing acceptor centres (12 in the former, 10 in the latter) leads to the possible formation of highly bent, bifurcated or trifurcated hydrogen bonds – which occurs more readily than for water in crystalline hydrates – or alternatively, to temperature dependent disorder of NH_4^+ groups. In a collection of ammonium salts of oxy-anions[63], examples in which tetrahedral coordination and therefore linear N—H—O bonds occur number only a fifth. In the other cases, the coordination number is in the range 5–9, with a progressive increase in the average N—O distance within a coordination polyhedron. In tetragonal $(NH_4)_2CuCl_4 . 2H_2O$, two sets of four Cl atoms related by an S_4 symmetry axis through the N atom provide eight almost equal length N—Cl hydrogen bonds[72a], giving two energetically comparable arrangements. Indeed, spectroscopic evidence for two non-equivalent orientations of the NH_4^+ ion at 77K has been provided[72b]. On the other hand, rapid reorientation of ammonium occurs in the fluoride perovskites, NH_4MF_3, M = Mn, Co, Zn, where the ammonium coordination number is 12[73].

In view of the reorientation or orientational disorder often encountered in ammonium salts, it is difficult in many instances to identify hydrogen bonding from crystallographic studies. In fact the problem originates in the definition of the hydrogen bond for such compounds. Several criteria have been suggested[74-78], including the presence of a combination band v_4 (deformation) + v_6 (libration) in infrared spectra, while the existence of K^+ or Rb^+ isomorphs of the NH_4^+ salt has been interpreted as indicative of no hydrogen bonding[75]. Whereas the validity of the latter yardstick has been questioned, the converse might be true, i.e. in cases where the ammonium salt is not isomorphous with the corresponding K^+, Rb^+ analogues, hydrogen bonding could be present[76]. Currently, the term hydrogen bonding as applied to the NH_4^+ ion is accepted as including those interatomic forces restricting its librational motion about an equilibrium position[78]. Broadening of the absorption band is expected if the lifetime in a given orientation is of the same order as the lifetime of a vibrationally excited state; such broadening has been observed in NH_4ClO_4 down to 22 K[79]. This salt has been the subject of extensive investigation by crystallographic techniques[80]: the ammonium ion is surrounded by eight hydrogen bond acceptors at similar distance, no four of which describe a regular tetrahedron. At 10 and 78 K, the root mean square amplitude of librational motion about one of the principal axes is particularly high, while at 298 K, rotational motion has been described as oscillatory[80b], and by a model containing a random spherical distribution of hydrogen around nitrogen[80f].

2.3.3 Solvation of the ammonium ion

The ammonium–ammonia system might be expected to show much similarity to the oxonium–water series, both in the extent of solvates of the $[NH_4(NH_3)_n]^+$ formed in the solid state, and in the richness of structural arrangements of the resulting hydrogen bonded ions. Disappointingly, the literature does not bear witness to this.

Solid phases reported over the last century are limited to the amines of the ammonium halides[81], and only the iodide comes near to demonstrating the range of phases characteristic of many acid-water binary systems. The proposed monoamines are particularly unstable, and only the triamines and $NH_4I.4NH_3$ have been studied crystallographically[82,83]. In $NH_4X.3NH_3$ discrete uncharged triamine tetrahedra are formed through hydrogen bonds N—H--N of length 2.65-3.05 Å, and

N—H—X[82]. The structure of $NH_4I.4NH_3$ at $-140\,°C$ is also based on ammonium–ammonia tetrahedra in which a single N—H—N distance is observed, 2.96 Å[83]. In neither series does hydrogen bond interaction occur between ammonia molecules. Similarly, $CH_3COONH_4.nNH_3$, $n = 1, 2$, are possible candidates for a discrete $N_2H_7^+$ ion; here the structures are made up of CH_3COO^-, NH_4^+ and NH_3 groups linked to form three-dimensional networks of hydrogen bonds. In both compounds, the NH_3 molecules accept a hydrogen bond of length 2.87 Å from the NH_4^+ ion[84]. It therefore seems that, as far as the second member is concerned, the ion $(NH_4.NH_3)^+$ has been reported only as $NH_4SbCl_6.NH_3$ and characterization limited to the determination of unit cell parameters[85]. The corresponding isostructural hydrate has been shown to consist of $SbCl_6^-$ ions and $(NH_4--OH_2)^+$ groups, in which the O--N distance is 2.72 Å.

Analogues of the $N_2H_7^+$ system are formed by heterocyclic bases (B) hydrogen bonded to their conjugate cations (HB). This family has stimulated much crystallographic and spectroscopic interest through the possibility of studying the NHN hydrogen bond in a formally symmetric environment[86].

In summary, although it is tempting to draw parallels between the ammonium amines and the isoelectronic oxonium–water series, this is simply not possible: the structural variety displayed by this latter family as hydration is increased finds no comparison, nor does the broad range of anions accompanying $[H(H_2O)_n]^+$ in the solid state find any counterpart.

2.3.4   Hydrazinium $N_2H_5^+$

Monoprotonated hydrazine $N_2H_5^+$ is an asymmetric species containing a basic, NH_2 moiety and a positively charged NH_3^+ fragment, with the N–N distance in the range 1.4–1.5 Å. All five N–H groups are capable of donating hydrogen bonds while the free electron pair on the $-NH_2$ site allows hydrogen bond acceptor behaviour as well as coordination to metal ions. These characteristics allow a division into four groups, based on the bonding arrangement of the hydrazinium ion:

(i) head-to-tail hydrogen bonded chains

$$--H_2H-NH_3^+--H_2N-NH_3^+--$$

(ii) side-on hydrogen bonds

$$
\begin{array}{ccc}
\text{H} & \text{NH}_3{}^+ & \text{H} \\
| & | & | \\
----\text{H--N}----\text{H--N}----\text{H--N}---- \\
| & | & | \\
\text{NH}_3{}^+ & \text{H} & \text{NH}_3{}^+
\end{array}
$$

(iii) discrete $N_2H_5{}^+$ ions
(iv) coordinated N_2H_5 groups in metal complexes

The number of neighbours around an N_2H_5 ion at possible hydrogen bonding distances is usually greater than six, indicating that bifurcated (possibly trifurcated) bonds may exist in these cases. Reorientation of the $NH_3{}^+$ fragment with respect to NH_2 has been observed spectroscopically; however, unlike the situation in ammonium compounds, crystallographic studies have not demonstrated any pronounced *orientational disorder* of the hydrazinium ion. As expected, the hydrogen bond acceptor environment is often more extensive and usually at shorter average distance for the $NH_3{}^+$ groups than for NH_2.

Linking of hydrazinium into infinite chains by head to tail hydrogen bonding is observed in a large number of salts having organic, simple oxy or halide counteranions[87]. In N_2H_5Br, the chains are linked through N–H--N hydrogen bonds of length 2.93 Å[88], which contract on cooling to 2.91(1) Å at $-60\,°C$[89]. The covalent N–N bond expands (1.45–1.47 Å) despite identical unit cell parameters at both temperatures. A recent examination of hydrazinium halides[89] indicates a range of conductivities (up to $10^{-5}\,\Omega^{-1}\,cm^{-1}$) depending on the counteranion, and high activation energies.

The alternative hydrogen bonding arrangement involves NH_2 groups hydrogen bonded 'side-on' into chains, as in $Li(N_2H_5)SO_4$[90]. The chains run through channels defined by LiO_4 and SO_4 tetrahedra, the oxygen atoms of which are hydrogen bonded to the $NH_3{}^+$ extremity of hydrazinium. The isomorphous fluoroberyllate differs in its hydrogen bonding arrangement, which involves single, bifurcated and trifurcated bonds[91]. Lithium hydrazinium sulphate was first shown to be a proton conductor in 1964 and since that time has been the subject of extensive crystallographic[90], spectroscopic[92,93] and conductimetric[94] studies. For a long time, it represented the only known example in which hydrazinium is involved in the proton transport process.

For example, analysis by low temperature polarized Raman spectroscopy and room temperature polarized infrared reflectivity[92] demonstrated temperature dependence of bandwidths and intensities in the region of the external modes that could be related to the onset of reorientation of the $-NH_3^+$ fragment at 125 K. Proton exchange in the H–N–H plane of NH_2 groups above 150 K was detected by NQR[95]. These groups are expected to participate in proton transport via the extended hydrogen bonded network, although the possible rôle of NH_3^+ is less well defined. So-called 'proton-injection' is an unusual new approach to the vibrational spectroscopic study of protonic conducting solids. The solid electrolytic properties of $Li(N_2H_5)SO_4$ have been exploited to electrochemically insert deuterium into the protonated form, and its diffusion subsequently followed by the evolution of N–H and N–D stretching vibrations as a function of the distance from the anode (D_2 source). With increasing proximity of the laser to the anode, the N–H intensity decreased and that of N–D increased, demonstrating a nonhomogeneous distribution of deuterium[96].

Apparently the majority of hydrazinium salts can be classified in one or other of the above hydrogen bond arrangements, which thus seem to be mutually exclusive. However, in a few examples, the classification cannot be applied: in certain series of halocomplexes, although NH_3 and NH_2 could not be identified with certainty, mixtures of head to tail and side-on chains appear to develop in the lattices[97].

Examples in which hydrazinium ions are isolated from each other are rarer[98] and seem to be characterized by a richly furnished hydrogen bond acceptor environment of oxygen atoms, leading to the possible formation of polyfurcated bonds. In $N_2H_5ClO_4 . \frac{1}{2}H_2O$ no less than eleven oxygen neighbours belonging to water molecules and perchlorate groups surround the hydrazinium ion. It would seem that in the limited series of hydrated hydrazinium compounds, $N_2H_5^+$ groups are often isolated from each other, but are hydrogen bonded to water molecules. This has led authors to consider the possibility of proton transfer to water to form an oxonium ion[99].

The coordination of hydrazinium to metal ions[100] could be regarded as being of little relevance with respect to proton transfer processes however, the general formulae $(N_2H_5)_2M(II)(SO_4)_2$[101] or $(N_2H_5)M(III)(SO_4)_2 . H_2O$[102] define classes of compounds, the distinctive characteristics of which are their low dimensional structure, and potential associated properties. Ions M(II) or M(III) are linked through bridging sulphate groups to form linear chains while hydrazinium groups coordinate through their basic amine moiety leaving the positively charged fragment projecting into the interchain voids giving distinctive thrusting of NH_3^+ between essentially anionic strands. That the structural arrangement is conducive to the possession of interesting physical properties is shown by the antiferromagnetic behaviour of the Fe, Mn, Co, Ni and Cu members of the former series[103]. Although no examination of proton conducting properties has been performed to date, the presence of regions of high protonic density in these families of compounds could be favourable to proton transfer under specific conditions.

2.3.5   Hydrazinium $N_2H_6^{2+}$

In the family of inorganic hydrogen bond donating cations, $N_2H_6^{2+}$ is the only member to carry a double charge. This distinction does not preclude protonic conduction properties, which have been reported for a range of hydrazinium$(2+)$ salts[89, 104, 105]. Structural data up to 1982 have been reviewed[106].

In its most symmetric conformation, all hydrogen atoms are equivalent and, in this respect $N_2H_6^{2+}$ represents the dicationic equivalent of NH_4^+. As the ammonium ion is usually classed within the series of alkali metal ions, with an ionic radius equal to that of Rb^+, so it is useful in crystal chemistry to ponder upon the similarity between $N_2H_6^{2+}$ and an alkaline earth metal cation. It is clear that the $N_2H_6^{2+}$ ion radically departs from spherical symmetry, with a shape that can be approximated by an ovoid of principal axis dimensions $1.1 \times 1.1 \times 2.9$ Å^3 (if the longest dimension is estimated as a function of the mean of the interatomic N–N distance[106]). The two shortest dimensions are approximately equal to the ionic radius of Sr^{2+}, while the average, 1.7 Å, corresponds more closely to Ba^{2+}. Such size considerations can be used in tentative rationalization of the structural types adopted by hydrazinium$(2+)$ salts[106]. For example, in $N_2H_6X_2$, X = F, Cl[89, 107], the environment of $N_2H_6^{2+}$ is distorted cubic, and is related to the fluorite arrangement. Furthermore, it would seem that the size of hydrazinium should favour a CsCl type structure for N_2H_6X, X = MF_6[108], BeF_4[109], SO_4[110] etc. That this is not always observed, as for example in $N_2H_6SiF_6$[106, 111], must undoubtedly emphasize the rôle of anion size as a determining factor.

Whatever the structural type, $N_2H_6^{2+}$ is most typically encountered isolated from other $N_2H_6^{2+}$ species, but hydrogen bonded to an acceptor framework of oxygen or fluorine atoms. Examination of the literature shows that $N_2H_6{}^{2+}$ rarely adopts maximum $\bar{3}m$ symmetry and that this is not compensated for by formation of fluxional hydrogen bonds, disorder not having been detected crystallographically. On the other hand, the flexibility is reflected in the diversity of hydrogen bonding arrangements, including highly bent and polyfurcated bonds, to such an extent that single, linear hydrogen bonds are rare features of hydrazinium(2+) compounds[89,107].

Few examples of hydrated hydrazinium(2+) salts are known, but the limited data available demonstrate the possibility of forming low dimensional hydrogen bonded arrangements involving the water molecule as acceptor site. Chains, as in $N_2H_6(GeF_6).H_2O^{112}$ or $N_2H_6[NbOF_5].H_2O^{113}$, or two-dimensional arrays, as observed in $N_2H_6Br_2.2H_2O^{89}$, are built up through hydrogen bonding between alternating $N_2H_6{}^{2+}$ and H_2O units. Such substructures may play a non-negligible rôle in proton transport through a crystal lattice and, indeed, the possibility of proton transfer to water to form H_3O^+ in $N_2H_6(ClO_4)_2.2H_2O$ and in $N_2H_6Br_2.2H_2O$ has been suggested[99].

Acknowledgements

This review has largely made use of results and ideas of some of those who have made significant contributions to this field. We particularly acknowledge I. Olovsson, J. M. Williams, D. Mootz, O. Knop and M. Falk, and their coworkers, as well as others too numerous to name explicitly.

2.4 References

1. J.-O. Lundgren and I. Olovsson in *The Hydrogen Bond: Recent Developments in Theory and Experiments*, Vol. II, Ch. 10, P. Schuster, G. Zundel and C. Sandorfy (eds) (Elsevier (1976)).
2. I. Taesler, *Acta Universitatis Upsaliensis* **591** (1981) Uppsala.
3. J. M. Williams in *The Hydrogen Bond: Recent Developments in Theory and Experiments*, Vol. II, Ch. 14, P. Schuster, G. Zundel and C. Sandorfy (eds) (Elsevier (1976)).
4. J. Rozière, Thesis University of Montpellier 1973.
5. J. Emsley, *Chem. Soc. Rev.* **9** (1980) 91.
6. J. Emsley, D. J. Jones and J. Lucas, *Rev. Inorg. Chem.* **3** (1981) 105–40.
7. C. I. Ratcliffe and D. E. Irish in *Water Science Reviews*, F. Franks (ed) (Cambridge University Press (1986)), 149–214.

Protonic species and their structures

8. H. Goldschmidt and O. Udby, *Z. Physik. Chem.* **60** (1907) 728.
9. A. Hantzsch, *Z. Physik. Chem.* **61** (1908) 257.
10. G. Mascherpa, Thesis University of Montpellier 1965.
11. (a) F. S. Lee and G. B. Carpenter, *J. Phys. Chem.* **63** (1959) 279–82;
    (b) C. E. Nordman, *Acta Cryst.* **15** (1962) 18–23.
12. I. Olovsson, *J. Chem. Phys.* **49** (1968) 1063–7.
13. J. Almlöf, J.-O. Lundgren and I. Olovsson, *Acta Cryst.* B**27** (1971) 898–904.
14. J. Almlöf, *Acta Cryst.* B**28** (1972) 481–5.
15. J. Almlöf, *Chem. Scripta* **3** (1973) 73–9.
16. D. Mootz, E. J. Oellers and M. Wiebcke, *J. Am. Chem. Soc.* **109** (1987) 1200–2.
17. M. L. Huggins, *J. Phys. Chem.* **40** (1936) 723–31.
18. M. Volmer, *Justus Liebigs Ann. Chem.* **440** (1924) 200–2.
19. See, for example, A. Potier, J. M. Leclerq and M. Allavena, *J. Phys. Chem.* **88** (1984) 1125–30; E. Kochanski, *J. Am. Chem. Soc.* **107** (1985) 7869; R. Remington and H. F. Schaefer, reported in L. I. Yeh, M. Okumura, J. D. Myers, J. M. Price and Y. T. Lee, *J. Chem. Phys.* **91** (1989) 7319–30.
20. J. Almlöf and U. Wahlgren, *Theoret. Chim. Acta* **28** (1973) 161–8; J. Angyan, M. Allavena, M. Picard, A. Potier and O. Tapia, *J. Chem. Phys.* **77** (1982) 4723–33.
21. C. O. Selenius and R. G. Delaplane, *Acta Cryst.* B**34** (1978) 1330–2; A. Bino and F. A. Cotton, *J. Am. Chem. Soc.* **101** (1979) 4150–4; P. Teulon and J. Rozière, *Z. anorg. allg. Chem.* **483** (1981) 219–24.
22. J. M. Williams and S. W. Peterson, *Acta Cryst.* A**25** (1969) S113–14; R. Attig and J. M. Williams, *Inorg. Chem.* **15** (1976) 3057–61; J.-O. Lundgren and R. Tellgren, *Acta Cryst.* B**30** (1974) 1937–47.
23. M. Eigen and G. Demaeyer in *The Structure of Electrolytic Solutions* (Wiley (1959)).
24. R. A. Bell, G. G. Cristoph, F. R. Fronczek and R. E. Marsh, *Science* **190** (1975) 151–2.
25. R. M. Izatt and B. L. Haymore, *J. Chem. Soc. Chem. Commun.* (1972) 1308–9.
26. G. S. Heo and R. A. Bartsch, *J. Org. Chem.* **47** (1982) 3557; R. Chênevert, A. Rodrigue, M. Pigeon-Gosselin and R. Savoie, *Can. J. Chem.* **60** (1982) 853; R. Chênevert, A. Rodrigue, P. Beauchesne and R. Savoie, *Can. J. Chem.* **62** (1984) 2293; R. Chênevert, A. Rodrigue, D. Chamberland, J. Ouellet and R. Savoie, *J. Mol. Struct.* **131** (1985) 187–200.
27. J. P. Behr, P. Dumas, D. Moras, *J. Am. Chem. Soc.* **104** (1982) 4540–3.
28. C. B. Shoemaker, L. V. McAfee, D. P. Shoemaker and C. W. DeKock, *Acta Cryst.* C**42** (1986) 1310–13.
29. Yu. A. Simonov, N. F. Krasnova, A. A. Dvorkin, V. V. Vashkin, V. M. Abashkin and B. N. Laskorin, *Sov. Phys. Dokl.* **28** (1983) 823–6.
30. R. Chênevert, D. Chamberland, M. Simard and F. Brisse, *Can. J. Chem.* **67** (1989) 32–6.
31. J. L. Atwood, S. G. Bott, A. W. Coleman, K. D. Robinson, S. B. Whetstone and C. M. Means, *J. Am. Chem. Soc.* **109** (1987) 8100–1.
32. R. Neumann and I. Assael, *J. Chem. Soc. Chem. Commun.* (1989) 547–8.

The hydrogen bond and protonic species

33. D. Brodalla and R. Kniep, Z. Naturforsch. **35b** (1980) 403–4.
34. G. M. Brown, M.-R. Noë-Spirlet, W. R. Busing and H. A. Levy, Acta Cryst. **B33** (1977) 1038–46.
35. K. Mereiter, Tschermaks. Min. Petr. Mitt **21** (1974) 216–32; J. Tudo, B. Jolibois, G. Laplace and G. Nowogrocki, Acta Cryst. **B35** (1979) 1580–3; I. Vencato, E. Mattievich, L. DeMoreira and P. Mascarenhas, Acta Cryst. **C45** (1989) 367–71; E. Husson, M. Durand Le Floch, C. Doremieux-Morin, S. Deniard and Y. Piffard, Solid State Ionics **35** (1989) 133–42.
36. D. Mootz, E. J. Oellers and M. Wiebcke, Acta Cryst. **C44** (1988) 1334–7.
37. A. N. Fitch, A. F. Wright and B. E. F. Fender, Acta Cryst. **B38** (1982) 2546–54.
38. J.-O. Lundgren and I. Olovsson, Acta Cryst. **23** (1967) 971–6; J.-O. Lundgren, Acta Cryst. **B26** (1970) 1893–9.
39. I. Taesler, R. G. Delaplane and I. Olovsson, Acta Cryst. **B31** (1975) 1489–92.
40. D. Mootz and A. Merschenz-Quack, Z. Naturforsch. **42b** (1987) 1231–6.
41. J. M. Williams and S. W. Peterson, J. Am. Chem. Soc. **91** (1969) 776–7; T. Gustafsson, Acta Cryst. **C43** (1987) 816–19.
42. H. Henke, Acta Cryst. **B36** (1980) 2001–5.
43. J.-O. Lundgren, Acta Cryst. **B34** (1978) 2432–5.
44. H. Bode and G. Teufer, Acta Cryst. **8** (1955) 611–14.
45. D. W. Davidson and S. K. Garg, Can. J. Chem. **50** (1972) 3515–20.
46. D. W. Davidson, L. D. Calvert, F. Lee and J. A. Ripmeester, Inorg. Chem. **20** (1981) 2013–16.
47. M. Wiebcke and D. Mootz, Z. Kristallogr. **177** (1986) 291–9.
48. M. Wiebcke and D. Mootz, Z. Kristallogr. **183** (1988) 1–13.
49. D. Mootz and M. Wiebcke, Inorg. Chem. **25** (1986) 3095–7.
50. D. Mootz and E. J. Oellers, Z. anorg. Chem. **559** (1988) 27–39.
51. D. Mootz, E. J. Oellers and M. Wiebcke, Z. anorg. Chem. **564** (1988) 17–25.
52. W. K. Thompson, Trans. Faraday Soc. **62** (1966) 2667–73.
53. K.-D. Kreuer, A. Rabenau, W. Weppner, Angew. Chem. Int. Ed. Engl. **21** (1982) 208–9.
54. H. Wise, J. Phys. Chem. **71** (1967) 2843–6.
55. A. Potier and D. Rousselet, J. Chim. Phys. **70** (1973) 873–8.
56. K.-D. Kreuer, A. Rabenau and R. Messer, Appl. Phys. **A32** (1983) 45–53; A. T. Howe, and M. G. Shilton, J. Solid State Chem. **28** (1979) 345–61.
57. E. Krogh Andersen, I. G. Krogh Andersen, E. Skou and S. Yde-Andersen, Solid State Ionics **18/19** (1986) 1170–4; S. Yde-Andersen, E. Skou, I. G. Krogh Andersen and E. Krogh Andersen, in Solid State Protonic Conductors for Fuel Cells and Sensors III, J. B. Goodenough, J. Jensen and A. Potier (eds) (Odense University Press (1985)).
58. M. A. Subramanian, B. D. Roberts and A. Clearfield, Mat. Res. Bull. **19** (1984) 1471.
59. D. R. Balasubramanian, S. V. Bhat, M. Mohan and A. K. Singh, Solid State Ionics, **28/30** (1988) 664–7.
60. S. Chandra, S. K. Tolpadi and S. A. Hashmi, Solid State Ionics **28/30** (1988) 651–5.
61. D. J. Jones and J. Rozière, to be published.

Protonic species and their structures

62. I. Brach and J. Rozière, to be published.
63. A. A. Khan and W. H. Baur, *Acta Cryst.* **B28** (1972) 683–93.
64. I. Knop, W. Westerhaus, J. Wolfgang, M. Falk and W. Massa, *Can. J. Chem.* **63** (1985) 3328–53, and references therein.
65. W. C. Hamilton and J. A. Ibers, in *Hydrogen Bonding in Solids* (W. A. Benjamin, New York (1968)) 222–6.
66. B. Morosin, *Acta Cryst.* **B26** (1970) 1635–7.
67. R. S. Seymour and A. W. Pryor, *Acta Cryst.* **B26** (1970) 1487–91.
68. R. C. Plumb and D. F. Hornig, *J. Chem. Phys.* (1953) 366–7.
69. E. O. Schlemper, W. C. Hamilton and J. J. Rush, *J. Chem. Phys.* **44** (1966) 2499–505.
70. E. O. Schlemper and W. C. Hamilton, *J. Chem. Phys.* **45** (1966) 408–9.
71. C. Knakkerguard Møller, *Acta Chem. Scand.* **8** (1954) 81–7.
72. (a) S. N. Bhakay-Tamhane, A. Sequeira and R. Chidambaram, *Acta Cryst.* **B36** (1980) 2925–9; (b) I. A. Oxton, O. Knop and M. Falk, *Can. J. Chem.* **54** (1976) 892–9.
73. O. Knop, I. A. Oxton, W. J. Westerhaus and M. Falk, *J. Chem. Soc. Faraday* II **77** (1981) 309–20.
74. S. D. Hamman, *Aust. J. Chem.* **31** (1978) 11–18.
75. T. C. Waddington, *J. Chem. Soc.* (1958) 4340–4.
76. J. T. R. Dunsmuir and A. P. Lane, *Spectrochim. Acta* **28A** (1970) 45–50.
77. J.-P. Mathieu and H. Poulet, *Spectrochim. Acta* **16** (1960) 696–703.
78. I. A. Oxton, O. Knop and M. Falk, *Can. J. Chem.* **53** (1975) 3394–400.
79. I. A. Oxton, O. Knop and M. Falk, *J. Mol. Struct.* **37** (1977) 69–78.
80. (a) H. G. Smith and H. A. Levy, *Acta Cryst.* **15** (1962) 1201–4; (b) C. S. Choi, H. J. Prask and E. Prince, *J. Chem. Phys.* **61** (1974) 3523–9; (c) G. Peyronel and A. Pignedoli, *Acta Cryst.* **B31** (1975) 2052–6; (d) C. S. Choi, H. J. Prask and E. Prince, *Acta Cryst.* **B32** (1976) 2919–20; (e) J.-O. Lundgren and R. Liminga, *Acta Cryst.* **B35** (1979) 1023–7; (f) J.-O. Lundgren, *Acta Cryst.* **B35** (1979) 1027–33.
81. G. W. Watt and W. R. McBride, *J. Am. Chem. Soc.* **77** (1955) 1317–20.
82. I. Olovsson, *Acta Chem. Scand.* **14** (1960) 1453–65.
83. I. Olovsson, *Acta Chem. Scand.* **14** (1960) 1466–74.
84. I. Nahringbauer, *Acta Chem. Scand* **22** (1968) 1141–58; 2981–92.
85. H. Henke, E. Buschmann and H. Bärnighausen, *Acta Cryst.* **B29** (1973) 2622–4.
86. D. J. Jones, I. Brach and J. Rozière, *J. Chem. Soc. Dalton Trans.* (1984) 1795–800, and references therein.
87. K. Sakurai and Y. Tomiie, *Acta Cryst.* **5** (1952) 293–4; J. W. Conant and R. B. Roof, *Acta Cryst.* **B26** (1970) 1928–32; P. Bukovec and L. Golic, *Acta Cryst.* **B32** (1976) 948–50; R. Liminga, *Acta Chem. Scand* **19** (1965) 1629–42; S. A. Hady, I. Nahringbauer and I. Olovsson, *Acta Chem. Scand.* **23** (1969) 2764–72; N. A. K. Ahmed, R. Liminga and I. Olovsson, *Acta Chem. Scand.* **22** (1968) 88–96; A. Nilsson, R. Liminga, I. Olovsson, *Acta Chem. Scand.* **22** (1968) 719–30.
88. K. Sakurai and Y. Tomiie, *Acta Cryst.* **5** (1952) 289.

89. C. Garcia, D. Barbusse, R. Fourcade and B. Ducourant, *J. Fluorine Chem.* **51** (1991) 245–56.
90. I. D. Brown, *Acta Cryst.* **17** (1964) 654–60; J. H. Van den Hende and H. Boutin, *Acta Cryst.* **17** (1964) 660–3; V. M. Padmanabhan and R. Balasubramanian, *Acta Cryst.* **22** (1967) 532–7; M. R. Anderson and I. D. Brown, *Acta Cryst.* **B30** (1974) 831–2.
91. M. R. Anderson, I. D. Brown and S. Vilminot, *Acta Cryst.* **B29** (1973) 2625–7.
92. S. H. Brown and R. Frech, *Solid State Ionics* **18/19** (1986) 1020–4; R. Frech and S. H. Brown, *Solid State Ionics* **35** (1989) 127–32.
93. S. H. Brown and R. Frech, *Spectrochim. Acta* **44A** (1988) 1–15.
94. J. Vanderkooy, J. D. Cuthbert and H. E. Petch, *Can. J. Phys.* **42** (1964) 1871–8; K.-D. Kreuer, W. Weppner and A. Rabenau, *Solid State Ionics* **3/4** (1981) 353–8.
95. R. N. Hastings and T. Oja, *J. Chem. Phys.* **57** (1972) 2139–46.
96. S. H. Brown and R. Frech, *Solid State Ionics* **28/30** (1988) 607–10.
97. J. Slivnik, J. Pezdic and B. Sejec, *Mh. Chem.* **98** (1967) 204–5; B. Kojic-Prodic, S. Scavnicar, R. Liminga and M. Sljukic, *Acta Cryst.* **B28** (1972) 2028–32; A. Braibanti and A. Tiripicchio, *Gazz. Chim. Ital.* **96** (1966) 1580–8.
98. R. Liminga and J.-O. Lundgren, *Acta Chem. Scand.* **19** (1965) 1612–28; R. Liminga, *Acta Chem. Scand.* **21** (1967) 1217–28; D. Gajapathy, S. Govindarajan, K. C. Patil and H. Manohar, *Polyhedron* **2** (1983) 865–73; A. Braibanti, G. Bigliardi, A. M. Manotti Lanfredi and A. Tiripicchio, *Nature* **211** (1966) 1174–5.
99. K. C. Patil, C. Nesamani and V. R. Pai Verneker, *J. Chem. Soc. Dalton Trans.* (1983) 2047–50.
100. D. B. Brown, J. A. Donner, J. W. Hall, S. R. Wilson, R. B. Wilson, D. J. Hodgson and W. E. Hatfield, *Inorg. Chem.* **18** (1979) 2635–41; W. Granier, S. Vilminot and H. Wahbi, *Rev. Chim. Min.* **22** (1985) 285–92; A. Taha, Thesis University of Montpellier 1989.
101. C. K. Prout and H. M. Powell, *J. Chem. Soc.* (1961) 4177–82.
102. S. Govindarajan, K. C. Patil, H. Manohar and P.-E. Werner, *J. Chem. Soc. Dalton Trans.* (1986) 119–23.
103. C. Cheng, H. Wong and W. M. Reiff, *Inorg. Chem.* **16** (1977) 819–22.
104. S. Chandra and N. Singh, *J. Phys.* **C16** (1983) 3081–97.
105. H. Barbès, G. Mascherpa, R. Fourcade and B. Ducourant, *J. Solid State Chem.* **60** (1985) 100–5.
106. T. S. Cameron, O. Knop and L. A. Macdonald, *Can. J. Chem.* **61** (1983) 184–8.
107. M. L. Kronberg and D. Harker, *J. Chem. Phys.* **10** (1942) 309–17; J. Donohue and W. Lipscomb, *J. Chem. Phys.* **15** (1947) 115–19.
108. B. Kojic-Prodic, B. Matkovic and S. Scavnicar, *Acta Cryst.* **B27** (1971) 635–7; B. Kojic-Prodic, S. Scavnicar and B. Matkovic, *Acta Cryst.* **B27** (1971) 638–44.
109. M. R. Anderson, S. Vilminot, I. D. Brown, *Acta Cryst.* **B29** (1973) 2961–2.

110. P.-G. Jönssen and W. C. Hamilton, *Acta Cryst.* **B26** (1970) 536–46; L. F. Power, K. E. Turner, J. A. King and F. H. Moore, *Acta Cryst.* **B31** (1975) 2470–4.
111. B. Frlec, D. Gantar, L. Golic and I. Leban, *Acta Cryst.* **B36** (1980) 1917–18.
112. B. Frlec, D. Gantar, L. Golic and I. Leban, *Acta Cryst.* **B37** (1981) 666–8.
113. Yu. E. Gorbunova, V. I. Pakhomov, V. G. Kuznetsov and E. S. Kovaleva, *J. Struct. Chem. USSR* **13** (1972) 154–5; V. A. Sarin, V. Ya. Dudarev, L. E. Fykin, Yu. E. Gorbunova, E. G. Il'in and Yu. B. Buslaev, *Dokl. Phys. Chem.* **236** (1977) 892–5.

3 Proton conductors: classification and conductivity

PHILIPPE COLOMBAN AND ALEXANDRE NOVAK

3.1 Introduction

The distinction between conductors and insulators is often not well-defined but is, to a certain extent, a matter of choice. Conductivity can vary by more than ten orders of magnitude, sometimes over a temperature interval of only a few degrees. The charge carriers can be either electrons (or holes) or ions (or vacancies). The occurrence of ionic conductivity implies diffusion of matter through the solid and/or at its surface, although the distinction between the latter two is not always clear-cut. In a densely packed structure, ionic diffusion is associated with a local perturbation of the structure, such as the presence of defects which facilitate the migration of ions or vacancies.

The most common defects in a compound AB are, using Kröger's notation:

$$A_A \rightleftharpoons V_A + A_i \text{ (Frenkel defect)}$$

where A_A is atom A on site A, V_A is vacant site A and A_i is atom A on an interstitial site,

$$A_A + B_B \rightleftharpoons A_B + B_A \text{ (Anti-structure defect)}$$

in which A_B represents atom A on site B and B_A is atom B on site A,

$$A_A + B_B \rightleftharpoons V_A + V_B \text{ (Schottky defect)}$$

The defects can be charged positively (˙), negatively (′) or they can remain neutral (x), giving rise to electronic defects, e.g.

$$V_A{}^x \rightleftharpoons V_A{}' + h^\cdot$$

where h represents an electron hole.

In materials containing ionic bonds, the defects (i.e. ions or vacancies) are charged naturally and therefore ionic transport is synonymous with ionic conduction.

Ionic compounds, such as halides, sulphides, oxides, nitrides and certain polymers, can be divided into three major groups: (i) insulators with a residual ionic conductivity lower than $10^{-10}\,\Omega^{-1}\,\text{cm}^{-1}$, where the electronic contribution is generally of the same order of magnitude, (ii) ionic conductors (IC) in which the presence of point defects leads to a conductivity of up to $10^{-5}\,\Omega^{-1}\,\text{cm}^{-1}$, and (iii) superionic conductors (SIC) with a conductivity of at least $10^{-4}\,\Omega^{-1}\,\text{cm}^{-1}$. The main difference between the last two groups of materials concerns the activation energy (E_a): in the case of SIC, E_a is lower than 0.4 or even 0.2 eV, while in IC, values varying between 0.6 and 1.2 eV are usually observed. The superionic conductors have thus a high conductivity far below the melting point. This fundamental difference is due principally to the particular structures of SIC. In ionic conductors, the defects allowing the diffusion of charge carriers must be created thermally, while in superionic conductors, the potentially mobile species are already numerous and the structure appears sufficiently loosely packed to facilitate dynamic disorder and diffusion of such entities. In fact, as shown by the Nernst–Einstein law,

$$\sigma = (DC\,e^2)/kT$$

the conductivity, σ, is proportional to the product of the diffusion coefficient, D, and the concentration, C, of the mobile species. Both D and C are thermally activated and σ can be expressed as follows:

$$\sigma T = [(D_0 C_0\,e^2)/k][\exp - (E_f + E_d)/kT] = \sigma_0 \exp - E_a/kT$$

where E_f and E_d are enthalpies of formation and diffusion of charge carriers, respectively, E_a is the activation energy for conduction and σ_0 is the associated preexponential factor. E_a and σ_0 are thus characteristic electrical properties of a material. For glasses and polymers other conductivity laws are often used (see Chapters 4, 20 and 30).

There are numerous publications concerning the theory and applications of superionic conduction[1-6] (see also the Preface). Table 3.1 gives the main electrical parameters of some typical superionic conductors. Superionic conductivity has been observed for a number of ions, Ag^+, Li^+, Na^+, K^+, Cu^+, Pb^{2+}, F^-, O^{2-}, NH_4^+, $H^+(H_2O)_n$; the lowest E_a has been found in AgI-based materials. Some compounds exhibit both high ionic and electronic conductivity. Glassy materials are usually poorer conductors and have higher activation energy than crystalline materials of similar composition.

Table 3.1. Non-protonic superionic conductors: main electrical parameters[1–6]

Compounds	Ionic charge carrier	Dimensionality	σ_T (Ω^{-1} cm^{-1})	T (°C)	E_a (eV)	E_G (eV)	σ_e (Ω^{-1} cm^{-1})
Na β-Al$_2$O$_3$	Na$^+$	2D	3×10^{-2}	RT	0.16	6	10^{-8}
Na$_3$Zr$_2$Si$_2$PO$_{12}$ (NASICON)	Na$^+$	3D	10^{-1}	300	0.15		
NASICON (glassy)	Na$^+$	3D	10^{-4}	600	0.65		
LiAlSiO$_4$	Li$^+$	1D	10^{-3}	RT	1		
LiAlSiO$_4$ (glassy)	Li$^+$	3D	10^{-9}	RT			
K β''-Al$_2$O$_3$	K$^+$	2D	10^{-1}	RT	0.21		
Ag$_4$RbI$_5$	Ag$^+$	3D	3×10^{-1}	RT	0.1	3.2	
AgI	Ag$^+$	3D	1.9	200	0.1		
0.75AgI 0.25Ag$_2$SeO$_4$ (glassy)	Ag$^+$	3D	6×10^{-2}	RT			
0.67AgI 0.25Ag$_2$O 0.08P$_2$O$_5$ (glassy)	Ag$^+$	3D	10^{-2}	RT			
Cu$_4$RbCl$_3$I$_2$	Cu$^+$	3D	5×10^{-1}	RT	0.15		
Li$_3$N	Li$^+$	2D	5×10^{-3}	RT	0.3	2.2	$<10^{-4}$
Pb β''-Al$_2$O$_3$	Pb^{2+}	2D	4×10^{-3}	RT			
PbF$_2$	F$^-$	3D	10^{-4}	100	0.45		
ZrO$_2$: Y$_2$O$_3$	O^{2-}	3D	0.12	1000	0.8		

Mixed conductors

Compound	Ion	Dim	σ_T	T	E_a	E_G	σ_e
Ag$_3$Si	Ag$^+$	3D	3	200	0.1	1.8	$\leq 10^{-4}$
Ag$_2$S	Ag$^+$	3D	4	200	0.1	0.9	
Cu$_{1.75}$Se	Cu$^+$	3D	3×10^{-2}	RT	0.05		~ 1
Ag$_{2x}$GeSe$_{2+x}$ (glassy)	Ag$^+$	3D					

Polymers

Compound	Ion	Dim	σ_T	T	E_a		
PPO–NaCF$_3$SO$_4$ (amorphous)	Na$^+$	3D	10^{-5}	37			
MEEP–LiSCN (amorphous)	Li$^+$	3D					
(MEEP)$_4$LiCF$_3$SO$_3$	Li$^+$	3D	10^{-4}	300	1.6		
PEO–LiClO$_4$	Li$^+$	3D	$10^{-6}/10^{-7}$	RT	0.31		
(PEO)$_4$–AgClO$_4$	Ag$^+$	3D	$10^{-3}/10^{-5}$	RT	1.02		

PEO: Poly(ethylene oxide); MEEP: $-(\text{N}=\text{P}(\text{OC}_2\text{H}_4\text{OC}_2\text{H}_4\text{OCH}_3)_2)_n$, Poly(bis-(methoxy ethoxy ethoxide) phosphazene; PPO: poly(propylene oxide). Dimensionality of the conduction: mono-(1D), bi-(2D) or tri-(3D) dimensional. E_a: activation energy of ρ_T; RT: room temperature; σ_T: total conductivity; σ_e: electronic conductivity; E_G: band gap.

3.2 Classification of protonic conductors

3.2.1 Definition

A material is usually defined as a solid proton conductor if protons can be transferred through the solid and converted to hydrogen gas at the cathode*. This process must be maintained over a long period of time provided there is a corresponding supply of some form of protons at the anode. The conducting species can be protons, oxonium, ammonium or hydrazinium ions and hydroxyl groups. A major problem is to obtain convincing proof that charge is transferred by a protonic species. Quantitative determination of the amount of hydrogen gas liberated at the electrode as a function of the current passed, according to Faraday's law, is a necessary but not sufficient criterion. A suitable method is to establish different charge-carrier concentrations at opposite faces of a sample of proton conductor, for instance by applying different pressures of H_2 gas. Alternatively, different materials, each supplying a fixed electrochemical potential of the proton (e.g. PdHx) may be used as the electrodes.

Space-charge polarization at the partially non-ohmic sample–electrode interfaces indicates the occurrence of ionic transport. The polarization arises from the accumulation of charge-carriers at the electrodes and is related to the different rates of arrival (injection) and transfer (discharge, diffusion) of charge carriers at the sample–electrode interfaces. A decay in current over an extended time period is thus observed.

The Hall effect is well-known for electronic conductors. Although it forms the basis of the best method to determine the sign and effective number of electronically conducting species, only a few attempts have been made to use it with ionic conductors and these have been limited to non-protonic conductors[7].

3.2.2 Classification

Protonic conductors can be classified according to the preparation method, structural dimensionality and conductivity mechanism. As far as the preparation method is concerned, there are two main ways: either direct synthesis or ion exchange. The first is usually used for hydrates while the second involves the substitution of conducting ions of a

---

* Note that metal hydrides, containing hydrogen ion, develop hydrogen at the anode when subjected to electrolysis.

superionic conductor by immersing the sample in a molten salt or an acid solution. If, in a particular compound, a mobile ion can be reduced easily (e.g. Ag^+ in β-alumina), then annealing at 300–500 °C under hydrogen, followed by hydration, allows ion exchange with H^+ to occur. Thus Ag^+ β-alumina may be converted into H^+ β-alumina by the following scheme[8]:

$$Ag^+ \text{ β-alumina} + H_2 \rightarrow H^+ \text{ β-alumina} + Ag^{\cdot}$$

Na^+/H^+ (D^+) ion exchange in a radio frequency plasma has also been carried out with success[9].

Direct synthesis methods have the advantage of the versatility of chemical synthesis and it is possible to prepare materials in a rich variety of forms (low-temperature sinterable powders, thin or thick film etc). Ion exchange methods, on the other hand, have numerous problems: (i) exchange by protonic species must frequently take place in hot concentrated acid, which may attack the rigid anionic framework; exchange in molten salts appears to be milder, (ii) substitution of an ion by a protonic species of different size brings about considerable constraints in polycrystalline material. Size differences associated with ion exchange may give rise to stresses/strains in the ceramics or even in the component crystals[10]. Moreover, in order to carry out the exchange without destroying the structure, the latter must be tridimensional (3D) or bidimensional (2D). The case of β-alumina shows that, with care, ceramics tolerate ion exchange without becoming porous. In particular, mechanical degradation can be minimized using successive exchange by ions of progressively increasing (or decreasing) size (Chapter 33).

The most usual classification criteria of ionic conductors concern the nature of the mobile ion and the dimensionality of the potentially mobile ion sublattice. These criteria have important practical consequences for the possible applications of the materials. A polycrystalline material with a structure containing 1D conduction pathways will have a very low overall conductivity. This is because the grain boundaries are effectively a barrier to long range conduction. Each grain boundary encountered along the macroscopic diffusion pathway is a wall. There is little chance that tunnels of the 1D structure may be aligned from one grain to another. On the other hand, for theoretical or analytical aspects of conduction mechanisms, the 1D and 2D structures are simpler to handle. Table 3.2 gives a classification according to the dimensionality[11–33].

Finally, protonic conductors and protonic superionic conductors can be distinguished using their σ_0 and E_a values as criteria. Fig. 3.1 and Table 3.3

Table 3.2. *Classification of typical protonic conductors according to the (approximate) dimensionality of the conduction path*

	References	Abbreviations
One dimension (1D)?		
$CsHSO_4$ below 318 K (phase I)	11	CsH
Imidazole	12	
Proteins	13	
Two dimensions (2D)		
$H_3OUO_2PO_4 . 3H_2O$	14	HUP
$H_3OFe(SO_4)_2 . 3H_2O$	15	HFeS
$1.25NH_4 . 11Al_2O_3$ (β-Al_2O_3)	16	$NH_4\beta$
$1.25H^+(H_2O)_n . 11Al_2O_3$ (β-Al_2O_3)	17	$H_3O^+\beta, n = 1$
$1.66H^+(H_2O)_n . 11Al_2O_3$ (ion-rich β-Al_2O_3)	17	irHn$\beta, n \sim 2$
$1.66(NH_4^+, H^+(H_2O)_n) . 11Al_2O_3$ (ion-rich β-Al_2O_3)	17	irN/Hβ
$1.66(NH_4^+, H^+(H_2O)_n) . 11Al_2O_3$ ($\beta''Al_2O_3$)	17	irN/Hβ''
$UO_2(H_2PO_4)_2 . 3H_2O$	18	$U(H_2P)_2$
Three dimensions (3D) – bulk properties		
$CsHSO_4$ above 417 K (phase III)	11, 9	CsH
$(NH_4)_3H(SeO_4)_2 - (NH_4)_3H(SO_4)_2$	15, 20, 21	$N_3HSe_2 - N_3HS_2$
$H_3PW_{12}O_{40} . nH_2O$ ($n = 21-29$)	20, 23	P21–P29
H_3OClO_4	24	PO
$H_2Sb_4O_{11} . nH_2O$ ($n = 2, 3$)	25	HSb-2
$HxWO_3$	26	
$H_3OZr_2(PO_4)_3$	27	
$C_6H_{12}N_2 . 1.5HSO_4$	28	TED $1.5H_2SO_4$
Transition aluminas (γ)	17	
Three dimensions (3D–S) – surface properties		
$V_2O_5 . nH_2O$	29	
$Zr(HPO_4)_2 . nH_2O$	30	αZrP
$SnO_2 . 2H_2O$	31	Sn-2
$ZrO_2 . 1.75H_2O$	31	Zr-1.75
Three dimensions (3D–P) – porous structures		
Nafion®	32	
Porous inorganic gels	33	

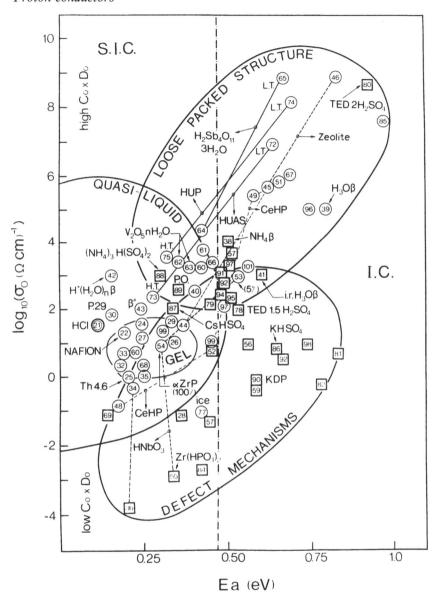

Fig. 3.1. Relationship between the pre-exponential factor σ_0 and the activation energy of conduction for protonic conductors. Squares and circles correspond to anhydrous and hydrated compounds, respectively (double circle: HCl solution). Dotted straight lines indicate the effect of varying water vapour pressure and solid lines the effect of temperature. The numbers are explained in Table 3.3 (with permission[34]).

Table 3.3. *Room temperature conductivity* (σ_{RT}) *and activation energy* (E_a) *of protonic conductors classified according to their method of synthesis*

Materials		Number[a]	σ_{RT} ($\Omega^{-1}\,cm^{-1}$)	E_a (eV)	Mechanisms (assumed)	References
Ionic exchanged materials						
β-Al$_{11}$O$_{16}\cdot(1+x)$M$_2$O						
M = NH$_4^+$	$x = 0.25$	38	1.5×10^{-6}	0.5	ion jump	16
M = H$_3$O$^+$	$x = 0.25$	39	$\sim 10^{-10}$	0.8	ion jump	16
M = M$^+$(H$_2$O)$_2$	$x = 0.25$	40	3×10^{-7}	0.4	ion jump + Grotthuss	17
β-Al$_{11-y}$O$_{17}$Mg$_y$M$_{1+y}$						
M = H$_3$O$^+$	$y = 0.66$	41	$\sim 10^{-10}$	0.6	ion jump	17
M = H$^+$(H$_2$O)$_2$	$y = 0.66$	42	6.5×10^{-6}	0.15	ion jump + Grotthuss	17
M = NH$_4^+$ + εH$^+$(H$_2$O)$_n$	$y = 0.66$		2.5×10^{-5}	0.18	ion jump + Grotthuss	17
β''-Al$_{11-y}$O$_{17}$Mg$_y$M$_{1+y}$						
M = H$^+$(H$_2$O)$_2$	$y = 0.66$	42	7×10^{-6}	0.17	ion jump + Grotthuss	17
M = NH$_4^+$ + εH$^+$(H$_2$O)$_n$	$y = 0.66$		3×10^{-5}	0.24	ion jump + Grotthuss	17
NH$_4$ β-gallate			10^{-4}			41
Surface acidic conductivity of H$^+$(H$_2$O)$_n$						
β''-Al$_2$O$_3$		43	$\geqslant 10^{-3}$	0.2	surface (liquid-like)	36
H$_3$OZr$_2$(PO$_4$)$_3$ (NASICON)		56	$< 10^{-6}$	0.56	ion jump	27
HZr$_2$(PO$_4$)$_3$		57	10^{-7}	0.44		
H$_2$Sb$_4$O$_{11}\cdot$3H$_2$O		64	2×10^{-3}	0.42	ion jump + Grotthuss	25
H$_2$Sb$_4$O$_{11}\cdot$2H$_2$O		67	2×10^{-6}	0.69	ion jump	25

$HTaWO_6 \cdot 0.5/1\ H_2O$	35	$10^{-8}/10^{-4}$	0.23		38
$HNbO_3 \cdot 0.4\ H_2O$	34	10^{-4}	0.21		43
$HNbO_3$	36	5×10^{-8}	0.20		89
$H_3OTi_{1.2}Nb_{0.7} \cdot H_2O$		8×10^{-3}	0.19		92
$HTiNbO_5$		4×10^{-4}	0.14		92
$HTiTaO_5$		4×10^{-4}	0.15		92
$HSbTeO_6 \cdot H_2O$	60	$10^{-2}/10^{-5}$	0.44		39
H-natrolite		$\leqslant 10^{-4}$			40
Zeolite (NH_4^+)	44-45	$0.5/2.5 \times 10^{-5}$	0.36/0.84		37
H-mordenite $(H_{4.3}(NH_4)_{4.4}(AlO_2)_{8.7}(SiO_2)_{39.3} \cdot 10.5H_2O)$				ion jump	40
NH_4-analcine $\cdot nH_2O$		$\leqslant 4 \times 10^{-5}$			37
NH_4-sodalite $\cdot nH_2O$		$\leqslant 2 \times 10^{-3}$	0.23		37
$NH_4\beta''$-gallate		$\leqslant 2 \times 10^{-4}$			41
$(H_3O)_5GdSi_4O_{12}$		3×10^{-6}	0.20		42
		10^{-6}	0.48		
$HSbO_3 \cdot 1\text{-}2H_2O$ (pyrochlore)	33	3×10^{-3}	0.20		43
$HSbO_3 \cdot 1\text{-}2.5H_2O$ ('layer-ilmenite')	32	2×10^{-3}	0.19		43
$HSbWO_6 \cdot H_2O$		$< 10^{-6}$			93
$HSbO_3 \cdot 0.5H_2O$ (cubic)	31	9×10^{-4}	0.39		43
H^+-montmorillonite		$\sim 10^{-4}$			44
$HSbP_2O_8 \cdot 4H_2O$		10^{-6}	0.2		45
$HSbP_2O_8 \cdot 10H_2O$		10^{-2}			45

(continued)

Table 3.3. *Continued*

Materials	Number[a]	σ_{RT} (Ω^{-1} cm^{-1})	E_a (eV)	Mechanisms (assumed)	References
$H_3Sb_3P_2O_{14} \cdot 6H_2O$		10^{-4}	0.43		45
$H_3Sb_3P_2O_{14} \cdot 10H_2O$		4×10^{-3}	0.33		45
$H_5Sb_5P_2O_{20} \cdot 2H_2O$		10^{-8}			45
$H_5Sb_5P_2O_{20} \cdot 7H_2O$		10^{-4}	0.4		45
$H_5Ti_4O_9 \cdot 1.2H_2O$		10^{-4}			46
Direct synthesis					
$Sb_2O_5 \cdot 5.4H_2O$	23	7.5×10^{-3}	0.16	surface (liquid-like)	47
$In(OH)_3 \cdot 1.4H_2O$	24	3.5×10^{-3}	0.24	surface (liquid-like)	47
$ThO_2 \cdot 4.6H_2O$	25	4×10^{-4}	0.20	surface (liquid-like)	47
$ZrO_2 \cdot 1.75H_2O$	26	2×10^{-5}	0.34	surface (liquid-like)	31
$SnO_2 \cdot 2H_2O$	27	4×10^{-4}	0.20	surface (liquid-like)	31
$H_3PW_{12}O_{40} \cdot 28/29H_2O$		$\sim 10^{-4}$	0.15/0.25		22, 23
$H_3PW_{12}O_{40} \cdot 21H_2O$		$\sim 6 \times 10^{-3}$	0.4		22, 23
$H_3SiW_{12}O_{40} \cdot 28H_2O$		$\sim 3 \times 10^{-2}$	0.40		22, 23
$H_3PMo_{12}O_{40} \cdot 29H_2O$	30	1.7×10^{-1}	0.15		
$Zr(HPO_4)_2 \cdot nH_2O$	53–54	10^{-3}–10^{-8}	0.3–0.5	surface and bulk	30, 48
$Ce(HPO_4)_2 \cdot 3.8H_2O$	48	2×10^{-4}	0.17	surface	49, 50
$Ce(HPO_4)_2 \cdot 1.2H_2O$	52	10^{-6}	0.45	surface	49, 50
γ-$Ti(HPO_4)_2 \cdot 2H_2O$	58	7×10^{-5}	0.1	surface	51
γ-$Ti(HPO_4)_2$	59	4×10^{-6}	0.6	surface	51
$H_3OUO_2AsO_4 \cdot 3H_2O$	72	5×10^{-3}	0.27	ion/water jump	14, 52

$H_{2x}Sb_{2x}W_{2-2x}O_6 \cdot nH_2O$	74–76	$<10^{-6}$	0.35	surface?	93
$H_3OUO_2PO_4 \cdot 3H_2O$	77	5×10^{-3}	0.19	surface?	52, 53
$H_3OFe(SO_4) \cdot 3H_2O$		10^{-3}	0.30	surface?	15
$H_8UO_2(IO_6)_2 \cdot 4H_2O$		6×10^{-3}	0.20	surface?	54
$UO_2(H_2PO_4)_2 \cdot 3H_2O$		5×10^{-5}	0.48		18
$NH_4H_2(IO_3)_3$	94	10^{-6}	0.36		55
H_3OClO_4	89	3.5×10^{-4}		H^+ jump + 'free' rotation	24
$HClO_4 \cdot 5 \cdot 5H_2O$ (liquid)	78–79	3	0.02		82
$C_6H_{12}N_2 \cdot 1 \cdot 5H_2SO_4$	82	10^{-7}	0.53		28
$(C_2H_5NO_2)_3H_2SO_4$	83	10^{-7}	0.80		56
$(NH_2)_2COHNO_3$	84	2×10^{-7}	1		57
$(NH_2)_2CO$	85	10^{-9}	0.4		58
$(COOH)_2 \cdot 2H_2O$	86	10^{-9}	1		59
$KHSO_4$	57	10^{-8}	0.64	defect	60
H_2O (ice) (260 K)		$10^{-7}/10^{-8}$	0.55		34
(KOH doped, 260 K)		$\leqslant 10^{-2}$			95
$CsHSO_4$		10^{-6}	0.3	defect	11, 19
$RbHSO_4$		10^{-8}	0.35	defect	61, 62
$(NH_4)_3H(SeO_4)_2$		10^{-3}	0.25		15, 20
$LiN_2H_5SO_4$	96	2×10^{-8}	0.75		63
$N_2H_6SO_4$		10^{-7}	0.23		76
KD_2PO_4	90	10^{-8}	0.56	defects	64

(continued)

Table 3.3. Continued

Materials	Number[a]	σ_{RT} $(\Omega^{-1}\,cm^{-1})$	E_a (eV)	Mechanisms (assumed)	References
$NH_4H_2PO_4$	91	10^{-8}	0.48	defects	65
KH_2AsO_4	92	10^{-8}	0.66	defects	66
KH_2PO_4	93	10^{-8}	0.55	defects	67
CsH_2PO_4		10^{-7}	1		94
$SnCl_2 \cdot 2H_2O$	97	10^{-6}	0.92		68
KHF_2	98	10^{-6}	0.74		69
$V_2O_5 \cdot 0.5H_2O$	63	5×10^{-5}	0.42		29
$V_2O_5 \cdot 1.6H_2O$	62	3×10^{-3}	0.35		29
$H_3OAl_3(SO_4)_2(OH)_6$		$10^{-4}/10^{-9}$			75
$(CH_3)_4NOH \cdot 5H_2O$		4.5×10^{-3}	0.32		96
Polymers					
$PEO-HSO_4$		2×10^{-4}			70
$PAA-NH_4HSO_4$		2.4×10^{-5}			70
$PEI-H_3PO_4(H_2SO_4)$		5×10^{-6}			70
$PVA-H_3PO_4$		$10^{-5}-10^{-3}$	1		71, 73
$PEO-H_3PO_4$		10^{-4}	1.26		71
PANI (pH > 2) polyaniline $((C_6H_4)-NH-(C_6H_4)-NH) \cdot nH_2O$ $(C_6H_4)-N-(C_6H_4)-N)$		1		mainly electronic conductivity	91
Nafion®	22	5×10^{-2}	0.22		32
High temperature conductors (T)					
γ-Al_2O_3 (700 K)		10^{-2}	0.7–1	H^+ jump?	17

Compound	σ	Mechanism	Ref.
CsHSO$_4$ (420 K)	10^{-2}	H$^+$ jump + free rotation	11, 19
Cs$_3$H(SeO$_4$)$_2$ (460 K)	10^{-3}		89
Rb$_3$H(SeO$_4$)$_2$ (460 K)	10^{-3}		94
NH$_4$HSO$_4$ (430 K)	10^{-3}	H$^+$ jump + free rotation	72
CsH$_2$PO$_4$ (505 K)	10^{-2}		94
SrCeO$_3$: Yb (500 K)	10^{-6}		77–79
BaCeO$_3$: Yb (500 K)	10^{-5}		78
CaHPO$_4$.2H$_2$O (600 K)	10^{-3}		80
H$_3$AlP$_3$O$_{10}$.nH$_2$O (600 K)	10^{-2}		81
NH$_4$$^+$ β-gallate/aluminate (500 K)	$> 10^{-2}$		17, 41
Insertion compounds			
MnO$_{1.78}$.nH$_2$O	10^{-5}	H$^+$ jump?	83
MnO$_2$			84
PbO$_2$			85
Miscellaneous			
KOH (520 K)	10^{-3}		86
NiOH			90
CsOH, RbOH			74
CsOH.H$_2$O			87, 88
HxWO$_3$			26

^{a} Number corresponds to those used in Fig. 3.1.

Abbreviations: PEO, poly(ethylene oxide); PEI, Poly(ethylenimine); PVA, Poly(vinylalcohol).

show the correlation between these two parameters for different types of proton conductors[11–88] classified according to the conductivity mechanism: i.e. defect mechanisms in anhydrous compounds, ion jumps in loosely packed structures and quasi-liquid state conduction[34–37]. Conductivity plots of some typical protonic conductors are shown in Fig. 3.2[34]. The σ_0 term is directly related to the concentration of conducting species. Low values are observed for extrinsic defect conductors while loosely packed structures yield high σ_0 values. Quasi-liquid like conduction occurs in gels and is characterised by intermediate σ_0 values. Similar behaviour is seen for instance in HCl solutions, at least as far as especially low activation energy (E_a) values are concerned. Fig. 3.3 shows the following types of material.

(i) Anhydrous protonic conductors, the behaviour of which is rather similar to that of alkali halide IC. The activation energy is high and the proton conduction is related to the presence of (intrinsic and extrinsic) defects. The conducting species are protons or proton vacancies; typical examples are KH_2PO_4 and ice (Fig. 3.3a).

(ii) Ionic conductors containing a loosely packed lattice with a high concentration of potentially mobile species (Fig. 3.3b). These have high σ_0 and high activation energy at low temperature. The conducting species are ions such as H_3O^+ or NH_4^+ as found in, for example, oxonium or ammonium β-alumina. As the temperature rises, the rigid lattice is not much affected but dynamic disorder of mobile species commences and consequently E_a is strongly diminished.

(iii) Compounds containing protonic species in a quasi-liquid state. Such states can exist either inside the structure (intrinsic or bulk conductors) or at the surface (gels or particle hydrates). Various mobile species (Fig. 3.3c) can move at different rates using different paths and some, such as the proton, may jump from one to the other (proton jump or Grotthuss mechanism).

In some cases, several mechanisms may act simultaneously. The overall conductivity consists of the intrinsic conductivity due to the lattice and that of the surface. The latter depends strongly on the water pressure partial which determines the quantity of adsorbed water. Protonic species can be associated in several hydrogen bonded layers as found in clays and other compounds; finally gels or even salts may be formed. In Fig. 3.4, the characteristic properties, such as conductivity (σ) and water content (n) are plotted against water pressure partial for three typical protonic conductors: hydrated uranyl phosphate ($H_3OUO_2PO_4 . 3H_2O$ or HUP) an intrinsic conductor which is not water pressure sensitive; $V_2O_5 . nH_2O$, a gel where both intrinsic and surface conductivity must be taken into account; and $Ce(HPO_4)_2 . nH_2O$, which is a typical surface conductor[49]. It should be pointed out that the behaviour of σ and

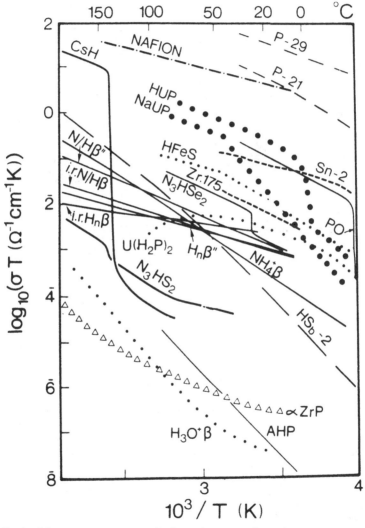

Fig. 3.2. Conductivity versus temperature plot for various protonic conductors. For explanation of symbols, see Table 3.2 (with permission[34]).

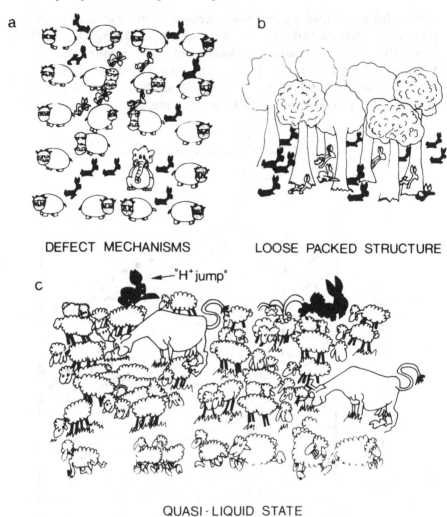

DEFECT MECHANISMS   LOOSE PACKED STRUCTURE

QUASI-LIQUID STATE

Fig. 3.3. Illustration of the main proton transfer mechanisms: (a) defect mechanism in a densely packed structure; (b) loosely packed structure with a high concentration of mobile species; (c) quasi-liquid state with a proton jump contribution[34]. In (a) the conductivity is favoured by intrinsic (interstitial rabbits) or extrinsic (impurity: elephant) point defects. An orientation defect (hippopotamus in the wrong orientation) can also favour disorder of rabbits (O_2^- for ZrO_2: CaO, H for $KHSO_4$); (b) the tree sublattice is a perfectly stable loosely packed structure and a high rabbit disorder can exist without affecting the host lattice (e.g. NH_4^+ in β-Al_2O_3); (c) only the mobile species sublattice is considered here; these entities are moving with different speeds in different directions and some are hopping: such may be the image of a quasi liquid or surface liquid ($V_2O_5 \cdot nH_2O$, HUP).

n for the above conductors is analogous to the I, II and III types of Brunauer, Emmett and Teller's adsorption isotherms.

The conductivity mechanism may change, within the same type of structure, as a function of temperature and water vapour partial pressure. In compounds such as HUP, HUAs and $H_2Sb_4O_{11} \cdot 3H_2O$, ion hopping dominates at low temperatures while at high temperature a quasi-liquid behaviour exists (Fig. 3.1). According to Hairetdinov[35], this is characteristic of structures where defect formation and migration are observed, i.e. for IC and SIC with $E_a > 0.4$ eV. In this correlation based on 40

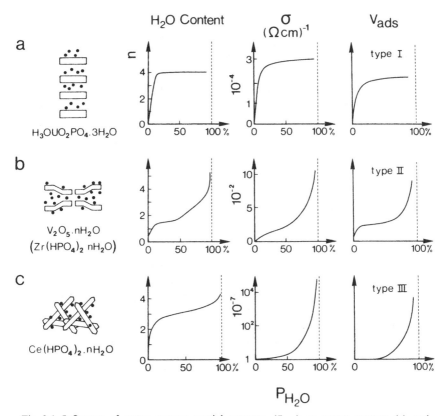

Fig. 3.4. Influence of water vapour partial pressure (P_{H_2O}) on water content (n) and conductivity (σ) for (a) an intrinsic conductor, a crystalline hydrate HUP; well-defined steps corresponding to various stoichiometries can be observed, e.g. in $H_xSb_xP_2O_{3x+5} \cdot nH_2O$ $(n = 2\text{--}10)$, $x = 1, 3, 5$[45]; (b) a mixed surface–bulk conductor $V_2O_5 \cdot nH_2O$ or $Zr(HPO_4)_2 \cdot nH_2O$; (c) surface conductor $CeHPO_4 \cdot nH_2O$. Brunauer adsorption isotherms (V_{ads}) are given for comparison[34, 49] (with permission).

non-protonic ionic conductors and 160 superionic conductors, a deviation from the linearity of the relationship $\log \sigma_0 = aE_a + b$ is observed only for low E_a values. This is related to the melting of the conducting ion sublattice in such structures. In the case of protonic conductors, the change in water content leads to a vertical shift indicating a variation in the number of charge carriers only.

In order to understand the mechanisms of proton conduction, it appears necessary to determine the structure of the rigid framework as well as that of the potentially mobile protonic species. X-ray and neutron diffraction methods are well-suited for an accurate determination of the rigid framework structure, but less so for the investigation of the exact nature of protonic species and their localization in superionic conductors. In particular, they do not distinguish easily between static and dynamic disorder. Vibrational spectroscopy, both neutron and optical, on the other hand, appears complementary and can give an 'instant' and not average image of different configurations of protonic entities.

Complex impedance spectroscopy, *quasi*-elastic neutron scattering and nuclear magnetic resonance spectroscopies may also contribute considerably to a good knowledge of proton dynamics.

3.3   References

1. G. D. Mahan and W. L. Roth (eds) *Superionic Conductors* (Plenum Press, New York (1976)).
2. P. Hagenmuller and W. Van Gool (eds) *Solid Electrolytes* (Academic Press, New York (1986)).
3. S. Geller (ed) *Solid Electrolytes* (Springer Verlag, Berlin (1977)).
4. T. W. Wheat, A. Ahmad and A. K. Kuriakose (eds) *Progress in Solid Electrolytes* (CANMET (Energy, Mines and Resources) Ottawa (1983)).
5. R. Collongues, D. Gourier, A. Kahn, J. P. Boilot, Ph. Colomban and A. Wicker, *J. Phys. Chem. Solid* **45** (1984) 981–1013.
6. R. Lascar and S. Chandra (eds) *Recent Developments in Superionic Conductors* (Academic Press, New York (1987)); J. R. MacCallum and C. A. Vincent, *Polymer Electrolyte Reviews* (Elsevier Applied Science, London (1987)).
7. B. A. Boukamp and C. A. Wiegers, *Solid State Ionics* **9/10** (1983) 1193–6; D. S. Newman, C. Frank, R. W. Matlack, S. Twining and V. Krishnan, *Electrochimica Acta* **22** (1977) 811–14.
8. Ph. Colomban, G. Lucazeau and A. Novak, *J. Phys. C (Solid State Physics)* **14** (1981) 4325–33.
9. J. P. Schnell, G. Velasco, M. Croset, D. Dubreuil, D. Dieumegard and Ph. Colomban, *Solid State Ionics* **9/10** (1983) 1465–8.

Proton conductors

10. P. Lenfant, D. Plas, M. Ruffo and Ph. Colomban, *Mat. Res. Bull.* **15** (1980) 1817–27, and see Chapter 33.
11. J. C. Badot and Ph. Colomban, *Solid State Ionics* **35** (1989) 143–9.
12. A. Kawada, A. R. McGhie and M. M. Labes, *J. Chem. Phys.* **52** (1970) 3121–5.
13. E. W. Krapp, K. Schulten and Z. Schulten, *Chem. Phys.* **46** (1980) 215–29.
14. K.-D. Kreuer, A. Rabenau and R. Messer, *Appl. Phys.* A32 (1982) 45–53; 155–8.
15. I. Brach, Thesis, University of Montpellier (1986).
16. Ph. Colomban, J. P. Boilot, A. Kahn and G. Lucazeau, *Nouv. J. Chim.* **2** (1978) 21–32.
17. N. Baffier, J. C. Badot and Ph. Colomban, *Solid State Ionics* **2** (1981) 107–13; *Solid State Ionics* **13** (1984) 233–6.
18. R. Mercier, M. Pham-Thi and Ph. Colomban, *Solid State Ionics* **15** (1985) 113–26.
19. A. I. Baranov, N. N. Shchagina and L. A. Shuvalov, *J.E.T.P. Lett.* **36** (1982) 459–62.
20. A. I. Baranov, I. P. Makarova, L. A. Muradyan, A. V. Tregubchenko, L. A. Shuvalov and V. I. Simonov, *Kristallografiya* **32** (1987) 682–94.
21. I. P. Aleksandrova, Ph. Colomban, F. Denoyer, N. Le Calvé, A. Novak and A. Rozycki, *Phys. Status Sol.* (a) **114** (1989) 822–34.
22. O. Nakamura, I. Ogino and T. Kodama, *Solid State Ionics* **3/4** (1981) 347–51.
23. K.-D. Kreuer, M. Hampele, K. Dolde and A. Rabenau, *Solid State Ionics* **28/30** (1988) 589–93.
24. A. Potier and D. Rousselet, *J. Chim. Phys.* **70** (1973) 873–8.
25. Y. Piffard, M. Dion, M. Tournoux and H. Aribart, *C.R. Acad. Sci. Ser. C* **290** (1980) 437–40; *Solid State Ionics* **7** (1982) 91–9.
26. See Chapter 7.
27. M. A. Subramanian, B. D. Roberts and A. Clearfield, *Mat. Res. Bull.* **19** (1984) 1471–8; M. Ohta, A. Ono and F. P. Okamura, *J. Mater. Sci. Letts* **6** (1987) 583–5.
28. T. Takahashi, S. Tanase, O. Yamamoto and S. Yamauchi, *J. Solid State Chem.* **17** (1976) 353–61.
29. Ph. Barboux, N. Baffier, R. Morineau and J. Livage, *Solid State Ionics* **9/10** (1983) 1073–80.
30. M. Casciola and U. Costantino, *Solid State Ionics* **20** (1986) 69–73.
31. W. A. England, M. G. Cross, A. Hamnett, P. J. Wiseman and J. B. Goodenough, *Solid State Ionics* **1** (1980) 231–49.
32. R. C. T. Slade, A. Hardwick and P. G. Dickens, *Solid State Ionics* **9/10** (1983) 1093–100.
33. Ph. Colomban and J. P. Boilot, *Rev. Chim. Minérale* **22** (1985) 235–55.
34. Ph. Colomban and A. Novak, *J. Mol. Struct.* **177** (1988) 277–308.
35. N. F. Uvarov and E. F. Hairetdinov, *J. Solid State Chem.* **62** (1986) 1–10.
36. G. C. Farrington and J. L. Briant, *Mat. Res. Bull.* **13** (1978) 763–73.
37. K.-D. Kreuer, N. Weppner and A. Rabenau, *Mat. Res. Bull.* **17** (1982) 501–9.
38. C. M. Mari, F. Bonino, M. Catti, R. Pasinetti and S. Pizzini, *Solid State*

Ionics **18/19** (1985) 1013–19; C. M. Mari, A. Arghileri, M. Catti and G. Chiodelli, *Solid State Ionics* **28–30** (1988) 642–6.

39. X. Turillas, G. Delabouglise, J. C. Joubert, T. Fournier and J. Muller, *Solid State Ionics* **17** (1985) 169–74.
40. M. Lal, C. M. Johnson and A. T. Howe, *Solid State Ionics* **5** (1981) 451–4.
41. T. Tsurumi, H. Ikawa, K. Urabe, T. Nishimura, T. Oohashi and S. Udagawa, *Solid State Ionics* **25** (1987) 143–53.
42. K. Yamashita and P. S. Nicholson, *Solid State Ionics* **20** (1986) 147–51.
43. U. Chowdhry, J. R. Barkley, A. D. English and A. W. Sleight, *Mat. Res. Bull.* **17** (1982) 917–33; C. E. Rice, *J. Solid State Chem.* **64** (1986) 188–9.
44. S. H. Sheffield and A. T. Howe, *Mat. Res. Bull.* **14** (1979) 929–35.
45. S. Deniard-Courant, Y. Piffard, P. Barboux and J. Livage, *Solid State Ionics* **27** (1988) 189–94.
46. E. Krogh-Andersen, I. G. Krogh-Andersen and E. Skou, *Solid State Ionics* **27** (1988) 181–7.
47. D. J. Dzimitrocwicz, J. B. Goodenough and P. J. Wiseman, *Mat. Res. Bull.* **17** (1982) 971–9.
48. G. Alberti, M. Casciola, U. Costantino and M. Leonardi, *Solid State Ionics* **14** (1984) 289–95.
49. Ph. Barboux, Thesis, University of Paris (1987); Ph. Barboux, R. Morineau and J. Livage, *Solid State Ionics* **27** (1988) 221–5.
50. Ph. Colomban, EEC Contract, Final Report (1989), EN3E-0062-DK(B).
51. G. Alberti, M. Bracardi and M. Casciola, *Solid State Ionics* **7** (1982) 243–7.
52. A. T. Howe and M. G. Shilton, *J. Solid State Chem.* **28** (1979) 345–61.
53. M. Pham-Thi and Ph. Colomban, *Solid State Ionics* **17** (1985) 295–306.
54. M. G. Shilton and A. T. Howe, in *Fast Ion Transport in Solids*, P. Vashista, J. N. Mundy and G. K. Shenoy (eds) (Elsevier, North Holland, New York (1979)) 727–30.
55. A. I. Baranov, G. F. Dobrzhauskii, V. V. Ilyukhin, V. S. Ryabkin, Yu.N. Sokolov, N. I. Sorokina and L. A. Shuvalov, *Kristallographiya* **26** (1981) 1259–63; P. Bordet, J. X. Boucherle, A. Santoro and M. Marezio, *Solid State Ionics* **21** (1986) 243–54.
56. G. Royal, B. Martin and G. Godefroy, *C.R. Acad. Sci. B* **775** (1972) 353.
57. T. R. N. Katty and A. R. U. Marthy, *Ind. J. Chem.* **11** (1973) 253–6.
58. S. N. Batt, *Curr. Sci.* **42** (1973) 198–201.
59. J. M. Poolock and A. R. Ubberhold, *Trans. Faraday Soc.* **52** (1956) 1112–17.
60. S. E. Rogers and A. R. Ubberhold, *Trans. Faraday Soc.* **46** (1950) 1051–61.
61. A. I. Baranov, R. M. Fedosyuk, N. M. Shchagina and L. A. Shuvalov, *Ferroelectrics Lett.* **2** (1984) 25–8.
62. M. Komukae, T. Osaka, Y. Makita, T. Ozaki, K. Itoh and E. Nakamura, *J. Phys. Soc. Japan* **50** (1981) 3187–8.
63. J. Van der Kooy, *Can. J. Phys.* **42** (1964) 1871–7.
64. V. H. Schmidt and E. A. Vehling, *Phys. Rev.* **126** (1962) 447–54.
65. J. M. Pollock and M. Sharan, *J. Chem. Phys.* **51** (1969) 3604–6.
66. C. T. Perrino, B. Blan and R. Alsdork, *Inorg. Chem.* **11** (1972) 571–4.

67. R. S. Bradley, D. C. Munro and S. L. Ali, *J. Inorg. Nucl. Chem.* **32** (1970) 2513–16.
68. F. R. Mognaschi, A. Chierico and G. Panavacini, *J. Chem. Soc. Faraday Trans.* **74** (1978) 2333–8.
69. J. Bruinink, *J. Appl. Electrochem.* **2** (1972) 239–49.
70. M. F. Daniel, B. Desbat and J. C. Lassègues, *Solid State Ionics* **28/30** (1988) 632–6; M. F. Daniel, B. Desbat, F. Cruege, O. Trinquet and J. C. Lassègues, *Solid State Ionics* **28/30** (1988) 637–41.
71. P. Donoso, W. Gorecki, C. Berthier, F. Defendini and M. Armand, in *Proc. Int. Conf. on Solid Electrolytes, St Andrews, Scotland* (1987) 17–23; A. Polak, S. Petty-Weeks and A. J. Beukler, *Chem. Energ. News* (1985) 28–33; P. Donoso, W. Gorecki, C. Berthier, F. Defendini, C. Poinsignon and M. B. Armand, *Solid State Ionics* **28/30** (1989) 969–74.
72. Yu. N. Moskvich, A. A. Sukhovski and O. V. Rozanov, *Ser. Phys. Solid State* **26** (1984) 21–5.
73. S. Petty-Weeks and A. J. Polak, *Sensors and Actuators* **11** (1987) 377–86; R. P. Singh, P. N. Gupta, S. L. Agrawal and U. P. Singh, *Mater. Res. Soc. Symp. Proc.*, vol. 135 (1989) 361–66; K. C. Gong and H. Shou-Cai, *Mater. Res. Soc. Symp. Proc.*, vol. 135 (1989) 377–82.
74. D. T. Amm and S. Segel, *Z. Naturforsch.* **41a** (1986) 279–82; J. Henning, H. D. Lutz, H. Jacobs and B. Moch, *J. Mol. Struct.* **196** (1989) 113–23.
75. Y. Wing, M. Lal and A. T. Howe, *Mat. Res. Bull.* **15** (1980) 1649–54.
76. S. Chandra and N. Singh, *J. Phys. C (Solid State Physics)* **16** (1983) 3081–97.
77. See H. Iwahara, Chapter 18.
78. T. Scherban, W. K. Lee and A. S. Nowick, *Solid State Ionics* **35** (1989).
79. N. Bonanos, B. Ellis and M. N. Mahmoud, *Solid State Ionics* **28/30** (1988) 579–84.
80. E. Montoneri, G. Modica and G. C. Pappalardo, *Solid State Ionics* **26** (1988) 203–7.
81. E. Montoneri, F. J. Salzano, E. Findl and F. Kulesa, *Solid State Ionics* **18/19** (1986) 944–1002.
82. T. H. Huang, R. A. Davis, U. Frese and U. Stimming, *J. Phys. Chem. Letts* **92** (1988) 6874–6.
83. J. C. Charenton and P. Strobel, *Solid State Ionics* **24** (1987) 333–41.
84. K.-D. Kreuer, in *Proc. Electrochem. Soc.* (1985), 85–4, MnO_2 Electrode Theory, Electrochem. Appl., 21–47.
85. J. R. Gavarri, P. Garnier, P. Boher, A. J. Dianoux, G. Chedeville and B. Jacq, *J. Solid State Chem.* **75** (1988) 251–62.
86. D. W. Murphy, J. Broadhead and B. C. H. Steele (eds) *Materials for Advanced Batteries, Nato Conference, ser. VI, Mater. Science* (Plenum Press, New York (1980)); B. Sh. El'Kin, *Solid State Ionics* **37** (1990) 139–48.
87. M. Stahn, R. E. Lechner, H. Dachs and H. E. Jacobs, *J. Phys. C (Solid State Physics)* **16** (1983) 5073–82.
88. J. Gallier, B. Toudic, M. Stahn, R. E. Lechner and H. Dachs, *J. Physique (France)* **49** (1988) 949–57.

The hydrogen bond and protonic species

89. A. I. Baranov, B. V. Merinov, A. B. Tregubchenko, L. A. Shuvalov and
    N. M. Shchagina, *Ferroelectrics* **81** (1988) 187–91.
90. V. A. Volynsskii and Yu.N. Chernykh, *Electrokhimiya* **13** (1977) 1070–4.
91. J. P. Travers and M. Nechtschein, *Synthetic Metals* **21** (1987) 135–41; J. P.
    Travers, C. Menardo, M. Nechtschein and B. Villeret, *J. Chim. Phys.* **86**
    (1989) 71–84.
92. C. Chr. Schüler, *Z. Phys. B* (Condensed Matter) **68** (1987) 325–8.
93. M. Rivière, J. L. Fourquet, J. Grins and M. Nygren, *Mat. Res. Bull.* **23**
    (1988) 965–75.
94. A. I. Baranov, B. V. Merinov, A. V. Tregubchenko, V. P. Khiznichenko,
    L. A. Shuvalov and N. M. Shchagina, *Solid State Ionics* **37** (1989) 279–82.
95. A. V. Zaretskii, V. F. Petrenko, A. V. Trukhanov and V. A. Chesnokov,
    Solid State Ionics **36** (1989) 225–6.
96. N. Kukiyama, T. Sakai, H. Miyamura, A. Kato and H. Ishikawa, *J.
    Electrochem. Soc.* **137** (1990) 335–6.

4 Defects, non-stoichiometry and phase transitions

PHILIPPE COLOMBAN AND ALEXANDRE NOVAK

4.1 Introduction

Protonic conduction can be considered as a particular case of ionic conduction; however, there are some similarities with electronic conduction because of the proton size. In both cases, defects can play an important role.

Ionic conductivity in a compact structure is generally very low and requires a high activation energy. The diffusion coefficient of oxygen in amorphous silica at 1000 °C, for instance, is 10^{-14} cm^2 s^{-1} and that in MgO is 10^{-20} cm^2 s^{-1}, with activation energies between 1 and 4 eV[1]. In single crystal and polycrystalline alumina at 1700 °C, the diffusion coefficient decreases to 10^{-15} and 10^{-12} cm^2 s^{-1}, respectively, and the activation energy increases to 6–8 eV. The diffusion coefficients of smaller cations such as Al^{3+} are of the order of 10^{-9}–10^{-10} cm^2 s^{-1} with activation energy of about 5 eV[1,2].

The diffusion coefficient of Al^{3+} ions in interstitial sites can be obtained from electrical conductivity measurements. Its activation energy is considerably lower (2.5 eV) than that of lattice diffusion. This applies not only to Al^{3+} in alumina but generally to all atoms occupying defect sites[2]. All compounds have naturally a certain number of intrinsic defects as well as those associated with impurities (extrinsic defects). Intrinsic defects can be thermally activated and are particularly important for non-stoichiometric compounds. In the latter, diffusion coefficients are generally higher and the activation energy lower than that in stoichiometric compounds. The diffusion coefficient of the O^{2-} ion in Fe$_{0.95}$O at 800 °C, for instance, is about 10^{-8} cm^2 s^{-1} with $E_a \sim 2$ eV and varies strongly with non-stoichiometry which, in turn, is determined by the oxygen pressure[1]. Certain low density structures can have a large quantity of vacant sites and/or potentially mobile species independently of external factors. Such structures have been found among superionic conductors such as β- and β″-aluminas, structures derived from quartz or cristobalite, e.g. LiAlSiO$_4$ and structures based on silver iodide. Before discussion of the specific behaviour of protonic conductors, the role of defects in the ionic and electronic conductivity of solids will be reviewed.

4.2   Ionic mobility and conductivity

The presence of mobile species in a particular medium can give rise to a macroscopic concentration gradient obeying *Fick's law*[1]. The flux of matter in a given x direction is given by $J = -D\,dC/dx$, where C is the concentration of mobile species and D the diffusion coefficient. The conductivity is given by $\sigma = CBe^2$, where B is the mobility and e the charge of the mobile ions. The Nernst–Einstein law which connects diffusion coefficient and conductivity is: $\sigma = DCe^2/kT$, where k is Boltzmann's constant and T temperature. Furthermore, the thermal activation of C and D must be taken into account, as follows:

$$C = C_0 \exp - \frac{E_f}{2kT} \quad \text{and} \quad D = fD_0 \exp - \frac{E_m}{kT}$$

E_f corresponds to the enthalpy of formation of a defect pair (*Frenkel* or *Schottky*, cf. Chapter 3). In an alkali halide, E_f is typically about 3 eV implying a very low defect concentration, of the order of 10^{-14} at room temperature. E_m is the enthalpy of defect migration, usually several eV in dense structures, and f is the correlation or Haven factor. Its value varies between 0 and 1 and takes into account unfruitful attempts at transport in a given direction, caused by random jumps of mobile species and the particular geometry of each site[3]. In other words, this factor takes into account the correlation effects ($f = 1$ when correlations are absent). Usually f is of the order of 0.5–0.8 and plays a role in determining the transport mechanism.

The conductivity can thus be written

$$\sigma = f\,\frac{C_0 D_0 e^2}{kT} \exp - \frac{E_a}{kT} \quad \text{where } E_a = \frac{E_f}{2} + E_m$$

The term $f\,C_0 D_0 e^2/k = \sigma_0$ is frequently called the prefactor.

The tracer or self-diffusion coefficient represents only a random walk diffusion process, i.e. in the absence of chemical potential gradients. The true chemical diffusion coefficient refers to diffusion in a chemical-potential gradient and its expression is more complicated[4,5].

In the case of polymers, different expressions for the conductivity are used, such as the *Vogel–Tamman–Fulcher (VTF) relationship*[6]

$$\sigma = \sigma_0 \exp - (B/K(T - T_0))$$

where σ_0 is a prefactor combining the charge, carrier concentration and some constants and is proportional to $T^{-\frac{1}{2}}$; B is a constant proportional to V^* (*Van der Waals volume* of mobile species) and is inversely proportional to expansivity; T_0 is a critical temperature close to T_g, the *glass transition temperature*. (In fact, polymers are 'good' conductors only above T_g.) An improved form of the *VTF* equation, including an activation energy term accounting for the production of free carrier ions, has also been used[9]. Models taking into account dynamical contributions are discussed in Chapter 30.

The *VTF* equation was initially proposed in order to explain the viscosity of undercooled liquids. Other models using the free volume concept are connected to the above relationship. Free volume theories are based on the assumption that the mobile species move only when, locally, a sufficiently large void opens[6]. These models are used for glasses which do not follow the Davies–Jones law linking the difference of thermal dilation and compressibility of the liquid state.

High ion mobility in solids is attained if the following conditions are met: (i) the number of vacant positions in the rigid framework should exceed that of ions available to occupy the vacancies and (ii) the activation energy of migration between these positions should not be too high, i.e. the dimensionless E_a/kT ratio should not exceed a value in the range 1 to 4 at temperatures far below the melting (decomposition) point. In terms of crystalline structure, this means that a channel of sufficient size must link the accessible sites in order that ion diffusion is not hindered; if channels are too large with respect to the ion radius, ions can be trapped. In polymers, the region of high conductivity is found in the homogeneous, elastomeric amorphous phase and the presence of partial crystallinity inhibits conduction. The motion of mobile ions is strongly linked to that of polymer segments and ionic conductivity drops to extremely low values below T_g where the chain segment motions are frozen. Both anions and cations are mobile in many polymer/salt electrolytes.

4.2.1   Glasses

The simplest model of glass structure, Zachariasen's continuous random network model, represents glasses as being formed by random

Table 4.1. *Typical values of the diffusion coefficient (D) in typical non-protonic conductors*

Compound	Mobile species	D (cm^2 s^{-1})	T (°C)
AgI	Ag$^+$	10^{-5}	250
Li$_3$N	Li$^+$		
Na β-Al$_2$O$_3$[a]	Na$^+$	10^{-5}	300
K β''-Al$_2$O$_3$[b]	K$^+$	10^{-5}	50
ZrO$_2$	O^{2-}	10^{-6}	1300
PEO: Li$^+$[c]	Li$^+$	$\leqslant 10^{-10}$	100
Na	Na	4×10^{-5}	100
Li-Hg	Li	10^{-5}	25

[a] $1.25(Na_2O).11Al_2O_3$. [b] $(Al_{11-y}Mg_yO_{16})(K_{1+y}O)$, $y \sim 0.6$. [c] Polyethylene oxide: Li$^+$: $[(CH_2—CH_2—O)_n]_6LiCF_3SO_3$.

orientational disorder of bond angles. Correspondingly, the simple *Rasch–Henderson equation*, log $\sigma = A - B/T$, has been used to describe conductivity[8]. There are, however, a series of reports indicating that glasses and glassy electrolytes in particular are inhomogeneous on a 1–10 nm scale. The nature of the conduction mechanism can vary from hopping in a disordered random network potential[9] to a percolation pathway involving higher mobilities in a particular region of an inhomogeneous glass, with slower motions in the intervening structure[10]. Similar situations may be found in plastic or rotator phases[11]. Analogous models to those applied to polymers are also used.

In good conductors, the diffusion coefficient can attain a value similar to that in aqueous solution or in the molten state ($\sim 10^{-5}$ cm^2 s^{-1}). The corresponding value of the ionic mobility is about 10^{-3} cm^2/Vs. This has been observed for many non-protonic conductors (the main examples are given in Table 4.1) and for some protonic materials (Table 4.2). If the number of mobile ions is sufficiently high, superionic conductivity occurs[12].

4.3 Electronic conduction and non-stoichiometry

The main interest of ionic conductors or solid electrolytes in comparison with liquid electrolytes lies in the enormous difference in mobility of ionic species of opposite charge. In the former, this difference is generally more than ten orders of magnitude. Electronic conductivity can be a serious

Table 4.2. *Typical values of proton (hydrogen) diffusion coefficient at various temperatures*

Compounds	T (K)	D_H (cm²/s)	E_a (eV)	References
$H^+(H_2O)_n\beta''Al_2O_3$	300	$\leqslant 10^{-5}$	0.14	53, 54
$H^+(H_2O)_n\beta''\text{-}Al_2O_3$	300	3×10^{-8}	0.2	55
$H_3O^+\beta\text{-}Al_2O_3$	300	10^{-13}		53
$NH_4^+\beta\text{-}Al_2O_3$	300	7×10^{-9}	0.05	56
$Li^+\beta\text{-}Al_2O_3(.H_2O)$	300	6×10^{-7}		57
$Pb^{2+}\beta\text{-}Al_2O_3(\varepsilon H_2O)$	620	10^{-7}	0.54	58
H_3OClO_4	300	2×10^{-9}	0.36	59
$N_2H_6SO_4$	300	10^{-8}		60
$H_4PSi_{12}O_{40}.28H_2O$	300	10^{-6}	0.39	61
$H_3PW_{12}O_{40}.28H_2O$	300	4.5×10^{-6}	0.27	61
$H_3PW_{12}O_{40}.21H_2O$	300	2×10^{-7}		61
$H_3OUO_2AsO_4.3H_2O$	290	3×10^{-8}	0.8	41
$H_{1.68}MoO_3$	295	4.8×10^{-6}		62, 63
$(H_2O)_{0.42}H_2Ta_2O_6$	300	10^{-9}		64
Zeolites				
X(synthetic)	310	10^{-5}	0.3	65
NaP	350	10^{-4}	0.35	65
Chabazite	310	10^{-7}	0.3	65
Natrolite	310	10^{-8}	0.6	65
Heulandite	310	10^{-7}	0.2–0.5	65
PbO_2	300	$10^{-6}/10^{-7}$		66
Y_2O_3	800	4×10^{-7}	1.7	19
SiO₂				
Quartz	700	$10^{-8}/10^{-14}$	0.6–0.8	65, 67
Obsidian	400	10^{-14}	0.9	65
Vermiculite	300	$10^{-5}/10^{-7}$		65
$CsOH.H_2O$	400	2×10^{-7}		68
TiO_2	620	3×10^{-8}	0.6–1.2	22, 69
Polyaniline	300	10^{-9}		70

source of short-circuiting since it can easily amount to $10^{-9}\,\Omega^{-1}\,cm^{-1}$, thereby reducing the advantages of pure ionic conductors. In certain cases a high electronic conductivity associated with a high ionic conductivity is sought for use in low polarization electrodes. *Intercalation* materials belong to this group.

The electronic properties of crystalline materials are described in terms of the well-known band theory of solids. The generation of electronic carriers by interband transitions in superionic conductors is somewhat more complicated than in usual semi-conductors because of the structural disorder[13].

Intrinsic and extrinsic defects do not play the same role. With the latter, impurities determine the electron (hole) concentration in the conduction (valence) band. A very important practical case occurs when non-stoichiometry depends on the atomic exchange between a crystal and the surrounding medium, usually gas. A simple binary non-stoichiometric compound MX_{n+x}, for instance, is an intrinsic semiconductor or dielectric if $x = 0$. This ideal stoichiometry at constant temperature is observed under a partial pressure of the most volatile element if the electron concentration (n_e) is balanced by that of holes (n_h); $n_{\text{intrinsic}} = n_e = n_h$.

Under these conditions, the composition change may be described by Kröger's quasi-chemical equations (see Chapter 3).

$X_X \leftrightarrows V_X + \tfrac{1}{2}X_2\uparrow$

$\tfrac{1}{2}X_2 \leftrightarrows X_M + V_M$

M_M: M atom on normal site M

X_X: X atom on normal site X

V_X: X vacancy

V_M: M vacancy

Frenkel equation: $X_X \leftrightarrows X_i + V_X$

Schottky equation: $V_X + V_M \rightleftarrows 0$ (see Fig. 4.1).

For proton defects, we have:

$\tfrac{1}{2}H_2O$ (gas) $\leftrightarrows H_i^{\boldsymbol{\cdot}} + e' + \tfrac{1}{4}O_2$ (gas)

H_i: interstitial proton

e: electron

The ionization of defects can take place but is not described here (see reference 1 for detailed analysis).

Electronic conductivity is frequently caused by impurities of variable valence which may or may not be related to non-stoichiometry. A plot of log pressure (or chemical potential) versus T^{-1} is divided into two domains: the electrolytic (or ionic) domain above which the ionic conductivity is much higher than the electronic conductivity, and the domain below where the opposite is true[14]. Examples are given by ThO_2, ZrO_2 and CeO_2, which depend strongly on pO_2[15]. Fig. 4.2 shows the domains for ZrO_2, AgI and β-Al_2O_3 where the conductivity is due to the mainly electronic mobility (n or p type) and mainly ionic conductivity (electrolytic domain), respectively[16]. There is a region near the borderline where both types of conductivity contribute significantly to the overall conductivity (mixed conductor). This intermediate region is characterized

Defects

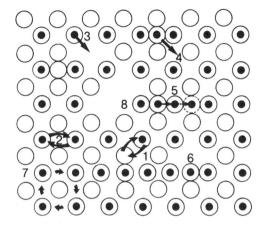

Fig. 4.1. Defects and associated diffusion mechanisms: $\bigcirc$, anion; $\odot$, cation. 1, 2, diffusion mechanism by direct exchange and 3, through vacancy; 4, direct interstitial mechanism; 5, indirect interstitial or caterpillar mechanism; 6, Frenkel defect; 7, indirect exchange; 8, Schottky defect.

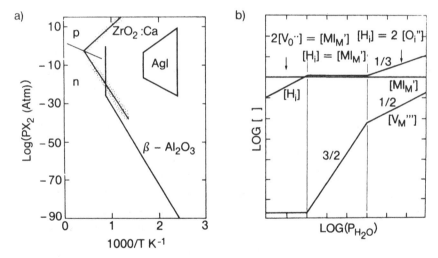

Fig. 4.2. (a) Schematic representation of electrolytic domain, i.e. relative electronic (for instance n and p type for ZrO_2: Ca) and ionic conductivity as a function of partial pressure pX_2 of the more volatile element (e.g. O_2 or I_2). Dotted zone corresponds to a mixed conduction domain where the ionic transport number (t_1) goes from 0 to 1[16] (with permission). The AgI area is limited by the α–β transition and by melting on the low and high temperature sides, respectively. (b) Schematic defect structure of an oxide M_2O_3 as a function of the water pressure. The oxide is dominated by anti-Frenkel defects and protons ($[H_i]$) and doped with MI^{2+} cations, concentration of which is assumed to be constant ($[MI'_n]$). Metal vacancies are shown as examples of minority defects.

The hydrogen bond and protonic species

by the *ionic transport number* t_i:

$$t_i = 1 - \sigma_e/(\sigma_i + \sigma_e)$$

where σ_e is electronic and σ_i ionic conductivity respectively.

The ionic transport number is directly related to the utilization of the given material as electrolyte or insertion electrode. The $\sigma_e/(\sigma_i + \sigma_e)$ ratio is usually of the order of 10^{-5} to 10^{-6} if the material is supposed to be used as electrolyte. Its measurement is not straightforward and generally has not been made for protonic conductors.

The nature and the concentration of ionic and/or electronic defects determining the conductivity domains may be modified by adding extrinsic defects by doping. If the dopant concentration is large, the structure usually changes. It must be pointed out that in electrochemical systems serving as the basis for various devices, strong redox pairs are sought. This contributes to the defect formation leading more or less rapidly to the deterioration of the system; i.e. strong redox couples enhance the defect formation. An electrolyte must be particularly stable to be used with such couples.

4.3.1 Protonic defects

Recently, it has become clear that hydrogen defects introduced from a hydrogen-containing atmosphere may influence the defect structure and defect-dependent properties of many oxides at high temperature[17]. Hydrogen defects can be present in metal oxides in the form of protons, neutral hydrogen, H or H^- hydride anions. It is generally assumed that there are protons present at high temperature and atmospheric oxygen pressure. They are bonded to oxygen ions on normal lattice sites and the corresponding defects are called either hydroxide ions or interstitial protons, $OH^{\cdot}$ and $H_i^{\cdot}$, respectively (Kröger–Vink notation). If the hydrogen source is water vapour, the following equilibrium can be assumed

$$\tfrac{1}{2}H_2O(gas) \leftrightarrows H_i^{\cdot} + e' + \tfrac{1}{4}O_2(gas)$$

$$K_1 = [H_i]nP_{O_2}^{\tfrac{1}{4}}P_{H_2O}^{\tfrac{1}{2}}$$

In stoichiometric oxides protons may determine charged defects and be counterbalanced by electrons or point defects with negative effective charges. When protons dominate, the defect-related properties such as conductivity, corrosion, sintering and creep become water pressure dependent.

68

Such defects have been reported for ZnO[18], Y_2O_3 and ZrO_2[17], CaO[19] Al_2O_3[20], SiO_2[21], TiO_2[22], $BaTiO_3$[23], Y, La, Sn-doped ThO_2[24], MoO_3[25] and Yb-doped $SrCeO_3$[26] (the last compound is described in Chapters 8 and 9). High conductivity, however, has been observed only for a few types of materials. If proton-dominated oxides are doped substitutionally with high-valent foreign cations, the protons disappear and the defect structure becomes independent of water vapour pressure.

The variation in *n*- or *p*-conductivity as a function of water pressure at constant oxygen activity appears to be a good method for determining the proton contribution to defect formation. When protons correspond to the majority defects, they can dominate the conductivity and the transport number may be determined.

4.3.2   Non-equilibrium processes

A material can contain both the sources and sinks of defects, most of which are found on the surfaces and at interfaces. It should be remembered that the partial pressure of any element on the surface of a material depends on the nature of this surface[1]. As soon as defects are created, they must migrate in order to reach an equilibrium distribution, which depends on temperature, concentration and electric field gradients. The time to reach equilibrium can be longer or shorter with respect to the preparation (annealing) time and the corresponding kinetics will depend on the density and distribution of sources and sinks. However, the protonic species in proton conductors can have a decomposition temperature not very far from the utilization temperatures. Since humidity can vary strongly, the defect formation will be particularly critical as far as the proton conducting properties are concerned and, in a phase diagram, water should always be considered.

4.4   Water vapour pressure

In the preceding chapter, protonic conduction has been classified according to the relation between electrical properties and water partial pressure. The stoichiometry and conductivity of 'intrinsic' conductors are almost water pressure independent while those of 'surface' conductors vary strongly with humidity. In the latter case, it appears as if a 'solid' solution (gel) between the skeleton compound (rigid framework) and the liquid water is formed. In the former case, on the other hand, several phases may exist, some of which have more or less non-stoichiometric domains (cf. Chapter 17 on $H_3OUO_2PO_4.3H_2O$). Here mainly intrinsic conductors will be discussed.

The role of defects in the conductivity process was already discussed in the first papers concerning protonic conductors, those of Bjerrum[27] on ice and of O'Keefe & Perrino on KH_2PO_4[28]. It was suggested that defects are formed following the reactions

$$2H_2O \leftrightarrows OH^- + H_3O^+ \qquad 2H_2PO_4^- \leftrightarrows HPO_4^{2-} + H_3PO_4$$

which can be displaced as a function of the water pressure.

The most recent studies of Sharon & Kalia of ^3H and ^{32}P diffusion coefficients confirm the role of phosphate ion rotation in the conductivity process[28]. The existence of extrinsic and intrinsic regions, however, has not been established with certainty, which leaves the discussion about the effective mechanism open.

In the case of ice, Bjerrum remarked that whether there is an excess or lack of protons, the proton jump must be followed by a structural reorganization since it creates a disorder. Moreover, proton migration in a chain induces a dipole moment in the opposite direction which is opposed to the jump. In order to counterbalance this blocking effect the orientational defects, D (two hydrogens within the same O . . . O distance) or L (lack of hydrogen) are proposed. They are (see Chapters 10 and 11) formed in pairs but migrate individually. The great possibility to have many orientations is related to the directionality of the hydrogen bond.

A similar interpretation has been suggested for KH_2PO_4[28], H_3OClO_4[29] and $CsHSO_4$[30]:

$$H_3OClO_4 \leftrightarrows H_2O + HClO_4 \qquad 2CsHSO_4 \leftrightarrows Cs_2S_2O_7 + H_2O$$

Experimental proof supporting the above equilibrium has been given for oxonium perchlorate: in the molten salt species corresponding to both sides of the equilibrium have been observed. The melting may be described by the reaction[31,32]

$$2(H_3OClO_4) \leftrightarrows H_5O_2^+ + ClO_4^- + HClO_4$$

Another type of defect has also been considered[29]:

$$2HClO_4 \leftrightarrows Cl_2O_7 + H_2O$$

In the case of $CsHSO_4$, which exhibits a high temperature superionic phase (cf. Chapter 11 on anhydrous compounds), the above reaction was supported by Raman spectra. Fig. 4.3 shows clearly a band at $320 \, \text{cm}^{-1}$ which was assigned to a bending mode of the S_2O_7 species. A detailed stoichiometric analysis leads to the formula $Cs(HSO_4)_{1-x}(S_2O_7)_{x/2}$, where $x \leqslant 0.05$[30].

4.4.1 Transition from crystalline to amorphous state

A vitreous phase can sometimes be obtained by quenching a liquid. Such a transition from the crystalline to the amorphous state may result from an orientational disorder, of SiO_4 tetrahedra in silica glasses, for instance, from 'condensation' of XO_4 to X_2O_7 groups as found in $CsHSO_4$[33] or from a positional disorder as in metal glasses. The vitreous state of cesium hydrogen sulphate (and selenate) in particular was studied recently and it was shown that the S_2O_7 (Se_2O_7) defects can explain the variation of electrical properties as a function of thermal history.

Another case of crystalline to amorphous transition has been observed for $Zr(HPO_4)_2 \cdot nH_2O$. This compound can be prepared in the crystalline

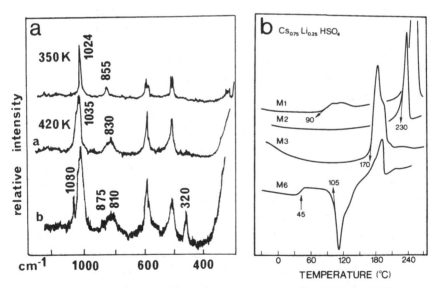

Fig. 4.3. (a) S_2O_7 defects in $CsHSO_4$: Raman spectra of $CsHSO_4$ at 350 and 420 K; lines a and b are the spectra at 420 K recorded after several minutes and hours, respectively[45] (with permission). (b) DSC traces, on heating, of $Cs_{0.75}Li_{0.25}HSO_4$ samples after successive heating–quenching cycles; M1, first cycle; M2, second cycle . . .; M6, sixth cycle. Formation of the vitreous phase is characterized by a glass transition kink (T_g point) at 45 °C (318 K), a crystallization peak at 105 °C (378 K) and a decrease of temperature of melting by 60°[33] (with permission).

state with $n \leqslant 1.6$ at 300 K and more or less amorphous materials can also be prepared. Less crystalline samples have a higher water content ($n \sim 6\text{--}12$) and there is a modification of protonic species[34]. In the crystallized compound there is an equilibrium between H_3O^+ and HPO_4^- entities and the sites occupied by H_2O and H_3O^+ are well defined. In a gel, on the other hand, there are P_2O_7 defects and OH groups as well as H_2O and H_3O^+ sites appear ill-defined, which leads to an increased number of water layers (2–12), surrounding the $Zr(HPO_4)_{2-x-y}(PO_4)_x(P_2O_7)_{y/2}$ skeleton. Moreover, this skeleton must be 'dissolved' at certain points giving rise to a highly ramified inorganic polymer.

4.4.2 Protonic defects and phase transitions

A possible origin of various defects, including orientational defects, in protonic materials must be sought in hydrogen bonding because of its relative weakness and directionality. There appears to be a relationship between the strength of OH . . . O hydrogen bonds, given by the O . . . O distance, thermal stability and conductivity as shown in Fig. 4.4. High conductivity values can be obtained when the strongest hydrogen bonds are broken. Differential scanning calorimetry of protonic conductors shows not only phase transitions but also the presence of specific defects which are directly related to proton solvation and hydrogen bonding. In doped ice, as in many hydrates, phase transitions near 100 and 230 K have been observed[35-37]. The hydrogen bond network formed either between various protonic species or between protonic species and host lattice depends strongly on the H^+/H_2O ratio. The protonated water trapped in the cavities or tunnels of a given structure could thus freeze at temperatures well below 0 °C (cf. antimonic acids, Chapter 18 and reference 38) if the hydrogen bonds are weaker than in pure water. Quenching can also induce the appearance of intermediate phases such as in ferro-paraelectric transition of HUP[39]. A substitution of half the protons by deuterons smears out the transition by hindering the *ferroelectric* order[40]. Phase transitions can also be modified by a partial substitution of cations: in $CsHSO_4$, the transition is spread out as soon as more than 10% of the Cs^+ ions are substituted by Li^+ ions[44] (cf. Chapter 30). The effect of annealing is similar[45,46]. External pressure, even a weak one, can easily modify the hydrogen bond strength (distance) and thus influence phase transitions. Fig. 4.5 shows such influence of pressure on

CsHSO$_4$[30] but similar effects have been observed also for HUP[41], NH$_4$HSO$_4$[42] and NH$_4$HSeO$_4$[43].

CsHSO$_4$ provides a good illustration of specific defects in protonic materials. The structure of the room temperature phase I consists of 'infinite' (HSO$_4{}^-$)$_n$ chains. However, it seems that numerous defects give rise to chains of different lengths containing H$_2$SO$_4$ or SO$_4{}^{2-}$ defects or even dimers (HSO$_4{}^-$)$_2$. The chain length may increase by annealing under high partial pressure of water and under these conditions, the enthalpy

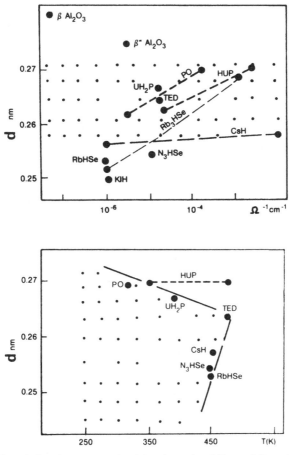

Fig. 4.4. Correlations between conductivity, thermal stability and O...O hydrogen bond distance for some protonic conductors: hydrogen β- and β″-alumina, oxonium perchlorate (PO), HUP, UO$_2$(H$_2$PO$_4$)$_2$.3H$_2$O (UH$_2$P), CsHSO$_4$ (CsS), RbHSeO$_4$ (RbSe), TED1.5H$_2$SO$_4$ (TED)[29] (with permission), Rb$_3$HSe: Rb$_3$H(SeO$_4$)$_2$, N$_3$HSe: (NH$_4$)$_3$H(SeO$_4$)$_2$. Dotted lines correspond to the change in O...O at phase transitions.

of the transformation ($T_c = 318$ K) of chains (phase I) into dimers (phase II) increases by more than a factor of 10.

The defect formation can be studied by measuring the characteristic infrared and Raman bands. Fig. 4.6 shows, for instance, the variation of the 476 cm^{-1} HSO$_4^-$ bending band: its intensity (and frequency) decreases with increasing temperature and changes suddenly at the I → II transition at 318 K. The intensity of this band, however, does not reach zero at the I → II transition but only at the subsequent II → III

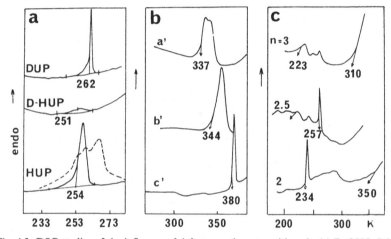

Fig. 4.5. DSC studies of the influence of defects on phase transitions in (a) D$_3$OUO$_2$PO$_4$. 3D$_2$O (DUP), mixed (0.5H$_3$O-0.5D$_3$O)UO$_2$PO$_4$ 1.5H$_2$O-1.5D$_2$O (D-HUP) and H$_3$OUO$_2$PO$_4$. 3H$_2$O (HUP); the dotted line is obtained after quenching[39,40]; (b) CsHSO$_4$[30] annealed at room temperature, a', after pressing, b', and grinding c' (with permission); (c) H$_2$Sb$_4$O$_{11}$. nH$_2$O[38] with various amounts of water stoichiometry (with permission).

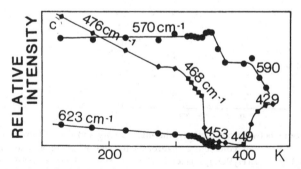

Fig. 4.6. Relative infrared intensity of some sulphate bending modes of CsHSO$_4$ as a function of temperature between 100 and 450 K[40.]

transition, thus indicating the conservation of chains in the II phase (Fig. 4.6).

Phase transition enthalpy values may also depend on the water partial pressure as shown by Lunden *et al.* for RbH_2PO_4 and $CsHSO_4$[47,48]. These enthalpies may vary by a factor of 2–5 and have been correlated with the defect content through surface stoichiometry modification[30].

4.4.3 *Electric field induced defects*

The application of an electric field of low (~ 2 V) voltage can modify an aqueous conductor and water can be reduced or oxidized, producing H_2 or D_2, with subsequent dehydration of the material. The situation can be more complicated in solid protonic conductors: a progressive proton transfer may occur between the protonated water layer and the framework, e.g. phosphate tetrahedra in HUP may form HPO_4^{2-} and even $H_2PO_4^-$ depending on the voltage ($\gg 10$ V). The voltage leading to irreversible structural change appears to be higher for a hydrate than for a liquid, doubtless because of lower kinetics in the former and because the reaction products remain in the area where the reaction takes place. Vibrational spectroscopy has been used in order to characterize the reaction in various compounds (HUP, $MoO_3 \cdot nH_2O$, WO_3, $N_2H_6Cl_2$, $(NH_4)_{10}W_{12}O_{41} \cdot 5H_2O$)[49–52]. The different stages of proton transfer have also been studied by simulating the action of the electric field using different solvents (alcohols, DMSO, acetones)[49,52].

4.5 *References*

1. W. D. Kingery, H. K. Bowen and D. R. Uhlmann, *Introduction to Ceramics*, Ch. 6 (J. Wiley and Sons, New York (1975)).
2. C. Monty, *Proceedings of the Summer School*, '*Défauts Ponctuels dans les Solides*' (Les Editions de Physique, Orsay (1978)) Ch. XII.
3. Y. Haven, Transport mechanisms and lattice defects, in *Solid Electrolytes*, P. Hagenmuller and W. Van Gool (eds) Ch. 5 (Mat. Sci. Ser., Academic Press, New York (1978)); C. P. Flynn, *Point Defects and Diffusion* (Clarendon Press, Oxford (1972)); G. E. Murch, *Atomic Diffusion Theory in Highly Defective Solids, Diffusion and Defect Monograph Series no. 6* (Trans Tech SA, Aedermannsdorf (Switzerland) (1980)).
4. G. Brebec, *Proceedings of the Summer School* '*Défauts Ponctuels dans les Solides*' (Les Editions de Physique, Orsay (1978)) Ch. V.
5. J. Philibert, *Diffusion et Transport de Matière dans les Solides* (Les Editions de Physique, Orsay (1985)).

The hydrogen bond and protonic species

6. M. A. Ratner, in *Polymer Electrolyte Review* 1, J. R. MacCallum and C. A. Vincent (eds) Ch. 7 (Elsevier Appl. Science London (1987)); C. A. Angell, *Solid State Ionics* **9/10** (1983) 3–16; **18/19** (1986) 72; C. A. Angell, *Chemistry in Britain*, April 1989, 391–5; C. A. Vincent, *Progress Solid State Chem.* **17** (1987) 145–71; W. Beier and G. H. Frischat, *J. Non-Cryst. Solids* **73** (1985) 113–33.

7. H. Cheradame, in *IUPAC Macromolecules*, H. Benoit and B. Rempp (eds) (Pergamon Press, Oxford (1982)).

8. D. Ravaine and J. L. Souquet, in *Solid Electrolytes*, P. Hagenmuller and W. Van Gool (eds) Ch. 17 Mat. Sci. Ser. (Academic Press, New York (1978)).

9. A. Pechenik, S. Susman, D. H. Whitmore and M. A. Ratner, *Solid State Ionics* **18/19** (1986) 403–9; M. A. Ratner and D. F. Schriver, *Chem. Rev.* **88** (1988) 109–24; M. A. Ratner and A. Nitzan, *Solid State Ionics* **28–30** (1988) 3–33.

10. J. P. Malugani, M. Tachez, R. Mercier, A. J. Dianoux and P. Chieux, *Solid State Ionics* **23** (1987) 189–96.

11. A. Lunden, *Solid State Ionics*, **28–30** (1988) 163–7; E. A. Secco, *Solid State Ionics*, **28–30** (1988) 168–72.

12. R. Collongues, D. Gourier, A. Kahn, J. P. Boilot, Ph. Colomban and A. Wicker, *J. Phys. Chem. Solids* **45** (1984) 981–1013; N. Baffier, J. C. Badot and Ph. Colomban, *Solid State Ionics* **2** (1981) 107–13.

13. Y. Yu Gurevich and A. K. Ivanov-Shits, *Semiconductor Properties of Superionic Materials in Semiconductors and Semimetals*, Vol. 26, Ch. 4 (Academic Press (1988)).

14. L. Heyne, in *Solid Electrolytes*, S. Geller (ed.) Topics in Appl. Phys. Vol. 21 (Springer Verlag (1977)) 169–221.

15. H. H. Fujimoto and H. L. Tuller, in *Fast Ion Transport in Solids*, P. Vashishta, J. N. Mundy and G. K. Shenoy (eds) (Elsevier, North Holland, Amsterdam (1979)) 649–52.

16. B. C. H. Steele and R. W. Shaw, Thermodynamic measurements with solid electrolytes, in *Solid Electrolytes*, P. Hagenmuller and W. Van Gool (eds), Ch. 28. Mat. Sci. Ser. (Academic Press, New York (1978)).

17. T. Norby and P. Kofstad, *J. Phys. Paris* **47–C1** (1986) 849–853. T. Norby and P. Kofstad, *J. Am. Ceram. Soc.* **67** (1984) 786–92; M. Yoshimura, T. Noma, K. Kawabata and S. Somiya, *J. Mater. Sci. Lett.* **6** (1987) 465–7; M. Yoshimura, T. Noma, K. Kawabata and S. Somiya, *J. Ceram. Soc. Jpn. Inter. Ed.* **96** (1988) 263–4.

18. D. G. Thomas and J. J. Lander, *J. Chem. Phys.* **25** (1956) 1136–42.

19. V. B. Balakireva and V. P. Gorelov, *Solid State Ionics* **36** (1989) 217–18.

20. S. K. Mohapatra, S. K. Tiku and F. A. Kröger, *J. Am. Ceram. Soc.* **62** (1979) 50–7.

21. G. O. Brunner, H. Wondratschek and F. Laves, *Z. Elektrochem. Ber. Bunsenges Phys. Chem.* **65** (1961) 735–40.

22. J. B. Bates, J. C. Wang and R. A. Perkins, *Phys. Rev.* **B19** (1979) 4130–9.

23. J. M. Pope and G. Simkovich, *Mat. Res. Bull.* **9** (1974) 1111–20.

24. D. A. Shores and R. A. Rapp, *J. Electrochem. Soc.* **119** (1972) 300–11.
25. W. G. Buckman, *J. Appl. Phys.* **43** (1972) 1280–1.
26. H. Iwahara, T. Esaka, H. Uchida and N. Maeda, *Solid State Ionics* **3/4** (1981) 359–63.
27. N. Bjerrum, *Science* **115** (1952) 385–91.
28. M. O'Keeffe and C. T. Perrino, *J. Phys. Chem. Solids* **28** (1967) 211–18; M. Sharon and A. K. Kalia, *J. Solid State Chem.* **21** (1977) 171–83.
29. M. Pham-Thi, Thesis, University of Montpellier (1985).
30. Ph. Colomban, M. Pham-Thi and A. Novak, *Solid State Ionics* **24** (1987) 193–203.
31. N. Bout and J. Potier, *Rev. Chem. Minerale* **4** (1967) 621–31.
32. M. Fournier, G. Mascherpa, D. Rousselet and J. Potier, *C.R. Acad. Sci. Paris* **269** (1969) 279–84.
33. T. Mhiri, A. Daoud and Ph. Colomban, *Phase Transition* **14** (1988) 233–42; Ph. Colomban and A. Novak, *J. Mol. Struct.* **177** (1988) 277–308.
34. Ph. Colomban and A. Novak, *J. Mol. Struct.* **198** (1989) 277–95.
35. T. Matsuo and H. Suga, *Rev. Inorg. Chem.* **3** (1981) 371–94.
36. H. Suga, *Pure Appl. Chem.* **55** (1983) 427–39.
37. M. Pham-Thi and Ph. Colomban, *Solid State Ionics* **17** (1985) 295–306.
38. Ph. Colomban, C. Dorémieux-Morin, Y. Piffard, M. H. Limage and A. Novak, *J. Mol. Struct* **213** (1989) 83–6.
39. M. Pham-Thi, Ph. Colomban and A. Novak, *J. Phys. Chem. Solids* **46** (1985) 565–78.
40. Ph. Colomban, M. Pham-Thi and A. Novak, *Solid State Comm.* **53** (1985) 747–51.
41. K.-D. Kreuer, A. Rabenau and R. Mesner, *Appl. Phys.* A**32** (1983) 45–53; 155–8.
42. A. I. Baranov, E. G. Ponyatovskii, V. V. Sinitsyn, R. M. Fedosyuk and L. A. Shuvalov, *Sov. Phys. Crystallogr.* **30** (1985) 1121–3.
43. Ph. Colomban, A. Rozycki and A. Novak, *Solid State Comm.* **67** (1988) 969–74.
44. T. Mhiri and Ph. Colomban, *Solid State Ionics* **35** (1989) 99–103.
45. Ph. Colomban, M. Pham-Thi and A. Novak, *Solid State Ionics* **20** (1986) 125–34.
46. A. T. Baranov, C. A. Shuvalov and N. M. Shchagina, *J.E.T.P. Lett.* **36** (1982) 459–62.
47. B. Baranowski, M. Friesel and A. Lunden, *Z. Naturforsch* **41a** (1986) 733–5; 981–3.
48. M. Friesel, B. Baranowski and A. Lunden, *Solid State Ionics* **35** (1989) 85–9; M. Friesel, A. Lunden and B. Baranowski, *Solid State Ionics* **35** (1989) 91–8.
49. M. Pham-Thi and Ph. Colomban, *J. Mater. Sci.* **21** (1986) 1591–600.
50. S. Chandra and N. Singh, *J. Phys. C (Solid State Phys)* **16** (1983) 3081–97.
51. S. Chandra, N. Singh and S. A. Hashmi, *Proc. Indian Nat. Sci. Acad.* **52A** (1986) 338–62; S. Chandra, N. Singh, B. Singh, A. L. Verma, S. S. Khatri and T. Chakravarty, *J. Phys. Chem. Solids* **48** (1987) 1165–71.

52. M. Pham-Thi and G. Velasco, *Rev. Chim. Min.* **22** (1985) 195–205; M. Pham-Thi and G. Velasco, *Solid State Ionics* **14** (1984) 217–20; M. Pham-Thi and Ph. Colomban, *Rev. Chim. Min.* **22** (1985) 143–60.
53. M. Anne, Thesis, University of Grenoble (1985).
54. A. R. Ochadlich Jr, H. S. Story and G. C. Farrington, *Solid State Ionics* **3–4** (1981) 79–84.
55. Y. T. Tsai, S. Smoot, D. H. Whitmore, J. C. Tarczon and W. P. Halperin, *Solid State Ionics* **9–10** (1983) 1033–9.
56. J. C. Lassègues, M. Fouassier, N. Baffier, Ph. Colomban and A. J. Dianoux, *J. Physique (France)* **41** (1980) 273–80.
57. N. J. Dudney, J. B. Bates and J. C. Wang, *Phys. Rev.* B**24** (1981) 6831–42.
58. J. B. Bates, N. J. Dudney and J. C. Wang, *J. Appl. Phys.* **58** (1985) 4587–93.
59. A. Potier and D. Rousselet, *J. Chim. Phys.* **70** (1973) 873–9.
60. S. Chandra and N. Singh, *J. Phys. C (Solid State Phys.)* **16** (1983) 3099–103.
61. K.-D. Kreuer, M. Hampele, K. Delele and A. Rabenau, *Solid State Ionics* **28–30** (1988) 589–93.
62. R. E. Taylor, M. M. Silva-Crawford and B. C. Gerstein, *J. Catal.* **62** (1980) 401–3.
63. R. C. T. Slade, P. R. Hirst, B. C. West, R. C. Ward and A. Magerl, *Chem. Phys. Lett.* **155** (1889) 305–12.
64. R. C. T. Slade, T. K. Halstead and P. G. Dickens, *J. Solid State Chem.* **34** (1980) 1983; P. G. Dickens and M. T. Weller, *Solid State Comm.* **59** (1986) 569–73.
65. R. Freer, *Contrib. Mineral. Petrol.* **76** (1981) 440–54; U. M. Gösele, *Ann. Rev. Mater. Sci.* **18** (1988) 257–82.
66. J. R. Gavarri, P. Garnier, P. Boher, A. J. Dianoux, G. Chedeville and B. Jacq, *J. Solid State Chem.* **75** (1988) 251–62; J. R. Gavarri, P. Garnier, P. Boher and A. J. Dianoux, *Solid State Ionics* **38–30** (1988) 1352–6.
67. J. Kirchkof, P. Kleinert, W. Radloff and E. Below, *Phys. Stat. Sol.* (a) **101** (1987) 391–401; A. K. Kronenberg and S. H. Kirby, *Am. Mineral.* **72** (1987) 739–47.
68. J. Gollier, B. Toudic, M. Stahn, R. E. Lechner and H. Dachs, *J. Phys. (France)* **49** (1988) 949–57.
69. O. W. Johnson, S. H. Poek and J. W. Deford, *J. Appl. Phys.* **46** (1975) 1026–33.
70. J. P. Travers and M. Nechtschein, *Synthetic Metals* **21** (1987) 135–41.

5 Structural studies of proton conductors

J. O. THOMAS

5.1 The structural situation

Let us here focus on the special structural situations we are likely to meet in studying crystalline proton conductors, and on how these can best be treated by conventional crystallographic refinement techniques. Reasonably, the same types of local situation occur in amorphous solids and in liquids, but their lack of translational symmetry renders them inaccessible to accurate study. We shall see that access to high quality single-crystal data (X-ray or neutron) gives no guarantee of a satisfactory result; careful thought must also be given to the method of refinement. Since many structural examples of proton conducting materials can be found elsewhere in this book, we shall here approach the problem from a general standpoint. Examples will subsequently be taken from the author's own work to illustrate the structural situations discussed.

What then is the structural feature most characteristic of a proton conductor system? It is generally believed that a proton is transferred through a solid in one of two distinct ways: by a *vehicular mechanism*, whereby the proton rides on a carrier molecule of type NH_4^+ or H_3O^+ ion, or by a *Grotthuss mechanism*, in which the proton jumps from a donor to a suitably placed acceptor molecule (typically, from H_3O^+ to H_2O, or from H_2O to OH^-). How then is such a process sensed in a conventional diffraction experiment? The nature of the diffraction method is such that we obtain a time- and space-average of the unit-cell content within the characteristic coherence length of the diffraction process (typically, hundreds of ångströms), and over the duration of the experiment (days to weeks). This follows from the extremely short photon–electron and neutron–nucleus interaction times ($\approx 10^{-18}$ s), which are significantly shorter than the characteristic time of the fastest of the dynamical processes in the structure ($\approx 10^{-13}$ s for the vibration of a covalently bonded atom). It follows then that some type of *structural disorder* must be a feature common to systems in which proton transfer occurs. Let us analyse the problems inherent in studying the different types of disorder commonly found in proton conductor systems.

5.2 Proton jumps

In its simplest form, as could occur in a hydrogen-bonded dimer, we are concerned with a proton jump process of type: $X–H + Y \rightarrow X' + H–Y'$, where (X, Y) and (X', Y') are the structural situations for molecules X and Y prior to and following the transfer of a proton from X to Y. A more detailed picture is that envisaged as a proton jump between situations 1 and 2 below.

Situation 1   *Situation 2*

Here, A and B represent the remainder of the molecules to which donor and acceptor atoms X and Y belong. The diffraction-determined picture will be:

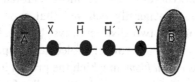

The protons $\bar{H}$ and $\bar{H}'$ occur at two mutually exclusive sites with (in the general case) occupations α and $(1 - \alpha)$; the corresponding pair of donor/acceptor geometries for each of the two H positions overlap such that $\bar{A} = \alpha . A + (1 - \alpha) . A'$ and $\bar{B} = \alpha . B + (1 - \alpha) . B'$. In this generalized situation, we will observe atom positions corresponding to $[\bar{A}, \bar{X}, \bar{H}, \bar{H}', \bar{Y}, \bar{B}]$, and the challenge in the refinement of a structural model is to separate out mechanistically meaningful coordinates $[(A, X, H, Y, B)$ and $(A', X', H', Y', B')]$ from the essentially meaningless set $[\bar{A}, \bar{X}, \bar{H}, \bar{H}', \bar{Y}, \bar{B}]$ – 'meaningless' in the sense that they do not represent true donor and acceptor coordinates, but rather a weighted average of the two hydrogen-bond situations. It can be expected that, apart from the grossly different hydrogen positions in the two cases, the separations of the donor/acceptor pairs [(X, X') and (Y, Y')] will be detectable, while the positions of the overlapping atom-pairs comprising A and B will be less dissimilar. The crystallographic challenge is thus to determine not only the positions of H and H', but also those of (X, X') and (Y, Y'). Indeed, in some instances, it is not even known initially whether the proton actually occupies two positions at all, or is situated on only one side of the hydrogen bond. The latter situation can be taken as structural evidence against a possible transfer of protons across the hydrogen bond.

By way of illustration, let us consider the real situation occurring in potassium bicarbonate ($KHCO_3$). A careful X-ray diffraction study provided a picture of the bicarbonate dimer $(HCO_3)_2{}^{2-}$ as shown in Fig. 5.1a. The less sensitive X-ray technique locates hydrogen atoms on only one side of the two hydrogen bonds, related by a centre of symmetry at the centre of the dimer[1]. A later neutron diffraction study[2], however, revealed that a vital feature had been missed; namely, that the hydrogen bonds were in fact disordered, with roughly 80% of the protons on one side of the O . . . O bond and 20% on the other; see Fig. 5.1b.

Recalling the earlier general discussion, this would imply that here we actually have an overlap of two dimer orientations related by a two-fold axis. The result of an attempt to refine such a model is indicated in Fig. 5.1c. This disorder could also be confirmed in a study of the deuterated compound $KDCO_3$, where the same type of structural situation was observed with a disorder ratio of 90%/10%. We can note that the diffraction study gives no indication of the nature of the disordering mechanism: correlated proton jumps, rotations of the whole dimer, etc. Other techniques, typically NMR, are more able to provide this type of information. This brings us to another type of structural disorder, and one which often figures as an important component in proposed Grotthuss-type mechanisms of proton transfer; namely, molecular reorientations.

5.3   *Proton jumps and molecular reorientations*

We here envisage a more complex type of proton transfer mechanism involving a combination of molecular reorientations and proton jumps. For the classic case of proton transfer from H_3O^+ to H_2O in a crystalline

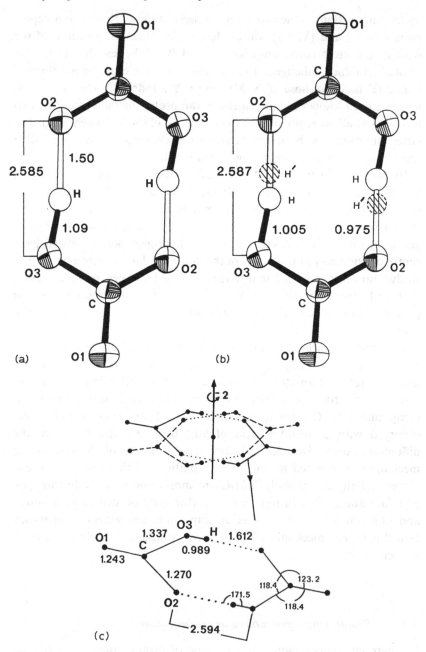

Fig. 5.1. The geometry of the hydrogen bond in the $(HCO_3)_2{}^{2-}$ dimer in $KHCO_3$ as obtained from single-crystal X-ray (a) and neutron (b) diffraction studies. A refined model involving two overlapping dimers related by a two-fold axis is shown in (c)[1,2].

solid, this could involve the steps represented schematically in the following way.

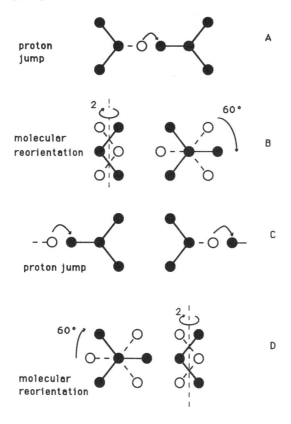

A neutron diffraction experiment will again sense this type of idealized overlap situation as a nuclear density involving a number of overlapping partially occupied sites. The refined occupations of each fractional site will give some indication of the mechanisms involved. In practice, we are faced with the problem of interpreting an overlap density in terms of some reasonable disordered model involving an overlap of the expected H_2O and H_3O^+ moieties. This interpretation is then incorporated into an overlap model which is then refined (including occupation parameters) in a conventional least-squares procedure, and the feasibility of the resulting parameters inspected.

The refinement of a single-crystal neutron diffraction data-set collected for the well-known proton conducting compound NH_4^+/H_3O^+

β''-alumina serves to illustrate this type of procedure[3]. Each apex of the hexagonal-shaped pathways of β''-alumina is occupied to $\approx 80\%$ by NH_4^+ ions. Following the scheme of the preceding discussion, the difference Fourier syntheses after the inclusion of NH_4^+ ions in the refinement were interpreted in terms of a complicated overlap of H_2O and H_3O^+ moieties, between which the protons of the system were assumed to be transferred; see Fig. 5.2a.

It is certainly conceivable that some other interpretation could have been made of the difference maps, but a satisfactory feature of this model was that the final refined occupations were physically reasonable, consistent with charge neutrality requirements, and provided a plausible

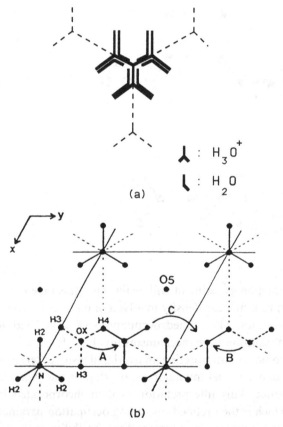

Fig. 5.2. (a) The refined model for the overlapped arrangement of H_3O^+ ions and H_2O molecules in the conduction plane of NH_4^+/H_3O^+ β''-alumina, as obtained from single-crystal neutron diffraction[3]. (b) The proposed proton transfer mechanism deduced from the refined overlap model.

proton transfer mechanism (Fig. 5.2b). Such considerations should be borne in mind in judging the correctness of a refined model. It should be noted that single-crystal neutron diffraction data of the highest quality are a prerequisite for this type of treatment to be at all feasible; the effective occupancies of the hydrogen-atom sites in the proposed proton conducting network are as low as 3%!

5.4   Short hydrogen bonds

A good proton conductor must have a low activation energy for proton transfer. This is the case across short hydrogen bonds of the O–H–O type, where the O . . . O distance is in the range 2.4–2.5 Å, and the hydrogen atom is either centred or disordered across the bond centre. Special problems can often be experienced in the treatment of this type of structural situation. By way of illustration, let us consider the case of the 2.457 Å O–H–O bond (of O–H/2· · ·H/2–O type) in hydrazinium hydrogen oxalate $(N_2H_5HC_2O_4)^4$: the X-ray and neutron diffraction determined models for the short hydrogen bond (Fig. 5.3) serve to indicate the nature of these difficulties.

Contrary to popular expectation, the X-ray study provides better resolution of the hydrogen atoms. The reason is clear: the hydrogen atoms in their two disordered sites are each polarized by the oxygen atoms to which they are covalently bonded (here O(2)s). This has the consequence of effectively shifting the charge centroid for each of the two hydrogen atoms by ≈ 0.2 Å towards their respective oxygen atoms. This results in a separation of the charge density centroids for the hydrogen atoms which is at least 0.4 Å greater than the separation of the proton density centroids to be determined by neutron diffraction. The latter distance can be as short as 0.25 Å in a 2.45 Å O . . . O bond: a distance not readily resolved by the most careful of ambient temperature neutron diffraction studies; a centrally placed proton refines equally well as two 'half-protons' in disordered positions straddling the bond-centre (Fig. 5.3b). The corresponding H/2 . . . H/2 separation to be resolved in the X-ray experiment is at least 0.65 Å, and as much as 1.04 Å in $N_2H_5HC_2O_4$ (Fig. 5.3a); a distinctly less demanding proposition.

This almost pathological consequence of the fundamental difference between X-ray and neutron diffraction should be borne in mind in choosing between X-ray and neutron diffraction. The latter is not necessarily always the more favourable technique.

5.5 Ordered and disordered networks

Let us return finally then to consider the crucial question which can often arise in diffraction studies of proton conducting systems: in an otherwise close to centrosymmetric non-hydrogen-atom network, how do we ascertain unequivocally whether a given hydrogen bond is genuinely disordered, and hence a potential structural source of proton mobility? The

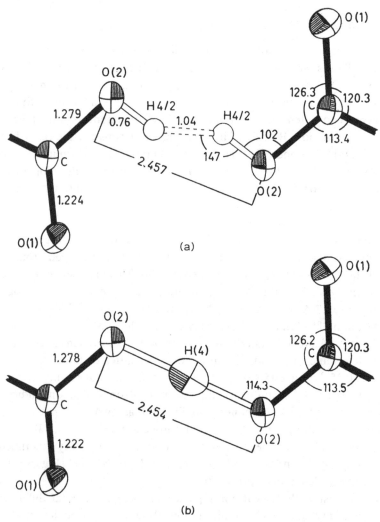

Fig. 5.3. The single-crystal X-ray(a) and neutron(b) diffraction determined models for the short O ... O hydrogen bond in $N_2H_5HC_2O_4$[4].

risk is that we are misled by the dominant symmetry of the rest of the structure and, without justification, force the hydrogen atoms to follow this same (higher) symmetry. We should not prejudge the symmetry of the hydrogen atoms of our structure by assuming it to be the same as that of the rest of the structure.

The literature provides many examples of this type of situation. The following is typical: the case of $Ca_2KH_7(PO_4)_4 . 2H_2O$. Studied first by X-ray diffraction[5], the structure was found to contain non-hydrogen atoms which were closely centrosymmetric ($P\bar{1}$), with the K^+ ion disordered across a centre of symmetry. Three of the O . . . O hydrogen bonds (lengths 2.48–2.51 Å) were refined with their respective hydrogen atoms (H11, H13 and H14) fixed at the centres of symmetry at the bond centres. The corresponding hydrogen-bond lengths were too long, however, for these to be truly symmetrical bonds: they had to be disordered or asymmetric. A single-crystal neutron diffraction study was therefore undertaken by the same authors to clarify the detail of the hydrogen-bond network. This led to the conclusion that the whole structure was actually non-centrosymmetric (P1), but that one of the hydrogen bonds [O(14) . . . H(14) . . . O(14)] was still found to be almost exactly centred; see Fig. 5.4a. This somewhat unlikely situation was later focussed upon by Baur & Tillmanns[6], who noted a number of telling features in the neutron refinement: chiefly, that the non-centrosymmetric refinement was unstable and gave unphysical bond distances throughout the structure. This was taken as evidence that the structure was actually centrosymmetric, and the hydrogen bonds disordered. What then are we to make of these two diverging pictures? Which, if either, is correct?

That neither is correct has been established (by the present author) by testing a third model against the previous two: one in which the non-hydrogen atoms were constrained to be centrosymmetric ($P\bar{1}$), while the hydrogen atoms were free to take up non-centric disordered sites (P1). The result was striking: the refinement of this latter model was stable, and the unhealthy symptoms of the earlier neutron refinement were now alleviated. All bond lengths (especially for P–O bonds) now had physically reasonable values and, above all, all hydrogen bonds (including that involving H14) were clearly asymmetric, with internal geometries entirely consistent with their observed O . . . O distances: see Fig. 5.4b.

How can the earlier inconsistencies be reconciled with this result? What was going wrong? Clearly, the structure is truly non-centrosymmetric, but the hydrogens are the only atoms which diverge significantly from

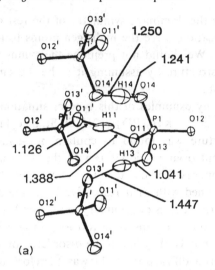

(a)

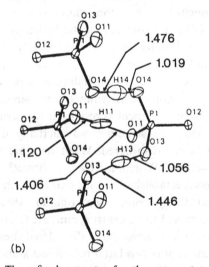

(b)

Fig. 5.4. The refined geometry for the near-centrosymmetric part of the structure of $Ca_2KH_7(PO_4)_4 \cdot 2H_2O$, as determined from a single-crystal neutron diffraction study by Prince *et al.*[5] (a), and from a later refinement of these same data made by the present author (b); (see text).

centrosymmetric. Earlier attempts to refine *all* atoms in a P1 space group resulted in the familiar instabilities characteristic of near-singularities in the least-squares matrix, which were brought about by part of the structure being close to centrosymmetric. When only the truly non-centrosymmetric atoms (the hydrogens) were refined as such, this problem disappeared. In the earlier refinement of the X-ray data, the non-centric hydrogens had contributed so weakly to the total scattering that they had exerted only a minor perturbation on the centrosymmetric refinement.

This last example has been discussed at considerable length since it serves to underline the subtlety of hydrogen bonded networks, and the extreme caution which must be exercised in ascertaining their detailed structure. Qualitatively incorrect conclusions can easily be drawn from intrinsically accurate data. The essential feature, indeed the very motivation, in the investigation of this or any other potential proton conductor is to discover the ordering – or otherwise – of the hydrogen atoms in the structure. It is worth noting just how challenging this task can be.

In view of the severity of the complications which can be experienced in assessing, by diffraction methods, the degree of order in a short hydrogen bond, it is important to state that vibrational spectroscopy (Raman, IR and neutron inelastic scattering) can provide definitive evidence as to the local effective symmetry of the bond; see, for example, Chapters 21, 23 and 24. Indeed, in situations of this type, an ideal requirement would be that a mutually consistent structural and vibrational description be achieved before our diffraction-obtained structural picture can be accepted unequivocally.

5.6   References

1. J. O. Thomas, R. Tellgren and I. Olovsson, *Acta Cryst.* **B30** (1974a) 1155–66.
2. J. O. Thomas, R. Tellgren and I. Olovsson, *Acta Cryst.* **B30** (1974b) 2540–9.
3. J. O. Thomas and G. C. Farrington, *Acta Cryst.* **B39** (1983) 227–35.
4. J. O. Thomas and R. Liminga, *Acta Cryst.* **B34** (1978) 3686–90.
5. E. Prince, S. Takagi, M. Mathew and W. E. Brown, *Acta Cryst.* **C40** (1984) 1499–502.
6. W. H. Baur and E. Tillmanns, *Acta Cryst.* **B42** (1986) 95–111.

6 Hydrogen in metals: structure, diffusion and tunnelling

IVAR SVARE

6.1 Introduction

Hydrogen ions can diffuse into the interstitial wells between atoms in many metals to form solid solutions MH_x or ordered metal hydride phases. Such materials are technically useful for the storage and purification of hydrogen, but absorbed H may also lead to unwanted brittleness in steel and other construction metals. The interesting material problems and the possible technical applications have led to large efforts over many years in studies of hydrogen absorbed in metals and its diffusion mechanisms. Lately, the promise of cheap energy from cold fusion of deuterium in metal hydrides has made most materials scientists aware of this field of research.

We will, in a simple way, review some properties of absorbed hydrogen in metals and especially the quantum mechanical tunnelling that often determines diffusion and other dynamic effects. For more details we refer to several reviews[1-3] and a vast literature.

6.2 Hydrogen absorption in metals

Dissolved hydrogen in metals may be thought of as single ions that are located on certain interstitial sites and that have given their electrons to the conduction bands. However, the electronic band picture is not very good for such solid solutions since protons distributed at random on sites destroy the periodicity of the lattice. There must also be a shielding electronic cloud near each H^+. Calculations of the electronic properties of the dissolved H and its heat of solution, ΔH, relative to that of $(\frac{1}{2})H_2$ in the gas phase, are therefore difficult and uncertain. Interactions between H^+ ions will also influence ΔH and make it concentration-dependent.

Experimentally, $\Delta H/k_B$ is negative and of order -5000 K for H in Pd, V, Nb, Ta, Ti, etc. Consequently, these metals readily absorb large quantities of hydrogen that can be driven off again at higher temperatures[4]. By contrast, ΔH is positive for H in Fe, Ni, Cu, etc., so these metals can absorb little hydrogen and only at high T. Single H^+ ions prefer octahedral (O) sites in the fcc metal Pd and tetrahedral (T) sites in the bcc metals V, Nb and Ta. Dissolved H causes an increase in electric resistance of metals and changes in magnetic susceptibility and other properties that depend upon filling of the conduction band.

Dissolved hydrogen ions interact with each other through shielded Coulomb forces, lattice distortions and the electronic clouds. The interaction is repulsive at small separations, so two H cannot be in the same well or even in neighbouring wells separated by, for example, only 1.17 Å in Nb. But, the interaction is attractive for larger separations with $d > 2$ Å, so that the dissolved hydrogen usually condenses at low temperatures to form ordered hydride phases. Such condensed phases may be in localized regions if there is insufficient hydrogen to fill the whole sample. As an example, part of the complicated phase diagram for Nb/H[5] is shown in Fig. 6.1. The α and α' phases are referred to as lattice gases with H on random T sites, while the β phase is face centred orthorhombic NbH. The ε phase is $NbH_{0.75}$, and the δ phase is the dihydride NbH_2.

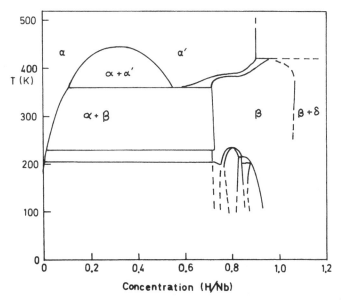

Fig. 6.1. Simplified phase diagram of the system H/Nb, after Schober & Wenzl[5] (with permission).

91

The most stable hydride of palladium is $PdH_{0.6}$, but hydrogen contents up to nearly PdH can be made with electrolytic charging.

A metal hydride, MH, has a hydrogen density equivalent to H_2 gas at about 1000 atmospheres at room temperature, and this is the basis for *storage* of hydrogen in metal hydrides instead of in heavy steel bottles. The individual H ions cause strain of the surrounding lattice and typically give a volume expansion corresponding to 20% for MH. Most metals therefore crack easily as they swell and also undergo phase changes when charged with hydrogen. Only PdH_x retains a reasonable mechanical integrity and can be used in the form of permeable foils to purify diffusing hydrogen. Iron and its alloys may absorb a little H during casting and welding. This diffuses to defects and microscopic cracks and reduces the mechanical strength, causing hydrogen embrittlement. Note that surface layers, slow diffusion, hydrogen trapping at impurities, etc. may prevent the establishment of a true thermodynamic equilibrium of H in a metal sample.

In Fig. 6.2, we show a simple model that can be used to correlate many experimental results for H in a metal. If we assume that the potential

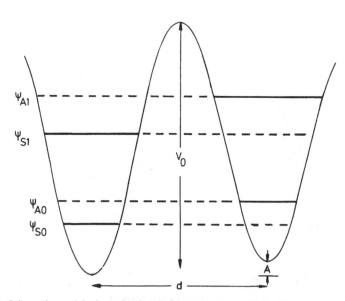

Fig. 6.2. Schematic model of nearly sinusoidal two-well potential with asymmetry A for hydrogen in metals, showing ground and first excited eigenstates. Heavy lines for the states show where the particle is likely to be.

barrier between neighbouring wells separated by distance d has height V_0 and approximate sinusoidal shape, the H oscillation energies in one of the wells are given by $E_n = a_n - a_0$ where[6,7]

$$a_n \simeq (V_0/k^2)[(2n + 1)k - \{n^2 + (n + 1)^2\}/4 + \cdots] \qquad (6.1)$$

with $k = (d/\pi\hbar)(2mV_0)^{1/2}$. The first excited state E_1/k_B of H in the α-phase of NbH_x is at 1240 K from inelastic neutron scattering[8], so here we find $V_0/k_B = 3000$ K from (6.1) with $d = 1.17$ Å; this will be used later to calculate the H motion. The isotope effects from H, D and T on the energies of solution and phase diagrams are moderate. Explanations are usually attempted in terms of the differences in the oscillation zero-point energies between the isotopes in the interstitial wells, Fig. 6.2, and in H_2, D_2 or T_2 molecules.

The interaction with defects and other H neighbours gives rise to an asymmetry A between the wells in Fig. 6.2, and the magnitude of A has a random distribution over the well pairs. The asymmetry gives a Boltzmann probability preference for H to be in the lowest well and thus a tendency for ordering. Asymmetry also strongly influences tunnelling rates.

6.3   Hydrogen diffusion and tunnelling

Hydrogen ions can diffuse classically by hopping over the barrier $E_a \simeq (V_0 - E_1/2)$ in Fig. 6.2 with the attempt frequency

$$f_a \simeq E_1/2\pi\hbar \simeq 2 \times 10^{13} \text{ s}^{-1},$$

so that the *classical hopping* rate is

$$R_{Class} \simeq f_a \exp(-E_a/k_B T) \qquad (6.2)$$

The prefactor f_a has also been associated with the Debye frequency of the lattice modes. The diffusion of H in the fcc metals Pd, Ni, Cu and Al, reduced to the elementary hop rate between the neighbouring sites, shows classical behaviour (6.2) above room temperature[9]. Note that many experimental methods, including neutron inelastic[10] and quasielastic scattering[11], NMR line narrowing and spin-lattice relaxation[12], internal friction and ultrasonic damping[13], can give important information about the H motion on various timescales, that complements that obtained from ordinary diffusion measurements.

Deviations from (6.2) due to tunnelling through the barrier appear as unreasonably small experimental values of E_a and f_a at low T. The observation of much slower rates for the isotopes D and T than for H is also a sign of tunnelling, as may be a much slower rate in alloys or in hydrides. These effects are seen in hydrogen diffusion in the bcc metals V, Nb and Ta[9,14]. When tunnelling occurs, the different experimental methods may give results for hopping rates that are so different that they are usually assumed to be unrelated effects. This is illustrated in Fig. 6.3, which shows the measured rates R for low concentrations of H in Nb, where we see three ranges of H motion: (i) very rapid transition rates $\simeq 10^{11}$ s^{-1} around 10 K for local H motion[15]; (ii) rapid ordinary diffusion above 100 K with rates $\simeq 10^{10}$–10^{12} s^{-1} and apparent activation energy $E_a{}^* < E_1$[14]; (iii) slow internal friction of H bound to impurities with rates 1–10^9 s^{-1} for $T \simeq 50$–300 K[13].

There are no generally accepted theories for the basic tunnelling motion of hydrogen in metals, which is surprising after so many years of great efforts. A good theory should be able to explain from a few basic assumptions all the effects in Fig. 6.3 that differ by so many orders of

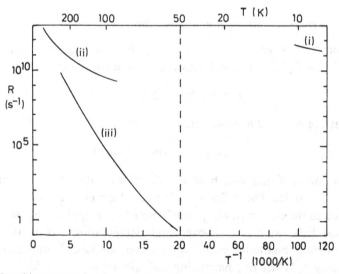

Fig. 6.3. Transition rates R between neighbouring wells for small amounts of H in Nb as a function of inverse temperature. Note change of scale at $T = 50$ K. The data are from: (i) localized motion of H bound to O impurities[15]; (ii) ordinary long-range diffusion[14]; and (iii) internal friction and ultrasonic attenuation from H bound to impurities[13].

magnitude. Here we give our understanding of the problem, and give formulae for tunnelling of hydrogen in metals that can be directly applied to other systems.

6.4   A model of tunnelling

The model in Fig. 6.2 has been widely used for two-level tunnelling systems in disordered glass structures, although the details of the tunnelling units in glass are unknown. For H in metals, the particle mass m and the distance d between the wells are known, so that the various dynamic effects can be checked for quantitative consistency much better than in glasses where crude averages have to be taken over most parameters.

If the barrier V_0 is fitted in Fig. 6.2, we can calculate the tunnel splittings of the oscillation states as approximately[6,7]

$$\Delta E_{Tn} \simeq 8\pi^{-\frac{1}{2}}2^{3n}k^{(n-\frac{1}{2})}\,e^{-2k}V_0/n!, \tag{6.3}$$

with $k = (d/\pi\hbar)(2mV_0)^{\frac{1}{2}}$ as in (6.1). For H in Nb with $d = 1.17$ Å and $V_0/k_B = 3000$ K, this gives $\Delta E_{T0}/k_B \simeq 2$ K, in agreement[16] with the tunnel splittings observed with INS[15], specific heat[17] and ultrasonic damping[18]. For H in Pd, the tunnelling distance is longer, $d = 2.85$ Å, and this gives the exponentially smaller splitting $\Delta E_{T0}/k_B = 0.35 \times 10^{-6}$ K, in agreement with the observed smaller diffusion.

The asymmetry A between the potentials in the two wells will partly localize the H to the left (L) or right (R) well in symmetrized ground eigenstates

$$\Psi_{S0} = [(\Delta E_0 + A)/2\Delta E_0]^{\frac{1}{2}}\Psi_{L0} + [(\Delta E_0 - A)/2\Delta E_0]^{\frac{1}{2}}\Psi_{R0} \tag{6.4a}$$

and

$$\Psi_{A0} = [(\Delta E_0 + A)/2\Delta E_0]^{\frac{1}{2}}\Psi_{R0} - [(\Delta E_0 - A)/2\Delta E_0]^{\frac{1}{2}}\Psi_{L0}. \tag{6.4b}$$

These states are split in energy by $\Delta E_0 = [\Delta E_{T0}^2 + A^2]^{\frac{1}{2}}$, and the minimum splitting ΔE_{T0} has in some cases been observed directly with inelastic neutron scattering or specific heat measurements. The excited eigenstates Ψ_{Sn} and Ψ_{An} are similarly partly localized and split by

$$\Delta E_n = [\Delta E_{Tn}^2 + A^2]^{\frac{1}{2}}.$$

The asymmetry A in (6.4) is caused by interactions with neighbouring hydrogens and with other impurities and defects. The values of A therefore have some distribution over the well pairs in a given sample, and the observed effects must be an average over A in the sample. The magnitude of A must, of course, be much smaller than V_0, and A/k_B may range from less than 1 K to about 100 K in pure metals with small concentrations of H. The binding energies of H to impurities or dislocations typically correspond to about 1000 K, and this A must be thermally overcome before the H can diffuse away. The binding to a condensed hydride phase is of the same order of magnitude.

Phonons and electrons dynamically perturb the potential in Fig. 6.2, induce transitions between Ψ_{S0} and Ψ_{A0} and take up the energy difference ΔE_0. Such transitions clearly take the H from one well to the other when the wave-functions are localized by $A > \Delta E_{T0}$. Motion from Ψ_{S0} to Ψ_{A0} is also possible via the excited states Ψ_{Sn} and Ψ_{An}.

We now briefly review the various transition processes that are made possible by the tunnelling and may contribute to the total rate, and how they are expected to vary with temperature T and asymmetry A.

The downward direct transition rate from Ψ_{A0} to Ψ_{S0} caused by the free conduction electrons in a normal metal, is[19]

$$R_{e\downarrow} = (\pi/\hbar)K(\Delta E_{T0}{}^2/\Delta E_0)/[1 - \exp(-\Delta E_0/k_B T)]$$

$$\rightarrow (\pi/\hbar)K(\Delta E_{T0}{}^2/\Delta E_0) \quad \text{spontaneous, for } \Delta E_0 \gg k_B T,$$

$$\rightarrow (\pi/\hbar)K(\Delta E_{T0}/\Delta E_0)^2 k_B T \quad \text{induced, for } \Delta E_0 \ll k_B T. \tag{6.5}$$

The asymmetry reduces the overlap between Ψ_{S0} and Ψ_{A0} and gives the reduction factor $(\Delta E_{T0}/\Delta E_0)^2$ in (6.5). The corresponding upward rate $R_{e\uparrow}$ from Ψ_{S0} to Ψ_{A0} is slower than $R_{e\downarrow}$ by the Boltzmann factor $\exp(-\Delta E_0/k_B T)$. The rate (6.5) and its modification for superconducting metals has been used for two-level tunnelling systems in metallic glasses, and for H in the two-well potential near O or N impurities in Nb where the coupling $K \simeq 0.06$ has been fitted[15, 18].

The downward rate for *direct one-phonon* transitions caused by a Debye spectrum of phonons, is[19]

$$R_{1p\downarrow} \simeq F_{1p}\Delta E_{T0}{}^2\Delta E_0/[1 - \exp(-\Delta E_0/(k_B T)] \tag{6.6}$$

This has been used for two-level tunnelling systems in insulating glasses. The coupling coefficient F_{1p} from the phonon deformation potential should be independent of T and A, because the density of phonon modes in the Debye model is proportional to ω^2 up to the maximum frequency ω_D, and this ω-dependence counteracts the smaller overlap for larger A. The electron rate R_e may therefore dominate the total rate at small values of A, while R_{1p} may be faster for large A up to the Debye energy $k_B T_D$.

Above 20 K the two-phonon tunnelling transitions may contribute to the total rate. Two-phonon transitions are well-known from electron spin-lattice relaxation[20], but they have hardly been discussed for two-level tunnelling systems. Such processes often have a T^5 or T^7 dependence for low temperatures, decreasing to a T^2 dependence for temperatures above T_D. A T^5 contribution with the approximate A-dependence $(\Delta E_{T0}/\Delta E_0)^2$ is expected if there is a direct matrix element connecting Ψ_{S0} and Ψ_{A0}, and this is also the condition for the one-phonon direct relaxation. The T^7 two-phonon relaxation is the result when the two E_1 states are used as intermediates. When one phonon is absorbed and another is emitted, we have approximately[16]

$$R_{2p(7)\downarrow} \simeq F_{2p}(E_{T1}^2/E_1^4)$$

$$\times \int_{\Delta E_0/\hbar}^{\omega_D} \frac{\omega^3(\omega - \Delta E_0/\hbar)^3 \, d\omega}{[\exp\{(\hbar\omega - \Delta E_0)/k_B T\} - 1][1 - \exp(-\hbar\omega/k_B T)]}$$

$$\rightarrow \propto T^7 \text{ for } \Delta E_0 < k_B T \rightarrow 0 \qquad (6.7)$$

Other processes where two phonons are emitted are also possible and they are important for large ΔE_0. The coupling F_{2p} must be treated as an adjustable parameter, but the A and T dependencies of the numerical integral (6.7) can be calculated. However, the Debye model for the phonons is of course uncertain for large ω.

At higher temperatures, transitions are possible through real excitations to the tunnel-split states E_n. The change from quantum tunnelling to seemingly classical hopping occurs when the tunnel-assisted transition rates are significant only for the states near the top of the classical barrier.

We have seen that the asymmetry A between wells strongly decreases the tunnelling transition rates. Large A will also lead to depopulation of the upper level at low T and thus a freezing of the motion. A is indeed the only parameter that can be varied in the calculations of tunnelling rates for hydrogen in a given metal, and in this way we can explain the data for H in Nb shown in Fig. 6.3: (i) the very rapid transition rates observed below 10 K for local H motion[15] goes as R_e with $A \simeq 0$; (ii) the rapid ordinary diffusion that is observed above 100 K is in samples with moderate A/k_B of the order 10–100 K, and it goes as $R_{2p(7)}$ below 250 K[16]; (iii) the slow internal friction observed from 50 to 300 K is from H bound to impurities with large binding energy $A/k_B \simeq 1000$ K $\gg T_D$ and it goes as R_{Class} with $E_a/k_B = 2400$ K at $T > 200$ K, and as $R_{e\uparrow}$ around 50 K, with a possible contribution from the resonance at $A \simeq E_1$[21].

The effect of a larger A that suppresses tunnelling is also seen in the greatly reduced diffusion of H in the Nb–V alloys[22] compared with H in either of the pure metals. Another case where the formulae above for the transition rates of H tunnelling in a two-well potential can quantitatively explain the observed effects, is the temperature and frequency dependence of the NMR spin-lattice relaxation peak around 60 K of H in Sc[23,24].

6.5 Other theories of tunnelling

The interference of Ψ_{S0} and Ψ_{A0} gives rise to adiabatic tunnelling motion, as is discussed in textbooks[25]. Asymmetry leads to an increase in the frequency of interference by $(\Delta E_0/\hbar)$ and a reduction in the amplitude by $(\Delta E_{T0}/\Delta E_0)^2$[26]. But the interpretation of this motion is uncertain, the uninterrupted interference cannot absorb the energy needed for it to be observed in experiments. Tunnelling interference in the excited states E_n is used in the old formula[27]

$$R_{En} \simeq (\Delta E_{Tn}/\hbar) \exp(-E_n/k_BT) \qquad (6.8)$$

which, although difficult to justify, seems to work in many cases[7,16]. It requires that the asymmetry A is smaller than the tunnel splitting ΔE_{Tn}, so, for wide barriers with small ΔE_{Tn}, we must go to higher states E_n where (6.8) transforms naturally into the classical rate (6.1).

The textbook formulae for barrier penetration to a free wave on the other side, give wrong results for the two-well potential with left and right states in near resonance.

The polaron, consisting of the deformed lattice and the electron cloud around a hydrogen, has to follow when the H moves to the other well, and this may add to the effective mass of the tunnelling particle. Most calculations of the polaron effects for hydrogen in metals predict the effective tunnel splitting $\Delta E_{T0\text{eff}}$ to be greatly reduced compared with what is expected for the 'bare' particle. However, the fitted effective barrier V_0 that we use may contain a contribution from the polaron, and the hydrogen mass m used in (6.1) and (6.3) gives good results for H, D and T in many metals[16].

The small polaron theories for hydrogen tunnelling in metals assume that the asymmetry A is mostly caused by the H self-trapping energy, which has to be thermally overcome by a dynamic tilting of the potential before instantaneous tunnelling can occur in the coincidence configuration of energy E_{ci}. That leads to the tunnelling transition rates[28]

$$R \propto \Delta E_{T0}{}^2 \exp(-E_{ci}/k_B T) \qquad \text{for } k_B T > \hbar\omega_D \qquad (6.9a)$$

and

$$R \propto E_{ci}{}^2 (\Delta E_{T0\,eff}{}^2/\omega_D{}^{10})T^7 \qquad \text{for } k_B T < \hbar\omega_D \qquad (6.9b)$$

But the magnitude of E_{ci} is uncertain, and there have been problems in fitting (6.9) to the experimental data[29]. Even recent refined calculations with this approach predict diffusion rates of H in Nb that do not fit case (ii) in Fig. 6.3 too well[30,31]. We believe that the self-trapping energy does not have to be explicitly included in the formulae for tunnelling, since it is the same whether the particle is in the left or the right well.

Some other theories of tunnelling assume that the two-level system is overdamped by the electrons, and that leads to the surprising predictions that the tunnelling transition rate and the diffusion of H in Nb should vary with T as $\propto T^{2K-1} = T^{-0.9}$ and be extremely fast at $T \to 0$[32]. However, the overdamping where $\hbar R_e > \Delta E_0$ from (6.5) occurs only for the few well pairs that have very small values of A, and these few pairs have limited effects in most experiments[33]. The electron-induced rate (6.5) has now been derived also with path-integral methods[34].

The tunnelling required for cold fusion is through the D–D repulsive barrier, which is some 10^5 times higher than the barrier against D diffusion in Pd, and the D–D separation in PdD_x is greater than in a D_2 molecule.

6.6 References

1. G. Alefeld and J. Völkl (eds) *Hydrogen in Metals I and II*, (Springer, Berlin (1978)).
2. L. Schlapback (ed) *Hydrogen in Intermetallic Compounds I*, (Springer, Berlin (1988)).
3. R. Lässer, *Tritium and Helium-3 in Metals* (Springer, Berlin (1989)).
4. R. Griessen and T. Riesterer, in [2], pp. 219–84.
5. T. Schober and H. Wenzl, in [1, *II*], pp. 11–71.
6. T. P. Das, *J. Chem. Phys.* **27** (1957) 763–81.
7. I. Svare, *Physica* **141B** (1986) 271–6.
8. A. Magerl, J. J. Rush and J. M. Rowe, *Phys. Rev.* B33 (1986) 2093–7.
9. J. Völkl and G. Alefeld, in [1, *I*] pp. 321–48.
10. T. Springer, in [1, *I*] pp. 75–100.
11. K. Sköld, in [1, *I*] pp. 267–78.
12. R. M. Cotts, in [1, *I*] pp. 227–65.
13. C. G. Chen and H. K. Birnbaum, *Physica status solidi* a **36** (1976) 687–92.

14. Z. Qi, J. Völkl, R. Lässer and H. Wenzl, *J. Phys. F* (Metal Phys.) **13** (1983) 2053–62.
15. H. Wipf, D. Steinbinder, K. Neumaier, P. Gutsmiedl, A. Magerl and A. J. Dianoux, *Europhys. Lett.* **4** (1987) 1379–84.
16. I. Svare, *Physica* **145B** (1987) 281–92.
17. H. Wipf and K. Neumaier, *Phys. Rev. Lett.* **52** (1984) 1308–11.
18. W. Morr, A. Müller, G. Weiss, H. Wipf and B. Golding, *Phys. Rev. Lett.* **63** (1989) 2084–7.
19. W. A. Phillips, *Rep. Prog. Phys.* **50** (1987) 1657–1708.
20. A. Abragam and B. Bleaney, *Electron Paramagnetic Resonance*, Ch. 10 (Oxford (1970)).
21. I. Svare, to be published.
22. D. T. Peterson and H. M. Herro, *Metall. Trans.* A17 (1986) 645–50.
23. R. L. Lichty, J.-W. Han, R. Ibanez-Meier, D. R. Torgeson, R. G. Barnes, E. F. W. Seymour and C. A. Sholl, *Phys. Rev.* **B39** (1989) 2012–21.
24. I. Svare, D. R. Torgeson and F. Borsa, *Phys. Rev.* **B43** (1991) 7448–57.
25. R. P. Feynman, R. B. Leighton and M. Sands, *Lectures on Physics*, Vol. 3, Ch. 8 (Addison-Wesley, Reading MA (1965)).
26. I. Svare, *Physica* **145B** (1987) 293–8.
27. J. A. Sussmann, *J. Phys. Chem. Solids* **28** (1967) 1643–8.
28. C. P. Flynn and A. M. Stoneham, *Phys. Rev* B1 (1970) 3966–78.
29. K. W. Kehr, in *Electronic Structure and Properties of Hydrogen in Metals*, P. Jena and C. B. Satterthwaite (eds) (Plenum Press, New York (1983)) 531–41.
30. A. Klamt and H. Teichler, *Phys. Stat. Sol.* **b 134** (1986) 103–14; 533–44.
31. H. R. Schober and A. M. Stoneham, *Phys. Rev. Lett.* **60** (1988) 2307–10.
32. J. Kondo, *Physica* B141 (1986) 305–11.
33. I. Svare, *Phys. Rev.* **B40** (1989) 11585.
34. U. Weiss and M. Wollensak, *Phys. Rev. Lett.* **62** (1989) 1663–6.

7 Structure and characterization of hydrogen insertion compounds of metal oxides

P. G. DICKENS AND A. M. CHIPPINDALE

7.1 Introduction

Most transition-metal oxides are unreactive towards molecular hydrogen below elevated temperatures. However, dissociated hydrogen reacts topotactically with a wide range of binary and ternary transition-metal and uranium oxides at ambient temperature to produce hydrogen insertion compounds of formula H_xMO_n or $H_xMM'O_n$ which are often referred to as 'hydrogen bronzes'. These compounds are, in general, non-stoichiometric, with biphasic regions containing solids of fixed composition in equilibrium, separated by single-phase regions of variable hydrogen content. Powder X-ray diffraction confirms that the lattice parameters of the parent oxide are little changed on hydrogen insertion, with the implication that the metal–oxygen framework is largely retained. (At high x values, however, amorphous products are sometimes found implying that a structural collapse has occurred.) The maximum hydrogen contents achieved under standard conditions are controlled by both structural factors and the redox characteristics of the metal oxidation states involved (Table 7.1).

The hydrogen insertion compounds H_xMO_n are formally mixed valence and, in marked contrast to the parent oxides, often have deep colours and behave as good electronic conductors, either metallic or semiconducting. The controllable variation in electronic properties has been exploited in electrochromic displays, which utilize the colour changes induced by insertion into e.g. WO_3 or IrO_2 films[34] and in other sensors, which

Table 7.1. *Hydrogen insertion compounds, H_xMO_n*

	Established phase ranges	$(x_{max})^a$	References
I Corner-sharing octahedra			
ReO_3	$0.15 < x < 0.5$	1.4	1–3
	$0.8 \ < x < 1.05$		
	$1.2 \ < x < 1.4$		
WO_3	$0.09 < x < 0.16$	0.4	4, 5
	$0.31 < x < 0.5$		
	$0.5 \ < x < 0.6$		
$Mo_yW_{1-y}O_3$	$0.1 \ < x < 0.2$	$1.24y + 0.36(1 - y)$	6, 7
	$x > 0.35$		
β-MoO_3		1.23	8, 9
δ-UO_3		1.0	10
hex-WO_3	$0.1 \ < x < 0.47$	0.5	11–16
II Edge-sharing octahedra/polyhedra			
α-MoO_3	$0.2 \ < x < 0.43$	1.7	17–21
	$0.85 < x < 1.03$		
	$1.55 < x < 1.72$		
	$x = 2.0$		
$VO_{2.5}(V_2O_5)$	$x < 0.25$	1.9	22–26
	$0.65 < x < 1.1$		
	$x > 1.9$		
$VO_{2.33}(V_3O_7)$		0.5	26
$VO_{2.17}(V_6O_{13})$		1.2	22
$VO_2(B)$		0.8	27
VO_2(rutile)	$0.13 < x < 0.37$	0.37	26, 75
CrO_2		0.8–1.0	28, 29
$UO_{2.67}(U_3O_8)$		0.8	30
α-UO_3		1.2	10
β-$MoO_3.H_2O$		1.0	31
$MoO_3.2H_2O$		1.0	31
$WO_3.H_2O$		0.2	32, 33
$WO_3.2H_2O$		0.2	32, 33

^{a} Maximum value at $p_{H_2} = 1$ atm (293 K $< T <$ 350 K).

respond to conductivity or optical changes in oxides on hydrogen incorporation[35]. The recovery of hydrogen on heating, $H_xMO_n = H_{x-y}MO_n + y/2H_2$ (for favourable systems at high x values), suggests a possible application as hydrogen storage materials[36]. The reactivity of inserted hydrogen towards neutral Lewis bases such as NH_3, organic amines etc, and the function of H_xMO_n as a hydrogen source in hydrogenation reactions have both been investigated[37,38].

Originally, emphasis was placed on the apparent similarities between the electronic and structural properties of the hydrogen-containing phases and those of the well known alkali-metal oxide bronzes in which alkali-metal ions reside in cavities in the parent-oxide matrix. This approach is justifiable in the case of the electronic properties, since the two classes of compound share a common redox insertion process

$$A \xrightarrow{\text{MO}_n} A_i^{\cdot} + e_M' \quad (A = H \text{ or alkali metal})$$

where $A_i^{\cdot}$ represents an ionised donor located at an interstitial site and e_M' an electron donated to a metal-based orbital. (The existence of a suitable acceptor orbital is assured if M in MO_n is in a high formal oxidation state.) This can lead to the progressive filling of a broad conduction band, as in A_xWO_3 and the development of the characteristic properties of metallic conductivity and reflectivity associated with quasi-free electrons. Alternatively, the electron can be localized at a particular metal site and lead to the formation of a semiconductor as in $A_xV_2O_5$. In either case, the particular outcome does not depend on the nature and location of $A_i^{\cdot}$ but is rather a property of the parent oxide MO_n and its band structure. Although the mode of attachment of hydrogen in H_xMO_n does not control electronic properties, it will, of course, be of importance in influencing the actual crystal structure adopted and in affecting those properties dependent on hydrogen mobility. In this respect, the analogies drawn between the structures of alkali-metal oxide bronzes and their hydrogen counterparts are misleading. The alkali metal in an oxide bronze has a definite ionic radius which requires a minimum cavity size for its central accommodation, usually of > 0.2 nm radius. This is not so for hydrogen which invariably attaches itself in oxide materials to oxygen as —OH or —OH_2, with a characteristic bond length of about 0.1 nm. This is found to be so for the many *oxy-hydroxides* and *oxide hydrates* which occur naturally including mixed valence compounds such as $V_4O_4(OH)_6$, häggite, and $V_3O_3(OH)_5$, protodoloresite. The same mode of attachment is found in those *hydrogen insertion compounds* H_xMO_n for which complete crystal structure determinations have been carried out (see below). A natural presumption therefore, should be to formulate *hydrogen 'bronzes'*, H_xMO_n, as *oxy-hydroxides* $MO_{n-x}(OH)_x$ rather than as $H_x^+MO_n(xe')$, which might seem to imply the existence of isolated protons and is, of course, chemically unrealistic.

Materials

Several recent reviews[39–42] cover different aspects of the hydrogen insertion compounds formed by oxides but in the present discussion attention is deliberately focussed on work which establishes the crystal structure and the mode and energetics of hydrogen attachment. These are static properties which relate most directly to the transport properties discussed in a later chapter.

7.2 Preparations

Insertion of hydrogen into oxides can be achieved by a variety of methods which produce dissociated hydrogen and avoid elevated temperatures, most of which were originated by Glemser and co-workers[17,18].

1. Heterogeneous reduction with $H_2(g)$ in the presence of a noble-metal catalyst – *'hydrogen spillover'*[30,43,44].

$$U_3O_8 + x/2H_2(g) \xrightarrow{\text{Pt}} H_xU_3O_8$$

2. Chemical reduction using metal/H^+ couples[45].

$$WO_3 \xrightarrow{\text{Zn/HCl}} H_xWO_3$$

3. Ion-exchange reactions from the corresponding alkali- or hydrated alkali-metal insertion compounds[46–48].

$$Li_{1+x}V_3O_8 + (1 + x)H^+ + yH_2O \rightarrow H_{1+x}(H_2O)_yV_3O_8$$
$$+ (1 + x)Li^+$$

$$Li_2MoO_3 + H_2SO_4 \rightarrow H_2MoO_3 + Li_2SO_4$$

In the second reaction, the product retains the Li_2MoO_3 structure[48,49] and is structurally different from H_2MoO_3 prepared from MoO_3 using Zn/HCl[19].

4. Hydrothermal synthesis[50].

$$x/2Mo + (1 - x/2)MoO_3 + x/2H_2O \rightarrow H_xMoO_3$$

5. Electrochemical reduction in aqueous or non-aqueous media[3,46].

$$ReO_3 + xH^+ + xe' \rightarrow H_xReO_3$$

The curves of EMF versus x of the cell Pt, $H_2|H^+$(sol)$|MO_n$, Pt typically consist of a series of slopes and plateaux corresponding to single- and two-phase regions respectively[5]. Insertion reactions are usually carried out galvanostatically and current reversal leads to de-intercalation[21]. Thermodynamic data and phase behaviour of these compounds can also be obtained from equilibrium (i.e. open circuit) data[15,32].

Preparation must usually be performed under an inert atmosphere as many of the products are readily oxidisable in air. (A metastable polymorph of MoO_3, β'-MoO_3, which is isostructural with m-WO_3, has been prepared by heating β-$D_{0.99}MoO_3$ in air at 200 °C[9].) The hydrogen content of the product is usually determined by redox titration, thermogravimetric analysis or directly from the curve of E versus x where this method of preparation is used.

7.3 *Thermodynamic aspects of hydrogen insertion*

Standard *enthalpies of formation* of pure phases H_xMO_n and enthalpy changes for the insertion reaction (I)

$$1/2H_2(g) + 1/xMO_n(s) = 1/xH_xMO_n(s) \qquad (I)$$

have been determined calorimetrically for a wide range of metal oxides (Table 7.2). From these data, accurate estimates of corresponding free-energy functions can be derived, since for reaction (I), $\Delta S^{\ominus} \sim -1/2S^{\ominus}$ $H_2(g)$, the small differences in entropies between the two solid phases are negligible in comparison with the molar entropy of gaseous hydrogen[53]. In a few cases, equilibrium (EMF) data have been obtained from suitable electrochemical cells and the values of $\Delta G_I^{\ominus}$ derived from the two approaches are in good agreement. For an initial two-phase region where $MO_n(s)$ and $H_xMO_n(s)$ coexist, the standard electrode potential for the process

$$H^+(a_{H^+} = 1) + 1/xMO_n(s) + e' = 1/xH_xMO_n(s)$$

is given by $-\Delta G_I^{\ominus} = E^{\ominus}F$.

The insertion process, I, is invariably exothermic but $-\Delta H_I$ decreases with increasing degree of reduction, x. For different systems, the magnitude of $-\Delta H_I$ parallels the oxidizing capacity of the parent oxide MO_n, as illustrated by the sequence $V_2O_5 > MoO_3 > WO_3$, for example.

These qualitative trends can be rationalized and put on a quantitative basis by using a simple bond model for the topotactic insertion of hydrogen into MO_n[55]. It is assumed that the product is an oxy-hydroxide

Table 7.2. *Standard thermodynamic data for pure phases of hydrogen insertion compounds* $H_x MO_n$ *at 298 K* $(kJ\ mol^{-1})$

	x	$\Delta H_f^{\ominus}$	$\Delta H_I^{\ominus}$	$\Delta G_I^{\ominus}$	References
$H_x VO_{2.5}$	0.23	-794.98 ± 0.87	-85.56 ± 5.21	-66.1	51
	0.72	-835.18 ± 0.94	-83.87 ± 1.77	-64.4	
	1.89	-890.55 ± 1.14	-61.14 ± 0.75	-41.7	
$H_x MoO_3$	0.34	-765.6 ± 1.0	-60.3 ± 2.9	-40.1	52
	0.93	-796.1 ± 1.4	-54.8 ± 1.3	-35.3	
	1.68	-819.0 ± 2.4	-44.0 ± 1.3	-24.5	
	2.00	-813.5 ± 2.9	-34.2 ± 1.4	-14.7	
$H_x WO_3$	0.35	-852.3 ± 1.1	-27.4 ± 2.3	-7.9	53
hex-$H_x WO_3$	0.30	-853.5 ± 1.8	-42.0 ± 5.7	-22.5	15
$H_x VO_2$ (rut.)	0.30	-729.17 ± 0.87	-44.53 ± 3.8	-25.1	26
$H_x VO_{2.33}$	0.21	-773.74 ± 1.27	-75.89 ± 7.68	-56.4	26
	0.47	-784.87 ± 0.67	-57.45 ± 2.51	-37.9	
$H_x VO_{2.17}$	0.97	-788.33 ± 1.32	-52.50 ± 1.75	-33.0	26
α-$H_x UO_3$	1.08	-1301.7 ± 2.26	-79.97 ± 1.7	-60.5	54
$H_x UO_{2.67}$	0.20	-1204.4^a	-64.45^a	-45.0	30

[a] Derived from galvanostatic measurements.

$MO_{n-x}(OH)_x$ (or oxide hydrate $MO_{n-x/2}(H_2O)_{x/2}$) in which

(i) xOH bonds are formed with bond enthalpy E_{OH},
(ii) the number of M–O bonds in $H_x MO_n$ and MO_n is the same, and
(iii) the M–O bond enthalpy contribution in $H_x MO_n$ varies continuously with the oxidation state $(2n - x)$ of M and is equal to that for a lower oxide $MO_{n-x/2}$ with the same local coordination of M.

From these assumptions of bond additivity, it follows that E_{OH} may be identified as the enthalpy change for the reaction

$$1/x\,H_x MO_n(s) = 1/x\,MO_{n-x/2}(s) + H(g) + \tfrac{1}{2}O(g)$$

which in turn can be expressed in terms of measured quantities as

$$E_{OH} = -\Delta H_I + (1/x\Delta H_f^{\ominus}MO_{n-x/2} - 1/x\Delta H_f^{\ominus}MO_n)$$
$$+ \tfrac{1}{2}\Delta H_f^{\ominus}O(g) + \Delta H_f^{\ominus}H(g)$$

where, for intermediate values of x, $\Delta H_f^{\ominus}MO_{n-x/2}$ is found by interpolation between values for MO_n and adjacent stoichiometric lower oxides

MO_{n-m}. A plot of $-\Delta H_I$ (for systems of different parent oxides and products which have comparable x values in $H_x MO_n$) against the term in brackets, denoted by ΔH_{ox}, a measure of the ease of reduction of the parent oxide, is illustrated in Fig. 7.1. The line drawn through the data has the theoretical slope of -1 and the intercept on the ordinate leads to a mean value of E_{OH} of ~ 492 kJ mol^{-1}. That this value is physically reasonable and consistent with the values found for a selection of stoichiometric oxy-hydroxides and oxide hydrates, is emphasized by reference to Table 7.3 in which values for individual compounds are listed, including those shown in Fig. 7.1.

Thermochemical data thus support the view that H in $H_x MO_n$ exists as —OH (or —OH$_2$) groups with H strongly bonded to oxygen. The relatively small values of enthalpies of insertion found for molecular

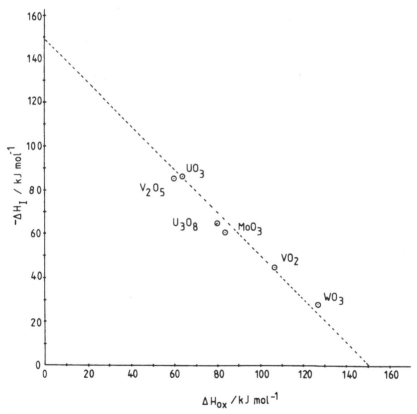

Fig. 7.1. Plot of ΔH_I versus ΔH_{ox} for $H_x MO_n$ phases.

Table 7.3. *Mean O–H bond enthalpies*
E_{OH} *(kJ mol^{-1})*

	E_{OH}		E_{OH}
$H_{0.23}VO_{2.5}$	494	$Al(OH)_3$	482
$H_{0.34}MoO_3$	487	$In(OH)_3$	497
$H_{0.35}WO_3$	497	$MoO_3.H_2O$	489
hex-$H_{0.30}WO_3$	506	$VO(OH)_2$	504
$H_{0.30}VO_2$ (rut.)	494	$H_2O_{(1)}$	486
α-$H_{0.83}UO_3$	493		
$H_{0.33}UO_{2.67}$	479		

hydrogen (Table 7.2) are a consequence not of an intrinsic weakness in the OH bond formed but rather of the high dissociation energy of H_2 and the smaller M–O bond enthalpy contributions made in the reduced product H_xMO_n. The insertion process involving molecular hydrogen is understandably a strongly activated one and it follows in practice that dissociated hydrogen is necessary for the formation of H_xMO_n at ambient temperature.

All the compounds in Table 7.2 are thermodynamically unstable towards oxidation

$$H_xMO_n + x/4O_2 = x/2H_2O + MO_n$$

since the corresponding free-energy change is large and negative. At ambient temperatures, however, the process is kinetically controlled and, at low x values especially, many of the compounds are insensitive to aerial oxidation.

There are two principal routes available for anaerobic decomposition

$$1/xH_xMO_n = 1/2H_2O + 1/xMO_{n-x/2} \tag{7.1}$$

and

$$1/xH_xMO_n = 1/2H_2 + 1/xMO_n \tag{7.2}$$

From the relations developed above it follows that

$$\Delta H_{7.1} = E_{OH} \text{ (in } H_xMO_n) - E_{OH} \text{ (in } H_2O)$$

and

$$\Delta H_{7.2} = -\Delta H_{7.1}$$

Reference to Tables 7.2 and 7.3 shows that $\Delta H_{7.1} \ll \Delta H_{7.2}$ at low x and also, in general, the disproportionation route (7.1) is thermodynamically

Hydrogen insertion compounds

preferred to route (7.2) and should proceed at a lower temperature. This is usually found to be so in practice except at high hydrogen loading. H_2MoO_3[52] decomposes by hydrogen evolution as does H_xWO_3 $(x > 0.3)$[36] but in general, it is not possible to recover by heating all the hydrogen in H_xMO_n as gaseous H_2. At low x values, little or no hydrogen is evolved on heating. This constraint limits the prospective value of hydrogen insertion compounds H_xMO_n as *hydrogen storage* materials.

7.4 Location of hydrogen in H_xMO_n

The complete structural characterization of compounds of this type is difficult to achieve. Firstly, the usual *ambient-temperature preparative routes* tend to produce poorly crystalline and frequently inhomogeneous products. Secondly, even when single-crystal materials are available, conventional X-ray diffraction methods are unsuited to the location of hydrogen atoms in the presence of much heavier metal atoms. Thirdly, compounds are often non-stoichiometric and a random or partially ordered arrangement of hydrogen amongst a surplus of available and near equivalent sites is a likely outcome. The most effective method to deal with the first two difficulties is by means of *powder neutron diffraction*. However, since only Bragg reflections are measured, the method can only determine sites and occupancies of the *average* unit cell. No information on the local ordering or clustering of hydrogen will be obtainable. To complete the structural characterization, other techniques which probe the local proton environment such as *NMR* or *incoherent inelastic neutron scattering* are necessary to supplement information gained about the average unit cell and its contents. Inelastic neutron scattering (*INS*) spectroscopy is particularly well suited to this purpose since it is applicable to metallic materials e.g. $H_{1.4}ReO_3$ and $H_6V_6O_{13}$ for which IR and Raman spectroscopies are not. It is also most sensitive to vibrational modes involving hydrogen atom displacement. The results of such a combination of methods are given below, starting with the few phases of H_xMO_n for which definition of the average unit cell has been achieved.

7.4.1 Phases of known structure

7.4.1.1 ReO_3-related phases

This group of compounds which, at high x values, have cubic structures, are based on MO_6 octahedra sharing vertices of two-connected oxygen

(Fig. 7.2). It includes $H_{0.5}WO_3$, $H_{1.36}ReO_3$, $H_xMo_yW_{1-y}O_3$, and β-$H_{0.99}MoO_3$, the last being an insertion compound formed from a meta-stable form of MoO_3 having the ReO_3 structure[9]. The compounds have common features and are classified as oxy-hydroxides $MO_{3-x}(OH)_x$. On passing from the idealized ReO_3 framework to H_xMO_3, a characteristic tilting of octahedra occurs (Fig. 7.2) which leads to the conversion of eight cavities in the parent compound to six smaller cavities, enclosing squares (or rectangles) of oxygens, together with two larger cavities. The eight sites within the squares of oxygen and lying approximately along the edges are occupied statistically by hydrogen. Only a small fraction of such sites are occupied at $x \leqslant 1$ in H_xMO_3, one per square corresponding to $H_{0.75}MoO_3$, and one per square edge to $H_3MO_3 = M(OH)_3$. The hydrogen sites lie at ~ 0.1 nm from O, and simultaneous occupation of adjacent sites along an edge is precluded on steric grounds.

The presence of —OH groups is confirmed by the appearance of an intense peak in the *INS spectrum* falling in the range 950–1300 cm^{-1}, which is characteristic of the

$$\begin{array}{c} H \\ | \\ M\!-\!O\!-\!M \end{array}$$

in-plane bending mode (Table 7.4). The out-of-plane bending mode is highly dependent on the M—O—M bond angle, falling to low frequencies as this angle approaches $180°$[62]. In these examples, for

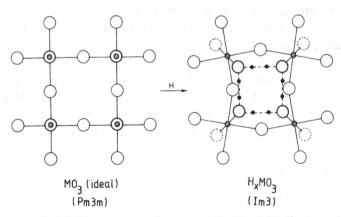

MO$_3$ (ideal)   H$_x$MO$_3$
(Pm3m)   (Im3)

Fig. 7.2. Tilting of the MO_6 octahedra on hydrogen insertion into the ReO_3-type framework. Symbols: ◉, metal; ○, oxygen; and ●, hydrogen atoms.

Table 7.4. Structural and vibrational data for cubic oxy-hydroxides $MO_{n-x}(OH)_x$

	$D_{0.53}WO_3$ [4,56]	$D_{0.56}Mo_{0.25}W_{0.75}O_3$ [7]	$D_{0.81}Mo_{0.69}W_{0.31}O_3$ [7,57,58]
M–O distance (Å)	1.926	1.930	1.949
M–O tilt (°)	11.0	11.8	13.7
O–O distance (Å)	3.264	3.219	3.136
O–D distance (Å)	1.10	1.11	1.12
O–H bend (cm^{-1})	1146		1081

	$\beta\text{-}D_{0.99}MoO_3$ [9]	$D_{1.36}ReO_3$ [59]	$HNbO_3$ [60]	$In(OH)_3$ [61]
M–O distance (Å)	1.940	1.904	1.988	2.163
M–O tilt (°)	12.08	10.1	16.0	22.8
O–O distances (Å)	3.190	3.09	2.90	2.733
	3.252	3.63	3.22	2.892
O–D distance (Å)	0.99, 1.11	0.96–1.09	0.99, 1.15	1.01
O–H stretch (cm^{-1})			3280	3200, 3400
O–H bend (cm^{-1})		1253	1060	1060

which (M—O—M) > 150°, the out-of-plane bend is predicted to occur below 300 cm^{-1} and is not easily identified from the spectra. That the —OH groups are isolated and distributed nearly randomly amongst the available sites is confirmed in the case of $H_{0.5}WO_3$ by the agreement found between calculated and measured values of the proton NMR second moment values (< 2 G^2)[63]. In the analogous β-$D_{0.99}MoO_3$ there is some evidence for an ordered arrangement of —OD groups[9]. The common skutterudite-type structure found for the non-stoichiometric reduced phases described above is also adopted by the hydroxides M(OH)$_3$ of In, Lu and Sc[61, 64, 65] and by the fully oxidized and stoichiometric compounds HNbO$_3$ and HTaO$_3$[60, 66].

An interesting feature of the hydrogen insertion compounds is that the MO$_6$ octahedra from which they are constructed become progressively more regular as x increases. The parent oxides MO$_n$, with formal d^0 electron configuration, are all built from strongly distorted octahedra as a result of $O_{p\pi}$–$M_{d\pi}$ interactions producing off-centre displacement of M (so-called ferroelectric distortion[67]). On formation of the corresponding insertion compound $H_x MoO_3$ (which at $x = 1$ becomes isoelectronic with the metallic and structurally regular ReO$_3$ (d^1)), electrons enter $M_{d(t_{2g})}$–$O_{p\pi}$ (antibonding π*) levels, hydrogen is attached to bridging oxygens and the distortion is suppressed through equalisation and lengthening of M–O bonds[68].

The adoption of the skutterudite structure is assisted by O–H . . . O hydrogen bonding but for these hydrogen insertion compounds, the O . . . O bond lengths formed (> 0.3 nm) imply only a weak interaction. The O—H groups sited at bridging oxygens are effectively isolated from those in neighbouring squares, a situation which is not conducive to facile hydrogen transport (Chapter 29).

7.4.1.2 α-MoO$_3$ and related phases

Of the known phases of α-$H_x MoO_3$ given in Table 7.1, average unit cells have been determined by *powder neutron diffraction* only for orthorhombic $H_{0.34}MoO_3$[69, 70] and for monoclinic $H_{1.7}MoO_3$[71]. Orthorhombic α-MoO$_3$ is the polymorph stable at room temperature and has a layered structure of double chains of edge-sharing MoO$_6$ octahedra linked through vertices to form corrugated layers. This structure contains two-connected (O$_2$), three-connected (O$_1$) and terminal (O$_3$) types of oxygen, and all phases of α-$H_x MoO_3$ have this same basic heavy-atom framework. In α-$H_{0.34}MoO_3$, hydrogen atoms are located within the

MoO_3 layers and attached to bridging two-connected oxygens (O_2) as —OH groups (Fig. 7.3)[69,70]. Again, as is the case for $H_{0.5}WO_3$, ferroelectric distortions present in MoO_3 are suppressed by hydrogen insertion[68]. These changes drive the phase separation of $H_x MoO_3$ into MoO_3 and $H_{0.23}MoO_3$ for ($0 < x < 0.23$). At the upper phase limit, $H_{0.4}MoO_3$, less than a quarter of the hydrogen sites available in the average unit cell are occupied. Confirmatory evidence that effectively isolated —OH groups are present is provided by the sharp peak in the *INS spectrum* (and a much weaker peak in the *IR*) observed at 1267 cm^{-1}, characteristic of a Mo_2O—H in-plane bend[69,72]. The small value of the proton NMR second moment[73] precludes the presence of —OH_2 groups but detailed analysis of data (Chapter 29) leads to the conclusion that some local ordering of H occurs amongst the available sites in the zig-zag oxygen chain (Fig. 7.4).

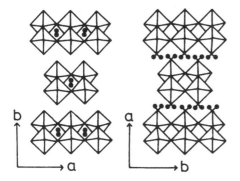

Fig. 7.3. Idealized structures of $H_{0.34}MoO_3$ (left) and $H_{1.7}MoO_3$ (right).

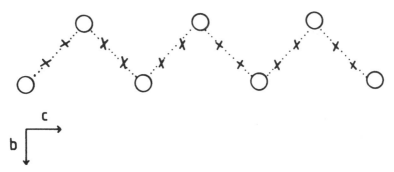

Fig. 7.4. The intralayer zig-zag of oxygen atoms in $H_{0.34}MoO_3$ (×, possible hydrogen site; ○, oxygen atom).

α-$H_{1.7}MoO_3$ has most, but not all, hydrogens located as —OH_2 groups at the terminal oxygen (O_3) (Fig. 7.3). This conclusion follows from a determination of the hydrogen sites for the average unit cell by powder neutron diffraction[71], the observation of an

$$
\begin{array}{c}
\text{H} \\
/ \\
—\text{O} \\
\backslash \\
\text{H}
\end{array}
$$

scissors vibration at 1652 cm^{-1} and a number of M—OH_2 libration modes below 1000 cm^{-1} in the *INS spectrum*[69,74] (Fig. 7.5), and the occurrence of an unusually large proton NMR second moment (>20 G^2) characteristic of protons placed in close two-spin pairs[73]. Attachment of two hydrogens at O_3 significantly reduces the amount of π-bonding between $O_{3p\pi}$ and $Mo_{d\pi}$ orbitals, causing a marked increase in the observed bond length[68]. This lengthening again provides a driving force for change of phase. Hydrogens now occupy inter- rather than intra-layer sites and their disposition is a favourable one for two-dimensional proton transport (Chapter 29). A concise structural formulation of this phase is $MoO_{3-x/2}(H_2O)_{x/2}$, an oxide hydrate. Its full structural characterization is less complete than for $H_{0.34}MoO_3$, however, largely because it is an intrinsically disordered phase containing some short Mo=O_3 and Mo—O_3H bonds in addition to a majority of long Mo—O_3H_2 bonds.

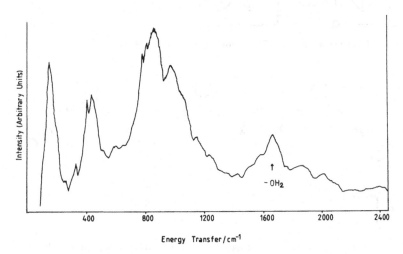

Fig. 7.5. INS spectrum of $H_{1.7}MoO_3$ at 20 K.

H_2MoO_3 prepared by Zn/HCl reduction also contains OH_2 groups[69]. In contrast, H_2MoO_3 prepared by ion exchange from Li_2MoO_3 which has a layered structure related to α-$HCrO_2$, contains only —OH groups[48,49].

7.4.2 *Phases of undetermined structure*

The structures of the remaining compounds in Table 7.1 are undetermined in the sense that the average unit cells have not been defined by diffraction methods. Crystallographic information is restricted to a listing of dimensions and types of Bravais lattice which may establish that the hydrogen insertion compounds in this table are topotactically related to the parent oxides, but give no indication of hydrogen location. In these cases, vibrational spectroscopy, using both neutrons and electromagnetic radiation (*IR* and Raman) and proton NMR play key rôles in obtaining information about the mode of attachment of hydrogen.

Given below are some of the phases for which partial structural information of this type is available.

7.4.2.1 H_xMO_2 (M = V, Cr)

Under ambient conditions, the parent rutile phases have, in general, a low reactivity towards hydrogen, presumably because, in these structures, all the oxygen atoms are bonded to three metal atoms. VO_2 and CrO_2, however, react by spillover to form $H_{0.36}VO_2$[26,75] and β-$H_{0.8}CrO_2$[28], both of which have an orthorhombically distorted rutile structure as found for a number of stoichiometric oxy-hydroxides MO(OH) (M = V, Cr, In[76]) prepared under high temperature/high pressure conditions. A powder neutron diffraction study of β-CrOOD prepared hydrothermally has located the hydrogen as M_3OH groups[77]. H_xVO_2 $(0.16 < x < 0.365)$ appears to be identical to the *oxy-hydroxide* $VO_{2-x}(OH)_x$, prepared previously by high pressure/high temperature decomposition of ammonium metavanadate[78]. The INS spectrum of $H_{0.3}VO_2$ further confirms the presence of —OH groups at three-connected oxygens (Fig. 7.6) with vibrations at 1135 and 917 cm^{-1}, corresponding to two orthogonal O—H bending modes[75]. Similar splitting, which can be predicted by simple valence force field analysis, is observed in the range 900–1400 cm^{-1} for a number of oxy-hydroxides containing M_3OH groups including α-AlO(OH) (diaspore)[79], GaO(OH)[80] and β-CrO(OH). In the latter, the bending modes, observed in both *IR and INS spectra*[81], occur at higher frequencies than in $H_{0.3}VO_2$, i.e. at 1391 and 1215 cm^{-1}, reflecting the greater strength of hydrogen bonding in this compound. The O ... O distance is 0.246 nm in β-CrOOH compared to 0.311 nm in $VO_{1.7}(OH)_{0.3}$[78].

7.4.2.2 $H_x VO_n$ ($n = 2.5, 2.33, 2.17$)

The parent oxides V_2O_5, V_3O_7 and V_6O_{13} contain two- and three-connected bridging oxygens and, in addition, V_2O_5 and V_3O_7 also contain terminal oxygens as vanadyl groups (V = O < 0.16 nm). To account for the experimentally observed maximum compositions in $H_x VO_n$ (Table 7.1), hydrogen must be located in more than one of these available sites. Hydrogen insertion proceeds topotactically over the range ($0 < x < 0.5$) for $H_x V_2O_5$ and up to the maximum compositions for the remaining phases. Evidence from vibrational spectra (INS only in the case of metallic $H_6V_6O_{13}$) suggests that for all these phases, hydrogen is present as —OH groups only with a number of OH bending modes in the range 800–1100 cm^{-1}. The vanadyl stretch, seen in the IR at ~ 1020 cm^{-1} is retained in $H_{0.33}V_2O_5$[22] (Fig. 7.7).

A number of high-hydrogen-content phases of $H_x V_2O_5$ exist (Table 7.1). The structures of these phases are unknown but it is suggested that the stacking sequence of the V_2O_5 layers is destroyed, although the integrity

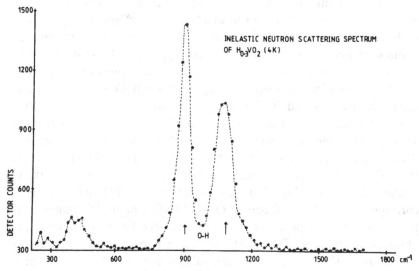

Fig. 7.6. INS spectrum of $H_{0.3}VO_2$ at 4 K.

of the layers themselves is partially preserved[23]. For $x > 3.0$, amorphous products are formed, implying that considerable structural rearrangement of the V_2O_5 lattice has taken place. An INS study of $H_{2.5}V_2O_5$ provides evidence for the formation of V—OH_2 groups[82]. This is further supported by a shift in frequency of the —OH_2 scissor mode in the IR of $H_{1.87}V_2O_5$ from 1560 to 1170 cm^{-1} on complete deuteration[51] (Fig. 7.7). The absence of a vanadyl stretching vibration at 1020 cm^{-1} in the IR spectra is consistent with, although not conclusive evidence for, hydrogen being coordinated to the terminal oxygen atoms, paralleling the behaviour of the H_xMoO_3 series. In contrast, the mineral häggite, $H_3V_2O_5$, contains —OH groups only located at three-connected oxygens which hold together layers of composition V_2O_5 by hydrogen bonding[83].

7.4.2.3 H_xUO_n ($n = 3.0, 2.67$)

α-U_3O_8 and α-UO_3 have closely related structures. The former contains layers of edge-sharing UO_5 pentagons connected by U—O—U—O chains. α-UO_3 can be formally derived from U_3O_8 by random removal of $\sim 12\%$ of the U atoms[84]. These U vacancies lead to $\sim\frac{1}{4}$ of the oxygen atoms in the U—O—U—O chains being bound to only one U atom. The

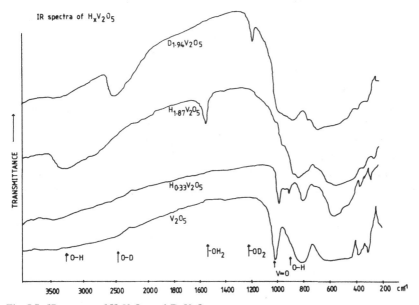

Fig. 7.7. IR spectra of $H_xV_2O_5$ and $D_xV_2O_5$.

resulting short uranium–oxygen linkages give rise to the characteristic uranyl stretching frequencies in the vibration spectra (930 and 890 cm^{-1}).

The nature of hydrogen attachment in the two oxides is rather different. In $H_{1.52}U_3O_8$, *IR and INS spectra* indicate that only OH groups are formed[74]. Modelling of the INS spectrum using normal coordinate analysis requires hydrogen to reside on both two-connected pillar and three-connected in-plane oxygen atoms[74]. The stoichiometric compounds H_2UO_4 and $H_2U_3O_{10}$ both contain two connected —OH groups and are correctly formulated as oxy-hydroxides $UO_2(OH)_2$[85] and $U_3O_8(OH)_2$[86] and not as oxide hydrates $UO_3 . nH_2O$.

In α-UO_3, hydrogen insertion proceeds topotactically up to α-$H_{1.17}UO_3$. Evidence for the formation of —OH_2 groups comes from the INS spectrum, which shows a large peak at 909 cm^{-1}[74]. In the IR spectrum, the uranyl bands decrease in intensity and eventually disappear at high hydrogen contents[10].

7.5   Conclusions

As regards their electronic properties, hydrogen insertion compounds of transition-metal and actinide oxides are analogous to the corresponding alkali-metal bronzes. They are, however, structurally distinct with hydrogen directly bonded to oxygen within the oxide framework. At low hydrogen contents, the structures adopted contain only —OH groups and can be formulated as mixed-valence oxy-hydroxides. Where the —OH groups are isolated, low proton mobility is anticipated and found. At higher hydrogen contents, however, —OH_2 groups are formed and the compounds may be regarded as oxide hydrates. Relatively few complete structure determinations have been carried out on hydrogen insertion compounds, but supplementary evidence for the nature of hydrogen attachment is provided by thermodynamic and spectroscopic measurements.

Acknowledgement

One of us (A.M.C.) wishes to thank New College, Oxford, for a Research Fellowship.

7.6 References

1. N. Kimizuka, T. Akahane, S. Matsumoto and K. Yukino, *Inorg. Chem.* **15** (1976) 3178.
2. P. G. Dickens and M. T. Weller, *J. Solid State Chem.* **48** (1983) 407.
3. M. T. Weller and P. G. Dickens, *Solid State Ionics* **9–10** (1983) 1081.
4. P. J. Wiseman and P. G. Dickens, *J. Solid State Chem.* **6** (1974) 374.
5. R. H. Jarman and P. G. Dickens, *J. Electrochem. Soc.* **128** (1981) 1390.
6. S. J. Hibble and P. G. Dickens, *Mat. Res. Bull.* **20** (1985) 343.
7. S. J. Hibble and P. G. Dickens, *J. Solid State Chem.* **63** (1986) 154.
8. E. M. McCarron, *J. Chem. Soc., Chem. Commun.* (1986) 336.
9. J. B. Parise, E. M. McCarron and A. W. Sleight, *Mat. Res. Bull.* **22** (1987) 803.
10. P. G. Dickens, S. V. Hawke and M. T. Weller, *Mat. Res. Bull.* **19** (1984) 543.
11. B. Schlasche and R. Schöllhorn, *Rev. Chim. Miner.* **19** (1982) 534.
12. M. Figlarz and B. Gérand in *9th Int. Symp. on Reactivity of Solids, Cracow, 1980.* p. 660.
13. B. Gérand, J. Desseine, P. Ndata and M. Figlarz in *Solid State Chem. 1982, Proc. of the Second European Conf., Veldhoven, The Netherlands, June 1982,* p. 457 (Elsevier, Amsterdam (1983)).
14. B. Gérand and M. Figlarz in *Spillover of Adsorbed Species,* G. M. Pajonk, S. J. Teichner and J. E. Germain (eds) (Elsevier, Amsterdam (1983)) 275.
15. P. G. Dickens, S. J. Hibble, S. A. Kay and M. A. Steers, *Solid State Ionics,* **20** (1986) 209.
16. M. Figlarz, *Prog. Solid State Chem.* **19** (1989) 1.
17. O. Glemser and G. Lutz, *Z. Anorg. Allg. Chem.* **264** (1951) 17.
18. O. Glemser, G. Lutz and G. Meyer, *Z. Anorg. Allg. Chem.* **285** (1956) 173.
19. J. J. Birtill and P. G. Dickens, *Mat. Res. Bull.* **13** (1978) 311.
20. P. G. Dickens and J. J. Birtill, *J. Electron. Mater.* **7** (1978) 679.
21. R. Schöllhorn, R. Kuhlmann and J. O. Besenhard, *Mat. Res. Bull.* **11** (1976) 83.
22. P. G. Dickens, A. M. Chippindale, S. J. Hibble and P. Lancaster, *Mat. Res. Bull.* **19** (1984) 319.
23. D. Tinet and J. J. Fripiat, *Rev. Chim. Miner.* **19** (1982) 612.
24. S. J. Hibble, A. M. Chippindale and P. G. Dickens, *J. Electrochem. Soc.* **132** (1985) 2668.
25. D. Tinet, M. H. Legay, L. Gatineau and J. J. Fripiat, *J. Phys. Chem.* **90,** (1986) 948.
26. A. M. Chippindale, D.Phil. Thesis, Oxford 1987.
27. C. Ellis, Chemistry Part II Thesis, Oxford 1990.
28. M. T. Weller, Chemistry Part II Thesis, Oxford 1982.
29. M. A. Alario-Franco and K. S. W. Sing, *J. Thermal Anal.* **4** (1972) 47.
30. S. D. Lawrence, Chemistry Part II Thesis, Oxford 1984.
31. S. Crouch-Baker and P. G. Dickens, *Mat. Res. Bull.* **19** (1984) 1457.
32. P. G. Dickens, S. J. Hibble and R. H. Jarman, *J. Electron. Mater.* **10** (1981) 999.
33. R. H. Jarman, D.Phil. Thesis, Oxford 1981.
34. B. W. Faughnan and R. S. Randall, in *Display Devices in Topics in App. Phys.* (Springer-Verlag, Berlin (1980)).

Materials

35. P. J. Shaver, *Appl. Phys. Lett.* **11** (1967) 255.
36. A. R. Berzins and P. A. Sermon, *Nature* **303** (1983) 506.
37. R. Schöllhorn, T. Schulte-Nölle and G. Steinhoff, *J. Less Common Metals* **71** (1980) 71.
38. J. P. Marcq, G. Poncelet and J. J. Fripiat, *J. Catal.* **87** (1984) 339.
39. R. Schöllhorn, *Angew. Chem. Int. Ed. Engl.* **19** (1980) 983.
40. P. G. Dickens, S. Crouch-Baker and M. T. Weller, *Solid State Ionics* **18–19**, (1986) 89.
41. C. Ritter, *Z. Physik. Chemie Neue Folge* **151** (1987) 51.
42. R. Schöllhorn in *Inclusion Compounds*, J. L. Atwood and L. E. D. Davies (eds) Vol. I (Academic Press, NY (1984)) 249.
43. G. C. Bond and P. A. Sermon, *Catal. Rev.* **8** (1973) 211.
44. G. C. Bond in *Spillover of Adsorbed Species*, G. M. Pajonk, S. J. Teichner and J. E. Germain (eds) *Studies in Surface Science and Catalysis* **17**, 1 (Elsevier (1983)).
45. O. Glemser and C. Naumann, *Z. Anorg. Allg. Chem.* **265** (1951) 289.
46. R. Schöllhorn in *Intercalation Chemistry* M. S. Whittingham and A. J. Jacobson (eds) (Academic Press, NY (1982)) 315.
47. R. Schöllhorn, F. Klein-Reesink and R. Reimold, *J. Chem. Soc., Chem. Commun.* (1979) 399.
48. J. Gopalakrishnan and V. Bhat, *Mat. Res. Bull.* **22** (1987) 769.
49. A. C. W. P. James and J. B. Goodenough, *J. Solid State Chem.* **76** (1988) 87.
50. J. J. Birtill, D.Phil. Thesis, Oxford 1977.
51. A. M. Chippindale and P. G. Dickens, *Solid State Ionics* **23** (1987) 183.
52. J. J. Birtill and P. G. Dickens, *J. Solid State Chem.* **29** (1979) 367.
53. P. G. Dickens, J. H. Moore and D. J. Neild, *J. Solid State Chem.* **7** (1973) 241.
54. S. V. Hawke, Chemistry Part II Thesis, Oxford 1983.
55. P. G. Dickens, R. H. Jarman, R. C. T. Slade and C. J. Wright, *J. Chem. Phys.* **77**(1) (1982) 575.
56. C. J. Wright, *J. Solid State Chem.* **20** (1977) 89.
57. S. J. Hibble and P. G. Dickens, *Ber. Bunsenges Phys. Chem.* **90** (1986) 702.
58. R. C. T. Slade, A. Ramanan, P. R. Hirst and H. A. Pressman, *Mat. Res. Bull.* **23** (1988) 793.
59. P. G. Dickens and M. T. Weller, *J. Solid State Chem.* **48** (1983) 407.
60. J. L. Fourquet, M. F. Renou, R. De Pape, H. Théveneau, P. P. Man, O. Lucas and J. Pannetier, *Solid State Ionics* **9–10** (1983) 1011.
61. D. F. Mullica, G. W. Beall, W. O. Milligan, J. D. Korp and I. Bernal, *J. Inorg. Nucl. Chem.* **41** (1979) 277.
62. G. M. Hawkins, Chemistry Part II Thesis, Oxford 1988.
63. P. G. Dickens, D. J. Murphy and T. K. Halstead, *J. Solid State Chem.* **6** (1973) 370.
64. D. F. Mullica and W. O. Milligan, *J. Inorg. Nucl. Chem.* **42** (1980) 223.
65. A. N. Christensen, N. C. Broch, O. Von Heidenstam and A. Nilson, *Acta Chem. Scand.* **21** (1967) 1046.
66. D. Groult, J. Pannetier and B. Raveau, *J. Solid State Chem.* **41** (1982) 277.
67. J. B. Goodenough, *Mat. Res. Bull.* **2** (1967) 165.

Hydrogen insertion compounds

68. S. Crouch-Baker and P. G. Dickens, *Solid State Ionics*, **28–30** (1988) 1294.
69. P. G. Dickens, J. J. Birtill and C. J. Wright, *J. Solid State Chem.* **28** (1979) 185.
70. F. A. Schroeder and H. Weitzel, *Z. Anorg. Allg. Chem.* **435** (1977) 247.
71. P. G. Dickens, A. T. Short and S. Crouch-Baker, *Solid State Ionics* **28–30** (1988) 1294.
72. P. G. Dickens, S. J. Hibble and G. S. James, *Solid State Ionics* **20** (1986) 213.
73. R. C. T. Slade, T. K. Halstead and P. G. Dickens, *J. Solid State Chem.* **34** (1980) 183.
74. A. V. Powell, D.Phil. Thesis, Oxford 1990.
75. A. M. Chippindale, P. G. Dickens and A. V. Powell, *J. Solid State Chem.* **93** (1991) 526.
76. A. F. Wells, *Structural Inorganic Chemistry*, 5th edn (Oxford University Press (1985)).
77. A. N. Christensen, P. Hansen and M. S. Lehmann, *J. Solid State Chem.* **19** (1976) 299.
78. K.-J. Range and R. Zintl, *Mat. Res. Bull.* **18** (1983) 411.
79. E. Hartert and O. Glemser, *Z. Electrochem.* **60** (1956) 746.
80. M. F. Pye, Chemistry Part II Thesis, Oxford 1975.
81. M. T. Weller, D.Phil. Thesis, Oxford 1984.
82. G. C. Bond, P. A. Sermon and C. J. Wright, *Mat. Res. Bull* **19** (1984) 701.
83. H. T. Evans and M. E. Mrose, *Am. Miner.* **45** (1960) 1144.
84. C. Greaves and B. E. F. Fender, *Acta Cryst.* **B28** (1972) 3609.
85. A. M. Deane, *J. Inorg. Nucl. Chem.* **41** (1979) 277.
86. S. Siegel, A. Viste, H. R. Hoekstra and B. Tani, *Acta Cryst.* **B28** (1972) 117.

8 High temperature proton conductors based on perovskite-type oxides

H. IWAHARA

8.1 Introduction

High temperature proton conducting solids are useful materials for many electrochemical applications such as high temperature fuel cells, hydrogen sensors and hydrogen gas separators. However, many protonic conductors decompose at temperatures above 300 °C. About ten years ago, the author found that certain perovskite-type oxide solid solutions exhibit protonic conduction in an atmosphere containing hydrogen or steam at high temperatures. In this chapter, proton conduction in perovskite-type oxides and their electrochemical properties are described.

8.2 Proton conducting solids at high temperature

Fig. 8.1 shows the conductivities of representative protonic conductors as a function of temperature. Although there are many good proton conducting solids at low temperature[1-3], they are unstable at temperatures above 300 °C since they decompose to liberate water. In earlier work, the existence of protons in Cu_2O, CuO, NiO, ZrO_2[4] and ThO_2[5] in a hydrogen-containing atmosphere at high temperatures was studied, and the possibility of protonic conduction in those oxides suggested. However, their protonic conductivities below 1000 °C were expected to be quite low compared to those of low temperature proton conductors. Furthermore, these studies did not provide a direct demonstration of protonic conduction. After the discovery of $SrCeO_3$-based protonic conductors[6], $KTaO_3$-based oxides[7] and Y_2O_3 ceramic[8,9] were reported to have protonic conduction at high temperatures, although the conductivities were not as high as those of the cerate-based perovskite-type oxide ceramics.

The high temperature proton conducting oxides described here are perovskites based on $SrCeO_3$ or $BaCeO_3$ in which some trivalent cations are partially substituted for cerium[6, 10]. The general formula is written as $SrCe_{1-x}M_xO_{3-\alpha}$ or $BaCe_{1-x}M_xO_{3-\alpha}$ where M is a rare earth element, x is less than 0.1 and α is the number of oxygen vacancies per perovskite-type unit cell. These oxides exhibit protonic conduction under a hydrogen-containing atmosphere at high temperatures. Doping by aliovalent cations is essential for the appearance of protonic conduction in these oxides. $SrCe_{0.95}Y_{0.05}O_{3-\alpha}$, $SrCe_{0.95}Yb_{0.05}O_{3-\alpha}$ and $BaCe_{0.9}Nd_{0.1}O_{3-\alpha}$ are examples of this type of conductor.

8.3 Preparation and properties of ceramics

These perovskite-type oxides are prepared by solid state reaction of cerium dioxide, strontium or barium carbonate and dopant oxide at about

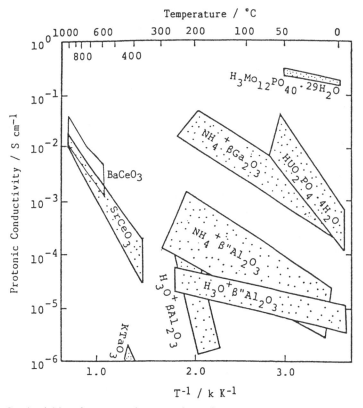

Fig. 8.1. Conductivities of representative protonic conductors.

Materials

1400–1500 °C and the ceramics are obtained by sintering at about 1500 °C for 10 h in air. The crystal structures of doped $SrCeO_3$ and $BaCeO_3$ are orthorhombic; the former is more asymmetric than the latter. Differential thermal analysis of $SrCe_{0.95}Yb_{0.05}O_{3-\alpha}$ showed that there was no significant phase change for temperatures up to 1000 °C[11].

The $SrCeO_3$- and $BaCeO_3$-based materials can be prepared as dense ceramics with moderate hardness, the colour of which is pale brownish-green except for the Nd-doped ones which are black. The porosity of the ceramics is less than 5%, without open pores to allow the penetration of gas. Dilatometric measurements on $SrCe_{0.95}Yb_{0.05}O_{3-\alpha}$ showed that the thermal expansion from room temperature to 1000 °C was about 1.0%[11]. These ceramics are soluble in some mineral acids at room temperature, especially in concentrated hydrochloric acid liberating chlorine gas, while they do not dissolve in aqueous alkaline solution[11].

8.4 Verification of protonic conduction

Protonic conduction in these materials was verified directly by means of electrochemical transport of hydrogen through the oxide ceramic. When one side of a $SrCeO_3$-based ceramic diaphragm attached with porous electrodes was exposed to hydrogen gas and a direct current passed through it, as shown in Fig. 8.2, hydrogen was observed to evolve at the cathode at a rate given by Faraday's law (Fig. 8.3). This indicated that, in a hydrogen atmosphere, the ceramic is almost a pure protonic conductor[12].

Also, in the same cell as described above, when steam is supplied to

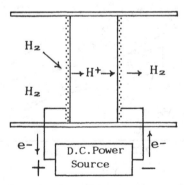

Fig. 8.2. Electrochemically induced transport of hydrogen through a protonic conductor.

the anode compartment and direct current is passed through the ceramic electrolyte, one can observe that hydrogen gas evolves at the cathode[6, 13, 14]. This is a kind of steam electrolysis, the principle of which is shown in Fig. 8.4. It means that the oxide ceramics behave as a protonic conductor not only in the presence of hydrogen but also in the presence of water vapour.

Protonic conduction in the presence of water vapour is also supported by the e.m.f. behaviour of a gas concentration cell using the specimen as a solid electrolyte (Fig. 8.5). The e.m.f.s of the cell obtained on applying various gases are given in Table 8.1 for the cases of $SrCe_{0.95}Yb_{0.05}O_{3-\alpha}$ and $SrCe_{0.95}Mg_{0.05}O_{3-\alpha}$ as typical examples[6, 10]. When dry oxygen gases of different partial pressure (gas I, air; gas II, pure oxygen) are introduced

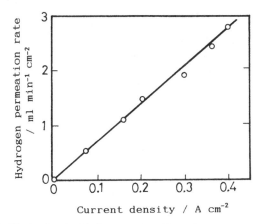

Fig. 8.3. Electrochemical hydrogen permeation rate.

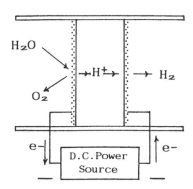

Fig. 8.4. Principle of steam electrolysis.

125

to each electrode compartment, only a very small and unstable e.m.f. is observed as shown for cell 1, Table 8.1, indicating that under these conditions, any conduction in the oxides is mainly electronic. The electronic conduction is p-type (hole conduction) since the conductivity of the ceramics in dry gas increases with increasing partial pressure of oxygen[6, 10].

However, when each gas of the above cell is moistened, a stable e.m.f.

Table 8.1. *E.m.f. of various gas cells, gas I, Pt|specimen oxide|Pt, gas II*

		E.m.f.(mV)[b]			
		$SrCe_{0.95}Yb_{0.05}O_{3-\alpha}$		$SrCe_{0.95}Mg_{0.05}O_{3-\alpha}$	
Cell no.	Cell type[a] gas I ‖ gas II	600 °C	800 °C	600 °C	800 °C
1	dry air‖dry O_2	0.5	1.0	0.1	−0.3
2	wet air‖wet O_2	25.0	14.0	16.0	12.0
3	wet air‖dry air	59.0	30.0	39.0	17.0
4	wet air‖dry O_2	69.5	30.5	46.8	19.5
5	dry air‖wet O_2	−43.5	−13.0	−25.0	−10.5

[a] Dry gas – dried with P_2O_5; wet gas – saturated with H_2O at room temperature (21–22 °C).
[b] Negative sign shows that the electrode of gas II is negative.

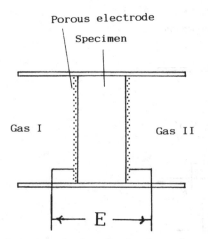

Fig. 8.5. A gas concentration cell using the specimen oxide as a solid electrolyte.

of the oxygen concentration cell is observed (cell 2 in Table 8.1), although the values are low compared with the theoretical values (29.5 mV at 600 °C and 35.7 mV at 800 °C).

A stable e.m.f. is also observed when air with different humidities is supplied to the electrode compartments (cell 3). In this case, the electrode with the higher humidity is the negative pole. This is a type of steam concentration cell using a proton conducting electrolyte. The principle of the *steam concentration cell* is shown in Fig. 8.6[6, 15]. The difference in water vapour partial pressure between the electrodes can be a driving force for the following electrode reactions.

Electrode with higher water vapour pressure:

$$H_2O \rightarrow 2H^+ + 1/2O_2 + 2e^- \tag{8.1}$$

Electrode with lower water vapour pressure:

$$2H^+ + 1/2O_2 + 2e^- \rightarrow H_2O \tag{8.2}$$

For this reason, this cell may give a stable e.m.f., with the electrode of lower vapour pressure being the cathode.

Generally, when gases I and II in the cell include oxygen and water vapour at different partial pressures, the e.m.f. E of the steam concentration cell can be given[6] as

$$E = \frac{RT}{2F} \ln \left(\frac{P_{H_2O}(I)}{P_{H_2O}(II)} \right) \left(\frac{P_{O_2}(II)}{P_{O_2}(I)} \right)^{1/2} \tag{8.3}$$

where P_{H_2O} and P_{O_2} are the partial pressures of water and oxygen, respectively, and R, F and T have their usual meanings. If the oxygen

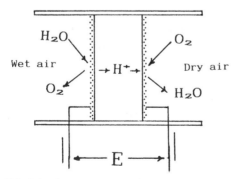

Fig. 8.6. Principle of steam concentration cell using a protonic conductor.

partial pressures of both electrodes are almost the same, the e.m.f. simplifies to[12]

$$E = \frac{RT}{2F} \ln\left(\frac{P_{H_2O}(I)}{P_{H_2O}(II)}\right) \qquad (8.4)$$

On the other hand, when the water vapour pressures are equal and oxygen partial pressures are different, the e.m.f. is given by

$$E = \frac{RT}{2F} \ln\left(\frac{P_{O_2}(II)}{P_{O_2}(I)}\right) \qquad (8.5)$$

which is the same expression as that for an oxygen concentration cell using oxide ion conducting electrolyte[15,16]. The e.m.f. behaviour of cells 4 and 5 in Table 8.1 can also be explained qualitatively by Eqn (8.3), provided the diaphragm is a proton conductor.

8.5 Conduction properties

From the experimental results described above and many other works[11–18], the conduction properties of $SrCeO_3$- or $BaCeO_3$-based ceramics may be summarized as follows.

(1) Sintered oxides exhibit p-type electronic conduction (hole conduction) in an atmosphere free from hydrogen or water vapour.

(2) When water vapour or hydrogen is introduced to the atmosphere at high temperature, electronic conductivity decreases and protonic conduction appears.

(3) When the ceramics are exposed to hydrogen gas, they become almost pure protonic conductors, the conductivities of which are at least $10^{-2}\,S\,cm^{-1}$ at 1000 °C and at least $10^{-3}\,S\,cm^{-1}$ at 600 °C, as shown in Figs 8.7 and 8.8.

(4) According to e.m.f. measurements on a hydrogen concentration cell using $SrCe_{0.95}Yb_{0.05}O_{3-\alpha}$ as electrolyte, the electronic conductivity of the ceramic in this condition is two orders lower than the proton conductivity[13].

(5) The conduction in $BaCeO_3$-based ceramics is partly protonic and partly oxide ionic, when they are used as a solid electrolyte for fuel cells above 800 °C[17].

8.6   Proton formation in oxides

Doping by aliovalent cations is indispensable for the appearance of protonic conduction in these oxides. It seems that electron holes and oxide ion vacancies formed by doping might play an important role in the formation of protons. For example, substitution of Yb^{3+} for Ce^{4+} in $SrCeO_3$ may provide oxygen vacancies V_0 as a means of charge compensation:

$$Yb^{3+} \rightarrow Yb^{\cdot}{}_{Ce} + 1/2V_0^{\cdot\cdot} \qquad (8.6)$$

The oxygen vacancies may be in equilibrium with electron holes as shown in Eqn (8.7) below. Possible equilibrium reactions and schematic defect chemistry for proton formation in $SrCe_{1-x}Yb_xO_{3-\alpha}$ are shown in Fig. 8.9.

Studies on electrical conductivity as a function of dopant content, or partial pressures of water vapour and oxygen have shown that the

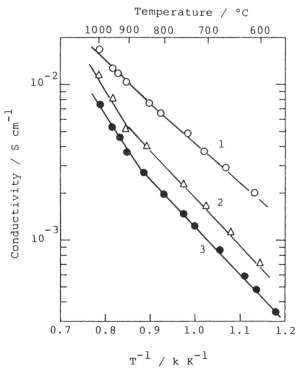

Fig. 8.7. Conductivities of $SrCeO_3$-based ceramics in hydrogen. 1, $SrCe_{0.95}Yb_{0.05}O_{3-\alpha}$; 2, $SrCe_{0.9}Y_{0.1}O_{3-\alpha}$; 3, $SrCe_{0.95}Sc_{0.05}O_{3-\alpha}$.

129

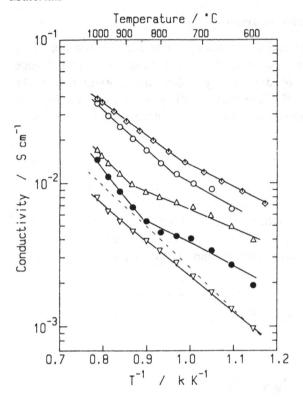

Fig. 8.8. Conductivity of $BaCe_{1-x}M_xO_{3-\alpha}$ in hydrogen. M(x value): $\diamond$, Y(0.10); $\bigcirc$, Nd(0.10); $\triangle$, La(0.10); $\bullet$, Nd(0.05); $\triangledown$, Ca(0.05); $----$, $SrCe_{0.95}Yb_{0.05}O_{3-\alpha}$ in H_2 [1].

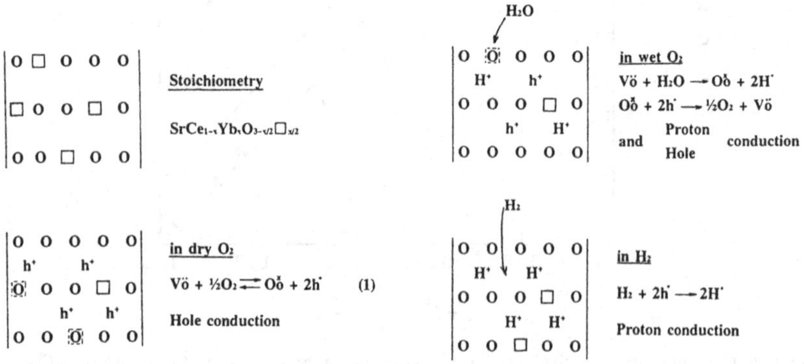

$$
\begin{array}{|ccccc|}
o & \square & o & o & o \\
 & & & & \\
\square & o & o & \square & o \\
 & & & & \\
o & o & \square & o & o \\
\end{array}
$$

Stoichiometry

$SrCe_{1-x}Yb_xO_{3-x/2}\square_{x/2}$

$$
\begin{array}{|ccccc|}
o & o & o & o & o \\
h^\cdot & & & & h^\cdot \\
\boxed{O} & o & o & \square & o \\
h^\cdot & & & & h^\cdot \\
o & o & \boxed{O} & o & o \\
\end{array}
$$

in dry O_2

$V\ddot{o} + \frac{1}{2}O_2 \rightleftharpoons O\ddot{o} + 2h^\cdot$ (1)

Hole conduction

H_2O

$$
\begin{array}{|ccccc|}
o & \boxed{O} & o & o & o \\
H^\cdot & & h^\cdot & & \\
o & o & o & \square & o \\
 & & h^\cdot & & H^\cdot \\
o & o & o & o & o \\
\end{array}
$$

in wet O_2

$V\ddot{o} + H_2O \rightarrow O\ddot{o} + 2H^\cdot$

$O\ddot{o} + 2h^\cdot \rightarrow \frac{1}{2}O_2 + V\ddot{o}$

and Proton Hole conduction

H_2

$$
\begin{array}{|ccccc|}
o & o & o & o & o \\
H^\cdot & & H^\cdot & & \\
o & o & o & \square & o \\
 & & H^\cdot & & H^\cdot \\
o & o & \square & o & o \\
\end{array}
$$

in H_2

$H_2 + 2h^\cdot \rightarrow 2H^\cdot$

Proton conduction

Fig. 8.9. Defects in $SrCe_{0.95}Yb_{0.05}O_{3-\alpha}$ and formation of protons.

following three equilibria are simultaneously established between the defects in the oxide and the atmosphere[18].

$$V_0^{\cdot\cdot} + 1/2O_2 \xrightarrow{K_1} O_0^x + 2h^{\cdot} \qquad (8.7)$$

$$H_2O + 2h^{\cdot} \xrightarrow{K_2} 2H^{\cdot} + 1/2O_2 \qquad (8.8)$$

$$H_2O + V_0^{\cdot\cdot} \xrightarrow{K_3} 2H^{\cdot} + O_0^x \qquad (8.9)$$

$$K_3 = K_1 K_2 \qquad (8.10)$$

where $V_0^{\cdot\cdot}$, O_0^x, $H^{\cdot}$, $h^{\cdot}$ and K are oxygen vacancy, oxide ion at normal lattice site, proton, hole and equilibrium constant, respectively. As is clear from Eqn (8.10), the equilibria can be expressed by using any two of the Eqns (8.7)–(8.9) since they are not independent of each other.

In order to clarify the process of dissolution of water vapour (or hydrogen) to generate protons in $SrCeO_3$-based oxides at high temperature, the evolution and the absorption of water vapour and/or oxygen were studied by changing the temperature or partial pressure in a flow of wet or dry gases[19,20].

Table 8.2 shows the behaviour of $SrCe_{0.95}Yb_{0.05}O_{3-\alpha}$ on changing the partial pressure of oxygen or water vapour[20]. Oxygen is released in response to an increase in the water vapour pressure, and is absorbed on decreasing P_{H_2O} (Nos 1, 2). This can be explained by Eqns (8.7) and (8.8). Conversely, when the oxygen partial pressure is increased suddenly, a small but distinct evolution of water vapour can be observed (No. 3) and vice versa. This can be also explained by Eqns (8.7) and (8.9).

The hydrogen concentration in $SrCe_{0.95}Yb_{0.05}O_{3-\alpha}$ which had been equilibrated with a certain water vapour pressure was measured by the SIMS method[21] and thermal desorption[20]. Fig. 8.10 shows the amount of evolved water vapour per mole of oxide measured by thermal desorption during heating from 600 °C.

In this experiment, undoped $SrCeO_3$ released little water vapour, but Yb-doped proton conducting oxides evolved a substantial amount of water vapour on raising the temperature. The 5% Yb-doped oxide released a larger amount of water vapour than the 3% doped specimen. This indicates that the higher the vacancy concentration produced by Yb-doping, the higher the proton concentration in the oxides.

As the water vapour evolution was almost complete at about 1300 °C,

Table 8.2. *Response of* $SrCe_{0.95}Yb_{0.05}O_{3-\alpha}$ *on changing* P_{O_2} *or* P_{H_2O} *at constant temperature and amount of gas evolved or absorbed at 800 °C. (Schematic illustrations for Eqns (8.9) and (8.7) are also shown.)*

No.	Change in atmosphere	Response	Schematic illustration
1	P_{H_2O} increase[a] ($\sim 10^{-4} \rightarrow 3.3 \times 10^{-2}$ atm)	O_2 evolution (6×10^{-4} mol%)	$H_2O + V_O^{\cdot\cdot} \rightleftharpoons 2H^{\cdot} + O_O^{x}$ (8.9)
2	P_{H_2O} decrease[a] ($3.3 \times 10^{-2} \rightarrow \sim 10^{-4}$ atm)	O_2 absorption (3×10^{-4} mol%)	
3	P_{O_2} increase[b] ($1.6 \times 10^{-4} \rightarrow 1$ atm)	H_2O evolution (4×10^{-3} mol%)	$V_O^{\cdot\cdot} + 1/2 O_2 \rightleftharpoons O_O^{x} + 2h^{\cdot}$ (8.7)
4	P_{O_2} decrease[b] ($1 \rightarrow 1.6 \times 10^{-4}$ atm)	H_2O absorption (3×10^{-3} mol%)	

[a] $P_{O_2} = 1.6 \times 10^{-4}$ atm.
[b] $P_{H_2O} = 7.2 \times 10^{-3}$ atm.

Fig. 8.10, the proton concentration $[H](T)$ at given temperature T can be calculated from

$$[H^{\cdot}](T) = 2[H_2O]_{limit} - [H_2O](T) \qquad (8.11)$$

where $[H_2O](T)$ is the amount of water vapour evolved during heating from 600 °C to a given temperature, T, and $[H_2O]_{limit}$ is the total amount of evolved water vapour at 1300 °C.

The hydrogen concentration in $SrCe_{1-x}Yb_xO_{3-\alpha}$ thus calculated is shown in Fig. 8.11 and is compared with that in other oxides. The solubility of hydrogen (or water vapour) is much higher than in other oxides. Typically, the proton concentration is 2.38 mol% at 600 °C and 0.7 mol% at 1000 °C in an atmosphere of $P_{H_2O} = 7.2 \times 10^{-3}$ atm and $P_{O_2} = 1.6 \times 10^{-4}$ atm. These values, obtained by thermal desorption, are close to those obtained by the SIMS method[21].

Lee et al.[7] reported the existence of the OH bond in Fe-doped $KTaO_3$ single crystal which showed protonic conduction. Shin et al.[22] measured the infrared absorption of Y-doped $SrCeO_3$ single crystal, and found that the proton is located in the interstitial sites between the oxygen ions.

8.7   Migration of protons

Since an interstitial proton associates strongly with a neighbouring oxygen ion, H_i is, in fact, better regarded as $(OH)_o$. As pointed out by Scherban & Nowick[23], the protons then migrate by a hopping mechanism, from one oxygen ion to a nearest neighbouring one.

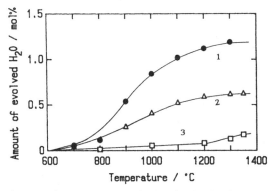

Fig. 8.10. Amount of water vapour evolved during heating from 600 °C. $P_{H_2O} = 7.2 \times 10^{-3}$ atm, specimen = $SrCe_{1-x}Yb_xO_{3-\alpha}$. 1, $x = 0.05$; 2, $x = 0.03$; 3, $x = 0.00$.

133

Materials

Some investigators have reported the isotope effect of migration by replacing protons with deuterons[7,22,24]. Scherban et al. observed non-classical behaviour in Yb-doped $SrCeO_3$; the ratio of conductivity σ_H/σ_D is about 2.5 which is significantly greater than the classical value of $\sqrt{2}$, although the reason for this is not yet clear.

Using the values of proton concentration [H] calculated as above, the mobility of protons can be evaluated from the Einstein equation:

$$\sigma_H = F\mu_H[H^{\cdot}]/\nu \qquad (8.12)$$

where σ_H, F, μ_H and $[H^{\cdot}]$ denote the proton conductivity, Faraday constant, proton mobility and proton concentration, respectively, and ν is the molar volume of $SrCe_{0.95}Yb_{0.05}O_{3-\alpha}$. The mobilities of the proton thus obtained are shown in Table 8.3[25]. The proton mobilities are of the order of $10^{-6}–10^{-5}$ cm^2 S^{-1} V^{-1} in the temperature range 600–1000 °C. Fukatsu et al. have also estimated the mobility in $SrCe_{0.95}Yb_{0.05}O_{3-\alpha}$ by means of a polarization method, and the values were close to those obtained from the thermal desorption method and Eqn (8.12)[26,27].

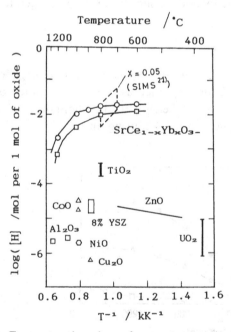

Fig. 8.11. Temperature dependence of proton concentration in Yb-doped $SrCeO_3$ under the conditions $P_{H_2O} = 7.2 \times 10^{-3}$ atm, $P_{O_2} = 1.6 \times 10^{-4}$ atm. The hydrogen contents in some other oxides are compared.

Fig. 8.12 shows an Arrhenius plot of proton mobility. Some data for other oxides are also presented in this figure. Various diffusion or mobility data for protons in some oxides have been reported [28–30]. Among these oxides, the proton mobility in Yb-doped $SrCeO_3$ seems to be of

Table 8.3. *Estimated values of proton concentration $[H^\cdot]$ and mobility μ_H in* $SrCe_{0.95}Yb_{0.05}O_{3-\alpha}$

$T(^\circ C)$	$[H^\cdot]$ (mol%)	$\mu_H(cm^2\ s^{-1}\ V^{-1})$
600	2.38	5.4×10^{-6}
650	2.34	7.9×10^{-6}
700	2.26	1.2×10^{-5}
750	2.20	1.5×10^{-5}
800	2.15	1.9×10^{-5}
850	1.9	2.3×10^{-5}
900	1.3	3.4×10^{-5}
950	1.0	4.1×10^{-5}
1000	0.7	6.2×10^{-5}

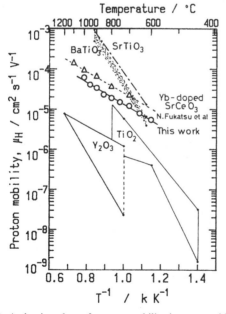

Fig. 8.12. Arrhenius plots of proton mobility in some oxides.

Materials

intermediate value. Therefore, the high protonic conductivity of $SrCeO_3$-based perovskite-type oxides is ascribed to the anomalously high content of protons compared with that for other oxides.

8.8   References

1. L. Glasser, *Chem. Rev.* **75** (1975) 21.
2. S. Chandra, Proton Conductors in *Superionic Solids and Solid Electrolytes*, S. Chandra *et al.* (eds) (Academic Press (1989)) 185–226.
3. F. W. Poulsen, Proton Conduction in Solids in *High Conductivity Solid Ionic Conductors* T. Takahashi (ed) (World Scientific (1989)) 166–200.
4. S. Stotz and C. Wagner, *Ber. Bunsenges. Physik. Chem.* **70** (1966) 781.
5. D. A. Shores and R. A. Rapp, *J. Electrochem. Soc.* **119** (1972) 300.
6. H. Iwahara, T. Esaka, H. Uchida and N. Maeda, *Solid State Ionics* **3/4** (1981) 359.
7. W. Lee, A. S. Nowick and L. A. Boatner, *Solid State Ionics* **18/19** (1986) 989.
8. T. Norby and P. Kofstad, *J. Am. Ceram. Soc.* **67** (1984) 786.
9. T. Norby and P. Kofstad, *Solid State Ionics* **20** (1986) 169.
10. H. Iwahara, H. Uchida, K. Ono and K. Ogaki, *J. Electrochem. Soc.* **135** (1988) 529.
11. H. Uchida, A. Yasuda and H. Iwahara, *Denki Kagaku* **57** (1989) 153.
12. H. Iwahara and H. Uchida, *Proc. Intern. Meet. Chemical Sensors (Fukuoka, Japan, 1983)* 227.
13. H. Iwahara, T. Esaka, H. Uchida, T. Yamauchi and K. Ogaki, *Solid State Ionics* **18/19** (1986) 1003.
14. H. Iwahara, H. Uchida and I. Yamasaki, *Int. J. Hydrogen Energy* **12** (1987) 73.
15. H. Uchida, N. Maeda and H. Iwahara, *J. Appl. Electrochem.* **12** (1982) 645.
16. H. Iwahara, H. Uchida and N. Maeda, *Solid State Ionics* **11** (1983) 109.
17. H. Iwahara, H. Uchida and K. Morimoto, *J. Electrochem. Soc.* **137** (1990) 462.
18. H. Uchida, N. Maeda and H. Iwahara, *Solid State Ionics* **11** (1983) 117.
19. H. Uchida, H. Yoshikawa and H. Iwahara, *Solid State Ionics* **34** (1989) 103.
20. H. Uchida, H. Yoshikawa and H. Iwahara, *Solid State Ionics* **35** (1989) 229.
21. T. Ishigaki, S. Yamauchi, K. Kishio, K. Fueki and H. Iwahara, *Solid State Ionics* **21** (1986) 239.
22. S. Shin, H. S. Huang, M. Ishigame and H. Iwahara, *7th International Conference on Solid State Ionics (Hakone, Nov. 1989) Extended Abs.*, p. 330, *Solid State Ionics* **40** (1990) 914.
23. T. Scherban and A. S. Nowick, *Solid State Ionics* **35** (1989) 189.
24. T. Scherban, W.-K. Lee and A. S. Nowick, *Solid State Ionics* **28–30** (1988) 585.
25. H. Uchida, H. Yoshikawa, T. Esaka and H. Iwahara, *Solid State Ionics* **36** (1989) 89.
26. N. Fukatsu, K. Yamashita, T. Ohashi and H. Iwahara, *J. Japan Inst. Metals* **51** (1987) 848.

High temperature protonic conductors

27. N. Fukatsu and T. Ohashi, *The 14th Symp. Solid State Ionics Japan, Extended Abstracts* (1987), p. 19.
28. R. Waser, *Ber. Bunsenges. Physik. Chem.* **90** (1986) 1223.
29. T. Norby and P. Kofstad, *Solid State Ionics* **20** (1986) 169.
30. G. J. Hill, *J. Phys.* **D1** (1968) 1151.

9 Highly ionic hydroxides: unexpected proton conductivity in Mg(OH)$_2$ and homologues

FRIEDEMANN FREUND

9.1 Introduction

Hydrogen bonding is considered a prerequisite for proton conduction. The underlying rationale is that protons can jump with relative ease only when the energy barriers between donor and acceptor sites are low. The corollary is that, when energy barriers are high, proton conductivity is not expected to occur.

Since the idea of low energy barriers is intuitively appealing, the search for proton conductors has concentrated worldwide on systems with recognizable H-bonding, both in inorganic systems and biological membranes[10]. This has created a bias which may have prevented the recognition of some important proton transport mechanisms that appear to be present in non H-bonded system.

In this chapter, I shall discuss highly ionic hydroxides that are probably the least likely candidates for proton conduction. However, on closer inspection, they reveal an interesting fundamental behaviour that may be important for understanding proton transport in other systems, for instance across bilayer membranes containing inner sections that are considered to be proton impermeable[1].

9.2 Non-hydrogen bonded systems

Inorganic hydroxides with the highest degree of ionicity are the binary hydroxides of the alkali and alkaline earth metals, Me^{I}OH and Me^{II}(OH)$_2$ with Me^I = Na$^+$, K$^+$ ..., and Me^{II} = Mg^{2+}, Ca^{2+} ..., respectively. They all contain well-defined, essentially non H-bonded OH$^-$ anions.

Most alkali hydroxides undergo a rotational transition prior to melting. At the transition temperature their low symmetry structures (usually monoclinic due to dipole–dipole interaction between the OH$^-$ anions) become dynamically disordered and cubic. The fact that the structurally bound OH^- can be thermally activated to fully rotate indicates how weak the crystal field is that determines the orientation of the OH^- anions. The transition to the rotator phase is accompanied by a sharp increase in the proton conductivity[5, 6, 11]. This proton conduction is believed not to be intrinsic but mediated by impurities, mainly H_2O. It probably proceeds by a series of proton jumps linked to a reorientation of the H_2O molecules, similar to a Grotthuss mechanism. LiOH adopts a layer structure with no reported H^+ conductivity but with an appreciable Li^+ conductivity of $10^{-4}\,\Omega^{-1}\,cm^{-1}$ at 385 °C[7].

The absence of H-bonding is clearly evident in $Mg(OH)_2$ and $Ca(OH)_2$. Both crystallize in the highest symmetry class, the CdI_2 type structure, D_{3d}^3-P3m. Their OH^- anions form a hexagonal close packing with the cations occupying the octahedral interstices in every other layer. The result is a sequence of layer packages stacked along the c axis, as shown by Fig. 9.1, separated by (0001) planes of perfect cleavage. Each package consists of two layers of OH^-, hexagonally densely packed in the a–b plane with the O–H vectors pointing along the $\pm c$ direction and the cation (symbolized by $\otimes$) sandwiched between them.

Each package satisfies the stoichiometry $Me^{II}(OH)_2$ and is electrostatically neutral. It may be regarded as a planar two-dimensional macromolecule. The forces across the cleavage planes are of Van der Waals type.

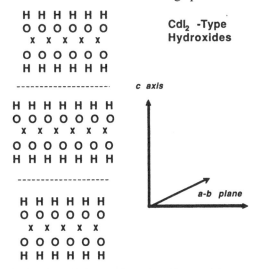

Fig. 9.1. Section through the two-dimensional layers of the CdI_2-type structure of $Mg(OH)_2$ and related hydroxides. Dashed lines indicate the perfect cleavage along (0001).

Materials

The large distance between adjacent OH^- anions, 3.22 Å, confirms the high degree of ionicity and makes the absence of H-bonding understandable. Protons in nearest OH^- neighbours are separated by at least 1.9 Å. Yet, these hydroxides undergo dehydration

$$Me^{II}(OH)_2 = Me^{II}O + H_2O \qquad (9.1)$$

The on-set of dehydration occurs below 400 K and 550 K for $Ca(OH)_2$ and $Mg(OH)_2$ respectively. In the case of $Al(OH)_3$ which has a somewhat more complicated, but related structure, dehydration starts around 450 K. Dehydration is a reaction which cannot do without proton transfer as an elementary step. It can be reduced to a simple proton transfer reaction

$$OH^- + OH^- = O^{2-} + H_2O \qquad (9.2)$$

Upon heating, the approach of the on-set of the dehydration is accompanied by subtle structural changes which have been recently reviewed[2]. Here only those questions will be addressed which specifically refer to proton mobilization and proton conductivity. The fundamental question is: how can protons become mobile that are tightly bound inside highly ionic OH^- and have no apparent H-bonding?

9.3   Potential energy curve of the O–H oscillator

Infrared (IR) spectroscopy confirms the absence of H-bonding in $Mg(OH)_2$. The O–H stretching band is very narrow with a width at half intensity of the order of only 4 cm^{-1}. It occurs near the highest end of the wavenumber scale known from any solid hydroxide, at 3700 cm^{-1}, indicating that the proton is tightly bound to its 'parent' O^{2-}. Coupling between OH^- ions is weak as indicated by the closeness to the Raman band (3655 cm^{-1}). The OH^- potential energy curve $V(r)$ can be determined from the optical transition energies E_{0n} of the O–H stretching mode from the zero level to the n-th excited level. We use the *Morse potential*

$$V(r) = D\{1 - \exp[-\alpha(r - r_0)/r_0]\}^2 \qquad (9.3)$$

where D is the dissociation energy for $OH^- = O^{2-} + H^+$, α an exponential factor, related to the force constant f by $f = 2d(\alpha/r_0)^2$, and r_0 the

Highly ionic hydroxides

equilibrium O–H distance. The eigenvalues E_n are given by

$$E(n) = D\gamma^{-2}[2\alpha\gamma(n + 1/2) - \alpha^2(n + 1/2)^2] \qquad (9.4)$$

where $\gamma = r_0/h(2\mu D)^{1/2}$, and μ is the reduced mass of the OH$^-$ group. The transition energies for the fundamental and the overtones are then given by $E_{0n} = E_n - E_0$ with $n = 0, 1, 2, 3 \ldots$. Knowing E_{0n} experimentally, the dissociation energy D and the factor α may be calculated.

The values for E_{0n} have been obtained from the overtones of the O–H stretching mode which occur at approximately twice, thrice, four times etc. the frequency of the O–H fundamental and in part consist of IR + Raman combination bands. Being forbidden by the dipole selection rules their intensities decrease rapidly, by about two orders of magnitude for each increment. As a consequence the higher overtone bands are very

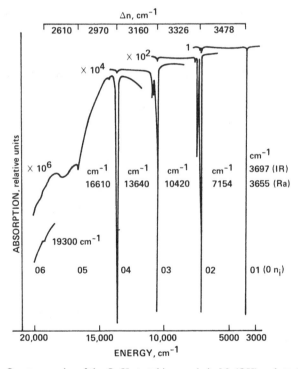

Fig. 9.2. Overtone series of the O–H stretching mode in Mg(OH)$_2$, plotted as a function of energy (wavenumbers), corresponding to the vibrational stretching excitation of the O–H oscillator. Note the appearance of an edge at the 05 transition (with kind permission of JAI Press).

141

weak. If we normalize the intensity of the E_{01} transition to unity, the relative intensities of the transitions 02, 03, 04, 05 etc. decrease as 10^{-2}, 10^{-4}, 10^{-6}, 10^{-8}, etc.

Despite the experimental difficulties associated with the measurement of such very weak bands, the complete overtone sequence extending from 3 μm in the near IR to 500 nm in the visible region was obtained, using $Mg(OH)_2$ single crystal and powder data[9]. The results are shown graphically in Fig. 9.2. From the first four bands we obtain the Morse potential

$$V(r) = 4.70\{1 - \exp[-2.294(r - r_0)/r_0]\}^2 \qquad (9.5)$$

The value of $D = 4.70$ eV agrees surprisingly well with the dissociation energy of the free OH^- as derived from thermodynamic data, 4.71 eV, attesting to the fact that, up to the fourth vibrationally excited level,

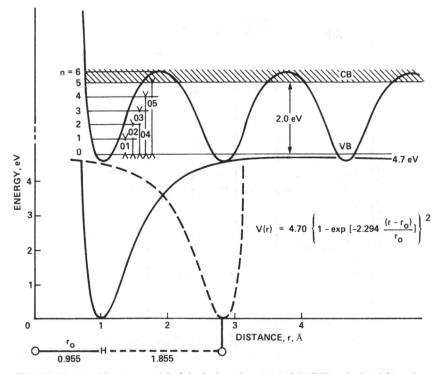

Fig. 9.3. Bottom: Morse potential of the hydroxyl proton in $Mg(OH)_2$ calculated from the O–H overtone series. Top: linear superposition to simulate the $Mg(OH)_2$ layer structure and formation of a proton conduction band (CB) above the fifth excited level (with kind permission of JAI Press).

the potential energy well of the hydroxyl proton in $Mg(OH)_2$ nearly coincides with that of free OH^-.

An interesting feature of Fig. 9.2 is that the O–H overtone sequence appears to come to an abrupt end above the 05 transition. In fact, the 05 transition appears as an edge, topped by a small band at $16\,610\ \text{cm}^{-1}$ or 602 nm. What this means becomes clear when we overlay the OH^- Morse potentials to give a linear array of OH^- ions separated by the same distances as in the $Mg(OH)_2$ structure.

The result, shown in Fig. 9.3. is a series of deep wells where the protons are localized on the 0 level. This defines the valence band (VB). However, the superposition also produces a continuum above the 05 transition, 2.0 eV above the VB. This energy region, hatched in Fig. 9.3, defines a proton conduction band (CB).

Fig. 9.3 leads to interesting possibilities for H^+ conductivity measurements. The existence of a CB implies that protons which have reached this high energy state cease to be bound to their parent O^{2-}. They become delocalized. At the same time, each proton injected into the CB leaves behind a deprotonated site, chemically an O^{2-}. Using Kröger's point

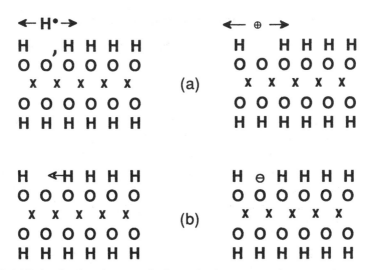

Fig. 9.4. (a) Delocalization of a proton in the conduction band and formation of an excess proton ($H^{\cdot}$) proton vacancy (O') pair. (b) Diffusion of a proton vacancy from left to right by successive hopping from right to left of protons from neighbouring OH^-, presumably using intermediate levels in the band gap.

defect notation[8], where superscript dot, prime and cross designate positive, negative and neutral charges respectively, we predict two types of defects: excess protons, $H^{\cdot}$, and proton vacancies, O'.

$$[OH]^{x} = H^{\cdot} + O' \qquad (9.6)$$

The excess protons are positive. Being delocalized they are potentially highly mobile charge carriers. Their formation is shown schematically in Fig. 9.4a. Proton vacancies carry a negative charge. They move by protons from neighbouring OH^{-} anions jumping onto the deprotonated sites as shown in Fig. 9.4b. The proton jumps necessary for this motion may occur on any of the four intermediate energy levels, separated by about 0.4 eV and marked in Fig. 9.3 by the 02, 03, and 04 transitions.

Eqn (9.6) and Fig. 9.4a, b represent the intrinsic case when excess protons and proton vacancies are produced in equal numbers: $n(H^{\cdot}) = n(O')$. The extrinsic case is also of direct relevance to the experiments described here. It arises when the number of deprotonated sites becomes very large, $n(H^{\cdot}) \ll n(O')$, for instance as a result of incipient dehydration. Note that, according to Eqn (9.2), each H_2O molecule forming in the $Mg(OH)_2$, leaves behind a deprotonated site, hence a proton vacancy.

9.4   Direct current proton conductivity measurements

If an electrical potential is applied to a conductor, a current I is produced

$$I = \sum_{ij} n_{ij}\mu_{ij} \qquad (9.7)$$

where n is the number and μ the mobility of all charge carriers i, j. In proton conductivity measurements the challenge is (i) to separate protonic charge carriers from all other possible charge carriers such as electrons or ions, and (ii) to distinguish between the two different types of protonic charge carriers defined above as excess and proton vacancies, $H^{\cdot}$ and O', respectively. Neglecting ionic contributions, the conductivity σ will contain four terms

$$\sigma = [n(e')\mu(e') + n(e^{\cdot})\mu(e^{\cdot})] + [n(O')\mu(O') + n(H^{\cdot})\mu(H^{\cdot})] \qquad (9.8)$$

where e' and $e^{\cdot}$ represent electrons and holes, respectively, in Kröger's defect notation[8]. In order to measure protonic contributions, special conditions must be met. Even if a current through a sample is carried by protons, in order to measure it by an external circuit the protons have to be converted to electrons. Hence, electrodes must be used that are capable of delivering protons to and accepting protons from the sample at the anodic and cathodic sides, respectively. With palladium (Pd) electrodes, the following reactions provide for the necessary electrochemical reactions at the anodic and cathodic sites, respectively, here written in simple chemical terms.

$$H_{diss} = H^+ + e^- \tag{9.9}$$

$$H^+ + e^- = H_{diss} \tag{9.10}$$

By contrast, Au or Cu are blocking electrodes that can neither inject nor dissolve protons. Hence, carrying out identical experiments with Au or Cu and with H_2-saturated Pd electrodes provides the means of separating protonic contribution, reducing Eqn 9.8 to $\sigma_{proton} = n(O')\mu(O') + n(H^{\cdot})\mu(H^{\cdot})$. For a thermally activated process we obtain

$$\sigma_{proton} = \sigma^{\cdot} \exp[-E^{\cdot}/kT\} + \sigma' \exp[-E'/kT] \tag{9.11}$$

where $E^{\cdot}$ and E' are the activation energies associated with the $H^{\cdot}$ and O' motions, T is the absolute temperature, and k the Boltzmann constant.

The current depends upon the number and the mobilities of the protonic charge carriers. In the intrinsic case, when $n(H^{\cdot}) = n(O')$, only the mobility μ determines which species carries the majority current. In the extrinsic case, for instance when $n(H^{\cdot}) \ll n(O')$, σ is determined by the product $n \cdot \mu$.

Note that, in an electric field, the two partial currents carried by $H^{\cdot}$ and O' run opposite to each other. However, the motion of O', say, from left to right depends upon a proton jump from right to left as illustrated in Fig. 9.4b. Therefore, under d.c. conditions, one actually measures two parallel proton fluxes moving on two different energy levels. Conductivity measurements cannot provide information about the nature of the charge carriers, in particular their signs. All that Eqn (9.11) provides is the sum of the $H^{\cdot}$ and O' contributions.

Conductivity measurements have still other limitations. Anodic proton injection requires high electric fields, and cathodic proton up-take requires dissolution of H in the electrode. If the anode does not supply enough protons, the current becomes injection-limited. If cathodic redissolution does not keep up with the arrival of protonic charge carriers, the cathode becomes polarized by the formation of H_2 gas. The performance can be improved by using Pd-black electrodes with small Pd particles which give high local electric field gradients for injection and a high surface area for redissolution.

To determine the signs of the charge carriers we may consider Hall effect measurements, i.e. apply a 90° magnetic field and allow the charge carriers to drift perpendicular to the direction of the electric field. This creates an electric potential across the sample which can be picked up with a pair of auxiliary electrodes. Easier and equally informative is the measurement of the thermopotential $U(T)$, i.e., the electric potential in a temperature gradient.

In the intrinsic case when $n(H^{\bullet}) = n(O')$, a temperature gradient across the sample leads to a concentration gradient: the concentration of both $H^{\bullet}$ and O' will be higher on the 'hot' side (T_2) than on the 'cold' side (T_1). As a result, both charge carriers diffuse in parallel in the same direction along the temperature gradient. This translates into two actual proton fluxes in opposite, i.e., antiparallel, directions, as explained above. The sign of the thermopotential is determined by the relative mobilities. If $\mu(H^{\bullet}) \gg \mu(O')$, $H^{\bullet}$ diffuse more rapidly from 'hot' to 'cold', so that the 'cold' side becomes positive. Assuming a linear temperature gradient $\Delta T = T_2 - T_1$, the concentration gradient across the sample is exponential. With the gap energy E_{gap} corresponding to the thermal activation energy, the concentration difference Δ between 'hot' and 'cold' sides becomes

$$\Delta[n(H^{\bullet})_2 - n(H^{\bullet})_1] = \exp[-E_{gap}/k(T_2 - T_1)] \qquad (9.12a)$$

$$\Delta[n(O')_2 - n(O')_1] = \exp[-E_{gap}/k(T_2 - T_1)] \qquad (9.12b)$$

The intrinsic case of greatest interest is when, due to the departure of H_2O molecules during incipient dehydration, protons are removed from the system and the number of deprotonated sites, i.e. O', increases rapidly. In this case $n(H^{\bullet}) \ll n(O')$. Since the sign of the thermopotential is determined by $n \cdot \mu$, the 'cold' side will eventually become negatively charged, even if $\mu(H^{\bullet}) \gg \mu(O')$.

Protonic thermopotentials can be established only if suitable electrodes are used, for instance H_2-saturated Pd. With blocking Au or Cu electrodes no protonic $U(T)$ is measured. The reason is that, even though thermopotentials should theoretically involve zero current, some charge carriers still need to react with the electrodes, extracting or injecting electrons, in order to establish $U(T)$. The absence of a measurable thermopotential with blocking electrodes allows us to differentiate between protonic effects and other contributions.

The zero current situation is never truly fulfilled because a certain number of charge carriers is needed to maintain a stable thermo-potential over time. Their number is determined by the internal resistance of the electrometer used in the outer circuit. Using an electrometer with a high resistance, for instance 10^{14} Ω, only a few charge carriers are drained from the sample to drive the measuring device. Thus, experimental difficulties due to space charge limitation at the anodic side or polarization at the cathodic side are minimized.

$U(T)$ measurements have the added advantage that they can also be modified to measure the conductivity, σ_{proton}, by increasing the charge carrier drainage. This is conveniently done by shunting the electrometer with calibrated resistances R_c. The lower the shunt resistances, the more charge carriers have to flow to the electrodes. The internal resistance of the sample, $R_s(T)$, can thus be determined from the effective thermo-potential $U_{eff}(T)$

$$U_{eff}(T) = U_0(T)[1 + R_s(T)/R_c] \qquad (9.13)$$

where $U_0(T)$ is the 'open circuit' thermopotential, i.e. the thermopotential measured with the highest electrometer resistance. If $R_s(T) \ll R_c$, $R_s(T)$ can be obtained with sufficient accuracy. Experimental details are given elsewhere[3].

9.5   Proton conductivity results

The d.c. conductivities of $Mg(OH)_2$, $Ca(OH)_2$ and $Al(OH)_3$ as a function of temperature are low when blocking Cu or Au electrodes are used, confirming that they are good insulators. The conductivity, however, increases significantly when H^+-injecting Pd-black electrodes are used. In addition, as exemplified for $Mg(OH)_2$ in Fig. 9.5, an anomaly appears between 450 and 500 K (solid curves) where the response is non-ohmic, indicating space charge limited currents. The very low conductivity measured with Au electrodes at the highest voltage, 120 V, is shown by the dashed curve. The increase and appearance of the anomalous region are indications of protonic charge carriers. Three regions of distinctly different behaviour may thus be distinguished prior to, during and above the anomaly. In the following, these regions will be called I, II and III, respectively.

Fig. 9.6a, b combines results for $Mg(OH)_2$, $Ca(OH)_2$ and $Al(OH)_3$, plotted linearly as σ versus T, and as Arrhenius plots according to Eqn

Materials

(9.11), i.e. $\log \sigma$ vs. $1/T$. A common mechanism is indicated by the straight line spanning the low temperature region I and the high temperature region III, prior to and after the 'anomaly', with activation energies of $\leqslant 1$ eV. Superimposed on this background, in the intermediate region II, rises the anomaly. Its initial portion gives a straight line, pointing to a second conductivity mechanism, with activation energies of 2.0, 2.2 and 1.65 eV for $Mg(OH)_2$, $Ca(OH)_2$ and $Al(OH)_3$, respectively.

Fig. 9.7 shows the thermopotentials measured for all three hydroxides with H_2-saturated Pd-black electrodes. With Cu electrodes $U(T)$ remains nil (not shown). Three regions are again noted: weakly negative in I, strongly positive in II and moderate negative in III. These regions of $U(T)$ coincide with I, II and III determined by d.c. measurements. Also included are the $U(T)$ curves obtained with shunt resistances of 10^9, 10^8, 10^7 and $10^6 \, \Omega$, leading to the breakdown of $U(T)$. Using Eqn (9.13) we can calculate $\sigma(T)$. In region II we find an Arrhenius behaviour extending over the entire region of positive $U(T)$ values, indicating no space charge limitation. The activation energies for $Mg(OH)_2$, $Ca(OH)_2$ and $Al(OH)_3$ are 2.1, 2.2 and 1.6 eV, respectively, in good agreement with those determined by d.c. measurements[3].

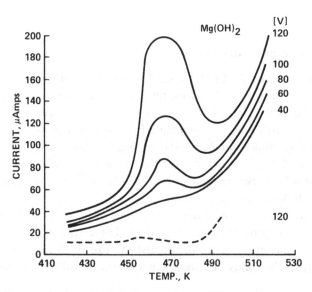

Fig. 9.5. D.c. conduction of $Mg(OH)_2$ powder at different voltages as a function of temperature approaching the on-set of dehydration, with H_2-saturated Pd electrodes (solid) and Cu electrodes (dashed) (with kind permission of Verlag Chemie).

148

Highly ionic hydroxides

Since $U(T)$ is positive in region II, this region can be assigned to excess protons, H·. The activation energy of 2.0–2.1 eV for $Mg(OH)_2$ agrees well with the 2.0 eV band gap determined from the O–H overtone series. This suggests that the mobile H· carriers are protons elevated to the CB. By inference, we conclude that the negative $U(T)$ values in regions I and III indicate proton vacancies, O'.

Though no spectroscopic data are available for $Ca(OH)_2$ and $Al(OH)_3$, the larger activation energy for the former and the smaller activation

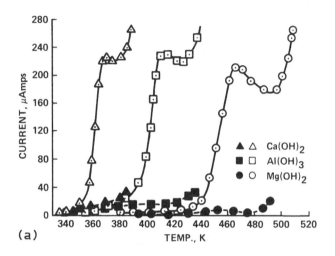

(a)

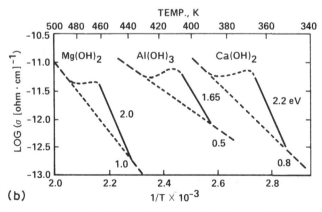

(b)

Fig. 9.6(a). D.c. proton conductivities of $Mg(OH)_2$, $Ca(OH)_2$ and $Al(OH)_3$ prior to their respective dehydration, using H_2-saturated Pd electrodes (open symbols) and Cu electrodes (solid symbols). (b) Corresponding Arrhenius plots with the activation energies (with kind permission of Verlag Chemie).

149

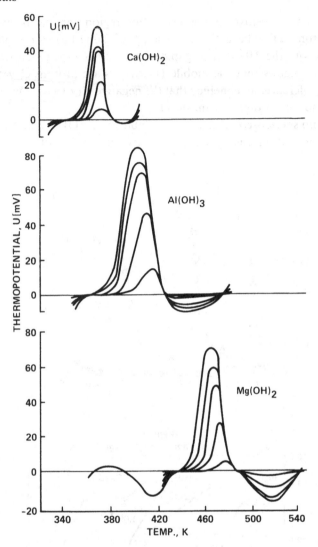

Fig. 9.7. Thermopotentials of $Mg(OH)_2$, $Ca(OH)_2$ and $Al(OH)_3$ measured with H_2-saturated Pd electrodes and an internal resistance of $10^{14}\,\Omega$ (outer envelope). Progressive break-down using shunt resistances of 10^9, 10^8, 10^7 and $10^6\,\Omega$. Note the regions of negative and positive $U(T)$ (with kind permission of Verlag Chemie).

energy for the latter appear reasonable. In $Ca(OH)_2$, the separation between OH^- anions is larger than in $Mg(OH)_2$, due to the larger ionic radius of Ca^{2+}. Superimposing Morse potentials as in Fig. 9.3 will cause the overlap to move to higher energies, giving a larger band gap in $Ca(OH)_2$. In $Al(OH)_3$, the repulsive Coulomb interaction between Al^{3+} and the protons in OH^- is stronger. This weakens the O–H bond and increases the asymmetry of the Morse potential. We therefore expect a lower energy barrier. The chemical consequence of the strong repulsion between Al^{3+} and the protons in the OH^- is a higher acidity of $Al(OH)_3$ compared to weakly basic $Mg(OH)_2$ and a higher basicity for $Ca(OH)_2$.

9.6   Proton carrier density in the conduction band

If the conductivity mechanism proposed here is correct, the number of excess protons that carry the current in region II must be very small. It would therefore be of interest to determine independently the charge carrier concentration.

A suitable technique is *thermally stimulated depolarization TSDp*. It has been successfully applied to the same $Mg(OH)_2$ which was used for the $\sigma(T)$ and $U(T)$ measurements described in the preceding section[12]. A layer of compressed $Mg(OH)_2$ powder, ~ 0.5 mm, was placed between the H_2-saturated Pd electrodes of a capacitor and heated. At selected temperatures, T_{pol}, a high voltage was applied, corresponding to $\approx 80\,000$ V/cm^{-1} *polarization field*. The field was kept on for a few minutes to allow positive and negative charge carriers to separate into charge clouds near the cathode and anode, respectively. Then the assembly was quenched under field to liquid nitrogen temperature. At 80 K the high voltage was disconnected and the voltage source exchanged for an electrometer. At this point, we expect two charge clouds to be frozen-in: H˙ at the cathode and O′ at the anode. During warm-up to 300 K these charge clouds should remobilize, probably one after the other, and depolarization currents may be measured.

In conventional TSDp experiments, when this technique is used to measure the unfreezing of dipoles, the depolarization currents always flow in the direction of discharging the capacitor. In the case of TSDp experiments with protonic charge carriers a seemingly paradoxical situation arises. Because electrons are consumed at the electrodes when protons discharge, the current in the external circuit flows in the 'wrong' direction: they apparently increase the charge on the capacitor plates. In the case

151

of proton conductors the TSDp maxima are therefore in a characteristic inverted form with respect to regular TSDp maxima.

Fig. 9.8 shows typical results with $Mg(OH)_2$. At the lowest T_{pol}, 390 K, two regular (positive) TSDp maxima are observed at 210 and 250 K. They are due to H_2O molecules adsorbed on the $Mg(OH)_2$ crystallites which regain their rotational freedom at these relatively high temperatures. At

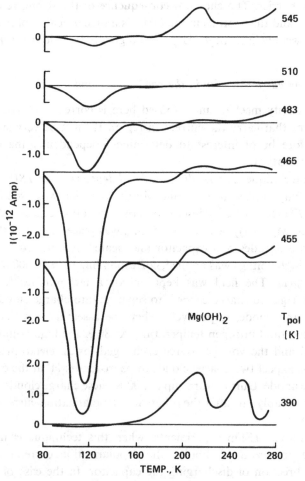

Fig. 9.8. Thermally stimulated depolarization (TSDp) curves of $Mg(OH)_2$, using H_2-saturated Pd electrodes, at different polarization temperatures, T_{pol}. The TSDp minimum at 118 K is caused by H^+ charge carriers, i.e. protons in the conduction band, reaching a density of $\sim 0.9 \times 10^{13}$ mol^{-1} in good agreement with the theoretical values of 1.5×10^{13} mol^{-1} (Eqn 9.14a), assuming $E_{gap} = 2$ eV (with kind permission of Verlag Chimie).

Highly ionic hydroxides

higher T_{pol}, $\geqslant 455$ K, two inverted (negative) TSDp maxima appear, a large one at 115–120 K, and a lesser one at ≈ 180 K.

The appearance of the large inverted TSDp maximum at 115–120 K coincides with region II, i.e. the temperature interval of positive $U(T)$ values where $H^{\cdot}$ charge carriers dominate. Its maximum intensity is reached at the same temperature at which the proton conductivity passes through its maximum in Fig. 9.5. To determine the concentration of excess protons $[H^{\cdot}]$ contributing to the 115–120 K inverted maximum we apply Fermi–Dirac statistics

$$[H^{\cdot}] = \frac{1}{V} \int_{E_{CB}}^{\infty} g^{\cdot}(E) \, dE \exp[(E - \eta)/kT + 1]^{-1} \qquad (9.14a)$$

$$[O'] = \frac{1}{V} \int_{\infty}^{E_{VB}} g'(E) \, dE \exp[(\eta - E)/kT + 1]^{-1} \qquad (9.14b)$$

Here $g^{\cdot}(E)$ and $g'(E)$ are the energy densities of levels in CB and VB, respectively, η is the chemical potential analogous to the Fermi level in electronic conductors, V is the volume and E_{VB} and E_{CB} are the band edge energies. With $(E_{CB} - E_{VB}) = E_{gap} = 2.0$ eV, we assume $(E_{CB} - \eta) \gg kT$ and $(\eta - E_{VB}) \gg kT$. Thus Eqns (9.14a, b) reduce to

$$[H^{\cdot}] \approx N^{\cdot} \exp[(E - \eta)/kT] \qquad (9.15a)$$

$$[O'] \approx N' \exp[(\eta - E)/kT] \qquad (9.15b)$$

with $N^{\cdot}(T) = 1/4[2m^{*\cdot}kT/\pi h^2]^{2/3}$ and $N'(T) = 1/4[2m^{*\prime}kT/\pi h^2]^{2/3}$, where $m^{*\cdot}$ and $m^{*\prime}$ are the effective masses of the excess protons and defect protons respectively. The dependence of the *chemical potential* vanishes

$$[H^{\cdot}][O'] = N^{\cdot}N' \exp[-E_{gap}/kT] \qquad (9.16)$$

Thus, for the intrinsic case, we obtain an expression for the charge carrier densities in terms of $(N^{\cdot}N')$ and T

$$[H^{\cdot}] = [O'] = (N^{\cdot}N')^{1/2} \exp[-E_{gap}/2kT] \qquad (9.17)$$

Assuming that the protonic space charges which give rise to the negative TSDp maxima are thermally generated, the transport of the charge carriers as a function of time t along the spatial coordinate x leads to the

153

Materials

current $J(t)$

$$J(t) = \int_0^x \{[H^\cdot](x, t)\mu^\cdot(T) + [O'](x, t)\mu'(T)\}F(x, t)\,dx \qquad (9.18)$$

where $\mu^\cdot$ and μ' are the mobilities and $F(x, t)$ the residual electric field.

Numerical solutions of this complex function can be found only by assuming specific conditions concerning the initial space charge distribution. The simplest case is that of a uniform sample with the distributed charge carriers concentrated in narrow layers close to both electrodes. With a linear heating program and bimolecular recombination kinetics, assuming $\mu^\cdot \gg \mu'$, symmetrical inverted current peaks are expected, one for each kind of charge carrier, as indeed shown by Fig. 9.8. (By contrast, positive TSDp maxima caused by the unfreezing of dipoles are theoretically predicted to be skewed as indeed observed for the two regular maxima at 210 and 250 K at the bottom of Fig. 9.8.)

If the space charge layers are close to the electrodes, half of the charges will reach the electrode and discharge, giving a recovery factor $R \approx 0.5$. If the $H^\cdot$ charge cloud spreads out first during warm-up, it will recombine with O' and thus consume part of the O' charge cloud. Therefore, R for O' will be much smaller than for $H^\cdot$.

$H^\cdot$ concentrations for different T_{pol} can thus be obtained from the TSDp experiments. The largest value at $T_{pol} = 465$ K is 1.1×10^{-9} C and occurs at the same temperature at which the $H^\cdot$ conductivity reaches its maximum (see Fig. 9.5). With an $Mg(OH)_2$ sample size of 50 mmol as was the case in this particular experiment, this gives a maximum $H^\cdot$ concentration of $\approx 1.5 \times 10^{-13}$ mol^{-1}, assuming $R \approx 0.5$.

Taking 2.0 eV as the protonic band gap in $Mg(OH)_2$, the concentration of $[H^\cdot]$ on the CB at 460 K is the order of 10^{-13} mol^{-1} according to a Boltzmann distribution[4]. This agrees very well with the value obtained from the intensity of the TSDp peak at 115–120 K. Agreement within one order of magnitude was also found for $Ca(OH)_2$, and $Al(OH)_3$[12]. The number of O' calculated from the TSDp peak around 180 K is much smaller, which appears reasonable in view of the fact that, if the $H^\cdot$ charge cloud spreads out around 115–120 K during warm-up, it must consume many O' charge carriers before they can unfreeze at 180 K. The fact that $H^\cdot$ remobilizes already at such a low temperature indicates that they are trapped in shallow energy levels.

The TSDp measurements confirm that the concentration of excess protons $H^\cdot$, $n(H^\cdot)$, is extremely small, only of the order of 10^{-13} mol^{-1}.

This confirms that they are thermally activated. It also allows us to comment on the mobility of the $H^{\cdot}$ charge carriers, $\mu(H^{\cdot})$, relative to the mobility of O' charges, $\mu(O')$. Only if $\mu(H^{\cdot})$ is extremely large can the product $n(H^{\cdot}) \cdot \mu(H^{\cdot})$ explain the proton conductivity results. Note that no absolute conductivity or mobility values are given in this report, because all d.c. measurements were carried out with powder samples where the grain–grain contact resistances remain unknown.

9.7 Summary

The most significant result which comes out of this work is that $Mg(OH)_2$, $Ca(OH)_2$ and $Al(OH)_3$ may be protonic semiconductors that can be described by highly localized protons occupying a valence band, VB, and the presence of a conduction band, CB, providing for proton delocalization. Between VB and CB lie intermediate levels, corresponding to the vibrationally excited states of the O–H oscillator, separated by ≈ 0.4 eV. The E_{gap} values are 2.0–2.1 eV for $Mg(OH)_2$, 2.2 eV for the more basic $Ca(OH)_2$ and 1.6–1.7 eV for the more acidic $Al(OH)_3$.

The large E_{gap} values mean that the number of protons that can be thermally injected into the CB is extremely small, of the order of 10^{-13} mol^{-1} at the maximum of the proton conductivity of $Mg(OH)_2$ at 465 K. Though the present experiments have not yielded absolute conductivity values, the $H^{\cdot}$ mobility can be ranked relative to the O' mobility. Since we observed both excess and defect proton conduction, we must have $n(H^{\cdot}) \cdot \mu(H^{\cdot}) \approx n(O') \cdot \mu(O')$. From the Boltzmann distribution we calculate that $n(H^{\cdot})$ is very small. If the motion of the O' charge carriers is based on proton jumps according to the schematic presentation in Fig. 9.4b, using some of the intermediate energy levels in the band gap, 0.5–1.0 eV above the VB, the occupancy of these levels is 10^6–10^7 larger than for CB density of states. Hence, in order to satisfy the condition for the intrinsic case $n(H^{\cdot}) \cdot \mu(H^{\cdot}) \approx n(O') \cdot \mu(O')$, via the O' mobility must be smaller by about the same factor. Therefore, the mobility of excess protons, delocalizing through the CB, should be 10^6–10^7 times larger than that of defect protons, moving by proton hopping through the more densely populated intermediate energy levels.

This order-of-magnitude estimate also explains why protonic thermo-potentials are unexpectedly large. Though ΔT is only a few degrees, $U(T)$ reaches values as high as 100 mV. Only if the mobilities of the two contributing charge carriers differ greatly can such large $U(T)$ values be expected. In electronic semiconductors where electrons, and defect electrons, mobilities normally differ by no more than a factor of $\approx 10^3$, the highest achievable thermopotentials fall in the range of 10–100 $\mu V\,K^{-1}$. Therefore, the high protonic $U(T)$ values obtained from the ionic hydroxides underline the unusual nature of the underlying proton conduction mechanism.

In conclusion it can be said that, though the highly ionic hydroxides discussed have probably no practical interest as high performance proton conductors, they represent a class of compounds which seem to have, by their own right, very interesting fundamental properties that open new aspects of proton behaviour.

Acknowledgements

Many of the data presented here were obtained in collaboration with Reinhard Martens and Heinz Wengeler whose enthusiasm for this experimentally demanding field of physics is highly appreciated. The Deutsche Forschungsgemeinschaft provided financial support.

9.8   References

1. F. Freund. The proton pump at work; Part I. The basic concept of excess proton/defect translocation. *J. Electroanal. Biochem.* **9** (1982) 61–77.
2. F. Freund and J. C. Nièpce. Protons in simple ionic hydroxides. *Adv. Solid States Chem.* **1** (1989) 26–64.
3. F. Freund and H. Wengeler. Proton conductivity of simple ionic hydroxides: Part I: The proton conductivities of $Al(OH)_3$, $Ca(OH)_2$, and $Mg(OH)_2$. *Ber. Bunsenges. Phys. Chem.* **84** (1980) 866–73.
4. F. Freund, H. Wengeler and R. Martens. Proton conductivity of simple ionic hydroxides: Part III. *J. Chim. Phys. (France)* **77** (1980) 837–41.
5. K.-H. Hass and U. Schindewolf. The electrical conductivity of solid and molten cesium hydroxide – A contribution to solid proton conductors. *Ber. Bunsenges. Phys. Chem.* **87** (1983) 346–8.
6. K.-H. Haas and U. Schindewolf. The electrical conductivity of solid alkali hydroxides. *J. Solid State Chem.* **54** (1984) 342–5.
7. R. T. Johnson, R. M. Biefield and J. D. Keck. *Mat. Res. Bull.* **12** (1977) 577–87.
8. A. F. Kröger and H. J. Vink. *Solid State Physics*, Ehrenreich, Seitz and Turnbull (eds.) (Academic Press, New York (1956)).
9. R. Martens and F. Freund. The potential energy curve of the proton and the dissociation energy of the OH^- ion in $Mg(OH)_2$. *Phys. Stat. Sol.* **37** (1976) 97–103.
10. P. S. O'Shea, S. Thelen, G. Petrone and A. Azzi. Proton mobility in

biological membranes: the relationship between membrane lipid state and proton conductivity. *FEBS Lett.* **172**(1) (1984) 103–8.

11. P. Stephen and A. T. Howe, Proton conductivity and phase relationships in solid KOH between 248 and 406 °C. *Solid State Ionics* **1** (1980) 461–71.

12. H. Wengeler, R. Martens and F. Freund. Proton conductivity of simple ionic hydroxides; Part II: In situ formation of water molecules prior to dehydration. *Ber. Bunsenges. Phys. Chem.* **84** (1980) 873–80.

10 Ice

I. A. RYZHKIN

10.1 Introduction

Vast masses of ice and snow on the Earth play an important part in our life. Nevertheless, ice is not a traditional topic in solid state physics. This is explained by difficulties of making controlled and reproducible experiments with ice for the following reasons.

First, ice has a large number of solid modifications: hexagonal or ordinary ice, I_h; cubic ice, I_c; ices II–IX; vitreous ice. Most of them exist at elevated pressure or need special formation conditions. Under ordinary conditions, only hexagonal ice is formed, which has therefore been investigated more frequently than other modifications. In this chapter we shall deal only with ordinary ice.

Second, ice is an unusual example of the solid state, since it consists of two very different parts: a crystalline, hard lattice of oxygen atoms and a disordered, *quasi-liquid* proton system. For this reason, the physical properties of ice are intermediate between those of a solid and a liquid.

Third, as a rule ice contains various impurities, whose distribution, homogeneity and concentration are very hard to control and which strongly affect the physical properties of ice.

In spite of such difficulties, by the beginning of the 1970s the essential principles of ice physics had been formulated. Using them it is possible to explain the unusual properties of ice and to predict the behaviour of ice under different conditions.

10.2 Structure of ordinary ice[1]

The oxygen lattice of ordinary ice is similar to that of the wurtzite structure, observed with $A^{II}B^{VI}$ semiconductors, in which both A and B atoms are substituted by oxygen ions. An important feature of this structure is that each oxygen ion has a regular tetrahedral environment of oxygens. The distance between the nearest oxygen ions equals 0.276 nm (see Fig. 10.1).

The protons are located in the bonds linking oxygen ions, forming hydrogen bonds. Inasmuch as the distance between an oxygen ion and a proton in a water molecule is about 0.1 nm, then there are two possible positions for a proton in each hydrogen bond, namely near to one or other of two neighbouring oxygen ions. So, the hydrogen bond may appear in one of two states and can be described by the pseudo-Ising variable $\sigma = \pm 1$. Consideration of Coulomb interaction between protons leads to an antiferromagnetic Hamiltonian for pseudo-Ising variables, localized in the middle of hydrogen bonds[2]. For the structure of hexagonal ice, the antiferromagnetic Hamiltonian has a degenerate ground state due to the impossibility of satisfying all the rules of antiferromagnetic ordering. Lattices possessing these properties are called *frustrated lattices*. Degeneration of the ground state implies that the proton distribution is disordered; two rules, however, should be fulfilled:

(i) there is one proton in each bond,
(ii) there are two protons near each oxygen ion.

These are known as the Bernal–Fowler rules, formulated empirically more than 50 years ago[3].

10.3   Defects and conduction mechanism

Ideally, ice has the structure described above and it can easily be seen that it is impossible for protons to move in it, since such motion would violate the Bernal–Fowler rules. However, at all real temperatures there

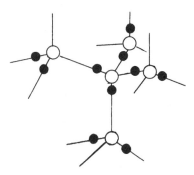

Fig. 10.1. Fragment of ice lattice. Open circles, oxygen atoms; solid circles, protons. The proton distribution satisfies the Bernal–Fowler rules.

are defects in the system, shown in Fig. 10.2, that make the proton system mobile (see Chapter 1).

Ionic defects H_3O^+ and OH^- involve violation of the second Bernal–Fowler rule. As shown in the figure, by way of motion of protons along the bonds, ionic defects can move across the oxygen lattice without forming new defects. It has to be emphasized that after a defect has passed along a bond, the latter appears in a quite definite orientation such that a second defect cannot pass in the same direction. So, the proton arrangement has a memory for whose description a new quantity is introduced, namely, the configuration vector. This vector is connected with defect fluxes or with the dipole moment of the hydrogen bonds. Ionic defects have effective charges $\pm 0.62e$ (e being the proton charge)[4a]. Their mobilities at temperature $T = 253$ K equal[4b],

$$\mu_{H_3O^+} = 9.0 \times 10^{-8}\, \mathrm{m^2 V^{-1}\, s^{-1}} \tag{10.1}$$

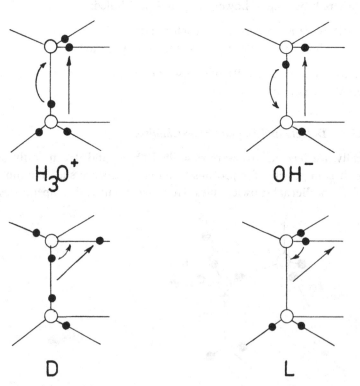

Fig. 10.2. Schemes of defect structures and defect movements in ice. Small arrows show the actual displacements of protons; large arrows show the resulting displacements of the defects.

$$\mu_{OH^-} = 2.5 \times 10^{-8} \, m^2 V^{-1} s^{-1} \tag{10.2}$$

The ionic defect concentrations are determined by the equation[1]

$$n = n_0 \exp(-E_{12}/2kT) \tag{10.3}$$

where $n_0 = 3 \times 10^{22} \, cm^{-3}$, $E_{12} = 0.98$ eV, k is the Boltzmann constant, and T is temperature. Taking account of conductivity results at the same temperatures, we obtain

$$\sigma_{H_3O^+} = 4.7 \times 10^{-8} \, (\Omega m)^{-1} \tag{10.4}$$

$$\sigma_{OH^-} = 1.3 \times 10^{-8} \, (\Omega m)^{-1} \tag{10.5}$$

The activation energy of conductivity equals the activation energy of defect creation which therefore indicates a nonactivated character for proton motion along the bonds.

Notwithstanding the fact that motion of ionic defects is connected with the motion of protons, it does not imply a real proton transfer. Actually there occur only successive local displacements of protons along the bonds. Due to the memory of the protonic system, after a certain amount of current has passed, all the bonds appear to be blocked, and further passage of current is impossible. It should be remembered however, that there are also defects of the second type in the system, namely, D and L defects, shown in Fig. 10.2, which represent violations of the first Bernal–Fowler rule. A D defect moving in the same direction as an H_3O^+ defect, polarizes the bonds in the opposite direction, that is, it unblocks them as seen in Fig. 10.2. In an analogous way the motions of OH⁻ and L defects are related. So, by a combined motion of all the defects (or of only an H_3O^+, D pair) a current may pass through ice indefinitely. D and L defects have effective charges $\pm 0.38e$. Their mobilities at a temperature $T = 253$ K equal[1]

$$\mu_D = 1.0 \times 10^{-10} \, m^2 V^{-1} s^{-1} \tag{10.6}$$

$$\mu_L = 1.0 \times 10^{-9} \, m^2 V^{-1} s^{-1} \tag{10.7}$$

The concentrations are determined by Eqn (10.3) with

$$n_0 = 6 \times 10^{22} \, cm^{-3},$$

$E_{12} = 0.68$ eV. Accordingly, for conductivities we obtain

$$\sigma_D = 6.0 \times 10^{-8} \, (\Omega m)^{-1} \tag{10.8}$$

$$\sigma_L = 6.0 \times 10^{-7} \, (\Omega m)^{-1} \tag{10.9}$$

Materials

These values of partial conductivities do not take into account the interaction via polarization of the bonds. During the motion of all four types of the defects such interaction leads to a frequency dependence of the overall conductivity

$$\sigma(\omega) = \sigma_0 - \frac{i\omega\tau(\sigma_\infty - \sigma_0)}{(1 - i\omega\tau)} \qquad (10.10)$$

where $\quad e^2/\sigma_0 = e_{12}{}^2/(\sigma_1 + \sigma_2) + e_{34}{}^2/(\sigma_3 + \sigma_4)$

$$\sigma_\infty = \sigma_1 + \sigma_2 + \sigma_3 + \sigma_4$$

$$\sigma_i = e_i \mu_i n_i$$

e_i, μ_i, n_i are effective charges, mobilities and concentrations respectively, σ_i are partial conductivities.

It is seen that at high frequencies, the conductivity is determined by the majority carriers, i.e. carriers with a maximum partial conductivity (at high frequencies the bonds have not enough time to be polarized). At low frequencies the conductivity is determined by minority carriers, i.e. carriers with smallest partial conductivities which unblock the bonds.

The processes of charge transfer described above involve protons. With the use of contacts made from electronic conductors, that do not permit the exchange of protons with ice, the passage of direct current should become impossible. However, Petrenko and co-workers have observed continuous currents in such systems without release of hydrogen or oxygen[5]. This phenomenon was attributed to the presence of two new types of defect which are complexes of ionic defects and electrons. Thus, H_3O^+ trapping electrons becomes $(H_3O^-)^{\cdot}$ with an effective charge $-0.38e$ and with polarization properties analogous to L defects. Similarly, OH^- giving up electrons becomes the $(OH^+)^{\cdot}$ defect with charge $+0.38e$; it is analogous to the D defect. The conduction mechanism is deduced to be the following: H_3O^+-defects move under the action of an electric field to the negative electrode, trap electrons at the interface and transform to $(H_3O^-)^{\cdot}$. The latter move to the positive electrode, giving up the electrons there. The current is defined by the motion of electrons that move not alone, but trapped on ionic defects. Thus, the new defects are capable of substituting D and L defects completely. Importantly, with the method proposed no proton transfer via ice takes place. This fact may be used in determination of the conduction mechanism for any particular case.

10.4 Electrical properties of doped ice[6]

The electrical properties of ice can be changed appreciably by doping with various impurities. Interesting results can be obtained on doping with the following materials: NH_3, HF, NH_4OH, NH_4F, KOH, which are related to their fairly extensive solubility in ice. Upon dissolution, fluorine and nitrogen occupy the sites of oxygen atoms, and hydrogen occupies protonic sites. During dissolution, depending on the amount of hydrogen, defects of all four kinds may be formed. Thus, doping of an NH_3 molecule leads directly to the formation of a D defect. Consequently, the dissolution of NH_3 leads to an increase of D type carriers and decrease of L carriers (since $n_D n_L = $ const.). In addition, NH_3 is prone to trap an extra proton and form $NH_4{}^+$. As a result, the concentration of OH^- carriers is increased and of H_3O^+ is decreased.

In a similar way, HF directly forms L defects, and by means of the $HF-F^-$ transition forms H_3O^+-defects. So, doping with HF leads to an increase of $n_{H_3O^+}$, n_L and to a decrease of n_{OH^-}, n_D. As a rule, the defects formed have a binding energy to an impurity centre identical to the ionization energy of impurity centres in semiconductors. Experimental studies have confirmed the effect of impurities on the electrical properties and yielded values for the ionization energy of defects.

10.5 Conclusion

In conclusion, we describe two phenomena between which there exists a certain relationship. The first phenomenon consists of a high *near-surface conduction* of ice: the near-surface layer, several nanometers in thickness, may possess a conductivity comparable with that of a macroscopic crystal. Detection of this phenomenon necessitates use of special experimental techniques (guard ring) for measuring conductivities and undoubtedly indicates a peculiar state of the protonic system in the near-surface layer[7, 8].

The second phenomenon is related to violation of the temperature dependence of concentration, described by Eqn (10.3). In fact, on formation of a pair of defects, H_3O^+ and OH^-, from two neutral water molecules a considerable energy is consumed in overcoming the Coulomb attraction between H_3O^+ and OH^-. With other defects present, this interaction is screened, which leads to a nonlinear dependence of the system energy on the defect concentration. As the temperature is increased the system may jump into a new state with an increased concentration of defects and decreased activation energy of their formation[9]. This transition leads to high conduction, and is a *superionic transition*. For it to occur, one has to increase both the temperature and pressure (the latter is needed for increasing the melting point of ice).

Another possibility for realizing the superionic transition is associated with modification of the atomic structure in the near-surface layer. In this case anomalously high near-surface conduction of ice would then acquire a natural explanation.

10.6   References

1. P. V. Hobbs, *Ice Physics* (Clarendon Press, Oxford (1974)).
2. I. A. Ryzhkin, *Solid State Commun.* **52** (1984) 49.
3. J. D. Bernal and R. H. Fowler, *J. Chem. Phys.* **1** (1933) 515.
4. (a) M. Hubmann, *Z. Physik* **B32** (1979) 127, 141. (b) V. F. Petrenko and N. Maeno, *J. Physique* **48** (1987) C1–11.
5. V. A. Chesnakov, V. F. Petrenko, I. A. Ryzhkin and A. V. Zaretskii, *J. Physique*, **48** (1987) C1–99.
6. N. H. Fletcher, *The Chemical Physics of Ice* (University Press, Cambridge (1970)).
7. J. M. Cavanti and A. J. Illingworth, *J. Phys. Chem.* **87** (1983) 4078–83.
8. J. Ocampo and J. Klinger, *J. Phys. Chem.* **87** (1983) 4325–8.
9. I. A. Ryzhkin, *Solid State Commun.* **56** (1985) 57.

11 Anhydrous materials: oxonium perchlorate, acid phosphates, arsenates, sulphates and selenates

PHILIPPE COLOMBAN and ALEXANDRE NOVAK

Proton-containing anhydrous materials are interesting proton conductors as they are stable to relatively high temperatures and contain fewer protons that do hydrates. Their conduction mechanism is thus expected to be easier to understand, and the proton diffusion may be investigated over a rather wide temperature range. However, the conduction mechanism can become complicated when the 'melting' temperatures of the protonic species and the non-protonic sublattice are close.

11.1 Oxonium perchlorate

Oxonium perchlorate H_3OClO_4 is the first solid proton conductor whose conductivity, $\sigma_{300\,K} = 3.4 \times 10^{-4}\,\Omega^{-1}\,cm^{-1}$, has been found to be comparable to that of a liquid electrolyte[1]. It is also one of the first materials in which the oxonium ion has been identified[2]. The compound is very hygroscopic and photosensitive. A glassy state can be obtained by quenching the molten salt ($T_m = 55\,°C$)[3]. The crystalline phase undergoes a first order transition on cooling, at 249 K, with a slight volume increase. The low temperature phase is monoclinic with $P2_1/n$ space group and four formula units in the unit cell. The structure consists of layers of oxonium and perchlorate ions hydrogen bonded by two short (O ... O = 0.263 and 0.261 nm) and one long (O ... O = 0.271 nm) hydrogen bond (Fig. 11.1). In the high temperature orthorhombic phase (Pnma, $Z=4$), the oxonium ions are disordered and each ion can have four different configurations[4]; the hydrogen bonds are considerably weaker with the shortest O ... O distance equal to 0.286 nm. The deuterated derivative, D_3OClO_4, shows two phase transitions at 245.5 and 251.9 K and the dynamic disorder increases in two steps.

165

Materials

This order–disorder transition in H_3OClO_4 is observed also by other techniques. The infrared absorption spectra of the low-temperature phase show three narrow and well-defined νOH bands at 3340, 3220 and 2820 cm^{-1} corresponding to three O . . . O distances while a single very broad absorption between 2000 and 3600 cm^{-1} has been observed for the high-temperature phase[5]. In the Raman spectrum there is a considerable broadening of the bands due to internal vibrations of ClO_4^- ions while the lattice modes due to translational and rotational motions of cations and anions merge into a broad wing near the Rayleigh line, characteristic of a plastic phase (Fig. 11.1). In the high temperature phase, the oxonium ions can thus be considered to be in a quasi-liquid state accompanied by an appreciable orientational disorder of ClO_4^- anions[5]. This assumption is supported by *NMR* measurements showing an 'isotropic' rotation of H_3O^+ ions[6-8] and by *quasi-elastic neutron scattering*[5,9]. Finally, the low-frequency neutron scattering ($P(\omega)$) indicates a quasi-liquid behaviour of the high temperature phase of oxonium perchlorate[5,10]. Fig. 11.2 shows the INS spectra of low and high temperature phases of H_3OClO_4 and $CsHSO_4$[5,11]. High temperature phases exhibit superionic conductivity. Comparison is also given for protonic β-aluminas in the low and high

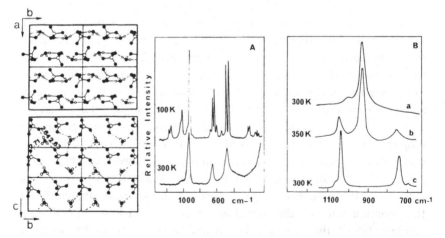

Fig. 11.1. Structure of the low-temperature phase (black circles: oxygen atoms)[4] of oxonium perchlorate (H_3OClO_4); A, Raman spectra[5]; and B, details of the $\nu Cl–O(H)$ stretching region are given for the crystal (line a), the molten salt (line b) and the aqueous solutions (line c)[5] (with permission).

Anhydrous materials

hydrated states, i.e. in the low and high conducting phases[11]. The low hydrated state of H_3O^+ β-alumina gives rise to a single narrow band at $130\ cm^{-1}$ due to the translational motion of cations. When the sample is hydrated forming a layer of protonated water, the spectrum is close to that of a Debye solid and a continuous distribution of modes exists. The same behaviour is observed for H_3OClO_4 (and $CsHSO_4$) when going from the low to the high temperature phase. The $P(\omega)$ function becomes similar to that of a liquid and the $P(\omega)$ value at $\omega = 0$ becomes different from zero because of the diffusion of mobile species[10].

In going from the low to the high temperature phase, the conductivity of oxonium perchlorate increases by a factor of 10 (while in $CsHSO_4$ the conductivity increase is close to 10^4) and the activation energy decreases from 0.37 to 0.29 eV (Fig. 11.3). These values are very similar to those of the glassy phase of perchloric acid hydrate (e.g. $HClO_4 \cdot 5.5H_2O$). On the other hand, the activation energy of liquid $HClO_4 \cdot 5.5H_2O$ is $\leqslant 0.05$ eV with a conductivity of about $3\ \Omega^{-1}\ cm^{-1}$ at $300\ K$[13]. Self-diffusion measurements show that hydrogen diffuses 1000 times faster than does oxygen[1]. Quasi-elastic neutron scattering indicates that 'nearly free' rotational motion of H_3O^+ occurs[9]. The correlation coefficient of the *Nernst–Einstein equation*, $f = 0.75$, suggests a vacancy mechanism and

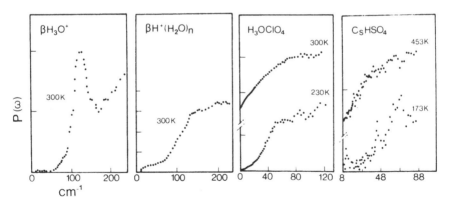

Fig. 11.2. Low-frequency inelastic neutron scattering spectra

$$P(\omega) = \lim_{q \to 0}(\omega^2 S_{inc}(Q, \omega)/Q^2)$$

of $11Al_2O_3 \cdot 1.3(H_3O_2)('\beta\text{-}H_3O^+')$, $11Al_2O_3 \cdot 1.3(H^+(H_2O)_n)_2$, $n \approx 3$ ('β-$H^+(H_2O)_n$'), H_3OClO_4 and $CsHSO_4$[10,11] in the low and high temperature (conductivity) phases (with permission).

167

Potier and Rousselet[1] proposed the following mechanism derived from Bjerrum[12]: there is a rotation of the H_3O^+ ion followed by proton transfer to a ClO_4^- anion, then a rotation of the resulting $HClO_4$ species and subsequently a proton transfer to H_2O. Such a mechanism implies the presence of 'proton *defects*' the formation of which can be expressed by two possible equilibria

$$H_3O^+ + ClO_4^- \leftrightarrows HClO_4 + H_2O \quad \text{and/or} \quad 2HClO_4 \leftrightarrows Cl_2O_7 + H_2O$$

The Raman spectra of molten oxonium perchlorate show bands characteristic of $HClO_4$ and support the former reaction.

Other perchlorates such as $H_5O_2ClO_4$, $H_3ONO_2(ClO_4)_2$ and $(H_3O)_2(NO_2)_9(ClO_4)_{11}$ are also good conductors and melt at higher temperature than H_3OClO_4[5,9]. Ammonium perchlorate has been extensively studied using conductivity[14], NMR[15] or neutron scattering[16]; an orthorhombic–cubic transformation is observed near 240 °C[14]. In the high temperature phase there is a 'nearly free' rotation of ClO_4^- and a rapid reorientation of NH_4^+ ions[15]. The conductivity is protonic as shown by electrolysis[14]. The activation energy varies from 0.5–1.5 eV – depending on the author – in the low temperature phase and increases to 2.5 eV in the high temperature phase. Such high values are consistent with

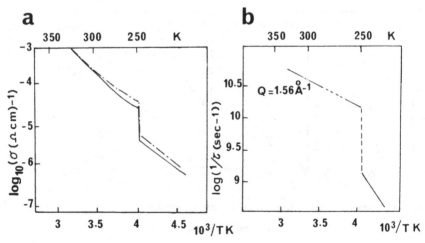

Fig. 11.3. (a) Conductivity plot σ^1 and (b) rotational diffusion coefficient $1/\tau^5$ as a function of the inverse temperature of oxonium perchlorate. The rotational diffusion coefficient has been measured by quasi-elastic neutron scattering[9].

Anhydrous materials

NH_4^+ ion diffusion although a possible proton defect mechanism has also been discussed[17].

11.2 Dihydrogen phosphates and arsenates, MH_2XO_4

A number of acid phosphates and *arsenates*, KH_2PO_4 (KDP) in particular, are interesting as ferroelectric crystals and have been intensively and extensively studied[18-20]. These materials are rather stable (≥ 500 K). The protons of $O-H \ldots O$ hydrogen bonds linking the phosphate groups must play a determining role in creating the ferroelectric phase since there is a very large isotope effect on the temperature (~ 100 K) of the paraelectric–ferroelectric phase transition when hydrogen is substituted by deuterium[21-23]. In the paraelectric phase, the protons are disordered, being statistically distributed between two equivalent sites off the centre of symmetry of the $OH \ldots O$ hydrogen bond, and they become ordered in the ferroelectric phase. The hydrogen bonds are usually short with $O \ldots O$ distances between 0.260 and 0.240 nm and thus very strong. The hydrogen bonds may be equivalent, as in KH_2PO_4[21], or different as in CsH_2PO_4[22], where there is a short, disordered H-bond of length 0.248 nm linking the phosphate tetrahedra into infinite chains and a long, ordered bond of length 0.254 nm joining the chains to form layers (Fig.11.4).

The room temperature conductivity of these MH_2XO_4 compounds is of the order of $10^{-8} \, \Omega^{-1} \, cm^{-1}$[10] and the activation energy varies between 0.5 and 0.7 eV. The conductivity increases to $10^{-6} \, \Omega^{-1} \, cm^{-1}$ at higher

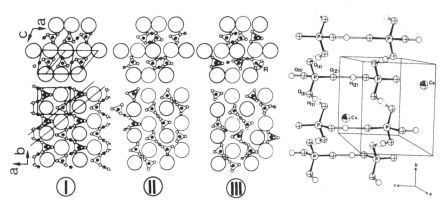

Fig. 11.4. Schematic representation of $CsHSO_4$ structures in phases I, II (I') and III, respectively (large circle, Cs ions; small black circle, S atoms; small white circle, oxygen atoms; small dashed circle, hydrogen atoms)[40,46] and of CsH_2PO_4[35.]

169

temperatures not far from the melting point and the activation energy increases to 0.9–1.0 eV[24]*. The conductivity is not appreciably altered by deuteration but increases proportional to the concentration of dopants such as sulphate. In the case of KH_2AsO_4, 0.01 mol% of sulphate ions increases the conductivity a hundred times and lowers the activation energy from 0.7 to 0.6 eV. Proton conduction is thus expected to be related to the presence of defects such as the Frenkel or Schottky defects in ionic conductors. In 1967, O'Keeffe & Perrino interpreted the KDP conductivity as an intrinsic property of the hydrogen bond network[25]. They considered an M defect which may arise due to proton translocation from a neighbouring hydrogen bond. The latter has thus no proton and is called an L defect by analogy with the Bjerrum defect in ice[12]. This can be described by the following reaction

$$2H_2PO_4^- \rightarrow HPO_4^{2-}[\ ] + (H_2PO_4^-)H^+, \text{ where } [\ ] \text{ is a proton vacancy}$$

Water non-stoichiometry associated with the presence of P_2O_7 defects is also possible[5]. Atmospheric humidity may also play a role as shown for RbH_2PO_4 by Baranowski *et al.*[26]. The enthalpy of a *phase transition* (352 K) in RbH_2PO_4 is directly related to the time of contact between the crystal (1–100 mm Hg) and the water vapour at various pressures. The mechanism is not well understood but the experimental evidence indicates the influence of defects.

Electrolysis of dihydrogen phosphate crystals yields hydrogen in the amount expected from Faraday's law and in agreement with proton conduction. Similar results are obtained for $NH_4H_2PO_4$ where ammonium ions can also play the role of charge carriers. The corresponding conductivity plot shows a knee near 360 K which was extensively discussed[17] and appears to be related to the formation of phosphoric acid at the sample surface according to the reaction $NH_4H_2PO_4 \rightarrow NH_3 + H_3PO_4$.

11.3   Hydrogen sulph*tes and selenates

Alkali hydrogen sulphates and selenates are frequently *ferroelectric* in their low temperature phases and may exhibit superionic behaviour and a *plastic* state in their high temperature phases, usually above 400–450 K.

---

* Recently A. I. Baranov *et al.* found that CsH_2PO_4 single crystal exhibits high proton conductivity ($>10^{-3}\ \Omega^{-1}\ cm^{-1}$) above 504 K with activation energy equal to ~ 0.3 eV[78].

Anhydrous materials

The transition temperature of the superionic phase depends on the alkali cation[27-31] and on external pressure[30] which may modify the hydrogen bond distance. A wider thermal stability range is observed for selenates than for sulphates and the high temperature phases are thus easier observed in the latter. The melting or decomposition temperature decreases in going from Cs^+ to Li^+ and for smaller cations, no superionic phases exist. In the case of NH_4HSO_4 ($RbHSO_4$), a superionic phase is found only under 0.4 GPa (0.3 GPa) hydrostatic pressure[30]. In general, hydrogen sulphate and selenate crystals can be grown from aqueous solutions or by cooling the melt.

$KHSO_4$ is the first crystal among hydrogen sulphates in which a pure though very low protonic conduction, $\sigma_{300\,K} = 10^{-11}\,\Omega^{-1}\,cm^{-1}$, was recognized[32]. The activation energy in a single crystal is 0.5 and 1.2 eV below and above 333 K, respectively, while lower values, 0.2 and 0.7 eV, have been reported for polycrystalline samples[33]. The conductivity of a frozen melt depends strongly on the nature of the sample and it is possible also to prepare a glassy state[28,34]. The melt is polymerized and the volume change on freezing is remarkably low, which is also usually observed for other good ionic conductors. High protonic conductivity in mixed organic-acid sulphates was also demonstrated by Takahashi *et al.* in 1976[31].

Unlike $H_2XO_4^-$ ions forming infinite layers or a three dimensional hydrogen bond network, HXO_4^- ions can be associated either in infinite chains or in cyclic dimers. $NaHSO_4$[35] and the room temperature phases of $RbHSO_4$[36], $CsHSO_4$[37], $RbHSeO_4$[38] and NH_4HSeO_4[39] contain infinite chains; the β-phase of $NaHSO_4$[35], $CsHSeO_4$[40,41] and the high temperature phase of $CsHSO_4$ ($CsDSO_4$) consist of cyclic dimers, while in $KHSO_4$ and $KHSeO_4$ both forms coexist[42,43]. The O—H ... O hydrogen bonds are usually short, as in phosphates with O ... O distances of 0.245–0.260 nm, and are weaker in cyclic *dimers* than in infinite *chains*. They are equivalent as in $CsHSO_4$ (O ... O = 0.257 nm) or different as in $KHSO_4$[42] or in the ferroelectric phase of NH_4HSeO_4 where nine non-equivalent $HSeO_4^-$ ions have been observed by NMR[44] and X-rays[54]. If there are no crystallographic data available, *vibrational spectroscopy* can be used to determine the hydrogen bond strength and the type of association. The main criterion consists of measuring the splitting of the XOH (donor) and X—O (acceptor) stretching frequencies of the —X—OH ... O—X— bonded form. The splitting is equal to zero for a truly symmetrical hydrogen bond, is small for very short bonds and increases with increasing

171

Materials

O ... O distance and OH stretching frequency. As illustrated in Figs 11.5 and 11.6, cyclic dimers correspond to a weak hydrogen bond whereas chains are observed for stronger bonds. Fig. 11.6 illustrates this behaviour for NH_4HSeO_4 containing several phases with different hydrogen bonding. The low temperature phase V' has the strongest OH ... O hydrogen bond characterized by the OH stretching frequency at 2220 cm^{-1} and the $\nu SeO-\nu SeO_4$ splitting of 60 cm^{-1}. There is a sudden increase of $\Delta\nu$ to 68 cm^{-1} in going to the ferroelectric phase IV' with an average $\nu OH = 2260$ cm^{-1} and a gradual increase of $\Delta\nu = 84$ cm^{-1} and $\nu OH = 2300$ cm^{-1} for the paraelectric phase II. The changes are much more important in the superionic phase I, similar to molten state containing dimers, with $\Delta\nu \sim 140$ cm^{-1} and $\nu OH = 3400$ cm^{-1}. In $KHSeO_4$, on the other hand, the room temperature phase contains equal proportions of infinite chains and cyclic dimers and the corresponding $\Delta\nu$ splittings are 87 and 136 cm^{-1}, respectively. Finally, the splitting of γOH *Raman* band can also be used to identify the association type[45].

Many hydrogen sulphates and selenates undergo structural phase transitions and the most interesting are $CsHSO_4$ and NH_4HSeO_4.

$CsHSO_4$. *Calorimetric* and spectroscopic measurements show that the phase transformations of $CsHSO_4$ depend considerably on the *thermal history* of the sample and on pressure[26,46]. On heating a freshly prepared

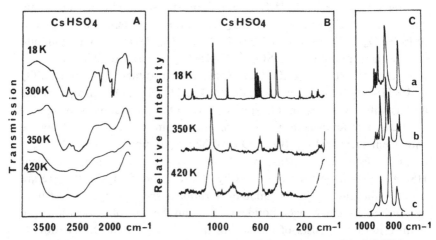

Fig. 11.5. Infrared spectra in the νOH region (A) and Raman spectra (B) of $CsHSO_4$ at various temperatures[11]. Details of the $\nu Se-O(H)$ stretching region are given in (C) for $CsHSeO_4$ (line a), $KHSeO_4$ (line b) and NH_4HSeO_4 (line c) at room-temperature[45] (with permission).

172

crystal, the room temperature phase I goes to phase II at 318 K and to phase III at 447 K, while on cooling phase II′ at 408 K and phase I′ at 353 K were obtained (Fig. 11.7). The plastic phase III exhibits unusually high conductivity of the order of $10^{-2}\,\Omega^{-1}\,cm^{-1}$ [27].

The phase I ($P2_1/m$, $Z = 2$) contains infinite hydrogen bonded $(HSO_4^-)_n$ chains with the O . . . O distance of 0.257 nm and the proton appears to be disordered between two equilibrium sites creating a statistically symmetric hydrogen bond[37]. On the whole the structure of $CsHSO_4$ is very similar to that of the paraelectric phase of CsH_2PO_4 apart from the long hydrogen bond of the latter[22]. There is also a similarity with the structure of ionic conductors such as Li_2SO_4, $LiNaSO_4$ and even $NASICON$[47,48].

Vibrational spectra are consistent with the X-ray diffraction results and confirm the statistical nature of the $P2_1/m$ symmetry[41]. The external

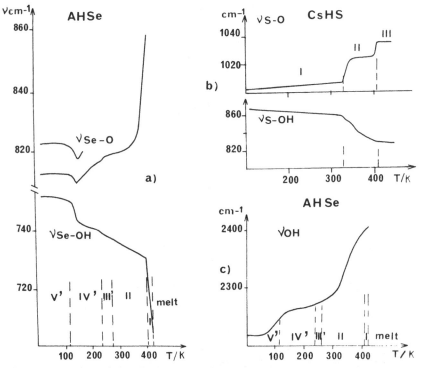

Fig. 11.6. Evolution of the νSe–0 and νSe–OH (a, NH_4HSeO_4), νS–O and νS–OH (b, $CsHSO_4$)[41] (adapted by permission of John Wiley & Sons Ltd) and νOH (c, NH_4HSeO_4) frequencies as a function of the temperature[44]. A = ammonium.

173

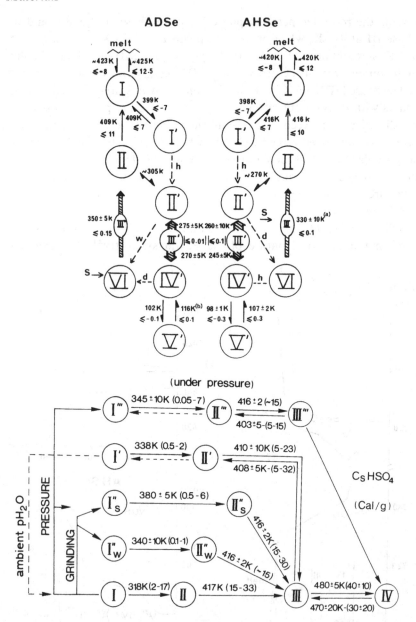

Fig. 11.7. Schematic representation of the phase relationship of $CsHSO_4$ (CsHS)[46], ND_4DSeO_4 (ADSe) and NH_4HSeO_4 (AHSe)[58] (both with permission). Metastable phases are designated with a prime ('). Dashed regions correspond to incommensurate or modulated phases. High humidity restores or imposes the I → II → III (superionic phase) sequence. For AH(D)Se: I, superionic phase; II, paraelectric phase (B2); III' and III", incommensurate

vibrations below 300 cm^{-1} obey the selection rules derived from the above symmetry while the internal vibrations see neither centres nor planes of symmetry: in particular, Raman bands due to νS–OH and νS–O stretching frequencies at 865 and 995 cm^{-1}, respectively, show that the O—H . . . O bond is asymmetric (Fig. 11.5).

The I → II phase transition is interpreted in terms of a conversion from infinite chains to cyclic dimers with a subsequent weakening of the hydrogen bonds and increased orientational disorder of HSO_4^- ions. The chain/dimer conversion leads to a νS–O frequency jump whereas the νS–OH frequency decreases smoothly (Fig. 11.6). This process is amplified in the II → III transition: the spectroscopic changes consist of broadening of the skeletal Raman bands and a diminishing of their number. This indicates a higher (average) symmetry and a highly disordered structure. The most dramatic changes, however, are observed for external modes where all the bands collapse into a broad wing near the Rayleigh line. This behaviour is characteristic of a plastic phase and implies a 'free' rotation of HSO_4^- ions on given sites. The Raman spectrum of phase III (I4$_1$/amd) can thus be considered as a spectroscopic manifestation of a *quasi-liquid* state in the superionic phase. In phase III, both protons and caesium ions are disordered while in phases II and I the disorder is essentially confined to the protons. The drastic increase in conductivity by almost five orders of magnitude on going from phase II to phase III is thus related to the high disorder of both sublattices. Pressure induces phase transitions associated with a partial conversion of chains into cyclic dimers (Fig. 11.7). The material is thus easily modified by grinding[46].

A 'quasi-liquid' mechanism of proton diffusion in $CsHSO_4$ is observed also by neutron scattering (Fig. 11.8). A detailed analysis of elastic and *quasi-elastic neutron scattering* as a function of the momentum transfer allows us to explore the proton (transfer) dynamics[40,49]. In the spectrum of $CsHSO_4$ in the 10 meV range, for instance, a quasi-elastic component about 0.1 meV wide corresponds to a rotation of about 0.22 nm consistent with $(HSO_4)_2$ dimer opening. At higher resolution (μeV) the elastic peak analysis can be used for determining the translational motion of proton, with a jump distance close to 0.1 nm[40]. This has been interpreted in terms

Caption for Fig. 11.7 (*cont.*)
or modulated phases; IV′, ferroelectric phase (B1); VI, orthorhombic phase (P2$_1$2$_1$2$_1$). Dashed lines indicate slow transitions (from hours to days). The sequence I → I‴ → II‴ ⇌ III‴ in $CsHSO_4$ is observed under pressure (∼400 MPa). Values in brackets correspond to the enthalpy changes (cal/g).

of a permutation of a proton between two sulphate tetrahedra of the same or of neighbouring dimers in $CsHSO_4$. Finally, a [133]Cs and deuteron NMR study of $CsDSO_4$ has shown that the Cs^+ (short range) and deuteron (long range) translational motion in phase II involves well defined lattice sites and directions and is not completely 'random' or '*liquid*'-like[50, 51]. On cooling, the dynamic disorder is frozen and phase I′ is obtained. Time is needed to reform phase I. Indeed, the rebuilding of

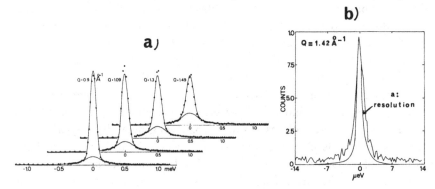

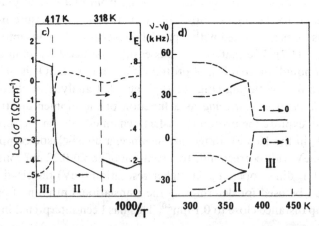

Fig. 11.8. Neutron scattering spectra: (a) quasi-elastic and (b) elastic scattering on $CsHSO_4$ samples recorded with IN6 time of flight and IN10 Doppler-back scattering spectrometers, respectively, at the Institut Laüe–Langevin[40]. (c) The intensity of the elastic component (I_E) for a fixed window is compared with the conductivity plot as a function of temperature[40] (with permission). (d) Temperature dependence of the deuteron quadrupole perturbed NMR spectra of $CsDSO_4$ at a general orientation near the II′ → III transition[50] (with permission).

infinite chains requires proton diffusion and rotation of tetrahedra. The phase transitions can be also regarded as chemical reactions between protonic species driven by protons the concentration of which depends on atmospheric water pressure: $n(HSO_4)_2 \rightarrow (SO_4—H—SO_4—H)_n$.

NH_4HSeO_4. Ammonium hydrogen selenate appears interesting since it undergoes several phase transitions from the high temperature superionic to the paraelectric, incommensurate, ferroelectric and a low temperature phase[44,52]. This compound and its deuterated derivative (ADSe) have been extensively studied by NMR[53], X-ray and neutron diffraction[54], dielectric measurements[52,55] and infrared and Raman spectroscopy[44]. Some controversy about the phase diagram has occurred[53-57]. Recent work[58] has shown that there is essentially the same phase sequence in both crystals, with a delicate equilibrium involving metastable and stable phases. The phase relationships are shown diagrammatically in Fig. 11.7. In the phase sequence, V′–IV′–III′–II–I, the orientational disorder of both ammonium and selenate ions increases progressively to reach a 'quasi-liquid' state (plastic phase) in the superionic phase I. The participation of $HSeO_4^-$ motion appears important at both ends of the sequence, i.e. in the V′ ↔ IV′ and II → I transitions. The VI–III″–II first order transition is that from an ordered to a disordered state via an incommensurate intermediate phase III″: the main mechanism is believed to consist of gliding motions of ammonium ions along the chain axis followed by a reorientation of $HSeO_4^-$ ions. The large number of phases encountered in ammonium hydrogen selenate may be related to the possibility of NH_4^+ ions having many positions (associated if necessary with distortion), as a kind of 'gear pinion' before reaching a full dynamic disorder. Consequently its electric dipole moment may be important[59].

The superionic phase I, above 409 K, has a conductivity of the order $10^{-2} \Omega^{-1}$ cm^{-1}, similar to that of $CsHSO_4$ and $CsHSeO_4$[27-30,59,60]. The $P2_1/b$ symmetry of the crystal is very close to the hexagonal system[43,54]. The vibrational spectra of phase I and of the melt are very similar but different from those of phase II. The internal vibrations show an almost C_{3v} symmetry of the $HSeO_4^-$ ion and Td symmetry of NH_4^+ while the external vibrations, as in $CsHSO_4$, indicate the existence of a plastic state in phase I. There is thus a high reorientational disorder with the formation of open dimers and weakening of OH . . . O hydrogen bonds (Fig. 11.7). The assumption of a quasi-liquid state in phase I is also supported by *enthalpy* measurements of the II → I phase transition; $\Delta H = 10$ K cal mol^{-1}, i.e. about the same as that of the I → melt transition.

11.3.1 Hydrogen disulphates ($M_3H(SO_4)_2$ and diselenates $M_3H(SeO_4)_2$

$M_3H(XO_4)_2$ type of compounds contain strongly hydrogen (~ 0.25 mm) bonded, one-proton dimers[61-63] $(O_3XO-H-OXO_3)^{3-}$. The OH . . . O hydrogen bond in $Na_3H(SO_4)_2$ is extremely short with an O . . . O distance of 0.243 nm but asymmetric and ordered[64]. Those in potassium, rubidium and ammonium derivatives are longer (0.248 and 0.254 nm for $Rb_3H(SO_4)_2$ and $(NH_4)_3H(SO_4)_2$, respectively) and crystallographically symmetric, the symmetry being of a statistical nature implying disordered protons in the room temperature phases. In the case of $(NH_4)_3H(SO_4)_2$, there is an additional orientational disorder of ammonium ions which appears to be responsible for numerous low-temperature phase transitions[65-69] which have no equivalent in potassium and rubidium analogues[70]. The structural disorder decreases with decreasing temperature: below 263 K the OH . . . O protons become ordered while the orientational disorder of NH_4^+ ions diminish gradually on passing through several phases[69,71]. The crystal becomes fully ordered below 78 K, giving rise to a ferroelectric phase[72] whose polarity is created mainly by ordering of dipoles of distorted ammonium tetrahedra. The high temperature phase of $(NH_4)_3H(SO_4)_2$ above 413 K has a protonic conductivity of about $10^{-5}\,\Omega^{-1}\,cm^{-1}$ which is anisotropic with a maximum along the c direction[65]. The activation energy is, however, high ($\geqslant 0.5$–0.7 eV). A marked dilation is also observed[73]. $Na_3H(SO_4)_2$, on the other hand, does not exhibit protonic conductivity[66].

$M_3H(SeO_4)_2$ acid salts undergo two phase transitions each and the high temperature phases above 300, 450 and 451 K for M = NH_4, Rb and Cs[74,75] have a high conductivity between 10^{-2} and $10^{-3}\,\Omega^{-1}\,cm^{-1}$ and a marked anisotropy with σ_a/σ_c varying from 15 to 60. Activation energy decreases from 0.6–1.2 eV in the low conductivity phases to 0.3–0.5 eV in the highly conducting phases[75]. The superionic phases of $M_3H(XO_4)_2$ compounds are characterized by an increased orientational disorder of XO_4 tetrahedra (plastic phase) and by a serious weakening and/or breaking of OH . . . O hydrogen bonds, as shown by vibrational spectra and X-ray diffraction. In the case of $Rb_3H(SeO_4)_2$, for instance, the O . . . O distance increases from 0.251 to 0.267 Å when going from room temperature to the superionic phase[74,75], while the Rb–O distances increase from 0.284 to 0.295 nm.

Finally, an acid salt of a different type $(NH_4)_2SeO_4 . 2NH_4HSeO_4$ containing short OH . . . O hydrogen bonds with O . . . O distances of

Anhydrous materials

0.254 Å[76], shows also a superionic phase above 378 K characterized by a high conductivity, $4.10^{-3} \Omega^{-1} cm^{-1}$ and a low activation energy, 0.11 eV[77]. The conductivity in the high temperature phase is isotropic while that in the low temperature phase is several orders of magnitude lower and strongly anisotropic for all three crystallographic directions. It should be pointed out finally that the superionic properties of the materials discussed in this chapter are directly related to the increase of the O—H ... O bond lengths.

11.4 References

1. A. Potier and D. Rousselet, *J. Chimie Physique (France)* **70** (1973) 873–8.
2. M. Volmer, *Liebigs Ann. Chem.* **440** (1924) 200–3.
3. M. Fournier, G. Mascherpa, D. Rousselet and J. Potier, *C.R. Acad. Sci. (Paris)* **269** (1969) 279–84.
4. F. S. Lee and G. B. Carpenter, *J. Phys. Chem.* **63** (1959) 279–82; C. Norman, *Acta Cryst.* **15** (1962) 18–23.
5. M. Pham-Thi, Thesis, University of Montpellier (1985); D. Rousselet, J. Rozière, M. H. Herzog-Cance, M. Allavena, M. Pham-Thi, J. Potier and A. Potier, in *Solid State Protonic Conductors I*, J. Jensen and M. Kleitz (eds.) (Odense University Press (1982)) 65–95.
6. P. E. Richard and J. S. A. Smith, *Trans. Faraday Soc.* **47** (1951) 1261–5; D. O. Reilly, E. M. Petersen and J. M. Williams, *J. Chem. Phys.* **54** (1971) 96–8; J. M. Janik and R. Rachwalska, *Physica* **72** (1974) 168–78.
7. M. H. Herzog-Cance, J. Potier and A. Potier, *Ad. Mol. Relax. Interaction Processes* **14** (1979) 245–67.
8. M. H. Herzog, M. Pham-Thi and A. Potier (eds) in *Solid State Protonic Conductors III*, J. B. Goodenough, J. Jensen and A. Potier (Odense University Press (1985)) 129–42.
9. M. Pham-Thi, J. F. Herzog, M. H. Herzog-Cance, A. Potier and C. Poinsignon, *J. Mol. Struct.* **195** (1989) 293–310; M. H. Herzog-Cance, M. Pham-Thi and A. Potier, *J. Mol. Struct.* **196** (1989) 291–305.
10. Ph. Colomban and A. Novak, *J. Mol. Struct.* **177** (1988) 277–308.
11. M. Pham-Thi, Ph. Colomban, A. Novak and R. Blinc, *Solid State Comm.* **55** (1985) 265–70; Ph. Colomban, G. Lucazeau and A. Novak, *J. Phys. C (Solid State Physics)* **14** (1981) 4325–33.
12. N. Bjerrum, *Science* **115** (1952) 385–91.
13. T. H. Huang, R. A. Davis, U. Frese and U. Stimmung, *J. Phys. Chem. Lett.* **92** (1988) 6874–6.
14. L. B. Harris and G. J. Vella, *J. Chem. Phys.* **58** (1973) 4550–7; P. W. M. Jacobs and H. M. Whitehead, *Chem. Rev.* **69** (1969) 551–90; V. V. Boldyrev and E. F. Hairetdinov, *J. Inorg. Nucl. Chem.* **31** (1969) 3332–7; P. W. M. Jacobs and W. L. Ng, *J. Phys. Chem. Solids* **33** (1972) 2031–9; G. P. Owens,

Materials

J. M. Thomas and J. O. Williams, *J. Chem. Soc. Faraday Trans. I,* **68** (1972) 2356–66.

15. R. Ikeda and C. A. McDowell, *Chem. Phys. Lett.* **14** (1972) 389–92; J. W. Riehl, R. Wang and H. W. Bernard, *J. Chem. Phys* **58** (1973) 508–16.
16. J. A. Janik, J. M. Janik, J. Mellor and H. Palevsky, *J. Phys. Chem. Solids* **25** (1964) 1091–8; J. M. Janik, *Acta Phys. Polonica* **27** (1965) 491–8; J. J. Rush, T. I. Taylor and W. W. Havens, Jr, *J. Chem. Phys.* **35** (1961) 2265–6.
17. L. Glasser, *Chem. Rev.* **75** (1975) 21–65; G. P. Owen, J. M. Thomas and J. O. Williams, *J. Chem. Soc. Dalton Trans.* (1972) 808–9; F. Croce and G. Ligna, *Solid State Ionics* **6** (1982) 201–2.
18. F. Jona and G. Shirane, *Ferroelectric Crystals* (Pergamon Press, London (1962)); W. C. Hamilton and J. A. Ibers, *Hydrogen Bonding in Solids,* W. A. Benjamin (ed) (New York (1968)); J. C. Burfoot and G. W. Taylor, *Polar Dielectrics and their Applications* (University of California Press, Berkeley, USA (1979)); Landolt-Börnstein, *Groupe III/16, Ferroelectrics and Related Substances* (Springer Verlag, Berlin (1981)).
19. T. F. Connaly and E. Turner, *Ferroelectric Materials and Ferroelectricity* (Plenum Press, New York (1970)).
20. R. Blinc and B. Žekš, *Soft Modes in Ferroelectrics and Antiferroelectrics* (North-Holland (1974)).
21. L. Tenzer, B. C. Frazer and M. Pepinsky, *Acta. Cryst.* **11** (1958) 505–9.
22. D. Semmingsen, W. D. Ellenson, B. C. Frazer and G. Shirane, *Phys. Rev. Lett.* **38** (1977) 1299–1302; B. Marchon, A. Novak and R. Blinc, *J. Raman Spectrosc.* **18** (1987) 447–55.
23. A. Levstik, R. Blinc, P. Kabada, S. Čižikov, I. Levstik and C. Filipič, *Solid State Comm.* **16** (1975) 1339–43.
24. A. L. de Oliveira, O. de O. Damasceno, J. de Oliveira and E. J. L. Schouler, *Mat. Res. Bull.* **21** (1986) 877–85.
25. M. O'Keeffe and C. T. Perrino, *J. Phys. Chem. Solids* **28** (1967) 211–18; J. Bruinink, *J. Appl. Electrochem.* **2** (1972) 239–49.
26. B. Baranowski, M. Friesel and A. Lunden, *Z. Naturforsch.* **41a** (1986) 733; 981–2; **42a** (1987) 565–71; M. Friesel, A. Lunden and B. Baranowski, *Solid State Ionics* **35** (1989) 84–9; 91–8.
27. A. I. Baranov, L. A. Shuvalov and N. M. Shchagina, *J.E.T.P. Letts.* **36** (1982) 459–62; A. I. Baranov, R. M. Fedosyuk, N. M. Shchagina and L. A. Shuvalov, *Ferroelectric Lett.* **2** (1984) 25–8.
28. T. Mhiri, A. Daoud and Ph. Colomban, *Phase Transitions* **14** (1988) 233–42.
29. N. G. Hainovsky and E. F. Hairetdinov, *Izvest Sibir. Otd. ANSSSR, Ser Khim. Nauk.* **8** (1985) 33–5; N. G. Hainovsky, Yu T. Pavlukhin and E. F. Hairetdinov, *Solid State Ionics* **20** (1986) 249–53.
30. A. I. Baranov, E. G. Ponyatovskii, V. V. Sinitsyn, R. M. Fedosyuk and L. A. Shuvalov, *Soc. Phys. Crystallogr.* **30** (1985) 1121–3; V. V. Sinitsyn, E. G. Ponyatowski, A. I. Baranov, L. A. Shuvalov and N. I. Bobrova, *Solid State Physics* **30** (1988) 2838–41.
31. T. Takahashi, S. Tanase, O. Yamamoto and S. Yamauchi, *J. Solid State Chem.* **17** (1976) 353–62.

Anhydrous materials

32. M. Sharon and A. J. Kalia, *J. Chem. Phys.* **66** (1977) 3051–5.
33. S. E. Rogers and A. R. Ubbelohde, *Trans. Faraday Soc.* **46** (1950) 1051–61.
34. T. Förland and W. A. Weyl, *J. Am. Ceram. Soc.* **33** (1950) 186–8.
35. E. J. Sonneveld and J. W. Wisser, *Acta Cryst.* **B35** (1979) 1975–7; **B34** (1978) 643–5; M. Matsunaya, K. Itoh and E. Nakamura, *J. Phys. Soc. Jpn* **48** (1980) 2011–14; J. A. Duffy and W. J. D. MacDonald, *J. Chem. Soc. A* (*Ing. Phys. Theor.*) (1970) 978–83.
36. F. A. Cotton, B. A. Frenz and D. L. Hunter, *Acta Cryst.* **B31** (1975) 302–4; F. Payen and R. Haser, *Acta Cryst.* **B32** (1976) 1875–9.
37. K. Itoh, T. Ozaki and Nakamura, *Acta Cryst.* **B37** (1981) 1908–9.
38. D. Jones, J. Rozière, J. Penfold and J. Tomkinson, *J. Mol. Struct.* **197** (1989) 113–21.
39. K. S. Aleksandrov, A. I. Kruglik, S. V. Misyul and M. A. Simonov, *Sov. Phys. Crystallogr.* **25** (1980) 654–6; A. I. Kruglik, S. A. Misyul and K. S. Aleksandrov, *Sov. Phys. Dokl.* **25** (1980) 871–4; A. Waskowska, S. Olejnik, A. Lukaszewicz and Z. Czapla, *Cryst. Struct. Comm.* **9** (1980) 663–89; A. Waskowska and Z. Czapla, *Acta Cryst.* **B38** (1982) 2017–20.
40. Ph. Colomban, J. C. Lassègues, A. Novak, M. Pham-Thi and C. Poinsignon, in *Dynamics of Molecular Crystals*, J. Lascombe (ed) (Elsevier, Amsterdam (1987)) 269–74.
41. M. Pham-Thi, Ph. Colomban, A. Novak and R. Blinc, *J. Raman Spectrosc.* **18** (1987) 185–94.
42. A. Goypiron, J. de Villepin and A. Novak, *J. Raman Spectrosc.* **9** (1980) 293–303.
43. J. Baran and T. Lis, *Acta Cryst.* **C42** (1986) 270–2.
44. I. P. Aleksandrova, Ph. Colomban, F. Dénoyer, N. Le Calvé, A. Novak, B. Pasquier and A. Rozicky, *Phys. Stat. Sol.* **a114** (1989) 531–44.
45. Ph. Colomban, M. Pham-Thi and A. Novak, *J. Mol. Struct.* **161** (1987) 1–14.
46. Ph. Colomban, M. Pham-Thi and A. Novak, *Solid State Ionics* **20** (1986) 125–34; **24** (1987) 193–203.
47. M. Vlasse, C. Parent, R. Salmon, G. Le Flem and P. Hagenmuller, *J. Solid State Chem.* **35** (1980) 318–24.
48. Ph. Colomban, *Solid State Ionics* **21** (1986) 97–115.
49. A. V. Belushkin, I. Natkaniec, N. M. Plakida, L. A. Shuvalov and J. Wasicki, *J. Phys. C.* (*Solid State Physics*) **20** (1987) 671–87.
50. R. Blinc, J. Dolinšek, G. Lahajnar, I. Zupančič, L. A. Shurlov and A. I. Baranov, *Phys. Stat. Sol.* **b123** (1984) K83; J. Dolinšek, R. Blinc, A. Novak and L. A. Shuvalov, *Solid State Comm.* **60** (1986) 877–9.
51. J. C. Lassègues and C. Poinsignon, *Solid State Protonic Conductors IV*, *Abstracts*, Exeter, 15–18 September 1988.
52. B. V. Merinov, A. I. Baranov, C. A. Shuvalov and B. A. Maksinov, *Kristallografiya* **32** (1987) 86–92.
53. I. P. Aleksandrova, Yu. N. Moskvich, O. V. Rozanov, A. F. Sadreev, I. V. Seryukova and A. A. Sukhovsky, *Ferroelectrics* **67** (1986) 63–84.
54. A. Rozycki, F. Dénoyer and A. Novak, *J. Physique* (*Paris*) **48** (1987) 1553–8; F. Dénoyer, A. Rozycki, K. Parlinski and M. More, *Phys. Rev.* **B39** (1989) 405–14.

55. Z. Czapla, O. Czupinski and L. Sobczyk, *Solid State Comm.* **51** (1984) 309–12.
56. R. Popravski, J. Dziedzic and W. Bronowska, *Acta Phys. Pol.* **A63** (1983) 601–4; R. Popravski and S. A. Taraskin, *Phys. Stat. Sol.* **a113** (1989) K31–5.
57. V. S. Krasikov, A. I. Kruglik, *Fiz. Tverd. Tela* **21** (1979) 2834–6.
58. Ph. Colomban, A. Rozycki and A. Novak, *Solid State Comm.* **67** (1988) 969–74.
59. J. C. Badot and Ph. Colomban, *Solid State Ionics* **35** (1989) 143–9.
60. Y. N. Moskvich, A. A. Sukhovskii and O. V. Rozanov, *Fiz. Tverd. Tela (Leningrad)* **26** (1984) 38–44.
61. W. Joswig, H. Fuess and G. Ivoldi, *Acta Cryst.* **B35** (1979) 525–9.
62. S. Suzuki and Y. Makita, *Acta Cryst.* **B34** (1978) 732–5.
63. S. Fortier, M. E. Fraser and R. D. Heyding, *Acta Cryst.* **C41** (1985) 1139–40.
64. W. Joswig, H. Fuess and G. Ferraris, *Acta Cryst.* **B38** (1982) 2798–801.
65. U. Syamaprasad and C. P. G. Vallashan, *J. Phys. C (Solid State Phys.)* **14** (1981) L571–4.
66. A. Devendar Reddy, S. G. Sathyanaragan and G. Sivarama Sastry, *Solid State Comm.* **43** (1982) 937–40.
67. M. Kamoun, A. Lautié, F. Romain, A. Daoud and A. Novak, in *Dynamics of Molecular Crystals*, J. Lascombe (ed) (Elsevier, Amsterdam (1987)) 219–24.
68. M. Kamoun, K. Chhor, C. Pommier, F. Romain and A. Lautié, *Mol. Cryst. Liq. Cryst.* **154** (1988) 165–77.
69. M. Kamoun, A. Lautié, F. Romain and A. Novak, *J. Raman Spectr.* **19** (1988) 329–35.
70. M. Damak, M. Kamoun, A. Daoud, F. Romain, A. Lautié and A. Novak, *J. Mol. Struct.* **130** (1985) 245–54.
71. M. Kamoun, A. Lautié, F. Romain, M. H. Limage and A. Novak, *Spectrochim. Acta* **44A** (1988) 471–7.
72. K. Gesi, *Jpn. J. Appl. Phys.* **19** (1980) 1051.
73. S. Suzuki, *J. Phys. Soc. Jpn* **47** (1979) 1205–9.
74. M. Ichikawa, *J. Phys. Soc. Jpn* **47** (1979) 681–2.
75. A. I. Baranov, I. P. Makarova, L. A. Muradyan, A. V. Tregubchenko, L. A. Shuvalov and V. I. Simonov, *Kristallografiya* **32** (1987) 683–94; F. E. Salmon, Cz Pawlacyzk, B. Hilczer, A. I. Baranov, B. V. Merinov, A. B. Tregubchenko, L. A. Shuvalov and N. M. Shchagina, *Ferroelectrics* **81** (1988) 187–91; V. Zelezny, J. Petzelt, Yu. G. Goncharov, G. V. Kozlov, A. A. Volkov and A. Pawlowski, *Solid State Ionics* **36** (1989) 175–8.
76. A. I. Kruglik and M. A. Simonov, *Kristallografiya* **22** (1974) 1082–5.
77. Cz. Pawlaczyk, F. E. Salman, A. Pawlowski and Z. Czapla, *Phase Transitions* **8** (1986) 9–16.
78. A. I. Baranov, B. V. Merinov, A. V. Tregubchenko, V. P. Khiznichenko, L. A. Shuvalov and N. M. Shchagina, *Solid State Ionics* **36** (1989) 279–82.

12 Hydrogen behaviour in graphite–nitric acid intercalation compounds

H. FUZELLIER AND J. CONARD

12.1 Graphite intercalation compounds

Graphite crystallizes in a structure which consists of stacked carbon atom layers in which each carbon atom is bonded to three neighbours in a hexagonal lattice. Foreign atoms or molecules can slip between graphite layers without completely disrupting the structure of the host material. During the past ten years, these *GICs* have been the subject of numerous studies as they present interesting application possibilities: electrical conductors, catalysts, high energy density batteries, lubricants and membranes. A good survey is given in the review *Intercalated Materials*[15] where the synthesis, structure and physical properties of GICs with metals, halogens, halogenides, oxides and acids are described.

In the last century, Brodie[4] and Schafhautl[29] reported the intercalation, between graphite layers, of a mixture of sulphuric and nitric acids. More recently, Rüdorff[27] determined the X-ray diffractograms of (*GNCs*) *Graphite–Nitric acid Compounds*. GNCs can be prepared by using dilute nitric acid[33] or pure HNO_3 as well as dinitrogen pentoxide mixed with, or without, nitric acid[10]. The best way to avoid swelling and distortion of pyrolytic graphite samples and to prepare high quality GNCs, is to intercalate, step by step, small amounts of pure nitric acid vapour.

The GNCs behave as good p-type conductors, which is the reason why Ubbelhode[34] described them as 'Synthetic Metals,' in agreement with a structural model in which the carbon hexagonal network acts as a macrocation with intercalation of nitrate ions and molecules of nitric acid. The intercalated layers of reagent are ordered in regular sequences along the graphite *c* axis. A stage s compound is defined by the number of graphite layers between two successive acid layers.

An order–disorder transformation was observed for a second stage GNC using ESR[18], electrical resistivity and thermoelectric power[34] as

Materials

well as DTA measurements[9] linked to X-ray diffraction experiments[22]. The complex crystal structure is described next.

Among the family of GIC phases, some contain hydrogen which is linked more or less strongly to the intercalated molecules; this ranges from physically absorbed hydrogen in KC_{24}[17] up to metal hydride graphites[14] for which hydrogen motion cannot be detected even by NMR experiments. In GNCs, hydrogen is weakly linked to the nitrate ions and as observed previously[7,10], only a small percentage of the hydrogens can be considered to be really mobile.

12.2   Crystal structure of GNCs

Pure nitric acid intercalates readily into graphite to form GNCs. The α form is characterized by a rounded-off formula: $C_{5s}HNO_3$ (stage, $s = 1$, 2, 3 . . .) and an interlayer distance, $d_i = 7.80$ Å. When these compounds are left in the room atmosphere, their composition changes to $C_{8s}HNO_3$ without any stage variation (except for the first one) but with a smaller interlayer distance, $d_i = 6.55$ Å; they were named β-GNCs by Fuzellier & Melin[12].

12.2.1   Structure of α-GNC

At room temperature, the intercalated layers of a stage two nitric acid compound are disordered in a liquid-like state modulated by the graphite lattice[13]. The intercalated nitric acid molecules are neither organized in double layers, as suggested earlier by Rüdorff[27], nor oriented parallel to the graphite c axis[32]. Instead, the molecule planes are tilted by about 30° to the c axis, as deduced from neutron spectroscopy[26] and from X-ray data[11].

Numerous studies have attempted to solve the low temperature structure by various techniques: X-ray and electron diffraction[5,23,28], neutron scattering[3,26]. These results may be summarized briefly as follows.

At 250 K, an organization of the intercalated layers in a large unit cell is observed which is completely incommensurate with respect to the graphite lattice.

At 240 K, an ordered state appears whose unit cell is commensurate with graphite in one direction.

184

Below 210 K, a weak superstructure appears corresponding to a large superlattice commensurate with graphite; the large unit cell ($A = 17a_G$, $B = 2a_G + 9b_G$) contains 60 nitric acid molecules, for which no diffusive motion can be observed.

The *order–disorder transition* exhibits a clear hysteresis of 5 K. A unique feature of the ordering during the phase transition is a pronounced sliding motion of the graphite layers below T_c away from the usual hexagonal stacking. The c axis sequence AIABIBCICA, where I indicates an intercalated layer and A, B, C the graphite layers shifted from (1/3, 2/3) proposed by Nixon at room temperature, becomes AIA'B'IB" below T_c. The relation between A'B' is the same as that between AB, and the AIA' sandwich corresponds to an A' shift of $0.2a_G + 0.05b_G$ relative to A according to Samuelsen[28]. This shift is in agreement with our evaluation of the orientation of the nitric acid molecules.

12.2.2 Structure of β-GNCs

During the past few years, some studies were carried out to determine the orientation of the intercalated molecules in relation to the graphite a or c axis. According to Clinard[6], nitric acid molecules lie nearly parallel (8°) to the graphite layers. Moreh[19,20] and Pinto[25,30] give values that are slightly higher (13°), but there is some uncertainty due to the presence of a mixture of stages. The intercalated molecules present a 2D organization which could be commensurate with the graphite lattice if one uses a large hexagonal superlattice ($a = 42.6$ Å) required by the presence of a small amount of water intercalated molecules. Along the c axis one can observe the usual sequence for a second stage compound: AIABIBCICA where no correlation can be observed between intercalated layers. According to Pinto[24], the hydrogen atoms are found to be above and below the layers of NO_3 groups, independently of the stage.

12.3 H mobility

The first suggestion of a specific H mobility in GNCs was published by Avogadro and Villa[1,2] and concerned the second, third and fourth stage in highly oriented pyrolytic graphite (HOPG)-based samples. By a detailed study, using mainly ^1H and ^2H NMR and DTA, they were able to demonstrate:

(1) a change in mobility of HNO_3 inserted in HOPG at 140 K and 250 K, with a 4 and 14 kcal g^{-1} HNO_3 endothermic effect;

(2) a $T_1(T)$ curve for ^1H, which showed a rapid increase of mobility near 250 K, followed by a T_1 anisotropy up to room temperature. They deduced, from their *2D* models for the dipolar relaxation mechanism, that 2D H diffusion occurs above 250 K, the *correlation time* for diffusion going from 7×10^{-4} s to 2×10^{-10} s on passing through the transition. Below the transition, they obtained evidence for a T_1 2D law, and above 250 K, from an angular dependence of a directly measured diffusion coefficient, $D = 0$ along the c axis and $D = 5.5 \times 10^{-6}$ cm^2 s^{-1} along the layer planes.

They then identified the slow organization process of the intercalated layers obtained after a week of sample storage at $-25\,°C$. When this ionomolecular structure ($C_{24}{}^+NO_3{}^-$. $3HNO_3$: Ubbelohde) is attained, the electrical field gradient measured by the $I = 1$ spin of ^2H NMR appears to be axially symmetric along c, which means either some static organization or a dynamic average of three molecules rotating around the observed nucleus and c axis. At the end of a careful discussion, they concluded that two motions were involved: one, of low activation energy, is the rotation of an individual molecule in its plane, and the other, three times more energy consuming, above $-21\,°C$, is a three molecules process able to transport the proton. At a lower temperature they proposed a ^1H hopping along the lattice of HNO_3 molecules which can be helped by the in-plane rotation of HNO_3 molecules.

Unfortunately, their hypothetical two-layer structure is not now believed to be correct for the HNO_3 intercalant of the α-compound. We think that a new interpretation of their experimental results would involve a lattice with groups of three canted molecules (relative to the c axis) around a proton with both kinds of rotation, depending on the temperature range. This local structure would be organized in strips or as a 2D lattice, in agreement with structural results of Moret[21,28].

More recently, the results of a detailed neutron study by Simon *et al.*[31] agree quite well with a model involving combination of two rotational motions.

(1) An in-plane HNO_3 molecular rotation, occurring above 210 K and temperature independent.

(2) An out-of-plane rotation, thermally activated at 75–100 meV, with an increasing number of molecules involved, depending on temperature above 210 K; the maximum ratio is 60% above 270 K. This combination of two rotational motions, one being temperature independent, has already been observed for water molecules around a small cation in lamellar silicates[8].

They see clearly (Fig. 12.1) a proton translational diffusion with $D = 4 \times 10^{-8}$ cm^2 s^{-1}, appearing at 250 K, i.e. when 30% of the protons undergo the two rotations. They conclude that no isolated protons are moving at this temperature. Nevertheless, various questions are still pending concerning the details of H mobility in GNCs.

(1) There probably exists a small water contamination, as indicated by appearance of the IR mode of H$_3$O$^+$ [26]. As a consequence, the proton vehicle could be H$_3$O$^+$ or a group of three HNO$_3$ molecules conically coordinated by this ion, as proposed by Clinard in β-compounds[6].

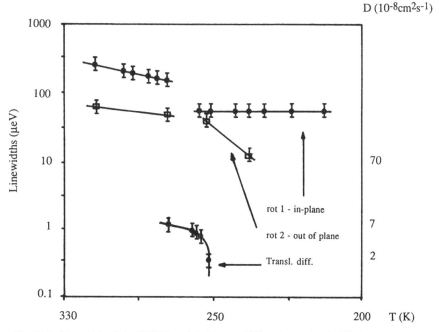

Fig. 12.1. Linewidth of the EISF (see definition, p. 332) and correlated diffusion coefficient as a function of temperature, deduced from neutron scattering experiments[31].

(2) An out-of-plane motion has been proposed involving a two-site rotation of the molecule around the *c* axis and which is clearly not related with other models.

But their elastic incoherent structure factor (EISF) curve is very favourable to a proton motion resulting, over 250 K, from two coordinated rotations, as for the very acidic water in clays[8].

In β-compounds (the so called residual compounds) where some water molecules have been intercalated during the formation process, Kunoff *et al.*[16] observe by ^{1}H NMR three critical temperatures, those at 210 and 250 K as for α-compounds and an additional one at 180 K for which they propose the break of a bimolecular H-bond.

12.4 Conclusion

As a whole, we conclude that in graphite–HNO_3 compounds two proton transfer mechanisms may be observed.

One, available between 210 and 250 K, concerns a limited number of protons and is assisted by the in-plane rotation of HNO_3 molecules

The other, above 250 K, uses either a group of HNO_3 molecules or H_3O^+ as proton vehicle, depending on the kind of compound (α or β). The latter (β) is stable for months at room atmosphere.

12.5 References

1. A. Avogadro, G. Bellodi, G. Borghesi and M. Villa, *Il. Nuovo Cimento* **38B** (1977) 22–7.
2. A. Avogadro and M. Villa, *J. Chem. Phys.* **70**(1) (1979) 109–15.
3. F. Batallan, I. Rosenman, A. Magerl and H. Fuzellier, *Phys. Rev.* **B32** (1985) 4810–13; *Synth. Met.* **23** (1988) 49–53.
4. J. Brodie, *Ann. Chem. Phys.* **45** (1855) 351–4.
5. R. Clarke, P. Hernandez, H. Homma and A. Montague, *Synth. Met.* **12** (1986) 27–32.
6. C. Clinard, D. Tchoubar, C. Tchoubar, F. Rousseaux and H. Fuzellier, *Synth. Met.* **7** (1983) 333–6.
7. J. Conard, H. Fuzellier and R. Vangelisti, *Synth. Met.* **23** (1988) 277–82.
8. J. Conard, H. Estrade-Szwarckopf, A. J. Dianoux and C. Poinsignon, *J. Phys.* **45** (1984) 1361–5.

Graphite intercalation compounds

9. A. Dworkin and A. R. Ubbelhode, *Carbon* **16** (1978) 291–6.
10. H. Fuzellier, Ph.D. Thesis Nancy (1974) C.N.R.S. A09580.
11. H. Fuzellier, M. Lelaurain and J. F. Mareche, *Synth. Met.* **34** (1989) 115–20.
12. H. Fuzellier, J. Melin and A. Herold, *Mat. Sci. Eng.* **31** (1977) 91–5.
13. J. E. Fischer, T. E. Thomson, G. M. T. Foley, D. Guerard, M. Hoke and F. L. Ledermann, *Phys. Rev. Lett.* **37** (1976) 769–72.
14. D. Guerard, N. Elalem, S. Elhadigui, L. Elansari, P. Lagrange, F. Rousseaux, H. Estrade-Szwarckopf, J. Conard and P. Lauginie, *J. Less Common Met.* **131** (1987) 173–80.
15. *Intercalated Materials*, F. Levy (ed) (D. Reidel Pub. Company, Dordrecht, Holland (1979)); *Intercalation Layered Materials*, M. S. Dresselhaus (ed) (Plenum NY (1986)).
16. E. Kunoff, *Phys. Rev. B* (to be published).
17. P. Lagrange, A. Metrot and A. Herold, *C.R. Acad Sci.* C**275** (1972) 765–8.
18. P. Lauginie, H. Estrade, J. Conard, D. Guerard, P. Lagrange and M. El Makrini, *Physica* **99B** (1980) 514–20; *Vth London Int. Carb. Conf.* (1978) 645–9.
19. R. Moreh and O. Shahal, *Sol. St. Comm.* **43**(7) (1982) 529–32.
20. R. Moreh, O. Shahal and G. Kimmel, *Phys. Rev.* B**33**(8) (1986) 5717–20.
21. R. Moret, R. Comes, G. Furdin, H. Fuzellier and F. Rousseaux in *Intercalated Graphite Mat. Res. Symp. Proc.* M. S. Dresselhaus, G. Dresselhaus, J. E. Fischer and M. J. Moran (eds) (North Holland NY (1983)) 27–32.
22. D. E. Nixon, G. S. Parry and A. R. Ubbelhode, *Proc. R. Soc. Lond.* A**291** (1966) 324–31.
23. G. S. Parry, *Mat. Sci. Eng.* **31** (1977) 99–106.
24. H. Pinto, M. Melamud, O. Shahal, R. Moreh and H. Shaked, *Physica B* **121** (1983) 121–6.
25. H. Pinto, M. Melamud, R. Moreh and H. Shaked, *Physica B* **156/157** (1989) 283–5.
26. I. Rosenman, C. Simon, F. Batallan, H. Fuzellier and H. J. Lauter, *Synth. Met.* **23** (1988) 339–44; *Synth. Met.* **12** (1985) 117–23.
27. W. Rüdorff, *Zeit. Phys. Chem.* **45** (1940) 42–68.
28. E. J. Samuelsen, R. Moret, H. Fuzellier, M. Klatt, M. Lelaurain and A. Herold, *Phys. Rev.* B**32** (1985) 417–27.
29. M. Schafhaeutl, *J. Prakt. Chem.* **21** (1841) 155–9.
30. H. Shaked, H. Pinto and M. Melamud, *Synth. Met.* **26** (1988) 321–5; *Phys. Rev.* B**35** (1987) 838–44.
31. C. Simon, I. Rosenman, F. Batallan, J. Rogerie, J. F. Legrand, A. Magerl, C. Lartigue and H. Fuzellier, *Phys. Rev.* B**41** (1990) 2380–91.
32. P. Touzain, *Synth. Met.* **1** (1979/80) 3–11.
33. A. R. Ubbelhode, *Carbon* **6** (1968) 177–82.
34. A. R. Ubbelhode, *Proc. R. Soc. Lond.* A**309** (1969) 297–311; A**304** (1968) 25–43; A**321** (1971) 445–60.

13 Proton-containing β- and β″-alumina structure type compounds

HIROYUKI IKAWA

The title compounds have received a considerable amount of attention because of their high protonic conductivities at 100–300 °C. However, because of their complex characters there remain many areas of uncertainty. The current status is reviewed here.

13.1 Synthesis

It is well known that the β- and β″-aluminas are sodium ion conductors with promising applications as an electrolyte for the sodium–sulphur secondary battery. They have been widely studied because of their technological and scientific interest[1-7].

The sodium ions in the conduction planes of β- and β″-alumina type structures are exchangeable with many other monovalent and divalent cations. The main host crystals are β- and β″-alumina, and the Ga, Fe analogues: β- and β″-gallate, β- and β″-ferrite. The guest ions which contain protons include H^+, H_3O^+ (in this review the use of H_3O^+ does not necessarily mean a hydronium ion), and NH_4^+. Many combinations of host crystals and guest ions have been prepared; details are given in Table 13.1.

The crystals, even of β-alumina, are attacked corrosively by hot sulphuric acid. Furthermore, crystals are inclined to cleave during ion exchange because the ionic radii of NH_4^+ and H_3O^+ are much larger than that of Na^+[14]. These factors have serious implications for the fabrication of high density, polycrystalline, flaw-free ceramics[17-20] (see Chapter 33).

During ion exchange, with the exception of those exchanges in which H^+ (D^+) β-alumina and NH_4^+ β- and β″-gallates are being prepared, it

Proton-containing β- and β″-aluminas

Table 13.1. *Synthesis of proton-containing β- and β″-alumina type compounds and polycrystalline bodies*

Product	Method, remark and reference

H^+ β-alumina: heating Ag^+ β- in H_2 at $\geqslant 300\ °C$; may be prepared from H_3O^+ and NH_4^+ β- by heating to liberate H_2O and NH_3[8,9] respectively

D^+ β-alumina: heating Ag^+ β- at $450\ °C$ in a sealed silica tube[8,10]

H_3O^+ β-alumina: repeated exposure of Na^+ β- to boiling conc. H_2SO_4[44]; ceramic attacked corrosively during ion exchange[11,12]

NH_4^+ β-alumina[a]: repeated exposure of Na^+ β- to molten NH_4NO_3, laminated[8,64]

H_3O β″-alumina and NH_4^+/H_3O^+ β″-alumina: the same methods mentioned above, the latter contains oxonium ion also[4,13,48]

H_3O^+ β-gallate and H_3O^+ β″-gallate: hydration of de-ammoniated NH_4^+ β- and NH_4^+/H_3O^+ β″-: easily attacked by acid

NH_4^+ β-gallate and NH_4^+/H_3O^+ β″-gallate: repeated exposure to molten NH_4NO_3 of K^+ β- and K^+ β″-[14,15]

H_3O^+ β-ferrite and NH_4^+ β-ferrite: repeated exposure of K^+ β- to boiling conc. H_2SO_4 and molten NH_3NO_3[16,57], respectively

Polycrystalline H_3O^+ β″/β-alumina: ceramic precursor of mixed alkali Na^+/K^+ β″/β- is ion exchanged in boiling conc. H_2SO_4 or in dilute acid under d.c. field[17,18] (see Chapter 33)

Polycrystalline NH_4^+ β″/β-gallate: ceramic precursor made by hot-pressing Rb^+ β″/β- is repeatedly exposed to molten NH_4NO_3[19,20] (see Chapter 33)

[a] NH_4^+ β-alumina made from a thin crystal is stoichiometric while that made from a large crystal is nonstoichiometric (N. Iyietal, *Solid State Ionics* **92** (1991) 578–96.)

is claimed that the stoichiometries of crystals, i.e. the numbers of exhangeable cations, are unchanged. There are reasons to doubt this, however, as discussed later.

13.2 Crystal structure and structural characteristics

The crystal structure of β-alumina may be described as an alternative stacking structure of 'spinel blocks' and 'conduction planes'. The spinel block is constructed by four layers of oxygen in a cubic close packed array, similar to that of spinel. Metal ions occupy a tetrahedral site, Al(2), and octahedral sites, Al(1) and Al(4). The conduction plane is a crystallographic mirror plane. In a unit cell section of the conduction plane (hereafter abbreviated as unit cell plane) of 'ideal' $NaAl_{11}O_{17}$, there is only one oxide ion, labelled O(5) and one Na^+ ion at a site named BR (Fig. 13.1). (Note in the β-alumina structure two BR sites are shown but these are different crystallographically. Occupied BR sites alternate with unoccupied aBR sites.) The BR site is coordinated trigonal prismatically by O(2) oxygens. Oxygen O(5) links two adjacent spinel blocks by acting as the common apex of two $Al(3)O_4$ tetrahedra. The *c*-parameter represents the length of two spinel blocks and two conduction planes.

The crystal structure of β″-alumina is very similar to that of β-alumina; however, the conduction plane is not a mirror plane. In a unit cell plane, there is one O(5) oxygen and two BR sites (Fig. 13.1). Strictly speaking, the BR sites are displaced slightly from the plane, alternately up and down, and are coordinated tetrahedrally by three oxygens O(2) and one O(4). The *c*-parameter reflects the thickness of three spinel blocks and conduction planes.

It is well-established that β-alumina prepared by the usual ceramic methods contains an excess of Na^+ ions over its ideal composition, and that the formula is expressed by $Na_{1+x}Al_{11}O_{17+x/2}$ ($x \simeq 0.3$). The parameter, $x/2$, represents the number of interstitial oxide ions, O_i, at mO sites – a middle point between two O(5) oxygens, Fig. 13.1 – per unit cell plane; each O_i coordinates to two Frenkel defects i.e. two interstitial Al_is which are associated with Al vacancies at the adjacent Al(1) site[21]. The situation in K^+ β-gallate is identical to that of β-alumina[22]. Local electroneutrality is not guaranteed in the ideal structure, since there is a

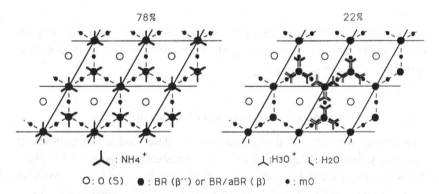

Fig. 13.1. Schematic probable representation of the disorder in the conduction plane of NH_4^+/H_3O^+ β″-alumina involving the superposition of a 78% NH_4^+ network and a 22% H_3O^+/H_2O network[13].

deficiency of positive charge around the conduction plane, and an excess of positive charges around the centre of the spinel block[23-25]. The above mentioned defect structures and nonstoichiometry are believed to diminish the electric imbalance. Defects are also present in β″-alumina in which the x value is as large as c. 0.6. The number of *Frenkel defects* is not enough to account for the large x values. Instead, single crystals of β″-alumina always contain 'stabilizing ions' – Li^+, Na^+ (only for β″-gallate), Mg^{2+}, Zn^{2+} – at the tetrahedral Al(2) site. The *charge compensation* mechanisms are topics of debate even today[26,27] and are related to the structure–property relations of β- and β″-alumina type crystals.

Several papers on the structures of proton-containing crystals, determined by X-ray and neutron diffraction have been reported as follows: D^+ β-alumina[10,28], H_3O^+ β-alumina[9,29], NH_4^+ β-alumina[9,30], H_3O^+ β″-alumina[31,32], NH_4^+/H_3O^+ β″-alumina[13,33], NH_4^+ β-gallate[22], and NH_4^+/H_3O^+ β″-gallate[34]. The structures of these proton-containing crystals are naturally very similar to those of the alkali-containing parent crystals. Accordingly, only limited structural characteristics of these compounds are discussed here, although small modifications always occur on ion exchange even to the centre of the spinel block.

The structure of D^+ β-alumina is interesting. Its composition is *stoichiometric*, $DAl_{11}O_{17}$; interstitial Al_i and O_i have disappeared during heating in D_2[10,28]. The O_i species must have been lost as D_2O whereas the Al_i atoms appear to return to their parent Al(1) sites[28]. The D^+ ion, one per unit cell plane, is located on the conduction plane and is connected to O(5) with an interatomic distance of c. 0.1 nm[10,28]. The orientation of this O(5)–D bond is consistent with spectroscopic data (see later). Occurrence of this bond accounts for the large c-parameter of D^+ β-alumina (2.262 nm, cf. 2.253 nm in Na^+ β-alumina). This is because the electrostatic bond strength between the Al(3) and O(5) is weakened by formation of the O(5)–D bond. The c-expansion is also encouraged by the repulsion between oxide ions in adjacent spinel blocks which have lost the screening provided by exchangeable cations.

A simple but powerful explanation of the fact that D^+ bonds exclusively to O(5) and not O(2), may be that the electrostatic bond strength sums (Pauling's second rule) of O(5) and O(2) are $+1.5$ and $+1.75$, respectively[23]. These values are smaller than the formally required $+2.0$. Such low values are the reason for the deficiency in positive charge around the conduction plane; clearly O(5) is more negatively charged than O(2) and is more likely to attract D^+.

Many researches have been concerned to identify the nature of the protonic ions and their configuration but the situation is not clear in general. A beautiful arrangement of atoms in the conduction plane of NH_4^+/H_3O^+ β''-alumina, as shown in Fig. 13.1[13], may be an exception. On the basis of this arrangement, Thomas & Farrington discussed the reason for the unusually high conductivity of this crystal as follows: the main hydrogen bonds are concerned with NH_4^+ ions which makes it easy to have a Grotthuss-type proton transfer involving rotational translation of H_3O^+ and H_2O, and proton transfer from H_3O^+ to H_2O[13].

There are many inconsistencies among reports on the structures of these protonic compounds. It may be possible that impurities and differences in preparation methods are responsible for the inconsistencies. Nevertheless, an understanding of the structures appears to be emerging. Some key experimental results are cited next; these also serve to make clear some characteristic features of these crystals.

1. The reported structures of D^+ β-alumina are that of a *stoichiometric* composition. It is said that hydrogen is effective in removing O_i[10,28]. Spectroscopic data change significantly with time on heating Ag^+ β-alumina in H_2 and D_2[35,36]. This change was attributed to a gradual loss of O_i[35,9,36].

2. The structure of H_3O^+ β-alumina is of a stoichiometric composition[9,29]. The thermogravimetric (TG) analysis by Udagawa *et al.*[37] is exactly consistent and that by Saalfeld *et al.*[11] is partially consistent with the stoichiometry.

3. The structure of NH_4^+ β-gallate is that of a stoichiometric composition[22]. Furthermore, the stoichiometry is consistent with TG results[14].

4. *Nonstoichiometric* NH_4^+ β-alumina (large crystal) transforms to stoichiometric H^+ β-alumina by heating at temperatures above c. 500 °C[4,5,9,30].

5. NH_4^+ β-ferrite is reported[38] to be stoichiometric from TG analysis.

6. The Na^+ ions which act as a 'stabilizer' of the β''-alumina structure, by occupying tetrahedral Ga(2) sites in Na^+ β''-gallate, are expelled from the spinel block during ion exchange in molten NH_4NO_3[15,34,39]. In order to compensate the charge imbalance caused by the missing stabilizing ions, O_i ions are introduced into mO sites of NH_4^+/H_3O^+ β''-gallate[34]. These O_i ions coordinate to three Ga atoms which enter interstitial sites[34].

Proton-containing β- and β″-aluminas

The structural changes cited above involving loss of O_i and compensating cations, lead to a decrease in conductivities because of a decrease in number of proton-containing ions. Thus, the stoichiometric structures of H_3O^+ and NH_4^+ β-alumina and β-gallate account for the large differences in conductivities between them and their non-stoichiometric β″-alumina analogues (see Table 13.3). These differences are orders of magnitude larger than the differences in conductivities between the alkali or silver β/β″ forms both of which are usually non-stoichiometric. The conductivity of NH_4^+ β-gallate may also be increased by doping with Mg, up to a maximum value ten times that of the undoped material[40]. The reason for this is that additional NH_4^+ ions enter the conduction plane at the same time as the spinel blocks are doped with Mg^{2+} ions.

The reason is not known why these characteristic structural changes occur but the following speculations may be made. A crystal of *nonstoichiometric* NH_4^+ β-alumina is significantly more stable than the stoichiometric form, because the former structure more closely approaches electroneutrality. Thus a stoichiometric crystal forms on ion exchange even though it appears energetically unfavourable for this to happen. The stoichiometric crystal must form because neither the stoichiometric nor the nonstoichiometric form is thermodynamically stable under the ion exchange conditions. Corrosion of crystals in conc. H_2SO_4 is noted. The energy necessary to break structural units may decrease in the sequence: spinel block, Al(3)–O(5)–Al(3) bridge, and Al_i–O_i–Al_i bridge, and therefore the *Frenkel defect* bridges are destined to be destroyed first during ion exchange. The *stoichiometric* phase may have some degree of metastability and be formed at a saddle point before complete destruction of the crystal. A similar explanation may be possible for the structural changes in NH_4^+/H_3O^+ β″-gallate in which a Na^+ ion at a Ga(2) tetrahedral site must be energetically unfavourable due to its large ionic size[34].

Spectroscopic methods – infrared (IR) and Raman spectroscopy, *NMR*, and inelastic neutron scattering – have been very effective to identify protonic species and determine their configurations[41–43]. The *IR* spectra of the O–H stretching mode should be strong (Fig. 23.2) when the electric vector of the incident beam is normal to the c-axis $(E \perp c)$[35, 36]. It is not clear in the literature, however, which are genuine spectra of H^+ β-alumina and which are spectra of hydrated H^+ β-alumina. This confusion is caused partly by difficulty in preparing pure and unhydrated specimens. It is almost impossible to make pure H^+ β-alumina free from

Materials

Ag contamination from Ag^+ β-alumina (Colomban, private communication). The proton-containing crystals are very hydroscopic after heating to temperatures higher than $c.$ 250 °C; an increase in weight by hydration is detectable in ordinary TG equipment as high as $c.$ 300 °C[14,15]. In addition the degree of hydration and aging affect the IR spectra (see p. 371).

Hayes *et al.* assigned two sharp peaks at 3548 and 3510 cm^{-1} to stretching modes of O_i–H and O(5)–H in unhydrated H^+ β-alumina, respectively[35]. The first peak decreased in intensity on heating in H_2 and is weak in de-ammoniated NH_4^+ β-alumina, which has been ion exchanged for 3 months and in which the interstitial oxide ions, O_i, are gradually lost. Further, they assign an intense line at 3508 and a shoulder at 3520 cm^{-1} to OH stretching modes of the H_3O^+ ion[9,31]. If we accept the assigments by Hayes *et al.*[35] as they are, although some were made without presenting enough data, we must also agree with the assumptions because their experimental methods are more persuasive. However, the presence of $H^+(H_2O)_n$ ions is conclusively documented by many reports[41-45] including data on hydrated Na^+ and K^+ β/β''-crystals[46,47]. The presence of the H_3O^+ ion, especially the hydrogen bond-free H_3O^+ ion in these crystals, is still an open problem.

The spectra of H_3O^+ and NH_4^+/H_3O^+ β''-alumina are broad with overlapping peaks[42,43] making definite assignments more difficult. However, a number of analyses have been carried out, with suggestions for the following: a wide variety of protonated species in protonated β- and β''-alumina[47,48]; occupation of mO sites by water molecules in hydrated Na^+ β''-alumina[49]; existence of three nonequivalent sites for NH_4^+ ions in NH_4^+ β-alumina[50]; two types of NH_4^+ reorientation mode in NH_4^+/H_3O^+ β''-alumina[51].

13.3   Thermal transformations

A number of studies have been reported in which thermal transformations of these compounds are one of the main themes: H_3O^+ β-alumina[11,12,37,52-54,57], NH_4 β-alumina[9,30,37,55-57], H_3O^+ and NH_4^+/H_3O^+ β''-alumina[13,18,33,53,54,56,57,64,75], NH_4^+ β-gallate and NH_4^+/H_3O^+ β''-gallate[14,15,40], and NH_4^+ β-ferrite and NH_4^+/H_3O^+ β''-ferrite[38,57]. Inconsistencies also exist in these reports which are not simply due to differences in composition related to the ion exchange processes. Many of the results and explanations are based on DTA (heating rate: 10 °C min^{-1}) and TG(A) (heating rate: 10, 2 or 1 °C min^{-1}) experiments. As many authors realize, liberation of H_2O and NH_3 from these compounds is very slow. Accordingly, such processes are sensitive to grain size and heating conditions. A scheme for the thermal transformations of these compounds is proposed below based mainly on this author's experience[14,15,37,40], (although they might be slightly dogmatic).

Thermal analysis curves of NH_4^+ β-gallate are shown in Figs 13.2 and 13.3[14]. The weight loss below 180 °C in Fig. 13.3 is attributed to water; accordingly, the stoichiometry of this specimen may be determined from the weight losses caused by de-ammoniation at *c.* 230 °C and dehydration at *c.* 480 °C. The fact that the crystal is stoichiometric, which is determined from the TG data, is fully consistent with structure analysis[22]. The stoichiometry of H_3O^+ β-alumina is also confirmed by a similar TG curve to Fig. 13.3[37]. However, NH_4^+ β-alumina of that experiment is not stoichiometric, and the results changed slightly from specimen to specimen. The phenomenon is now attributable to compositional differences which

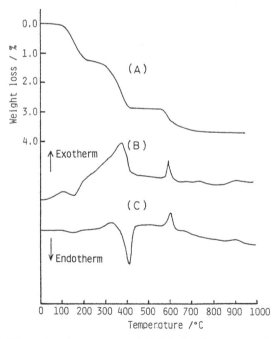

Fig. 13.2. Thermal analyses curves of NH_4^+ β-gallate measured at a heating rate of 10 °C min^{-1}[14]. (A), TG curve; (B), DTA curve measured in air; (C), DTA curve measured in a flow of nitrogen gas (reproduced with permission).

197

are caused by differences in the effective time of ion exchange. The proposed transformation sequences of stoichiometric compounds are given in Table 13.2. All except one have plateau regions with ideal compositions of H^+ β-alumina and H^+ β-gallate before the final, destructive dehydration at higher temperature. Only the case of NH_4^+ β-ferrite is different in that the de-ammoniation and the destructive dehydration occur simultaneously[38]. In general, specimens retain their β-alumina type structure until the final destructive dehydration; changes before destruction of the β-alumina and β″-alumina structures are reversible. Liberation of gases from the β″-alumina type compounds are also reversible as long as the β″-alumina structure is retained.

It has been reported[9, 30, 58] that *nonstoichiometric* NH_4^+ β-alumina transforms to a *stoichiometric* form when heated, by liberation of NH_3 and H_2O (O_i makes H_2O). The present author suggests that the nonstoichiometric specimens remain nonstoichiometric until the final destructive dehydration, and that the changes described as a nonstoichiometric to stoichiometric transformation[9, 30, 58] may be caused by a loss of water. Further studies are needed to resolve this problem.

The *thermal transformations* of NH_4^+ β-alumina[56, 57] and NH_4^+ β-gallate[40] doped with MgO have also been investigated. With increasing dopant concentration, the amounts of both de-ammoniation and destructive

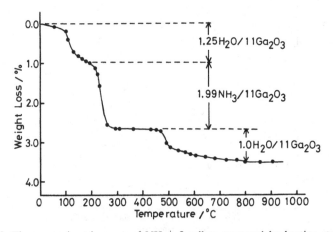

Fig. 13.3. Thermogravimetric curve of NH_4^+ β-gallate measured by heating stepwise at 20–30 °C intervals and holding constant at each temperature for 2.5 h[14]. The species of gases for each weight loss and their ratios (in mol per 11 mol of Ga_2O_3) are shown (reproduced with permission).

Table 13.2. *Thermal transformation sequence of proton-containing β-alumina type compounds*

H⁺ β-alumina	H₃O⁺ β-alumina	NH₄⁺ β-alumina	NH₄⁺ β-gallate	NH₄⁺ β-ferrite
$HAl_{11}O_{17}$	$H_3OAl_{11}O_{17}\cdot0.3H_2O$	$NH_4Al_{11}O_{17}\cdot0.23H_2O$	$NH_4Ga_{11}O_{17}\cdot0.63H_2O$	$NH_4Fe_{11}O_{17}\cdot H_2O$
	↓ dehydration, 60–270 °C; sharp at 210 °C	↓ dehydration, 60–270 °C; sharp at 310 °C	↓ dehydration, 60–180 °C; sharp at 130 °C	↓ dehydration, 100–200 °C
	$HAl_{11}O_{17}$	$NH_4Al_{11}O_{17}$	$NH_4Ga_{11}O_{17}$	
		↓ deammoniation, 270–430 °C; sharp at 310 °C	↓ deammoniation, 180–290 °C; sharp at 230 °C	
		$HAl_{11}O_{17}$	$HGa_{11}O_{17}$	
↓ dehydration, 650–880 °C, sharp at 690 °C	→	→	→ dehydration, 450–800 °C; sharp at 480 °C	→ deammoniation and dehydration, 350–430 °C
$Al_2O_3{}^a$ → $Al_2O_3{}^a$	$Al_2O_3{}^a$	$Al_2O_3{}^a$	$Ga_2O_3{}^a$	α-Fe_2O_3

Some reactions are estimated. The phase changes are those observed under almost equilibrium conditions except that of NH_4^+ β-ferrite[38]. The numbers of water molecules in the formulae change easily.

a Intermediate phase giving diffuse diffraction spots similar to β-alumina. Exothermic peaks in DTA curves at temperatures *c.* 830 and 590 °C of aluminate and gallate are caused by crystallization of the intermediate phases.

dehydration increase and the plateau region occurs at lower temperature[40]. From these changes it appears as if there may be a continuity in thermal transformation behaviour between the β and β″ forms.

The thermal transformations of β″-alumina type compounds are more complicated having no plateaux with which to identify intermediate phases, although some articles do report the occurrence of plateaux[13,33,64]. Tsurumi *et al.*[15] have studied thermal transformations of two kinds of NH_4^+/H_3O^+ β″-gallate: A-type, prepared from Na-doped gallate having O_i atoms at mO sites; and B-type, prepared from Zn-doped gallate and having no O_i atoms. The TG curves of the A- and B-types are shown in Fig. 13.4[15]. The A-type retains its β″-alumina structure up to *c.* 500 °C and the gases liberated between 170 and 230 °C are N_2O, H_2O and a trace of NH_3. The B-type loses its β″-alumina structure as low as 275 °C to transform into a γ-Ga_2O_3 type structure which contains a considerable amount of protonic species. Further, the gases liberated from 170 to 230 °C are NH_3, H_2O and trace of N_2O. It was not possible to estimate chemical compositions from the TG analyses without special quantitative measurements. The differences between the two types of behaviour are attributed to the presence or absence of O_i atoms[15,56]. Similar phenomena are expected for the proton containing β″-aluminas, although the thermal changes occur at higher temperatures than those of gallates. For example, NH_4^+/H_3O^+ β″-alumina doped with MgO transforms into a transition spinel structure at 320 °C according to DTA results[56,57].

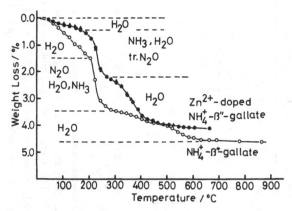

Fig. 13.4. Thermogravimetric curves of NH_4^+ β″-gallate (prepared from Na-doped Na^+ β″-gallate) and Zn-doped NH_4^+ β″-gallate measured by heating stepwise at 20–30 °C intervals and holding constant at each temperature for 2 h[15].

Proton-containing β- and β″-aluminas

It is said that the typical formula of NH_4^+/H_3O^+ β″-alumina is $(NH_4)_{1.00}(H_3O)_{0.67}Mg_{0.67}Al_{10.33}O_{17}$[59]. The H_3O^+ ions are incorporated into the crystal and the ratio of the two protonic ions. $NH_4^+:H_3O^+$, changes with preparation conditions[59]. According to DeNuzzio & Farrington[64] the content of NH_4^+ ion decreases and converges to the above value for times of exchange longer than seven days. The β″-alumina having O_i atoms in the conduction plane will be formed when the amount of a stabilizer (MgO) is small[34, 60].

13.4   Conductivity

It is necessary to make clear the following on an ionic conducting material: conductivity at a given temperature, species of moving ion(s) and transference number(s), and those changes with temperature. Further, we are interested in the conduction mechanism, and the Haven ratio is an important parameter in understanding that mechanism. In spite of several studies on this theme, it remains unclear. The first cause of these uncertainties may exist in the difficulty in sample preparation. It is almost impossible to find single crystal plates of aluminates with a thickness greater than *c.* 0.1 mm that are free from latent cleavage, although as for gallates, the situation is less critical owing to the longer Ga(3)–O(5) interatomic distance[14].

Conductivity data are listed in Table 13.3 together with some remarks. The data show considerable scatter, but we can estimate the conductivities to within an order of magnitude. The conductivities of NH_4^+/H_3O^+ β″-alumina type compounds are of the order $10^{-4}\,\Omega^{-1}\,cm^{-1}$ at room temperature, and their values are the highest among these compounds.

Conductivities of the β-alumina type compounds increase with temperature until the liberation of gases associated with exchangeable ions begins to occur, estimated as *c.* 200 °C for H_3O^+ β-alumina, 270 °C for NH_4^+ β-alumina, and 200 °C for NH_4^+ β-gallate. Conductivities then increase again with temperature through the TG plateau regions. If the measurements are made under stable conditions it is possible to evaluate the Arrhenius parameters for 'proton' conduction, since in the plateau regions the proton is thought to be the only moving species.

No such increases in conductivity with temperature are observed for β″-alumina type compounds.

The d.c. conductivities of polycrystalline specimens are important for applications, but there is only a limited amount of information available

Table 13.3. *Ionic conductivities of proton-containing β- and β''-alumina type compounds*

Compound	Conductivity[a]/ S.cm^{-1}	E_a/eV	Temp. range/°C	Reference, method, single (S) or polycrystal (P), other remark
H$_3$O$^+$ β-alumina	6.5×10^{-6}	0.15	25–400	53, AC, S, ion rich $y \approx 0.66$
	2×10^{-10}	0.6	400–500	same, after heating at 400 °C
	3×10^{-7}	0.4	25–400	61, AC, S, $x \approx 0.25$
	10^{-10}	0.8	<800	same, after heating
	1.7×10^{-11}	0.78	20–200	62, DC, S
	10^{-6} (500 °C)	1.3	300–500	same, partially dehydrated
NH$_4^+$ β-alumina	1.5×10^{-6}	0.5	25–400	9, AC, S, $x = 0.25$
	2.5×10^{-5}	0.18	25–400	56, AC, S, ion rich $y \approx 0.66$
	1.3×10^{-6}	0.47	25–300	55, AC, S
		0.66	400–700	same, after deammoniation
NH$_4^+$ β-gallate	8×10^{-6}	0.2	100–180	14, AC, S
	9×10^{-5} (390 °C)			same, after deammoniation
NH$_4^+$ β-ferrite	4.3×10^{-6}	0.21	25–150	38, AC, S
H$_3$O$^+$ β''-alumina	8.9×10^{-3}	0.24	20–100	63, AC, P, this σ is too high
	10^{-5}			59, AC, S of Mg-doped
	4×10^{-6}	0.20	25–300	53, AC, S of Mg-doped

NH$_4^+$/H$_3$O$^+$ β″-alumina	1×10^{-4}	0.3	25–200	59, AC, S of Mg-doped
	3×10^{-5}	0.24	25–200	56, AC, S of Mg-doped
	1.3×10^{-3} (30°C)	0.27	25–170	64, AC, S, $y = 0.67$
	10^{-6}	0.45	25–200	same, after heating at 280 °C
	2.3×10^{-4}	0.31	25–190	65, AC, S, $y = 0.68$
	3.6×10^{-4}	0.25	50–190	65, AC, S, $y = 0.85$
NH$_4^+$/H$_3$O$^+$ β″-gallate	1.1×10^{-4}	0.23	30–170	65, AC, S, estimated as $y = 0.0$
	2.9×10^{-4}	0.20	25–260	15, 66, AC, S, $y = 0.0$
	1.0×10^{-3}	0.22	25–190	15, 66, AC, S, $y = 0.68$ (Zn-doped)
	8.6×10^{-4}	0.35	14–200	67, AC, S, $y = 0.9$ (Zn-doped)
	6.0×10^{-3} (120 °C)			67, AC, P of Zn-doped
NH$_4^+$/H$_3$O$^+$ β″-ferrite	3×10^{-4}	0.3	20–190	20, AC, P of Zn-doped
	2.7×10^{-5}	0.15	25–150	38, AC, S of Cd-doped

[a] Conductivity at room temperature unless stated otherwise. Abbreviations: AC: a.c. method, DC: d.c. method, S: single crystal, P: polycrystal, same: the same as mentioned above. x and y are defined by the following: $M_{1+x}Al_{11}O_{17+x/2}$ and $M_{1+y}Mg_yAl_{11-y}O_{17}$ where M is exchangeable monovalent cation(s). (S is equivalent to Ω^{-1}.)

on this. The difficulties in sample preparation and selection of suitable electrodes[68] are cited as reasons. Studies on $NH_4{}^+/H_3O^+$ β''-gallate ceramics fabricated by hot pressing indicate that the d.c. conductivity is very low owing to the large grain boundary resistance[20,67]. But no large boundary resistances were reported for H_3O^+ β''-alumina ceramics prepared from normally sintered precursors[17,18,63,75]. The large grain boundary resistance may therefore be characteristic of those hot pressed materials which have poor intergranular connectivities.

There is a report[51] that conductivity is affected exponentially by humidity. It is thought to be caused by a change of grain boundary conductivity because (unpublished data) limited changes of bulk conductivity with environment, including water vapour, were seen.

Most reports on conductivities discuss moving ions mainly on the basis of activation energies. However, many of the activation energies listed in Table 13.3 may not be real values, even if the heating and cooling curves coincide, because specimens change their compositions with temperature (see TG curves). Studies by means of NMR[51,65,69-71] and *quasielastic neutron scattering*[31,45,72] are also effective for characterizing the moving species. Such measurements, however, may not give direct information about the protonic conductivity. In this context, the report by Smoot *et al.*[65] is noteworthy because of their effort to correlate conductivities and diffusion coefficients. Because these compounds are very complicated, transport number measurements are required to determine the nature of the conducting species. (See also Chapter 3.)

13.5 Applications

Application studies on these compounds have been made by two groups. One group uses a hot pressed $NH_4{}^+/H_3O^+$ β''/β-gallate ceramic and the other uses a H_3O^+ β''/β-alumina ceramic; details of the latter are described later in this book (see Chapters 33, 34, 36 and 39).

Tsurumi *et al.* have assembled a *hydrogen concentration cell* using an $NH_4{}^+/H_3O^+$ β''/β-gallate ceramic to measure e.m.f. values; the results agree well with a thermodynamically estimated value[19]. The group have also assembled a steam concentration cell and claim that the e.m.f. values cannot be explained by means of a simple equation[20]. Both studies found that the large resistance of the electrolyte, which is much greater than expected for the bulk resistance makes the measurement of e.m.f. difficult.

The H_3O^+ β''/β-alumina ceramics have been prepared[17,18,73,74,75] by

Proton-containing β- and β″-aluminas

various techniques – use of Na^+/K^+ mixed-ion precursor, gas-phase ion exchange, vapour-phase ion exchange, electric field assisted ion exchange, making a composite with ZrO_2. The resulting electrolytes have been demonstrated to have applications in a pH meter (HCl concentration cell)[18], an H_2/O_2 *fuel cell*[75] and *steam electrolysis cells*[18, 74, 75]. The steam electrolysis cell was operated at 100 °C for 2500 h and evolved hydrogen of volume 7×10^4 cm^3[74].

The proton-containing β- and β″-alumina type compounds, especially the latter are attractive for their high conductivities at intermediate temperatures (100–300 °C). However, the difficulties in making a ceramic electrolyte and subsequent cracking of the disc sample[76], caused by change of chemical composition with temperature and/or atmosphere coupled with large anisotropic character of the crystal, are serious drawbacks preventing applications. Approaches from different view points look necessary therefore before applications of these compounds become reality.

13.6   References

1. J. T. Kummer, β-Alumina electrolytes, *Prog. Solid State Chem.* 7 (1972) 141–75.
2. P. Vashishta, J. N. Mundy and G. K. Shenoy (eds) *Fast Transport in Solids* (North-Holland, New York (1979)).
3. B. C. Tofield, The intercalation chemistry of β-alumina, in *Intercalation Chemistry*, M. S. Whittingham and A. J. Jacobson (eds) (Academic Press, New York (1982)) 181–228.
4. Ph. Colomban and A. Novak, Hydrogen containing β and β″ alumina, in *Solid State Protonic Conductors*, J. Jensen and M. Kleitz (eds) (Odense University Press (1982)) 153–201.
5. R. Collongues, D. Gourier, K. Kahn, J. P. Boilot, Ph. Colomban and A. Wicker, β-Alumina, a typical electrolyte, *J. Phys. Chem. Solids* 45 (1984) 981–1013.
6. R. Stevens and J. G. P. Binner, Structure, properties and production of β-alumina, *J. Mater. Sci.* 19 (1984) 695–715.
7. B. Dunn, G. C. Farrington and J. O. Thomas, Frontiers in β″-alumina research, *MRS Bull.* (1989) 22–30.
8. Y.-F. Y. Yao and J. T. Kummer, Ion exchange properties of and rates of ionic diffusion in β-alumina, *J. Inorg. Nucl. Chem.* 29 (1967) 2453–75.
9. Ph. Colomban, J. P. Boilot, A. Kahn and G. Lucazeau, Structural investigation of protonic conductors: NH_4^+ β-alumina and stoichiometric H_3O^+ β-alumina, *Nouv. J. Chimie* 2 (1978) 21–32.

Materials

10. J. M. Newsam, A. K. Cheetman and B. C. Tofield, Anhydrous deuterium β-alumina: powder neutron diffraction studies at 4.2, 298, 573 and 720 K, *J. Solid State Chem.* **60** (1985) 214–29.
11. H. Saalfeld, H. Matthies and S. K. Datta, Ein Neues Aluminiumoxid-Hydrat mit β-Alumina Struktur, *Ber. Dtsch. Keram. Ges.* **45** (1968) 212–15.
12. M. W. Breiter, G. C. Farrington, W. L. Roth and J. L. Duffy, Production of hydronium β-alumina from sodium β-alumina and characterization of conversion products, *Mat. Res. Bull.* **12** (1977) 895–906.
13. J. O. Thomas and G. C. Farrington, Protonic solid electrolytes: a single-crystal neutron diffraction study of ammonium-hydronium β''-alumina, *Acta Cryst.* **B39** (1983) 227–35.
14. H. Ikawa, T. Tsurumi, K. Urabe and S. Udagawa, Electrical conductivity and thermal decomposition of β-alumina type NH_4^+-gallate, *Solid State Ionics* **20** (1986) 1–8.
15. T. Tsurumi, H. Ikawa, K. Urabe, T. Nishimura, T. Oohashi and S. Udagawa, Ionic conductivity and thermal decomposition of β''-alumina type NH_4^+-gallate, *Solid State Ionics* **25** (1987) 143–53.
16. S. Ito, N. Kubo, S. Nariki and N. Yoneda, Ion exchange in alkali layers of potassium β-ferrite $((1 + x)K_2O \cdot 11Fe_2O_3)$ single crystal, *J. Am. Ceram. Soc.* **70** (1987) 874–9.
17. P. S. Nicholson, M. Nagai, K. Yamashita, M. Sayer and M. F. Bell, Polycrystalline H_3O β''/β-alumina-fabrication characterization and use for steam electrolysis, *J. Am. Ceram. Soc.* **15** (1985) 317–26.
18. Y. Sheng, B. Cobbledick and P. S. Nicholson, The synthesis of polycrystalline H_3O^+ β''/β-alumina ceramics from conventional precursor Na/K β''/β-alumina powders, *Solid State Ionics* **27** (1988) 233–41.
19. T. Tsurumi, H. Ikawa, M. Ishimori, K. Urabe and S. Udagawa, Experimental assembling of gas cells using β/β''-alumina type NH_4^+-gallate electrolyte, *Solid State Ionics* **21** (1986) 31–5.
20. H. Ikawa, T. Ohashi, M. Ishimori, T. Tsurumi, K. Urabe and S. Udagawa, Fabrication of NH_4^+-gallate ceramic with β''-alumina structure and emf of gas cells using the electrolyte, in *High Tech Ceramics*, P. Vincenzini (ed) (Elsevier Science Pub. (1987)) 2137–46.
21. W. L. Roth, F. Reidinger and S. LaPlaca, Studies of stabilization and transport mechanisms in beta and beta'' alumina by neutron diffraction, in *Superionic Conductors*, G. D. Mahan and W. L. Roth (eds) (Plenum Press (1976)) 223–41.
22. H. Ikawa, T. Tsurumi, M. Ishimori, K. Urabe and S. Udagawa, Chemical composition and crystal structure of β-alumina type R^+-gallate ($R^+ = K^+$, NH_4^+), *J. Solid State Chem.* **60** (1985) 51–61.
23. A. R. West, Local structure in the β-alumina structures, *Mat. Res. Bull.* **14** (1979) 441–6.
24. H. Sato and Y. Hirotsu, Structural characteristics and nonstoichiometry of β-alumina type compounds, *Mat. Res. Bull.* **11** (1976) 1307–18.
25. T. Kodama and G. Muto, The crystal structure of T1 β-alumina, *J. Solid State Chem.* **17** (1976) 61–70.

Proton-containing β- and β″-aluminas

26. G. Collin , R. Comes, J. P. Boilot and Ph. Colomban, Structure, ion–ion correlation and compensation mechanisms in β- and β″-alumina, *Solid State Ionics* **28–30** (1988) 324–32.
27. G. K. Duncan and A. R. West, The stoichiometry of β″-alumina phase diagram studies in the system $Na_2O-MgO-Li_2O-Al_2O_3$, *Solid State Ionics* **28–30** (1988) 338–43.
28. J. M. Newsam, B. C. Tofield, W. A. England and A. J. Jacobson, Anhydrous deuterium beta alumina, in *Fast Ion Transport in Solids*, P. Vashishta, J. N. Mundy, G. K. Shenoy (eds) (North-Holland (1979)) 405–8.
29. K. Kato and H. Saalfeld, Alkalifreies wasserhaltiges β-Alumina *Acta Cryst.* B33 (1977) 1596–8.
30. J. M. Newsam, A. K. Cheetham and B. C. Tofield, Thermal evolution of the structure of ammonium β-alumina, *Solid State Ionics* **8** (1983) 133–9.
31. W. L. Roth, M. Anne and D. Tranqui, Protonic charge transport in hydronium β″-alumina, *Revue Chimie Min.* **17** (1980) 379–96.
32. J. O. Thomas, G. C. McIntyre and J. DeNuzzio, An alternative fabrication route of H_3O^+ β″-alumina via NH_4^+/H_3O^+ β″-alumina, *Solid State Ionics* **18/19** (1986) 642–4.
33. J. O. Thomas, K. G. Frase, G. C. McIntyre and G. C. Farrington, Decomposition and structure of NH_4^+/H_3O^+ β″-alumina, *Solid State Ionics* **9/10** (1983) 1029–32.
34. T. Tsurumi, H. Ikawa, T. Nishimura, K. Urabe and S. Udagawa, Crystal structure and charge compensation mechanism of β″-alumina type R^+-gallate ($R^+ = K^+$, NH_4^+), *Solid State Ionics* **71** (1987) 154–63.
35. W. Hayes, L. Holden and B. C. Tofield, Infrared studies of hydrogen beta alumina, *J. Phys. C* **13** (1980) 4217–28.
36. Ph. Colomban, G. Lucazeau and A. Novak, Vibrational study on hydrogen beta alumina, *J. Phys. C* **14** (1981) 4325–33.
37. S. Udagawa, S. Kimura and H. Ikawa, *Studies on the Materials Used for Efficient Production of Hydrogen, Research on Effective Use of Energy*, Vol. 3 (Reports of special project grant in aid of Sci. Res. of MESC Japan) (1982) 763–71.
38. K. Uchinokura, S. Nariki, S. Ito and N. Yoneda, Thermal decomposition process and electrical conductivity of single crystals of NH_4^+ β- and β″-ferrites, *Solid State Ionics* **35** (1989) 207–12.
39. H. Ikawa, K. Shima and K. Urabe, MAS-NMR spectra of ^{23}Na in spinel block of β and β″-alumina structures, *Chem. Letters* (1988) 613–14.
40. T. Tsurumi, H. Ikawa, K. Urabe and S. Udagawa, Ionic conductivity of Mg^{2+}-doped NH_4^+ β-gallate, *J. Ceram. Soc. Jpn* **97** (1989) 1308–10.
41. G. Lucazeau, Infrared, Raman and neutron scattering studies of β- and β″-alumina: a static and dynamic structure analysis, *Solid State Ionics* **8** (1983) 1–25.
42. A. Novak, M. Pham-Thi and Ph. Colomban, Vibrational study of some solid protonic superionic conductors, *J. Mol. Struc.* **141** (1986) 211–18.
43. Ph. Colomban and A. Novak, Proton transfer and superionic conductivity in solids and gels, *J. Mol. Struct.* **177** (1988) 277–308.

Materials

44. Ph. Colomban, G. Lucazeau, R. Mercier and A. Novak, Vibrational spectra and structure of $H^+(H_2O)_n$ β-alumina, *J. Chem. Phys.* **11** (1977) 5244–51.
45. J. C. Lassègues, M. Fouassier, N. Baffier, Ph. Colomban and A. J. Dianoux, Neutron scattering study of the proton dynamics in NH_4^+ and OH_3^+ β-alumina, *J. Physique* **41** (1980) 273–80.
46. J. B. Bates, N. J. Dudney, G. M. Brown, J. C. Wang and R. Frech, Structure and spectra of H_2O in hydrated β-alumina, *J. Chem. Phys.* **77** (1982) 4838–56.
47. J. B. Bates, D. Dohy and R. L. Anderson, Reaction of polycrystalline Na β″-alumina with CO_2 and H_2O and the formation of hydroxyl groups, *J. Mater. Sci.* **20** (1985) 3219–29.
48. Ph. Colomban and A. Novak, Protonic species, conductivity and vibrational spectra of β/β″-alumina, *Solid State Ionics* **5** (1981) 241–4.
49. J. B. Bates, J. C. Wang, N. J. Dudney and W. E. Brundabe, Hydration of β″-alumina, *Solid State Ionics* **9/10** (1983) 237–44.
50. J. B. Bates, T. Kaneda, J. C. Wang and H. Engstrom, Raman scattering from NH_4^+ and ND_4^+ in beta-alumina, *J. Chem. Phys.* **73** (1980) 1503–13.
51. Y. Furukawa, Y. Nakabayashi, S. Kawai and O. Nakamura, Proton magnetic relaxation and ionic conductivity of NH_4^+/H_3O^+ β″-alumina, *Solid State Ionics* **7** (1982) 219–23.
52. W. L. Roth, M. W. Breiter and G. C. Farrington, Stability and dehydration of alumina hydrates with the β-alumina structure, *J. Solid State Chem.* **24** (1978) 321–30.
53. N. Baffier, J. C. Badot and Ph. Colomban, Conductivity of β″ and β alumina. I. $H^+(H_2O)_n$ compounds, *Solid State Ionics* **2** (1980) 107–13.
54. G. C. Farrington and J. L. Briant, Hydronium beta″ alumina: a fast proton conductor, *Mat. Res. Bull.* **13** (1978) 763–73.
55. A. Hooper and B. C. Tofield, Characterization of NH_4^+ exchanged single crystal beta-alumina, in *Fast Transport in Solids*, P. Vashishta, J. N. Mundy and G. K. Shenoy (eds) (North-Holland, New York (1979)) 409–12.
56. N. Baffier, J. C. Badot and Ph. Colomban, Protonic conductivity of β″ and ion-rich β-alumina. II: ammonium compounds, *Solid State Ionics* **13** (1984) 233–6.
57. P. Lenfant, D. Plas, M. Ruffo and Ph. Colomban, Ceramiques d'alumine β et de ferrite β pour sonde a protons, *Mat. Res. Bull.* **15** (1980) 1817–27.
58. K. G. Frase and G. C. Farrington, Proton transport in the β/β″-aluminas, *Ann. Rev. Mater. Sci.* **14** (1984) 279–95.
59. G. C. Farrington and J. L. Briant, Ionic conductivity in β″-alumina, in *Fast Transport in Solids*, P. Vashishta, J. N. Mundy and G. K. Shenoy (eds) (North-Holland, New York (1979)) 395–400.
60. F. Harbach, High resolution X-ray diffraction of fully and partially magnesium stabilized β″-alumina ceramics, *J. Mater. Sci.* **18** (1983) 2437–52.
61. Ph. Colomban, J. P. Boilot, P. Chagnon and G. Guilloteau, Stabilité des conducteurs protoniques de type alumine β, *Bull. Soc. Fran. Ceram.* **111** (1978) 3–12.

Proton-containing β- and β″-aluminas

62. G. C. Farrington, J. L. Briant, M. W. Breiter and W. L. Roth, Ionic conductivity in H_3O^+ beta alumina, *J. Solid State Chem.* **24** (1978) 311–19.
63. A. Teitsma, M. Sayer, S. L. Segel and P. S. Nicholson, Polycrystalline hydronium β/β″-alumina: a ceramic fast proton conductor, *Mat. Res. Bull.* **15** (1980) 1611–19.
64. J. D. DeNuzzio and G. C. Farrington, Protonic solid electrolytes: thermal stability and conductivity of ammonium/hydronium β″-alumina, *J. Solid State Chem.* **79** (1989) 65–74.
65. S. W. Smoot, D. H. Whitmore and W. P. Halperin, Influence of stoichiometry and the nature of the spinel-block stabilizing element on proton transport behavior in solid electrolyte with the β″-alumina structure, *Solid States Ionics* **18/19** (1986) 687–93.
66. H. Ikawa, T. Tsurumi, T. Ohashi, K. Urabe and S. Udagawa, On the ionic conductivity of β″-alumina type NH_4^+-gallate, *J. Ceram. Soc. Jpn* **92** (1984) 473–4.
67. H. Ikawa, K. Shima, T. Tsurumi and O. Fukunaga, Protonic conductivities in single crystal and polycrystalline of β″-alumina type NH_4^+-gallate, in *Solid State Ionic Devices*, B. V. R. Chowdari and S. Radhakrishna (eds) (World Scientific, Singapore (1988)) 497–502.
68. H. Ikawa, K. Shima, T. Taniguchi, K. Urabe, O. Fukunaga and J. Mizusaki, Conductivity measurement of β″-alumina type NH_4^+gallate by dc four-terminal method, *Solid State Ionics* **35** (1989) 217–22.
69. D. Gourier and B. Sapoval, Jump diffusion and rotational tunnelling of ammonium in β-alumina by NMR, *J. Phys. C* **12** (1979) 3587–96.
70. A. R. Ochadlick, Jr and H. S. Story, Paramagnetic impurities and proton motion in NH_4^+ beta and NH_4^+/H_3O^+ beta″-alumina, *Solid State Ionics* **5** (1981) 257–60.
71. A. R. Ochadlick, Jr, H. S. Story and G. C. Farrington, Proton motion in NH_4^+ beta and NH_4^+/H^3O^+ beta″-alumina, *Solid State Ionics* **3/4** (1981) 79–84.
72. J. D. Axe, L. M. Corliss and J. M. Hastings, Neutron scattering study of NH_4^+ motion in NH_4^+ β-alumina, *J. Phys. Chem. Solids* **39** (1978) 155–9.
73. M. Nagai and P. S. Nicholson, Ion exchange kinetics of polycrystalline Na/K β/β″-alumina, *Solid State Ionics* **15** (1985) 311–16.
74. P. S. Nicholson, M. Z. A. Munshi, G. Singh, M. Sayer and M. F. Bell, Polycrystalline H_3O^+ β/β″-alumina: a designed superionic composite for steam electrolysis, *Solid State Ionics* **18/19** (1986) 699–703.
75. M. Z. Munshi and P. S. Nicholson, Demonstration of medium temperature, one atmosphere reversible H_2/O_2 fuel and steam electrolysis cell operation using polycrystalline H_3O^+ β/β″-alumina, *Solid State Ionics* **23** (1987) 203–9.
76. J. Gulens, T. H. Longhurst, A. K. Kuriakose and J. D. Canaday, Hydrogen electrolysis using a Nasicon solid protonic conductor, *Solid State Ionics* **28–30** (1988) 622–6.

14 Proton conduction in zeolites

ERIK KROGH ANDERSEN, INGE G. KROGH
ANDERSEN AND ERIK SKOU

14.1 Introduction

Zeolites are crystalline aluminosilicates. They are composed of cations (generally metal ions, but also ammonium ions or hydrogen ions) and an aluminosilicate anion framework. The formula for zeolites – usually written as the content of one unit cell – is

$$M_{x/m}[(AlO_2)_x(SiO_2)_y] \cdot n H_2O$$

In this formula M is a cation with valence m. The M cations are usually referred to as 'extra framework cations' (in the following they are called EFC-ions). The square brackets contain the formula for the anion framework. This is a three-dimensional network of silicon and aluminium atoms, both tetrahedrally coordinated to oxygen atoms. The anion framework forms cages and channels in which the EFC-ions and absorbed molecules (water) are located.

Zeolites are *absorbents* for gases and liquids, and they are cation exchange materials. The absorption and *ion exchange* takes place through the channels and cages. The walls of the cages and channels consist of rings formed by connection of silicon and aluminium tetrahedra. Such rings contain from 4 to 12 tetrahedral atoms (Si/Al) interconnected by 4 to 12 oxygen atoms. These rings have maximum free openings from 0.15 to 0.80 nm. The EFC-ions and absorbed molecules are however always situated so that they never have to pass windows with apertures smaller than 0.28 nm (six rings).

Excellent compilations of structural properties of zeolites are found in publications by Mortier[1] and Meier & Olson[2].

It has been known for many years that zeolites conduct electricity, and it has been assumed without experimental verification that the conduction was by migration of ions and not by electrons. The investigations of the conductivity of zeolites fall into three periods.

Proton conduction in zeolites

The early period with works by Günther–Schulze (1920)[3], Weigel (1923)[4] and by Rabinowitsch & Wood (1933)[5]. These investigations were made by *d.c. techniques* under application of very high voltages (70–500 V). Whether metal ions or protons (or hydroxyl ions) were the migrating species was a matter of debate in this period.

In a later period the conductivity measurements were made by a.c. techniques. In works by Beattie & Dyer (1957)[6] and Stamires (1962)[7] the conductivity of zeolites as a function of hydration state was studied. In this period it was generally assumed that the migrating ions were the EFC-ions.

In recent years it has been shown by d.c. methods that hydrated zeolites are proton (or hydroxyl ion) conductors.

The d.c. technique used in some of these experiments is described in detail in Chapter 27. Here only the information necessary for understanding results and judging the conclusions is reviewed. The electrodes in the conductivity cells were of platinum or palladium and were used in two different modes – either blocking for all species except electrons or blocking for all species but hydrogen and electrons. In the first mode, the electrodes were flushed with a gas of water-saturated argon or nitrogen; in the latter the electrodes were flushed with a gas of water-saturated hydrogen.

If – in experiments extended in time – there is observed conductivity with Ar/N_2 flushed electrodes, only electrons could be the conducting species. Such conduction was never observed, and therefore it can be concluded that zeolites are not electronic conductors.

If – in extended experiments – there is observed conductivity with hydrogen flushed electrodes, there may be contributions from three sources:

(1) from electronic conductivity (excluded, see above)
(2) from protonic species with electrode reactions
   anode: $H_2 \rightarrow 2H^+ + 2e$
   cathode: $2H^+ + 2e \rightarrow H_2$
(3) from hydroxylic species with electrode reactions
   anode: $H_2 + 2OH^- \rightarrow 2H_2O + 2e$
   cathode: $2H_2O + 2e \rightarrow 2OH^- + H_2$

It has been found in such long term d.c. experiments that hydrated zeolites are conductors for protonic and/or hydroxylic species.

Materials

In the following, a survey of the materials investigated is given and the conductivity measurements are reviewed.

14.2 Materials and materials modification

Most of the zeolites for which conductivity has been determined are listed in Table 14.1. Some zeolites which have only been measured in the dehydrated state have been excluded and so have others which have been reported only briefly in patents. The full list of zeolites which have been investigated as ionic conductors or which have been modified for such investigations are: Analcime[6]; Zeolite A, X and Y[7]; Natrolite and Mordenite[8]; Analcime, Phillipsite, Gismondine, Sodalite, Zeolite omega, Zeolite A, X, and Y[9, 10]; Zeolite A, X, and Y, Chabazite, Clinoptilolite and Mordenite[11–17].

Most of the zeolites for which the ionic conductivity has been reported have been subject to ion exchange. In principle, the EFC-ions of any zeolite can be exchanged by up to 100% with another cation. Such exchange has often been claimed. In practice, however, 100% exchange is extremely difficult and time consuming. For this reason, most zeolites contain more than one kind of EFC-ion.

Many of the zeolites which have been reported to possess conduction have been exchanged in part to give the ammonium form. This has been performed by treatment with ammonium salt solutions at temperatures between 25 °C and 100 °C. After the treatment, the solids are washed until the washing water has a low and constant conductivity.

During the ion exchange and washing procedures, the zeolites may undergo *dealumination*.

For this review, it is particularly relevant to mention the reaction between zeolites and water. With Linde type A zeolite as example, this reaction can be written

$$Na_{12}[Al_{12}Si_{12}O_{48}]aq_{(solid)} + H_2O_{(liquid)} \rightleftarrows$$
$$Na_{12-x}H_x[Al_{12}Si_{12}O_{48}]aq_{(solid)} + xNa^+_{(aq)} + xOH^-_{(aq)}$$

Therefore, there may be appreciable amounts of protons in zeolites in the form of hydroxyl bridges (Townsend (1986)[18, 19]).

Another type of hydrogen containing zeolite is obtained by calcination of ammonium zeolites. Ammonia is removed and the protons become attached to the anion framework. Such materials, usually called H-zeolites, have been studied by Lal *et al.*[8] and by Kreuer *et al.*[10].

212

Because of the possibility of more than one kind of EFC-ion and also the possibility of dealumination during modification of the zeolites, it is necessary to make full chemical analyses of the material in order to be sure of the composition. The majority of the materials dealt with in this section have, however, not been fully analysed.

The *hydration state* of zeolites can vary continuously between the dry state and the maximum hydration state, depending on the relative humidity (RH) of the surroundings. While the few milligrammes of zeolites used in thermogravimetric measurements are in equilibrium with the RH in the laboratory within a few minutes, pressed pellets for conductivity measurements take 1–2 weeks to equilibrate with the RH in new surroundings. Therefore, the hydration state deserves attention. The conductivity of zeolites depends strongly upon the hydration state.

14.3 Protonic conduction in alkali metal zeolites

Hydrated alkali metal zeolites conduct protons (or hydroxyl ions). This has been shown by long term conductivity measurements[13]. A typical data set of such an experiment is shown in Fig. 14.1. The material under investigation was a natural zeolite – chabazite – partially exchanged to the lithium form (composition $Li_{6.1}Na_{1.2}K_{0.15}Ca_{2.3}Al_{12}Si_{24}O_{72}$, 33 H_2O). The conductivity measurements were made at 22 °C. The electrodes were sputtered platinum flushed either with a gas – argon – that made them blocking for all species, or with a gas – hydrogen – that made them blocking for all species but hydrogen. In both cases, the gases were saturated with water vapour, i.e. the experiments were made on samples at 100% RH. The measuring cell was connected to a constant voltage of 0.75 V (provided by a potentiostat). During the first 10 days, the conductivity decreased from 5.5×10^{-7} to $3.3 \times 10^{-8}\ \Omega^{-1}\ cm^{-1}$. It can be concluded therefore that the maximum possible electronic conductivity is $3.3 \times 10^{-8}\ \Omega^{-1}\ cm^{-1}$ (impurity electrode reactions may sustain a low conductivity for a long time). As soon as hydrogen is introduced into the measuring cell the conductivity rises to a value around $10^{-5}\ \Omega^{-1}\ cm^{-1}$ and then later stabilizes at $3.1 \times 10^{-6}\ \Omega^{-1}\ cm^{-1}$. Since the electrodes are blocking for species other than hydrogen, it can be concluded that hydrogen ion or hydroxyl ion conduction takes place.

Similar experiments made with the synthetic zeolites X and A showed the same progress[15]. The conductivity found by such d.c. methods is in agreement with that determined by a.c. methods.

Materials

Table 14.1. *Conductivity of zeolites*

Compound	Reference	Chemical analysis	Phase analysis
Ag, Na-Permutite	3	Ag	not mentioned
Ag, K-Permutite	3	Ag	not mentioned
Natural zeolites	4		single crystals
Chabazite	5	H_2O	single crystals
Chabazite	5	H_2O	single crystals
Analcite	6	H_2O	not mentioned
Zeolite Na-A	7		not mentioned
Zeolite Na-X	7		not mentioned
H-natrolite	8	Al, NH_3, Na, H_2O	yes
NH_4-natrolite	8		yes
H-mordenite	8		yes
NH_4-mordenite	8		yes
Synthetic and natural	9	Na, NH_4, Al, Si	not mentioned
NH_4-zeolites	10	H_2O	
Zeolite NH_4-A	11	Na	yes
Zeolite NH_4-X	11	Na	yes
Zeolites NH_4-A, NH_4-X NH_4-Y	12	NH_4, Na, K, H_2O	yes
Zeolites Li-X and Li-A	13	Li, Na, K, Cl, H_2O	yes
Zeolites NH_4-A, NH_4-X, NH_4-Y NH_4-clinoptilolite	14	NH_4, Na, K, H_2O	yes
Li-chabazite	15	Li, Na, K, H_2O	yes
Tin-mordenites	16	Sn, Na, K, H_2O	yes
Tin-mordenites	17	Sn, Na, K, H_2O	yes

[a] In equilibrium with air of unknown RH. [b] Wet pellets. [c] Fully hydrated. [d] In equilibrium with air of 100% RH. [e] Electrodes flushed with water saturated hydrogen.

Method	Conductivity range ($\Omega^{-1}\,cm^{-1}$)	Temperature (°C)
isoconductometric	4.6×10^{-4}–6.2×10^{-4}	18
isoconductometric	4.1×10^{-4}–8.9×10^{-4}	18
d.c.	5.8×10^{-4a}–6.5×10^{-9a}	15–24
d.c.	2.0×10^{-8} (14% H_2O)	26
d.c.	2.4×10^{-7} (14% H_2O)	100
a.c.	2.6×10^{-6} (4.9% H_2O)–9.3×10^{-5} (0% H_2O)	140
a.c.	10^{-8} (0% H_2O)–5×10^{-4} (19% H_2O)	25
	8×10^{-8} (0% H_2O)–5×10^{-4} (26% H_2O)	25
a.c.	4×10^{-10a}–5×10^{-5b}	20
a.c.	6×10^{-8a}–7×10^{-4b}	20
a.c.	6×10^{-8a}–6×10^{-5b}	20
a.c.	3×10^{-5a}–8×10^{-4b}	20
a.c.	5×10^{-5c}–2×10^{-3c}	20
a.c.	5.3×10^{-6a}–1.8×10^{-4d}	20
d.c.	2.1×10^{-4e}	20
a.c.	3.9×10^{-6a}–1.7×10^{-4d}	20
d.c.	4.1×10^{-4e}	20
a.c.	3.2×10^{-6a}–1.8×10^{-4d}	21
d.c.	2.4×10^{-5e}–3.2×10^{-5e}	21
a.c.	1.6×10^{-6a}–5.4×10^{-6a}	22
d.c.	9×10^{-7e}–2.6×10^{-5e}	
d.c.	6.3×10^{-7e}–3.3×10^{-5e}	20
a.c.	1.1×10^{-6a}	22
d.c.	3.1×10^{-6e}	22
a.c.	10^{-3d}–10^{-2d}	20
a.c.	0.6×10^{-1d}	120
d.c.	5×10^{-4e}–1×10^{-4e}	20

215

14.4 Protonic conduction in ammonium zeolites and in hydrogen zeolites

Bell *et al.*[9] and Kreuer *et al.*[10] measured the conductivity of 10 different ammonium zeolites. Only approximate compositions of the starting materials were reported. Full exchange to the ammonium form was not attempted, but at least half of the sodium in the materials was replaced by ammonium.

The conductivity measurements were made by a.c. methods (silver paint electrodes) and by d.c. methods (hydrogen molybdenum *bronze electrodes*). The measurements were made in the temperature range from −10 °C to 85 °C. In the lower part of the temperature range, the Arrhenius plots show linear behaviour. In the upper part (30–85 °C), the Arrhenius law was not obeyed.

The conductivities were in the range 7×10^{-5} to $1 \times 10^{-3}\,\Omega^{-1}\,cm^{-1}$. The highest values were found for the ammonium forms of zeolite A, X and Y. Methyl ammonium zeolite A was also investigated. It had a conductivity similar to the ammonium zeolites.

Lal *et al.*[8] investigated the conductivity of the ammonium forms of natrolite and a synthetic mordenite. Their results resemble the results already reported above for ammonium zeolites. Conductivities measured on 'wet pellets' were $8 \times 10^{-4}\,\Omega^{-1}\,cm^{-1}$ for mordenite and $6 \times 10^{-5}\,\Omega^{-1}\,cm^{-1}$ for natrolite. For pellets measured in air (RH not specified) the conductivities were 3–4 orders of magnitude lower.

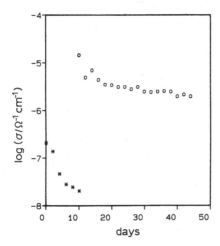

Fig. 14.1. The d.c. conductivity of lithium exchanged chabazite versus time. Platinum electrodes, and different purging gases[15] (with permission). Symbols: ∗, argon; ○, hydrogen.

Kreuer *et al.* assume that the proton transport in zeolites is by a 'vehicle-mechanism'[21]. By such a mechanism protons are transported attached to a vehicle, for example, ammonia or water molecules.

The question about the conduction mechanism was also studied by Krogh Andersen *et al.*[14] They made long term d.c. experiments on four ammonium zeolites. The cell they used is shown in Fig. 14.2. The electrodes were platinum sputtered on to the pellets. In one type of experiment, the measuring cell was flushed with hydrogen gas saturated with water. The outlet from the cell passed through a vessel with an

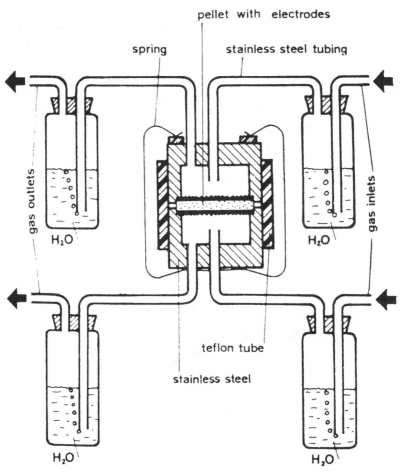

Fig. 14.2. The d.c. conductivity cell used for measurements of ammonium zeolites[15] (with permission).

accurately measured amount of 0.05 M sulphuric acid. During the d.c. experiment, ammonia was liberated and absorbed in the sulphuric acid. The amount of ammonia was determined by back titration. In Fig. 14.3 various amounts of ammonia are shown. Columns marked 'a' show the total amount of ammonia in the pellets at the start of the d.c. experiment, columns 'b' show the amount of ammonia liberated during the d.c. experiment and columns 'c' show the total charge passed through the cells during the d.c. experiment. In experiments where the total charge passed was larger than the total charge on the ammonium ions present in the pellet (zeolite 5A, clinoptilolite and zeolite Y), almost all ammonium was liberated as ammonia. In experiments where the charge passed was less than the charge on the ammonium ions present in the pellet, not all the ammonium was liberated, but only an amount equivalent to the charge passed. The experiments showed that the current is carried to a large extent by ammonium ions. When the ammonium ions reach to the cathode they are reduced to gaseous hydrogen and ammonia. The latter is to a large part liberated and absorbed in the sulphuric acid solution.

The experiments with ammonium zeolite Y showed that conductivity remains after all ammonia is removed. Therefore, conduction other than by ammonium ions must be present. In d.c. experiments, where the pellets were exposed to different dry and wet gases, current versus time curves

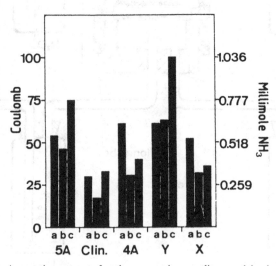

Fig. 14.3. Ammonia amounts for the ammonium zeolites used in d.c. experiments. Total amount of ammonia in the pellet (a), liberated ammonia during the experiment (b), and total charge passed through the cell (c)[14] (with permission).

were recorded. Part of such a curve is shown in Fig. 14.4. The curve is divided into seven parts marked on the figure with Roman numerals. During period I the electrodes were exposed to dry argon. The electrode reactions are blocked, and the current decreases. During period II the electrodes were exposed to dry hydrogen and ammonia. The electrode reaction

$$NH_3 + \tfrac{1}{2}H_2 \rightarrow NH_4^+ + e$$

provides ammonium ions. The current flows again. During period III the electrodes were exposed to dry hydrogen. The ammonium ions are gradually removed, and the current decreases. During period IV the electrodes were exposed to dry argon. The electrode reactions are blocked, and the current decreases. During period V the electrodes were exposed to dry hydrogen. The electrode reactions are reestablished, the current increases, but to a lower level than before (ammonium ions are still being removed from the anode side). During period VI the electrodes were exposed to dry hydrogen and ammonium. The anode reaction supplies ammonium ions again, and the current increases. During period VII the electrodes were exposed to hydrogen saturated with water. The ammonium vehicle is gradually removed, but a water-based conduction mechanism takes over. The conductivity decreases only slightly. The various periods were repeated in different order for six months, a proof of the long term stability of the material.

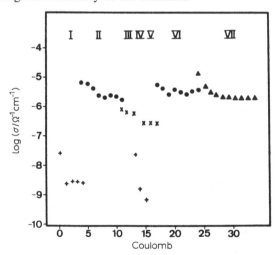

Fig. 14.4. Conductivity of ammonium Y zeolite exposed to various gases. Symbols: +, Ar; ●, hydrogen + ammonia; ×, hydrogen; ▲, hydrogen + water[14] (with permission).

Materials

Lal *et al.*[8] measured the conductivity of deammoniated ammonium natrolite and ammonium mordenite. These so-called H-zeolites had, in the hydrated state, a lower conductivity than the corresponding ammonium zeolites. The same behaviour was observed by Kreuer *et al.*[10] for a hydrated H-zeolite A.

14.5   Protonic conduction in tin zeolites

Small particle tin(IV) oxide hydrate (particle size 2.5 nm) has a room temperature conductivity of $4.1 \times 10^{-4} \, \Omega^{-1} \, cm^{-1}$ (England *et al.* (1980)[20]). Knudsen *et al.* (1988, 1989)[16,17] prepared tin-containing mordenites. By the preparation procedure (heat treatment of mixtures of tin salts with sodium or hydrogen mordenites), mixtures of tin(IV) oxide and tin-exchanged mordenite were created. Although the particle size of the tin(IV) oxide (around 10 nm) in the tin mordenites was larger than in the tin(IV) oxide hydrate prepared by England *et al.*, the former had the highest conductivity ($1 \times 10^{-3} \, \Omega^{-1} \, cm^{-1}$ at room temperature). The paper from 1989[17] presented results from the preparation of ten tin mordenites. The preparations differed in the ratio by weight R between tin(II) chloride dihydrate and hydrogen mordenite in the mixtures which were heat treated to give the *tin mordenites*. R was in the range from 0.1 to 1.3. For low values of R, no separate tin(IV) oxide phase was formed and only tin ion exchange took place. With increasing R the amount of tin(IV) oxide phase increased.

The tin mordenites were analysed for sodium, potassium, soluble tin, and tin(IV) oxide. With certain assumptions (Si/Al ratio not changing during the preparations, all tin in the zeolite associated with two positive charges, but not necessarily in oxidation state (II)) the formula for the $R = 0.2$ preparation was calculated to $H_{2.7}Na_{0.5}K_{0.05}Sn_{2.4}Al_8Si_{40}O_{96}$ $27.5 \, H_2O$. For R values less than 0.2, the amount of separate tin(IV) oxide was negligible.

The conductivities of the tin mordenites have been measured with a.c. and d.c. techniques. The electrodes were platinum black mixed with the electrolyte. The measurements were made at 25 °C and the electrodes were flushed with hydrogen gas saturated with water at 20 °C. The progress of such measurements is shown in Figs 14.5 and 14.6. The relative exchange along the abscissa is defined as 'moles of unit charge passed through the sample per mole of tin exchanged into the sample'. For both samples, the conductivity increased by one order of magnitude during an initial phase

220

Proton conduction in zeolites

of the experiment. During this period, the zeolites came to equilibrium with water in the purging gas. After a relative exchange of 40 in Fig. 14.5 and of 12 in Fig. 14.6, the cells were purged with dry hydrogen for one week. The figures show that the conductivity dropped by one to two

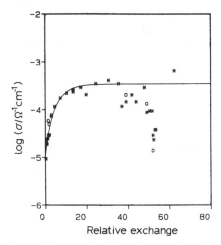

Fig. 14.5. A.c. and d.c. conductivities of tin-exchanged mordenite with $R = 0.2$. Platinum black electrodes. The cell is purged with hydrogen saturated with water at 20 °C. Measuring temperature 25 °C. Symbols: *, d.c. conductivity; ○, a.c. conductivity. At a relative exchange of 40 the cell was purged with dry hydrogen for one week[17] (with permission).

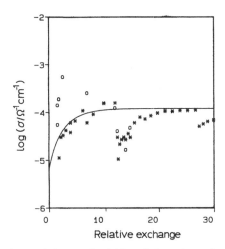

Fig. 14.6. A.c. and d.c. conductivities of tin-exchanged mordenite with $R = 0.4$. Experimental conditions as in Fig. 14.5. Dry hydrogen was applied at a relative exchange of 12[17] (with permission).

221

orders of magnitude, but recovered when humid hydrogen was applied to the cells again.

In order to evaluate the influence of the amount of tin in the samples on the conductivity, four samples with different R were measured at temperatures from 35 °C to 130 °C. The cells were closed and 1 ml of water was placed in them. In this manner, the measurements were made at 100% RH. The results are shown in Fig. 14.7. The conductivity, $0.6 \times 10^{-1} \, \Omega^{-1}$ cm^{-1} at 120 °C for material with $R = 0.7$, is the highest conductivity measured for zeolites.

14.6 Summary

From the conductivity measurements made until now, it is to be expected that all hydrated zeolites conduct protons (or hydroxyl ions). The conduction mechanism is most likely to involve a vehicle, either water or ammonia. In some experiments, small molecule amines and alcohols may have acted as vehicles.

Conductivities in the range from 10^{-14} to $10^{-2} \, \Omega^{-1} \, \text{cm}^{-1}$ have been observed; most zeolites tested for proton conduction have conductivities in the range 10^{-6} to $10^{-4} \, \Omega^{-1} \, \text{cm}^{-1}$.

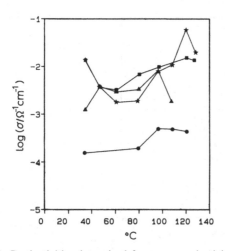

Fig. 14.7. Conductivities determined from a.c. conductivity measurements for a series of tin-mordenites. The measurements were performed at 100% RH. $SnCl_2 \, 2H_2O$/mordenite ratio $R = 0.1$ (●), 0.2 (▲), 0.4 (■) and 0.7 (∗)[17] (with permission).

14.7 References

1. W. J. Mortier, *Compilation of Extra Framework Sites in Zeolites* (Butterworth & Co. (1982)).
2. W. M. Meier and D. H. Olson, *Atlas of Zeolite Structure Types* (Butterworth (1987)).
3. A. Günther-Schulze, *Zeitschr. Elektrochem.* **25** (1919) 330–3.
4. O. Weigel, *Zeitschr. Krist.* **58** (1923) 183–202.
5. E. Rabinowitsch and W. C. Wood, *Zeitschr. Elektrochem.* **39** (1933) 562–6.
6. I. R. Beattie and A. Dyer, *Trans. Faraday Soc.* **53** (1957) 61–6.
7. D. N. Stamires, *J. Chem. Phys.* **36** (1962) 3174–81.
8. M. Lal, C. M. Johnson and A. T. Howe, *Solid State Ionics* **5** (1981) 451–4.
9. M. Bell, K.-D. Kreuer, A. Rabenau and W. Weppner, German Pat. DE 3127821 A 1.
10. K.-D. Kreuer, W. Weppner and A. Rabenau, *Mat. Res. Bull.* **17** (1982) 501–9.
11. E. Krogh Andersen, I. G. Krogh Andersen, K. E. Simonsen and E. Skou in *Solid State Protonic Conductors II*, J. B. Goodenough, J. Jensen and M. Kleitz (eds) (Odense University Press (1983)) 155–60.
12. S. Yde-Andersen, E. Skou, I. G. Krogh Andersen and E. Krogh Andersen in *Solid State Protonic Conductors III*, J. B. Goodenough, J. Jensen and A. Potier (eds) (Odense University Press (1985)) 247–57.
13. E. Krogh Andersen, I. G. Krogh Andersen, E. Skou and S. Yde-Andersen in *Solid State Protonic Conductors III*, J. B. Goodenough, J. Jensen and A. Potier (eds) (Odense University Press (1985)) 100–11.
14. E. Krogh Andersen, I. G. Krogh Andersen, E. Skou and S. Yde-Andersen, *Solid State Ionics* **18–19** (1986) 1170–4.
15. E. Krogh Andersen, I. G. Krogh Andersen, E. Skou and S. Yde-Andersen in *Occurrence, Properties and Utilization of Natural Zeolites*, D. Kalló and H. S. Sherry (eds) (Akadémiai Kiadó, Budapest (1988)) 275–81.
16. N. Knudsen, E. Krogh Andersen, I. G. Krogh Andersen and E. Skou, *Solid State Ionics* **28–30** (1988) 627–31.
17. N. Knudsen, E. Krogh Andersen, I. G. Krogh Andersen and E. Skou, *Solid State Ionics* **35** (1989) 51–5.
18. R. P. Townsend in *New Developments in Zeolite Science and Technology*, Y. Murakami, A. Iijima and J. W. Ward (eds) (Kodansha Ltd (1986)) 273–82.
19. K. R. Franklin, R. P. Townsend and S. J. Whelan in *New Developments in Zeolite Science and Technology*, Y. Murakami, A. Iijima and J. W. Ward (eds) (Kodansha Ltd (1986)) 289–96.
20. W. A. England, M. G. Cross, A. Hamnett, P. J. Wisemann and J. B. Goodenough, *Solid State Ionics* **1** (1980) 231–49.
21. K.-D. Kreuer, A. Rabenau and W. Weppner, *Angew. Chem. Int. Ed. Engl.* **21** (1982) 208–9.

15 Proton containing NASICON phases

ABRAHAM CLEARFIELD

15.1 Triphosphate phases

Tetravalent metals readily form triphosphates of the type $M(I)M(IV)_2(PO_4)_3$, where $M(I)$ may be any alkali metal and $M(IV)$ is most commonly Zr[1]. A large number of such compounds have been synthesized by high temperature solid state methods[2-4]. These compounds are of interest because they are low expansion materials[5-7] and form the basis for an extensive family of solid electrolytes[8]. Single crystals of $NaZr_2(PO_4)_3$ have been grown from melts and the structure determined.[8, 9] A schematic drawing of a portion of the structure is shown in Fig. 15.1. The crystals are rhombohedral $a = 8.8043(2)$ Å, $c = 22.7585(9)$ Å, space group $R\bar{3}c$. The structure consists of zirconium octahedra and phosphate tetrahedra which are linked by corners. A basic unit consists of two octahedra and three tetrahedra separated by Na^+ ions which form columns parallel to the c axis. Thus, the structure consists of $[Zr_2(PO_4)_3]^-$ anions and Na^+ cations. The anions are linked together through bridging phosphate groups to form a rigid framework which creates cavities. The cavity labelled A in Fig. 15.1 contains the Na^+ which is coordinated by phosphate oxygens in a trigonal anti-prism. Only half the structure in the a-direction is shown in the figure. On adding the second half, three more cavities are created. Six such cavities, three above and three below, termed type II, surround the A type (type I) and are connected through narrow passageways. The narrowness of the passageways largely confines the Na^+ to the type I cavities so that $NaZr_2(PO_4)_3$ is a poor conductor $(\sigma_{300} \simeq 2 \times 10^{-5} \, \Omega^{-1} \, cm^{-1})$[10].

The corresponding silicate, $Na_4Zr_2(SiO_4)_3$, has essentially the same structure as the phosphate but with all the cavities filled[11-15]. Single crystals were obtained hydrothermally[13, 14] as well as from melts[11]. Sizova et al.[12] indicated a non-centrosymmetric space group R3c whereas Tranqui refined their data in $R\bar{3}c$. Since all the cavities are filled with Na^+, the silicate is only a moderately good conductor (σ_{300} = $3.5 \times 10^{-4}\,\Omega^{-1}\,cm^{-1})^{15}$.

Hong[8] found that the trisilicate and triphosphate form a complete solid solution series. The phase relationships are readily seen by reference to the diagram of Fig. 15.2. If we focus on the line connecting the triphosphate and trisilicate, the replacement of a negative three charge by a negative four charge requires additional sodium ions for charge balance. Therefore, the type II cavities become increasingly occupied as the silicate content increases. The general formula for the series is $Na_{1+x}Zr_2Si_xP_{3-x}O_{12}$. As x increases, the passageways connecting the cavities become enlarged since the Si–O bond length (1.62 Å) is larger

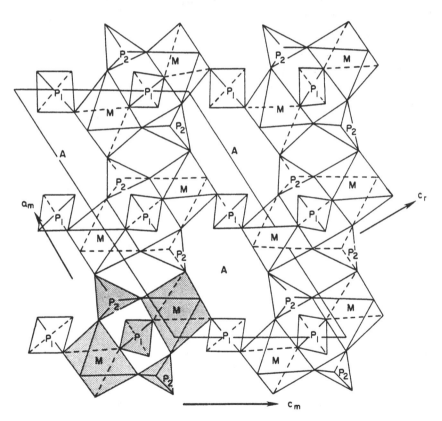

Fig. 15.1. Projection onto the ac plane of the $NaZr_2(PO_4)_3$ structure showing the buildup of $Zr_2(PO_4)_3^-$ species to create type I cavities labelled A. Only $\frac{1}{2}$ the unit cell in the b-direction is shown[8] (with permission).

225

than the comparable P–O value (~ 1.54 Å). As a result, the conductivity increases[8, 13, 16]. In the vicinity of $x = 1.8$ a phase transition occurs from rhombohedral to monoclinic and again reverts to rhombohedral for $x \geqslant 2.3$. The highest conductivities, equal to those of β-alumina, are achieved by these monoclinic phases[8, 16]. However, in most cases a small amount of monoclinic ZrO_2 is also present in the solids with monoclinic structure[16-18]. The trivial name NASICON (*Na* super *i*on *con*ductor) has been assigned to these highly conducting silicophosphates.

In an interesting development it was shown[19] that $ZrNaH(PO_4)_2$, on heating to 500 °C, formed $NaZr_2(PO_4)_3$. Subsequently, attempts to obtain single crystals of the sodium exchanged layered zirconium phosphates by hydrothermal methods yielded triphosphate type phases[20]. Therefore a systematic examination was carried out, by treating α-zirconium phosphate, $Zr(HPO_4)_2 . H_2O$, containing increasing amounts of sodium ion, hydrothermally in Teflon lined steel bombs at 200–300 °C for varying lengths of time. At 200 °C and low loadings (less than half-exchanged) the products were layered phases termed Na-I and Na-II[21] of composition $ZrNa_{0.2}H_{1.8}(PO_4)_2$ and $ZrNa_{0.8}H_{1.2}(PO_4)_2$, respectively. However, if a small amount of HF was added as a solubilizer, $NaZr_2(PO_4)_3$ formed. The triphosphate was also obtained without the use of HF at higher

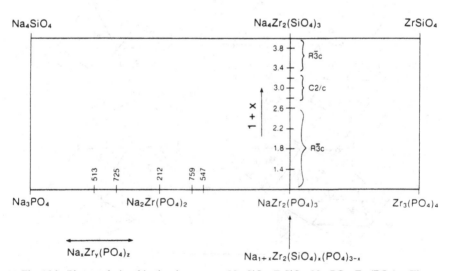

Fig. 15.2. Phase relationship in the system Na_4SiO_4–$ZrSiO_4$–Na_3PO_4–$Zr_3(PO_4)_4$. The solid line represents the solid solution $Na_{1+x}Zr_2Si_xP_{3-x}O_{12}$ (from *Solid State Ionics* **5** (1981) 301, with permission).

temperatures. Whenever acid was present, the triphosphate contained hydrogen ion. In such media, as much as 70% of the Na^+ could be replaced by H^+, yielding $Na_{0.3}H_{0.7}Zr_2(PO_4)_3$. A similar hydrothermal procedure was utilized to prepare $NH_4Zr_2(PO_4)_3$[22]. The reactions were carried out in Teflon lined Parr bombs at temperatures of 250–300 °C. The crystals are rhombohedral with hexagonal axes $a = 8.676(1)$ Å and $c = 24.288(5)$ Å. Ono[23] succeeded in preparing a cubic phase of $(NH_4)Zr_2(PO_4)_2$ using sealed tube reactions either with or without water present. Temperatures of 335 °C or higher and long heating times (72 h) were required. These crystals have the Langbeinite structure with $a = 10.186(3)$ Å.

The ammonium ion is tightly held in $(NH_4)Zr_2(PO_4)_3$. Treatment of this phase with 2M HCl at 300 °C under hydrothermal conditions did not yield the corresponding protonated phase[22]. Therefore, the ammonia was removed thermally[24, 25]. At about 440 °C a mole of NH_3 splits out to yield a triclinic phase $HZr(PO_4)_3$. However, if the ammonium zirconium phosphate is heated rapidly to 600 °C, a rhombohedral phase (hexagonal cell dimensions $a = 8.80(5)$ Å and $c = 23.23(4)$ Å) is obtained. Both protonated phases could be hydrated to $(H_3O)Zr_2(PO_4)_2$ ($a = 8.760(1)$ Å, $c = 23.774(4)$ Å) by refluxing in water for 12 h or treating hydrothermally for shorter periods of time. It should be remarked that the low temperature phase may be monoclinic rather than triclinic. The complex X-ray diffraction pattern obtained for this phase could not be indexed in the monoclinic system but it may be a mixed pattern containing lines of the high temperature phase also. This has been found to be the case for $LiZr_2(PO_4)_3$[26]. $(H_3O)Zr_2(PO_4)_3$ loses its water below 200 °C[24] to yield the low temperature phase of $HZr_2(PO_4)_3$. Further heating results in loss of $\frac{1}{2}$ mole of water above 650 °C to yield ZrP_2O_7 and ZrO_2. However, Ono[23, 25] found that cubic $(NH_4)Zr_2(PO_4)_3$ forms a phase of composition $Zr_4P_6O_{23}$ on loss of ammonia followed by water loss. This phase, when treated hydrothermally at 350 °C in a gold tube for 15 h, yielded rhombohedral $(H_3O)Zr_2(PO_4)_3$.

$HZr_2(PO_4)_3$ has also been prepared by treatment of $LiZr_2(PO_4)_3$ with acid[27]. Once this exchange has been made it was then found possible to exchange Na^+ and Ag^+ for the protons. However, direct exchange of H^+ or H_3O^+ for Na^+ in $NaZr_2(PO_4)_3$ does not take place. Thus, some slight rearrangement of the structure must occur when the process proceeds through the lithium ion phase.

Conductivity measurements have been carried out on both $HZr_2(PO_4)_3$

and $(H_3O)Zr_2(PO_4)_3$ by the impedance method[24]. At all temperatures the plots consisted of a single semicircle whose centre is slightly below the Z' axis. The off-centring of the semicircle plots could be explained on the basis of the dispersion of relaxation times which commonly occurs in polycrystalline samples[17]. A plot of log σ versus $1/T$ is shown in Fig. 15.3. This Arrhenius plot exhibits several distinct regions. Curve (a) represents data for the compound $(H_3O)Zr_2(PO_4)_3$ from which an activation energy of 0.56 eV was obtained. Region (b) follows and shows a decrease in the conductivities, due to the loss of water. Region (c) of Fig. 15.3 shows the Arrhenius plot for $HZr_2(PO_4)_3$ as obtained from the dehydrated pellet. An activation energy of 0.44 eV was calculated from the slope of this curve. When the impedance measurements on $HZr_2(PO_4)_3$ were made below 200 °C, the experimental points fell on the same straight line region (d). However, it is to be noted that at all temperatures between room temperature and 150 °C, $(H_3O)Zr_2(PO_4)_3$ is a better conductor than $HZr_2(PO_4)_3$ by about one order of magnitude.

The dielectric properties of $HZr_2(PO_4)_3$ and its hydronium analogue have been measured[28]. The ε' and ε'' values decrease with temperature increase, which is attributed to water loss. Above 200 °C both the dielectric constant and dielectric loss increase sharply, which is attributed to a phase

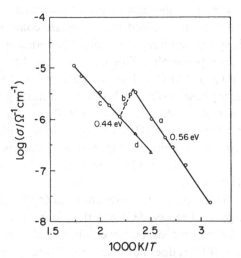

Fig. 15.3. Arrhennius plot of log σ versus $1000/T$ for $(H_3O)Zr_3(PO_4)_3$ (region marked a). The region indicated as b represents the temperature range over which water is lost to form $HZr_2(PO_4)_3$ and region c is a continuation of these conductivity data. Region d represents the measured conductivity of $HZr_2(PO_4)_3$ on cooling the sample[24] (with permission).

transition at 221 °C and the conduction behaviour of the proton. Ohta *et al.*[29] have also measured the protonic conduction of $HZr_2(PO_4)_3$. Cold pressed compacts were used with blocking (Pt) electrodes. Two semicircles were obtained in the impedance plots near room temperature indicative of a grain boundary conductance as well as conductance through the grains. The conductivity at 300 °C was determined to be $1.3 \times 10^{-6} \, \Omega^{-1}$ cm^{-1} which is slightly smaller than the value observed by Subramanian *et al*[24]. However, their activation energy for protonic conduction was exactly the same (0.44 eV). Ohta *et al.* considered the formula of their preparation to be $HZr_2(PO_4)_3 \cdot 0.4H_2O$ with the protons in the type I cavities and water in the type II cavities. At room temperature the conduction mechanism was considered to be of the Grotthuss type mainly through the grain boundary. As the temperature was increased water was lost and grain boundary conduction decreased. After the water loss further conduction was thought to occur by proton hopping through both type I and II sites. It should be remarked that the structure of $HZr_2(PO_4)_3$ is stable to about 640 °C[24,28] at which temperature water begins to split out. However, the framework does not collapse until about 900 °C.

The crystal structures of the ammonium and hydronium triphosphates have been determined by Rietveld refinement of their neutron diffraction patterns obtained at 15 K[30]. In both cases, the ions are confined to the type I cavities but even at 15 K are disordered. Thus, at room temperature they must be even more disordered and exhibit high thermal motion. These ions are larger than the passageways leading to the empty type II cavities and thus must overcome a high activation barrier to become mobile. This explains the low observed conductivity, the thermal stability of the ammonium phase and the inability to exchange H_3O^+ for NH_4^+. Similar conclusions were drawn from an NMR study[31] of the two compounds. In the case of $HZr_2(PO_4)_3$ the low conductivity must be attributed to the inherent slowness of the lattice hopping process. These structural findings are at odds with Ohta's proposal[29] that the water resides in the type II cavities. Since these cavities are isolated from each other, it is hard to visualize a Grotthuss type mechanism. However, the presence of free water in the grain boundaries could permit such a mechanism in that region. Within the grains a mixed hopping and vehicle mechanism is more likely.

A large number of triphosphate phases have now been prepared in single crystal form by hydrothermal methods[32-34]. A partial listing is given in Table 15.1. The experiments were carried out in Morey type

Table 15.1. *Cell parameters for new triorthophosphates*

Compound	System	Cell parameters (Å)			Axial angles (deg)	Volume (Å^3)
		a	b	c		
NaCu$_2$Zr(PO$_4$)$_3$	Tetragonal	5.04	—	3.57	—	90.68
NaNi$_2$Zr(PO$_4$)$_3$	Monoclinic	10.59	8.93	12.19	$\beta = 83.7$	560.80
NaMn$_2$Zr(PO$_4$)$_3$	Orthorhombic	9.29	7.17	6.05	—	402.96
Na$_2$LaZr(PO$_4$)$_3$	Triclinic	11.26	6.08	4.73	$\alpha = 84, \beta = \gamma = 90$	293.77
Na$_2$(La, Al)Zr(PO$_4$)$_3$	Triclinic	10.07	6.43	5.57	$\alpha = 90, \beta = 87, \gamma = 84$	291.11
Na$_2$(La, Al)Ti(PO$_4$)$_3$	Monoclinic	10.93	9.23	12.61	$\beta = 79$	310.65
Na$_2$(La, Cr)Zr(PO$_4$)$_3$	Orthorhombic	11.57	6.96	3.64	—	293.98
Na$_2$(La, Co)Zr(PO$_4$)$_3$	Triclinic	10.00	6.95	6.21	$a = \beta = 90, \gamma = 85$	413.44
Na$_2$(La, Co)Ti(PO$_4$)$_3$	Triclinic	8.50	7.17	6.73	$\alpha = \beta = 90, \gamma = 88$	410.72
Na$_2$(Ce, Co)Zr(PO$_4$)$_3$	Triclinic	10.39	7.01	5.62	$\alpha = \beta = 90, \gamma = 84$	409.93
Na$_2$(Nd, Co)Zr(PO$_4$)$_3$	Triclinic	9.62	6.94	6.08	$\alpha = \beta = 90, \gamma = 85$	405.31
Na$_2$(La, Fe)Zr(PO$_4$)$_3$	Orthorhombic	9.28	7.22	3.97	—	263.98

From reference 32, with permission.

autoclaves fitted with teflon liners. Temperatures were in the range of 200–300 °C and pressures ranged from 100–250 atm. Usually 10 days at the soaking temperature was required to obtain crystals up to 3 mm in length. Unusually high conductivities have been measured for some of these phases[34] as compared to triphosphates prepared by solid state high temperature reactions. It would be interesting if the ammonium and protonic analogues of these crystals, as well as of the $A_3M(III)_2(PO_4)_3$ type, could be prepared for single crystal conductivity studies.

15.2 Silicophosphate phases

Very little work has been carried out on proton phases of NASICON. Rather most of the research effort has been concentrated on sodium ion conductivity studies, on establishing the stoichiometry of NASICON and structural studies.

In Hong's paper[8] on NASICON he proposed a model structure for the monoclinic phase which was later shown to be essentially correct[35, 36]. In this model there are three cation sites. The type I sites remain essentially the same with a trigonal prism coordination of Na(1). However, the type II sites are split into one which is like the rhombohedral Na(2) site, in which the sodium is eight-coordinate, and another which is in a general position and can contain a maximum of two sodium ions per unit cell. This site has been designated the Na(3) or the mid-sodium site.

We have already indicated that the two end members of the NASICON solid solution have high resistivities[10, 13, 37, 38]. With increasing silicon content the conductivity increases reaching a maximum at about $x = 2$ and thereafter decreasing as the cavities fill up. It was therefore reasonable to assume that replacement of all the Na^+ in $Na_3Zr_2Si_2PO_{12}$ by protons would lead to a much better proton conductor than is $HZr_2(PO_4)_3$ with perhaps little loss in framework stability. Only a few attempts in this direction have been made.

Treatment of $Na_3Zr_2Si_2PO_{12}$ with acid does not remove all of the sodium. Bell[39] obtained higher replacements of Na^+ by electrolysing out the sodium in acid solution for a series of preparations containing 0.3–1.5 Si. However, the conductivities of these samples at 300 °C were relatively constant at ~ 5–$11 \times 10^{-4} \, \Omega^{-1} \, cm^{-1}$. The samples exhibited Arrhenius type behaviour with activation energies of 0.44–0.50 eV. Heating the samples above 450 °C resulted in a reduction in the conductivity attributed to the dehydration of the samples.

Materials

Our own work[40] involved measurement of potentials developed by exposure of NASICON to H_2 at different pressures in a gaseous concentration cell. The NASICON was prepared both by the sol–gel method[41] and the hydrothermal route[42]. Pressed sintered disks were then prepared which were approximately 1.27 cm ($\frac{1}{2}''$) in diameter and 1 mm thick. The disk separated two chambers, one of which contained N_2 at 1 atm and the other H_2. The potential of this cell was measured on open circuit as a function of hydrogen pressure and the results at 25 °C are plotted in Fig. 15.4. The slope of the line is close to that expected for Nernst type behaviour. At elevated temperatures, the voltage increased considerably but large deviations from Nernstian behaviour were observed presumably due to conduction of the sodium ions. These results are comparable to those obtained with β-alumina[43,44]. By applying a potential across the

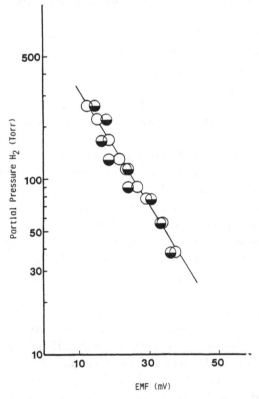

Fig. 15.4. Plot of log P_{H_2} versus EMF of an electrochemical cell with a NASICON solid electrolyte and Pt electrodes. The slope of the line indicates near Nernstian behaviour ($T = 25 \pm 2$ °C).

electrodes, hydrogen was transported across the electrolyte disk (Fig. 15.5)[40]. In these experiments a flow of H_2 at a rate of 25 ml min^{-1} was maintained in the anode chamber and 75 ml min^{-1} of N_2 in the cathode chamber. At 200 °C the rate of transport of H_2 across the NASICON membrane was ~ 0.95 ml h^{-1} cm^{-2} at a potential of 2 V. This compares to a rate of 3.1 ml h^{-1} cm^{-2} when AMP was used in a similar cell at 56 mV and 25 °C[45]. At high temperatures crystals formed on the disks at the cathode side. These may be sodium hydride.

One of the unpublished ways in which NASICON[40,41] was prepared was to produce a gel from $ZrOCl_2 . 8H_2O$ plus Na_2SiO_3 and H_3PO_4 in the correct ratios to yield $Na_3Zr_2Si_2PO_{12}$. However, if the gel was prepared at near neutral or acid pH and washed, the Na^+ was largely removed from the gel as soluble NaCl. This gel may be quite similar to ion exchange materials labelled zirconium silicophosphates[46] since they do possess ion exchange properties. However, if sodium is not washed out, heating the dried gel results in copious fumes of HCl being evolved starting at about 200 °C and continuing up to about 600 °C depending upon the heating rate. Above 900 °C crystallization sets in with the formation of NASICON. The original gel contained NaCl as shown by X-ray diffraction, but after heating to form NASICON, the salt was no longer present. These results are interpreted on the basis that the zirconium

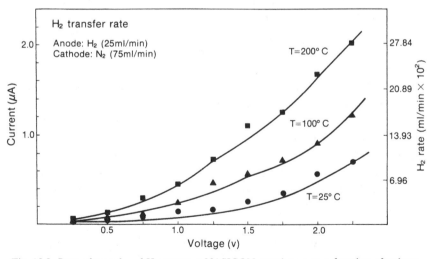

Fig. 15.5. Rate of transfer of H_2 across a NASICON membrane as a function of voltage and temperature.

silicophosphate gel has a disordered NASICON framework with H_3O^+ in the cavities. These ions are responsible for the ion exchange properties of the gel. On heating the mixture, exchange of Na^+ for H^+ takes place with evolution of HCl. The gel then crystallizes as NASICON. That ion exchange can take place between two solids without the presence of water has been shown previously for zirconium phosphate[47]. Crystallization of the washed gels in the presence of H_3O^+, perhaps by hydrothermal methods, may prove to be a route by which protonic NASICONS can be made.

An interesting paper[48] describes the preparation of

$$H_{0.5}Cu(I)_{0.5}Zr_2(PO_4)_3$$

by hydrogen reduction of $Cu(II)_{0.5}Zr_2(PO_4)_3$. No conductivity data were presented in this paper but rather certain optical properties were examined. However, this technique of metal reduction should be examined with silver ion NASICON to prepare more fully protonated phases.

15.3 Structural considerations

In order to prepare successfully NASICON based proton conductors, it is necessary for us to better understand the nature of the sodium ion phase. The structural variations which occur as the composition of the NASICON increases in silicon content were examined by means of neutron diffraction studies on powders[16,49]. It was found that the Na-I site was always totally occupied and the type II sites were progressively filled as the sodium ion level increased. The filling of the type II sites caused a rotation of the Si/P tetrahedra up to $x = 2$ followed by a return towards the original orientation from $x = 2$ to 3. This rotation resulted in an initial increase in the c axis length up to $x = 2$ with an accompanying increase in the oxygen distances blocking the conduction pathways. Hong[8] had earlier proposed that at low silicon contents, the bottlenecks were too narrow to allow diffusion between the two cavity types, but as the c-axis increased less of a barrier to ion transport resulted. Thus the neutron diffraction study appeared to substantiate Hong's hypothesis.

The situation is actually more complicated. Further neutron diffraction studies[36] on NASICONS with $x = 1.85–1.9$ revealed that the Na-I site is about $\frac{1}{4}$ occupied, the Na-II site is filled and the Na-III site, which lies along the conduction pathway displaced towards Na-I, contains the remainder ($\sim 1.7–1.8$) of the sodium ions. The distance between Na(1) and Na(3) in these compounds is 1.7–1.8 Å. This is too close a distance of approach for the sodium ions. It was therefore proposed that when one of these sites was occupied the adjacent other type of site was empty.[36] However, when the Na^+ content increases so that some of the adjacent Na-I and Na-III sites must both be occupied, the Na^+ in the Na-III site recedes to a position ~ 2.8 Å from Na-I.

The essential correctness of the situation was confirmed by single crystal studies[50,51]. The crystals were grown over a four-month period at 1200 °C and therefore reached an equilibrium composition very close to $Na_3Zr_2Si_2PO_{12}$. The maximum in conductivity is thus correlated with maximum occupancy of the mid-sodium site. The conduction process is thought to be accompanied by a breathing-like motion of the framework in agreement with the dynamic disorder observed in the crystals[51]. Space does not permit a detailed discussion of this dynamic disorder or the controversy concerning the stoichiometry of NASICON but the reader is referred to references 41 and 51 for more information on these subjects.

15.4   Conclusions

It is evident from our discussion that very little work has been carried out on the proton phases of NASICON. However, with the background given here it should be apparent that such phases are extremely interesting as solid electrolytes and deserve further study.

15.5   References

1. N. G. Chernorukov, I. A. Korshunov and T. V. Profokeva, *Sov. Phys. Crystallogr.* **23** (1978) 475.
2. R. J. Cava, E. M. Vogel and D. W. Johnson, Jr, *J. Am. Ceram. Soc.* **65** (1982) C157.
3. B. E. Taylor, A. D. English and T. Berzins, *Mat. Res. Bull.* **12** (1977) 171.
4. B. Makovic, B. Prodic and M. Sljukic, *Bull. Soc. Chem. (Fr)* (1968) 1777.
5. J. Alamo and R. Roy, *J. Am. Ceram. Soc.* **67** (1984) C78.
6. G. E. Lenain, H. A. McKinstry, S. Y. Limaye and A. Woodward, *Mat. Res. Bull.* **19** (1984) 1451.
7. T. Oota and I. Yamai, *J. Am. Ceram. Soc.* **69** (1986) 1.
8. H. Y.-P. Hong, *Mat. Res. Bull.* **11** (1976) 173.
9. L. Hagman and P. Kierkegaard, *Acta Chem. Scand.* **22** (1968) 1822.
10. C. Delmas, J.-C. Viala, R. Olazcuaga, G. LeFlem, P. Hagenmueller, F. Cheraoui and R. Brochu, *Solid State Ionics* **3/4** (1981) 209.
11. R. G. Sizova, A. A. Voronkov, N. G. Shumatskaya, V. V. Ilyukhin and N. V. Belov, *Sov. Phys. Dokl.* **17** (1973) 618.

Materials

12. R. G. Sizova, V. A. Blinov, A. A. Voronkov, V. V. Ilyukhin and N. V. Belov, *Sov. Phys. Crystallogr.* **26** (1981) 165.
13. D. Tranqui, J. J. Capponi, M. Gondrand, M. Saib, J. C. Joubert and R. D. Shannon, *Solid State Ionics* **3/4** (1981) 219.
14. F. Genet and M. Barj, *Solid State Ionics* **9/10** (1983) 891.
15. D. Tranqui, J. J. Caponi, J. C. Joubert and R. D. Shannon, *J. Solid State Chem.* **39** (1981) 219.
16. L. J. Schioler, D.Sc. Dissertation, Mass. Inst. Tech., Feb. 1983.
17. M. L. Bayard and G. G. Barna, *J. Electroanal. Chem.* **19** (1978) 201.
18. D. H. H. Quon, T. A. Wheat and W. Nesbitt, *Mat. Res. Bull.* **15** (1980) 1533.
19. A. Clearfield, W. L. Duax, A. S. Medina, G. D. Smith and J. R. Thomas, *J. Phys. Chem.* **73** (1969) 3423.
20. A. Clearfield, P. Jirustithipong, R. N. Cotman and S. P. Pack, *Mat. Res. Bull.* **15** (1980) 1603.
21. A. Clearfield and S. P. Pack, *J. Inorg. Nucl. Chem.* **37** (1975) 1283.
22. A. Clearfield, B. D. Roberts and M. A. Subramanian, *Mat. Res. Bull.* **19** (1984) 219.
23. A. Ono, *J. Mat. Sci. Lett.* (1985) 936.
24. M. A. Subramanian, B. D. Roberts and A. Clearfield, *Mat. Res. Bull.* **19** (1984) 1471.
25. A. Ono, *J. Mater. Sci.* **19** (1984) 2691.
26. F. Sudreau, D. Petit and J. P. Boilot, private communication.
27. G. Alberti in *Inorganic Ion Exchange Materials*, A. Clearfield (ed) (CRC Press, Boca Raton, Fl. (1982)), 98.
28. M. Ohta, F. P. Okamura, K. Hirota and A. Ono, *J. Mater. Sci. Lett.* **5** (1986) 511.
29. M. Ohta, A. Ono and F. P. Okamura, *J. Mater. Sci. Lett.* **6** (1987) 583.
30. P. R. Rudolf, M. A. Subramanian, A. Clearfield and J. D. Jorgensen, *Solid State Ionics* **17** (1985) 337.
31. N. J. Clayden, *Solid State Ionics* **24** (1987) 117.
32. K. Byrappa, G. S. Gopalakrishna, V. Venkatachalapathy and A. B. Kulkarni, *J. Less-Common Met.* **110** (1985) 441.
33. K. Byrappa, A. B. Kulkarni and G. S. Gopalakrishna, *J. Mat. Sci. Lett.* **5** (1986) 540.
34. K. Byrappa, A. B. Kulkarni and G. S. Gopalakrishna, *J. Cryst. Growth* **79** (1986) 210.
35. M. A. Subramanian, P. R. Rudolf and A. Clearfield, *J. Solid State Chem.* **60** (1985) 337.
36. P. Rudolf, A. Clearfield and J. D. Jorgensen, *Solid State Ionics* **21** (1986) 213.
37. J. B. Goodenough, H. Y.-P. Hong and J. A. Kafalas, *Mat. Res. Bull.* **11** (1976) 203.
38. J. P. Boilot, J.-P. Salanié, G. Desplanches and D. LePotier, *Mat. Res. Bull.* **14** (1979) 1469.
39. M. F. Bell and M. Sayer, *The Development of Ceramic Proton Conductors*, Report No. OSU 81-00440, Queen's University, Kingston, Ontario, Canada, 1983.

40. H. Nguyan and A. Clearfield, unpublished results.
41. A. Moini and A. Clearfield, *Adv. Ceram. Mat.* **2** (1987) 173.
42. A. Clearfield, M. A. Subramanian, W. Wang and P. Jerus, *Solid State Ionics* **9/10** (1983) 895.
43. J. S. Lundsgaard and R. J. Brook, *J. Mater. Sci. Lett.* **9** (1974) 2061.
44. J. Jensen and P. McGeehin, *Silicates Ind.* **9** (1979) 191.
45. N. Muira, Y. Ozawa, N. Yamazol and T. Seiyama, *Chem. Soc. Jpn, Chem. Lett.* (1980) 1275.
46. V. Vesely and V. Pekarek, *Talanta* **19** (1972) 219.
47. A. Clearfield and J. M. Troup, *J. Phys. Chem.* **74** (1970) 2578.
48. G. Le Polles, A. El Jazouli, R. Olazcuaga, J. M. Dance, G. LeFlem and P. Hagenmuller, *Mat. Res. Bull.* **22** (1987) 1171.
49. B. J. Wuensch, L. J. Scioler and E. Prince in *Proc. Conf. High Temp. Solid Oxide Electrolytes* (1983) F. J. Salzano (ed.) Brookhaven, N.Y.
50. J. P. Boilot, G. Collin and Ph. Colomban, *Mat. Res. Bull.* **22** (1987) 669.
51. J. P. Boilot, Ph. Colomban and G. Collin, *Solid State Ionics* **28/30** (1988) 403.

16 Phosphates and phosphonates of tetravalent metals as protonic conductors

GIULIO ALBERTI AND MARIO CASCIOLA

16.1 Introduction

For many years it has been known that amorphous precipitates having a composition of $M(IV)(HPO_4)_{2-x}(OH)_{2x} \cdot nH_2O$ (x usually ranging from 0 to 0.4) are easily formed when a salt solution of a tetravalent metal ($M(IV) = Zr, Ti, Ce, Th, Sn$ etc.) is mixed with a phosphoric or arsenic acid solution[1]. Due to their peculiar properties (high ion exchange capacity, thermal stability, insolubility in concentrated acids, resistance to radiation, selectivity for some cations of interest in nuclear energy etc.) these amorphous precipitates were intensely investigated between 1954–1964 especially in nuclear centres[1]. During this period Hamlen[2] and Alberti[3] also discovered that amorphous *zirconium phosphate* exhibits a good protonic transport at room temperature ($0.1–6 \times 10^{-3} \Omega^{-1} cm^{-1}$).

After 1964, the discovery that layered crystalline compounds (general formula $M(IV)(HO—PO_3)_2 \cdot H_2O$) can be obtained by refluxing amorphous materials[4] or prepared directly by slow precipitation in the presence of HF[5] gave rise to an increased interest in these inorganic ion-exchangers.

It was found that a tetravalent metal can be bonded to the oxygens belonging to XO_4 and/or $R—XO_3$ groups ($X = P, As; R = $ inorganic or organic radical such as $—H, —OH, —CH_3OH, —C_6H_5$ etc.) forming two different layered structures usually called α and γ. Some fibrous structures (e.g. cerium and thorium phosphates) or tridimensional structures (e.g. $HZr_2(PO_4)_3$ and $Zr(R(PO_3)_2)$) can also be obtained. As a result of intense research, the chemistry of these compounds is now well understood (synthesis, structure, thermal behaviour, phase transitions during ion-exchange and/or intercalation processes etc.).

Phosphates and phosphonates

In recent years, these compounds have begun to find practical applications not only as ion-exchangers but also as intercalating agents, molecular sieves, catalysts and ionic conductors. The reader interested in synthetic and structural details and/or general aspects and applications of these materials is referred to books and recent reviews[1, 6–9]. In this chapter, our attention is focused on proton transport in α-zirconium phosphate and its derivative phases which are presently the most investigated materials, while only a brief account of proton transport in other materials of this class is given. Further details on the conductivity of other acid salts of tetravalent metals can be found in the literature[10]. The relation between thermal behaviour, *IR* spectroscopic properties and a possible conduction mechanism in α-zirconium phosphate has recently been discussed by Colomban[11, 12] (see pp. 280–1).

16.2 Layered α-zirconium phosphate and its modified and intercalated phases

16.2.1 Structural data

If a tetravalent metal is octahedrally coordinated with six oxygens belonging to six different R–PO_3 groups, an α-type planar macromolecule is formed and layered crystals having a general formula $M(IV)(RPO_3)_2 . nS$ (S = solvent and/or other molecules intercalated in the interlayer region) are obtained.

The first and most extensively investigated member is the compound with $M(IV) = Zr$; $R = OH$; $n = 1$; $S = H_2O$, that is α-$Zr(HOPO_3)_2 . H_2O$ (hereafter indicated as ZrP). Its structure has been elucidated by Clearfield and collaborators[13]. Two adjacent layers are schematically drawn in Fig. 16.1a, while the arrangement of the —OH groups (and hence of any R group) on one side of the α-layer is shown in Fig. 16.1b. Note that these groups can be regarded as arranged at alternate vertices of a hexagonal array and that the distance between the centres of adjacent groups is 0.53 nm. The layers are stacked so that each P–OH is directly above and below the Zr of adjacent layers; the projection of —OH groups, on an intermediate plane, of two adjacent layers is shown in Fig. 16.1c. The distance between the —OH groups of adjacent layers depends upon the interlayer distance (see Fig. 16.2). In ZrP ($d = 0.756$ nm) the layer packing creates small zeolitic-type cavities (one for each zirconium atom) in the interlayer region where one water molecule for each cavity can be accommodated.

It must be pointed out that the number of surface —POH groups in microcrystals with a high degree of crystallinity is usually a small fraction (0.01–1%) of the total number of —POH groups. However, since their hydration may be different from that of the groups present in the interlayer regions, the contribution of hydrated surface groups to the total proton

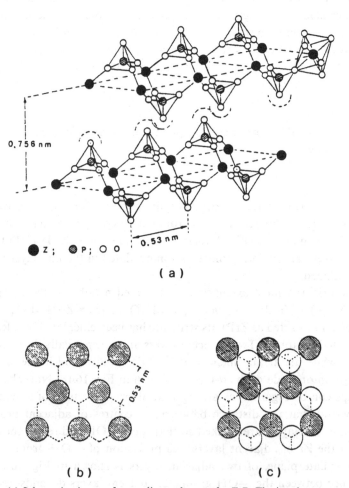

Fig. 16.1. (a) Schematic picture of two adjacent layers of α-ZrP. The interlayer water and protons are not shown. (b) Arrangement of OH groups on one side of an α-layer in ZrP. (c) Projection on an intermediate plane of the OH groups of adjacent layers.

Phosphates and phosphonates

transport may be important. It is therefore useful to examine proton transport in hydrated and anhydrous materials separately.

16.2.2 Proton transport in hydrated materials

16.2.2.1 α-Zirconium phosphate

The electrical transport properties of crystalline ZrP were studied in different laboratories using isoconductance[14], *a.c.-impedance*[15–18] and *d.c.-conductance*[19] techniques. All these investigations indicated that ZrP possesses a protonic conductivity largely dominated by *surface* transport, despite the small fraction of surface protons.

Specifically, isoconductance measurements[14] showed that surface proton mobility was higher than that of the bulk protons by a factor of at least 10^4, while the conductivity of microcrystalline ZrP pellets, determined by a.c.-impedance, was dependent on the size of the crystals[15,18] and the relative density of the pellets[16]. A systematic investigation was also carried out on pellets at *relative humidities* (RH) between 5 and 90%[17]. With increased surface hydration the conductivity turned out to be enhanced by about two orders of magnitude (Fig. 16.3a) and the activation energy, determined between 20 and $-20\,°C$, decreased from $0.52\,eV$ to $0.26\,eV$ (Fig. 16.3b). Based on the reported data[14–19], it is concluded that the conductivity of crystalline ZrP at $20\,°C$ can vary between 10^{-4} and $10^{-8}\,\Omega^{-1}\,cm^{-1}$ depending on hydration and surface area. However, at $20\,°C$ and 90% RH, the conductivity of highly

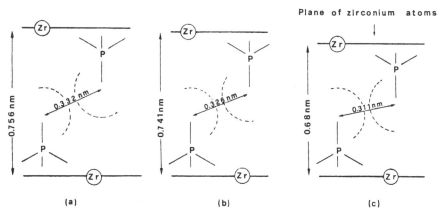

Fig. 16.2. Distance between the oxygens of OH groups belonging to adjacent layers for: monohydrated ZrP (a), anhydrous ZrP below 220 °C (b), anhydrous ZrP above 220 °C (c).

241

crystalline ZrP usually lies in the range 10^{-5}–$10^{-6}\ \Omega^{-1}\ \mathrm{cm}^{-1}$. Due to the almost linear dependence of conductivity on relative humidity (Fig. 16.3a), ZrP has been suggested for use as a humidity sensor[18].

Recent investigations were carried out on polyhydrated monolithium and monosodium salt forms of ZrP ($\mathrm{ZrHLi(PO_4)_2 . 3 . 5H_2O}$ and $\mathrm{ZrHNa(PO_4)_2 . 5H_2O}$) with combined a.c./d.c. measurements as a function of temperature and relative humidity[20]. In this case too, a mixed ionic–protonic conductivity was found arising from surface transport.

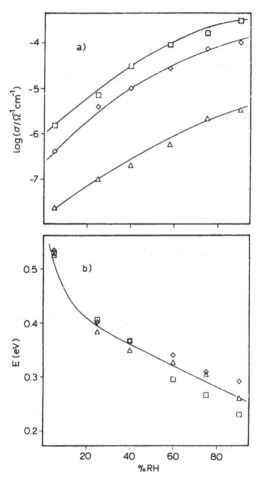

Fig. 16.3. Conductivity (σ) at 20 °C and activation energy (E) as a function of relative humidity for 'single layer' $\mathrm{ZrP_{(sl)}}$ ($\square$), pellicular $\mathrm{ZrP_{(p)}}$ ($\diamond$) and ZrP ($\triangle$). See text for explanation of terms.

Phosphates and phosphonates

The mechanism of protonic transport in ZrP is not yet known. Nevertheless, the fact that the conductivity is dominated by surface transport may be explained considering that, due to steric effects, the diffusion and/or reorientation of protonic species on the surface should be easier than in the bulk; in addition the ionogenic groups of the surface can be more hydrated than the inner ones, thus facilitating their dissociation and water protonation.

The structure of the α-layer accounts for the strong dependence of conductivity and activation energy upon surface hydration. The distance between neighbouring phosphate groups (0.53 nm) is too large to permit proton diffusion in the absence of water molecules forming bridges between the POH groups. Consequently, the number and arrangement of these molecules determine the conduction characteristics. To prove this hypothesis, an investigation was carried out on α-layered zirconium pyrophosphate[21] where protons were present only on the surface of the crystals. The conductivity of α-ZrP_2O_7 pellets, at 20 °C and 90% RH, was similar to that of the parent hydrogen form under the same experimental conditions. On the other hand, when the pellets of pyrophosphate were heated to make the surface of the microcrystals anhydrous, their conductivity decreased by more than three orders of magnitude, thus confirming that protons can hardly diffuse along the anhydrous surface of an α-layer.

16.2.2.2 Modified α-zirconium phosphate

The *intercalation* of a suitable amount of propylamine in ZrP (from 40 to 60% of the ion exchange capacity) leads to the formation of *colloidal* dispersions containing delaminated zirconium phosphate[22]. Each microcrystal is separated into single layers or in packets of a few layers without an appreciable alteration of the covalent structure and geometric size of the layers. The hydrogen form of ZrP can be regenerated by treating the intercalation compound with acids. Depending on the procedure followed, materials with different characteristics can be obtained.

Simple regeneration with 1M HCl leads to the formation of an aqueous suspension of ZrP microcrystals consisting of few layers with respect to the parent material. Pellicles or membranes made up entirely of ZrP can be obtained by filtering the suspension: this particular type of ZrP has been called '*pellicular*' *zirconium phosphate* (hereafter $ZrP_{(p)}$).

A slightly different regeneration procedure[23] consists of putting the colloidal dispersion under sonication before treatment with HCl, in order to break up the aggregates of lamellae. The suspension of delaminated hydrogen form is then subjected to sudden freezing in liquid nitrogen followed by liophilization to try to maintain, in the solid state, the situation present in the solution. Hereafter this material is called $ZrP_{(sl)}$. Compared to crystalline ZrP, it is characterized by a larger interlayer distance (0.105 nm) and surface area (15–30 $m^2 g^{-1}$) as well as a greater water content (3.5 molecules per unit formula).

The conductivity of $ZrP_{(p)}$[24] and $ZrP_{(sl)}$[23] was measured at relative humidities between 5 and 90% in the temperature range 20 to $-20\,^{\circ}C$ by applying the electric field both parallel and perpendicular to the flat surfaces of the pellets. Since the particles of both materials have one dimension (the thickness) which is much smaller than the other two dimensions, the pellets of $ZrP_{(p)}$ and $ZrP_{(sl)}$ are made up of microcrystals oriented prevalently with the layers parallel to the pellet surfaces. The a.c.-conductivity of both materials, measured with the parallel field, is six to eight times higher than that obtained in the perpendicular direction, thus indicating that protonic transport occurs, for the most part, parallel to the layers.

At 20 °C and 90% RH the parallel conductivity of $ZrP_{(sl)}$ ($\sim 10^{-3} \Omega^{-1}$ cm^{-1}) is higher than that of $ZrP_{(p)}$ and crystalline ZrP by one and three orders of magnitude, respectively. However, the dependence of both conductivity (Fig. 16.3a) and activation energy (Fig. 16.3b) on relative humidity is essentially the same for all three materials, so that $ZrP_{(p)}$ and $ZrP_{(sl)}$ effectively behave as crystalline ZrP but with a surface area much larger than usual.

16.2.2.3 Intercalation compounds

As already mentioned, in crystalline ZrP the number of surface ionogenic groups is about 0.01–1% of the total ion exchange capacity. Thus any modification in the structure and/or composition leading to an environment of the bulk fixed charges similar to that of the surface should increase the conductivity up to a limiting value of $\sim 10^{-2} \Omega^{-1} cm^{-1}$.

Polyhydrated compounds with a large interlayer spacing are expected to be endowed with these characteristics, since each layer could act as a hydrated surface. Highly hydrated materials, even at relative humidities less than 100%, can be obtained by intercalation of amines in ZrP. In the case of propylamine, compounds containing up to five water molecules per unit formula and having an interlayer distance up to 0.17 nm were obtained with an amine loading between 40 and 60%: $ZrP . x PrN . nH_2O$, $0.8 \leqslant x \leqslant 1.2$ ($PrN = C_3H_7NH_2$). In these compounds the amines are intercalated as alkylammonium ions and evenly distributed among all the interlayer regions[25].

Membranes made up entirely of highly oriented particles were obtained by filtering the colloidal dispersion which resulted from propylamine intercalation. Pellets were prepared by pressing a suitable number of membranes and their a.c.-conductivity was measured by applying the electric field both parallel and perpendicular to the membranes at temperatures between 20 and $-50\,^{\circ}\mathrm{C}$[25].

At $20\,^{\circ}\mathrm{C}$ and with increased amine loading, the parallel conductivity of the pentahydrated compounds ranged from 1.2×10^{-3} to $6 \times 10^{-4}\,\Omega^{-1}$ cm^{-1}; it decreased by about one order of magnitude when three water molecules were lost and was about $10^{-8}\,\Omega^{-1}$ cm^{-1} for the monohydrated compounds (Fig. 16.4). These results clearly indicate that the conductivity of the intercalation compounds cannot be accounted for by proton jumping between amines and phosphate groups, the intercalated/adsorbed water being the main vehicle for proton diffusion. Parametrization of the conductivity data, based on the Arrhenius equation, showed that the changes in both hydration and amine loading influenced the pre-exponential factor and the activation energy which goes from $0.74\,\mathrm{eV}$, for $n = 2.3$ and $x = 1.1$, to $0.4\,\mathrm{eV}$, for $n = 4.6$ and $x = 0.8$.

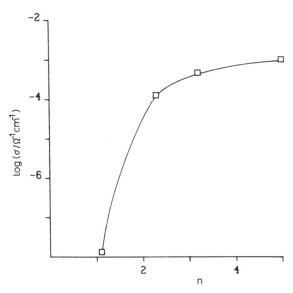

Fig. 16.4. Conductivity (σ) at $20\,^{\circ}\mathrm{C}$ of ZrP . PrN . $n\mathrm{H_2O}$ as a function of the number (n) of water molecules per unit formula.

Materials

The conductivity and activation energy of the intercalation compounds are rather different from those obtained under the same experimental conditions for microcrystalline ZrP. In both cases, proton transport requires intercalated/adsorbed water; the differences in activation energy could therefore arise to a great extent from a more or less strong coordination of water around charged and polar groups. It can be observed that the high conductivity of the intercalation compounds is due exclusively to an unusually high *pre-exponential factor*, which, according to the *hopping model*[26], is dependent upon both carrier concentration and activation entropy. Conductivity measurements do not allow these factors to be separated; however, considering that the pre-exponential factor of ZrP 0.8PrN.5H$_2$O at 90% RH ($\sim 10^6 \, \Omega^{-1} \, cm^{-1}$ K) is five orders of magnitude higher than that of microcrystalline ZrP under the same conditions, it may be reasonably admitted that the effective carrier concentration in the intercalation compounds with propylamine is higher than in the hydrogen form, thus indicating bulk transport.

16.2.3 Anhydrous materials

16.2.3.1 Crystalline and modified zirconium phosphate

Since protons cannot diffuse on the anhydrous surface of an α-layer, the conductivity of anhydrous ZrP is expected to come from protons jumping between oxygens belonging to the side faces of adjacent layers (see Fig. 16.2). The conductivity values of anhydrous ZrP[27], ZrP$_{(p)}$[21] and ZrP$_{(sl)}$[23], determined by applying an electric field perpendicular to the flat surfaces of the pellets, are shown in Fig. 16.5; in the case of ZrP$_{(p)}$, the parallel conductivity is also reported.

In the Arrhenius plot of ZrP and ZrP$_{(p)}$ a slope variation is observed around 220 °C; it is associated with a phase transition[28] which causes a discontinuous change of the interlayer distance from 0.74 nm (below 220 °C) to 0.68 nm (above 220 °C). The decrease of the interlayer distance also causes a decrease of the distance between the centres of neighbouring oxygens of adjacent layers and hence reduces the length of the proton jump (see Fig. 16.2). In agreement with these considerations, the activation energy of the 0.68 nm phase (0.35 eV for ZrP and 0.51–0.45 eV for ZrP$_{(p)}$) is lower than that of the 0.74 nm phase (0.65 eV for ZrP and 0.81–0.75 eV for ZrP$_{(p)}$). Similar values for ZrP were also found by Jerus *et al.*[29] No change of slope is seen in the Arrhenius plot of ZrP$_{(sl)}$ (*activation energy*

0.59 eV), in agreement with DTA and high temperature XRD data which do not show any phase transition in the range 120–300 °C.

The effect of the partial orientation relative to the electric field is much more evident here than in the hydrated materials: the 'parallel' conductivity of $ZrP_{(p)}$ ($10^{-4}\,\Omega^{-1}\,cm^{-1}$ at 300 °C) is more than two orders of magnitude higher than the 'perpendicular' conductivity. This has been attributed to the fact that, in hydrated materials, the conduction occurs essentially through the interparticle liquid phase surrounding the microcrystals in all directions.

16.2.3.2 Intercalation compounds of ZrP with diamines

The results obtained from the study of layered ZrP_2O_7[21] can be generalized by saying that proton diffusion in the layered structure of ZrP requires the presence of groups or molecules making bridges between phosphate groups of the same α-layer; the acid–base properties of these species are expected to affect proton transport. An investigation was thus undertaken to see how the conductivity of ZrP is modified by the presence of α–ω diamines having different base strengths[30].

The conductivity of anhydrous intercalation compounds of formula $ZrP \cdot xDAn$ (DAn being $N_2H_4(CH_2)_n$) with $n = 0, 2, 3, 4$ and 5 was

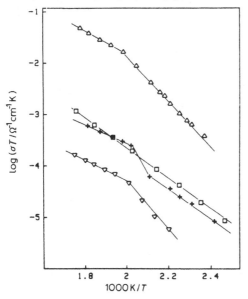

Fig. 16.5. Arrhenius plots for anhydrous ZrP ($+$), $ZrP_{(sl)}$ ($\square$) and $ZrP_{(p)}$ ($\triangle$: 'parallel' conductivity, ∇: 'perpendicular' conductivity).

Materials

Table 16.1. *Parameters of the Arrhenius equation (activation energy (E) and pre-exponential factor (A)) for the a.c.-conductivity of ZrP.xDAn anhydrous compounds*

n	x	E(eV)	$\log(A/\Omega^{-1}\,cm^{-1}\,K)$
0	0.80	0.74	5.13
2	0.93	0.94	4.38
3	0.96	1.14	6.04
4	0.85	1.07	5.45
5	0.99	1.19	6.21

measured up to the upper limit of the phase stability range (200 °C for $ZrP.0.8DA_0$ and 300 °C for the alkyldiamine intercalates) in an argon atmosphere. Activation energies and pre-exponential factors of the Arrhenius equation are listed in Table 16.1. As a general trend, an increase in base strength of the guest molecule is associated with an increase in activation energy; this is consistent with a transport mechanism which arises from protons jumping between the amine nitrogen and a phosphate oxygen.

The hydrazine intercalate exhibits the highest pre-exponential factor and the lowest activation energy, so that its conductivity at 180 °C $(1.8 \times 10^{-6}\,\Omega^{-1}\,cm^{-1})$ is more than one order of magnitude higher than that of ZrP and about three orders higher than that of the alkyldiamine intercalates. These results prove that it is possible to modulate the electrical conductivity of anhydrous ZrP by selecting the base-strength of the guest molecule.

16.3 Other phosphates and phosphonates of tetravalent metals

16.3.1 γ-M(IV) phosphates

Tetravalent metals may also be bonded by XO_4 and/or to $R{-}XO_3$ groups so as to build up a layered structure different from the α-structure. These compounds are usually called γ-*layered M(IV) phosphates*. Unfortunately, single crystals large enough for X-ray determination have not been obtained and the structure of γ-compounds is still unknown. Recent

Table 16.2. *Conductivities of some acid phosphates and phosphonates of tetravalent metals*

Compound	$\sigma(\Omega^{-1}\,cm^{-1})$	Temperature	Ref.
γ-$Zr(HPO_4)_2$	4×10^{-8}	180 °C	34
γ-$Ti(HPO_4)_2$	2×10^{-7}	180 °C	33
$Ce(HPO_4)_2$	7×10^{-8}	180 °C	39
$Zr(O_3P$—C_6H_4—$SO_3H)_2$	2×10^{-6}	180 °C	36
$Zr(O_3P$—C_6H_4—$SO_3H)_{2-x}$	2×10^{-5}	180 °C	36
$(O_3P$—$(CH_2)_2$—$COOH)_x$	$(\sim 10^{-2}$	20 °C, 90% RH)	42
γ-$Ti(HPO_4)_2 \cdot 2H_2O$	7×10^{-5}	20 °C, 90% RH	33
$Ce(HPO_4)_2 \cdot 3 \cdot 3H_2O$	3×10^{-5}	20 °C, 90% RH	39

investigations with solid state ^{31}P m.a.s. NMR spectroscopy[31] on γ-*zirconium phosphate* and X-ray powder diffraction methods on γ-*titanium phosphate*[32] gave evidence that, in the γ-layer, two different types of phosphate groups are present, one with all four oxygens bonded to the tetravalent metal and a second bonded with only two oxygens. Layered γ-M(IV) phosphates must therefore be formulated as $M(IV)PO_4H_2PO_4$. They are usually obtained as dihydrated compounds. Some conductivity data of hydrated and anhydrous γ-materials are reported in Table 16.2; it can be seen that, in spite of their different structure, they differ very little from those reported for α-zirconium phosphate. In addition, in this case also the conductivity of the hydrated Ti-compound seems to be dominated by surface transport[33].

16.3.2 Zirconium phosphonates

As we have seen, the pendant R group of α-$Zr(RPO_3)_2 \cdot nS$ compounds may be an organic radical. When polar groups with exchangeable protons (e.g. —COOH, —SO$_3$H etc.) are present in the R radical[35], the compounds obtained may be considered to be *organic ion-exchangers* possessing a crystalline inorganic backbone (see Fig. 16.6). These materials have very interesting electrochemical properties because protons of the acid groups can move on the surface and through the bulk. To date, research on these materials has been very limited; some preliminary data[36] on their conductivity are reported in Table 16.2.

16.3.3   *Fibrous acid phosphates of tetravalent metals*

While studying the Ce(IV)/phosphoric acid system, a fibrous material of composition $Ce(HPO_4)_2 . 2H_2O$ was obtained[37]. Similar results were obtained with thorium[38]. The structure of these compounds is unknown. The fibrous materials are of interest for electrochemical devices because they can be used to obtain very thin, autoconsistent membranes. The conductivity of anhydrous and trihydrated cerium phosphate[39] is reported in Table 16.2. The conductivity of the hydrated compound was investigated as a function of temperature at different relative humidities and parametrized on the basis of the Arrhenius equation. The dependence of both activation energy and pre-exponential factor on relative humidity was similar to that of ZrP. This suggests that in cerium phosphate also the proton conduction is due, for the most part, to the hydrated surface of the particles.

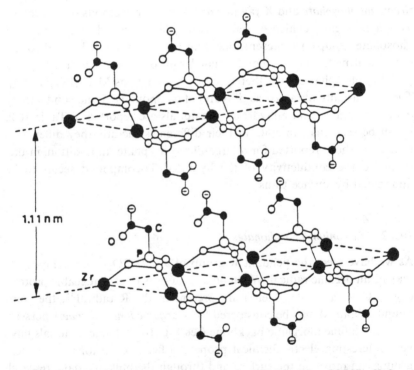

Fig. 16.6. Schematic picture of two adjacent layers of α-Zr(PO$_3$—CH$_2$—COOH)$_2$. H$_2$O, a layered zirconium phosphonate in which the organic radical possesses an acid group. The interlayer water and protons are not shown.

16.4   Applications and future perspectives

Phosphates and phosphonates of tetravalent metals are a large class of materials, some of which exhibit good protonic transport. Some can be obtained as colloidal dispersions and used to cover solid electrodes with a very thin film, or pellicle, of the protonic conductor. This peculiar property, together with good protonic transport, even under anhydrous conditions, makes pellicular phosphates very suitable materials for the fabrication of gas microsensors. Two sensors with good characteristics, one for H_2 working in the temperature range 0–300 °C[40] and another for O_2 at room temperature[41], have already been made in our laboratory by using a solid TiH_x electrode covered with $ZrP_{(p)}$.

Today, there is great interest in fuel cells which can operate at moderately high temperatures (200–350 °C). Since it is almost impossible to find a low-temperature oxide ion conductor below 500–600 °C, efforts must be directed towards proton conductors with high conductivity in the absence of water. The results already obtained with ZrP intercalated with diamines seem to be a promising way to obtain good protonic conductors with stability up to about 200 °C.

Zirconium phosphonates with long pendant aliphatic chains which carry some acid groups at their end are of interest since, due to thermal vibrations of the long chains, the terminal acid groups could approach each other at distances shorter than those between adjacent PO_3 sites (0.53 nm). Thus, protons could jump from one acid group to another with a mechanism that we jokingly call the 'Tarzan mechanism'. Furthermore, some phase transitions at relatively high temperature (150–200 °C) seem to indicate that the long pendant aliphatic chains behave as a melt between crystalline inorganic layers; thus, the liquid-like organization of the pendant groups in the interlayer region could further facilitate proton transport at these temperatures. Researches on the synthesis and protonic conductivity of pure zirconium phosphates with —SO_3H groups and on mixed zirconium phosphate with —SO_3H and —COOH groups attached to long aliphatic chains are in progress in our laboratory[36].

16.5   References

1. C. B. Amphlett, in *Inorganic Ion Exchangers* Ch. 5 (Elsevier, Amsterdam (1964)).
2. R. P. Hamlen, *J. Electrochem. Soc.* **109** (1962) 746.
3. G. Alberti and E. Torracca, *J. Inorg. Nucl. Chem.* **30** (1968) 1093.

Materials

4. A. Clearfield and J. A. Stynes, *J. Inorg. Nucl. Chem.* **26** (1964) 117.
5. G. Alberti and E. Torracca, *J. Inorg. Nucl. Chem.* **30** (1968) 317.
6. G. Alberti, *Accounts Chem. Res.* **11** (1978) 163.
7. A. Clearfield, in *Ion Exchange Materials*, A. Clearfield (ed.) Ch. 1 (CRC Press, Boca Raton, Fl (1982)); G. Alberti, in *Ion Exchange Materials*, A. Clearfield (ed.) Ch. 2 (CRC Press, Boca Raton, Fl (1982)).
8. G. Alberti and U. Costantino, in *Intercalation Chemistry*, M. S. Whittingham and J. A. Jacobson (eds) Ch. 5 (Academic Press, New York (1982)).
9. G. Alberti, *La Chimica e l'Industria* **70** (1988) 26.
10. G. Alberti, M. Casciola and U. Costantino, in *Solid State Protonic Conductors III*, J. B. Goodenough, J. Jensen and A. Potier (eds) (Odense University Press (1985)) 215.
11. Ph. Colomban and A. Novak, *J. Mol. Struct.* **198** (1989) 277.
12. Ph. Colomban and A. Novak, *J. Mol. Struct.* **177** (1988) 277.
13. A. Clearfield and G. D. Smith, *Inorg. Chem.* **8** (1969) 431.
14. G. Alberti, M. Casciola, U. Costantino, G. Levi and G. Ricciardi, *J. Inorg. Nucl. Chem.* **40** (1978) 533.
15. E. Krogh Andersen, I. G. Krogh Andersen, C. Knakkergaard Moller, K. E. Simonsen and E. Skou, *Solid State Ionics* 7 (1982) 301.
16. D. Bianchi and M. Casciola, *Solid State Ionics* **17** (1985) 7.
17. M. Casciola and D. Bianchi, *Solid State Ionics* **17** (1985) 287.
18. Y. Sadaoka, M. Matsuguchi, Y. Sakai and S. Mitsui, *J. Mater. Sci.* **22** (1987) 2975.
19. S. Yde-Andersen, J. S. Lundsgaard, J. Malling and J. Jensen, *Solid State Ionics* **13** (1984) 81.
20. E. Skou, I. G. Krogh Andersen, E. Krogh Andersen, M. Casciola and F. Guerrini, *Solid State Ionics* **35** (1989) 59.
21. G. Alberti, M. Casciola, U. Costantino and M. Leonardi, *Solid State Ionics* **14** (1984) 289.
22. G. Alberti, M. Casciola and U. Costantino, *J. Colloid Interface Science* **107** (1985) 256.
23. G. Alberti, M. Casciola, U. Costantino and F. Di Gregorio, *Solid State Ionics* **32/33** (1989) 40.
24. M. Casciola and U. Costantino, *Solid State Ionics* **20** (1986) 69.
25. M. Casciola, U. Costantino and S. D'Amico, *Solid State Ionics* **22** (1986) 127.
26. R. A. Huggins, in *Diffusion in Solids*, A. S. Nowick and J. J. Burton (eds) Ch. 9 (Academic Press, New York (1975)).
27. G. Alberti, M. Casciola, U. Costantino and R. Radi, *Gazz. Chim. Ital.* **109** (1979) 421.
28. A. La Ginestra, M. A. Massucci, C. Ferragina and N. Tomassini, in *Thermoanalysis*, Vol. 1, *Proc. IV I.C.T.A.*, Budapest (1974) 631.
29. P. Jerus and A. Clearfield, *Solid State Ionics* **6** (1982) 79.
30. M. Casciola, U. Costantino and F. Marmottini, *Solid State Ionics* **35** (1989) 67.
31. N. J. Clayden, *J. Chem. Soc. Dalton Trans.* **8** (1987) 1877.

Phosphates and phosphonates

32. A. Norlund Christensen, E. Krogh Andersen, I. G. Krogh Andersen, G. Alberti, M. Nielsen and M. S. Lehmann, *Acta Chem. Scand.* **44** (1990) 865.
33. G. Alberti, M. Bracardi and M. Casciola, *Solid State Ionics* **7** (1982) 243.
34. G. Alberti, M. G. Bernasconi and M. Casciola, *Reactive Polymers* **11** (1989) 245.
35. G. Alberti, U. Costantino and M. L. Luciani Giovagnotti, *J. Chromatogr.* **180** (1979) 45.
36. G. Alberti and S. d'Amico, unpublished data.
37. G. Alberti, U. Costantino, F. Di Gregorio, P. Galli and E. Torracca, *J. Inorg. Nucl. Chem.* **30** (1968) 295.
38. G. Alberti and U. Costantino, *J. Chromatogr.* **50** (1970) 482.
39. M. Casciola, U. Costantino and S. d'Amico, *Solid State Ionics* **28/30** (1988) 617.
40. G. Alberti and R. Palombari, *Solid State Ionics* **35** (1989) 153.
41. G. Alberti, M. Casciola and R. Palombari, Italian Patent 22663, 12 Dec. 1989.
42. G. Alberti, M. Casciola, V. Costantino, A. Peraio and E. Montoneri, *Solid State Ionics* (in press).

17 Hydrogen uranyl phosphate, $H_3OUO_4PO_4 . 3H_2O$ (HUP), and related materials

PHILIPPE COLOMBAN AND ALEXANDRE NOVAK

17.1 Introduction

Uranyl phosphates and arsenates appear interesting as uranium ores[1-4]. There are about 200 known uranyl phosphate containing minerals which can be divided into three groups according to the UO_2/PO_4 ratio[5]. The first group with $UO_2/PO_4 = 1$ comprises autunites and uranium phosphate micas; the second has $UO_2/PO_4 > 1$, most frequently 3/2 and 4/3, and is illustrated by phospharanglite; the third with $UO_2/PO_4 < 1$ is represented by parsonite. The first group of minerals contains usually Cu, Mn, Fe and H while in all groups Pb, Ca and Al are present. The formulae of these minerals have not been established accurately and errors in the determination of the UO_2/PO_4 ratio often lead to incorrect descriptions. Among the above minerals *autunite*, $CaUO_2PO_4 . 3-6H_2O$, appears to be the most extensively studied.

As far as *hydrogen uranyl phosphate* (*HUP*) is concerned, Beintema noticed – as early as 1938 – that 'true vagabond ions are present in the water layer of the structure'[6]. Thirty years later, Sugitami et al. concluded on the grounds of *NMR* measurements that water molecules in $KUO_2PO_4 . 3H_2O$ are highly mobile[7]. Much the same conclusions were reached by Wilkins et al.[2] who observed a high mobility of oxonium ions in $H_3OUO_2AsO_4 . 3H_2O$. Thus oxonium and water protons diffuse rapidly in the structure of uranyl phosphate or arsenate. In spite of the experimental evidence the question of the true and bulk conductivity has been discussed for a long time. First evidence of the high conductivity $(4 \times 10^{-3} \Omega^{-1} cm^{-1}$ at 290 K) in HUP was given by Howe & Shilton[8] and independently by Kobets et al.[9]. Sodium and ammonium uranyl phosphate trihydrates and hydrogen uranyl arsenate trihydrate are also good conductors[10,11]. All these materials, like most ionic conductors, are remarkable inorganic or organic ion exchangers[12,13].

Hydrogen uranyl phosphate

Major contributions to the understanding of the structure and properties of HUP and related compounds have been made by the groups in Leeds[8,11], Paris[14–16] and Stuttgart[17]. Simultaneously, many all-solid-state devices such as hydrogen sensors[18], electrochromic displays[19,20], batteries[21] and supercapacitors[22], using the fast diffusion of water and oxonium ions between $(UO_2PO_4)_n$ slabs were developed. Some applications are given in Part V of this book.

17.2 Preparation and chemistry

The preparation of uranyl phosphates has also been a subject of lively discussion. The presence of the heavy uranium atom hinders the observation of some structural modifications and the fact that there are many different complexes in solution complicates the situation[5]. Schreyer & Baes[23], studying the solubility of phosphate in phosphoric acid in the 0.001–15 M range, have identified three stable phases of uranyl phosphate

$(UO_2)_3(PO_4)_6 . 6H_2O$: less than 0.014 M, yellow,
$(UO_2)HPO_4 . 4H_2O$ (HUP): 0.014 M–6.1 M, yellow-green,
$UO_2(H_2PO_4)_2 . 3H_2O$: more than 6.1 M, green.

The color is related to the presence of UO_2 chromophore ions and can be modified by a very small change of an ion's coordination[24].

The existence of these phases was confirmed by chemical analysis, differential scanning calorimetry, vibrational spectroscopy and scanning electron microscopy[25–28]. Moreover, the presence of gels of approximate composition $(UO_2)_{1.5-x}PO_4H_{2x} . 2H_2O$ and of syntaxy between the above materials was demonstrated[26,27]. HUP, $UO_2(H_2PO_4) . 3H_2O$, $(UO_2)_3(PO_4)_2 . 4H_2O$ and the gel can be identified easily by the DSC signal shape; however, the temperature of the peak maxima as well as the onset temperature depend strongly on the heating rate[26]. Useful techniques for the identification of various phases are vibrational spectroscopy[26,29] and electronic microscopy[28] while the differentiation by X-ray powder patterns is often difficult. Typical electron microphotographs are shown in Figs 17.1 and 17.2. HUP and $UO_2(H_2PO_4)_2 . 3H_2O$ grow usually in the form of small cubic and large hexagonal platelets, respectively. $(UO_2)_3(PO_4)_2 . 4H_2O$ is obtained as small needles and gel as very small

255

Materials

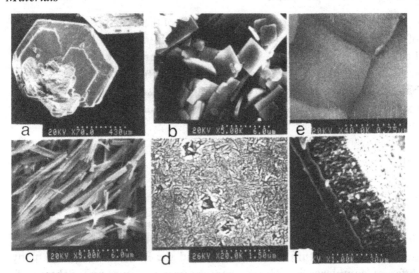

Fig. 17.1. Scanning electron micrographs of HUP related materials: (a) $UO_2(H_2PO_4)_2 \cdot 3H_2O$, (b) $H_3OUO_2PO_4 \cdot 3H_2O$ (HUP), (c) $(UO_2)_3(PO_4)_2 \cdot 4H_2O$, (d) optically clear gel $(UO_2)_{1.5-x}H_{2x}PO_4 \cdot 2H_2O$, (e) pressure-sintered HUP pellet and (f) the same pellet after heating for 1 h at 323 K and rehydrating in air. Note that only the surface layer is modified; reproduced with permission[26, 27].

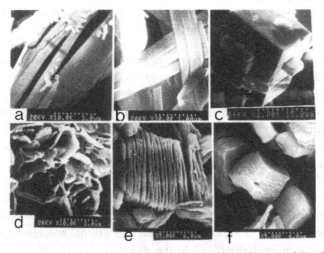

Fig. 17.2. Changes in the structure of HUP during washing with water [(a), after several days, exfoliation is observed, (b) after a week the composition reaches that of the gel], during slow (c) or rapid (d) rehydration in air of materials dipped in acetone for a few hours. Photomicrographs of HUP intercalated by ethylammonium (interslab distance ~0.79 nm) and by octadecylammonium ions (interslab distance ~3.27 nm) are shown in (e) and (f) respectively[26, 27], with permission.

256

particles which, pressed together, give rise to optically clear bodies. If water is present, HUP crystals are altered and progressively transform to a gel as shown in Fig. 17.2.

The strong Raman band due to the UO_2^{2+} symmetric stretching mode is a good probe to distinguish various phases and the characteristic frequencies of 835, 844, 866 and 868 cm^{-1} are observed for $UO_2(H_2PO_4)_2 \cdot 3H_2O$, HUP gel and $(UO_2)_3(PO_4)_2 \cdot 4H_2O$, respectively. In infrared spectra, the absorption pattern of strong P–O stretching bands in the 1050–1200 cm^{-1} region can be used for analytical purposes (Fig. 17.3).

HUP exchanges very easily its H_3O^+ ions with Na^+, NH_4^+, K^+, Rb^+, Cs^+ and some divalent cations[12]; the presence of these ions as impurities in reagents must be checked, therefore, in order to avoid crystalline stoichiometry modifications. The presence of these ions, in solid solution formation, can also lead to a broadening and smoothing of phase transitions[30].

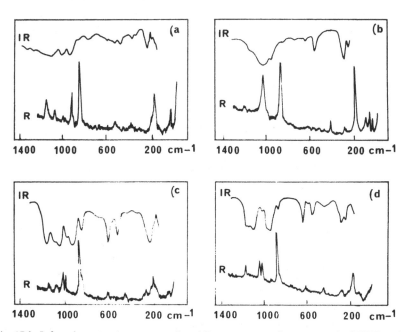

Fig. 17.3. Infrared spectra (upper curves) and Raman spectra (lower curves) of HUP and related materials at room temperature: (a) $UO_2(H_2PO_4)_2 \cdot 3H_2O$, (b) $H_2OUO_2PO_4 \cdot 3H_2O$ (HUP), (c) $(UO_2)_{1.5x}H_{2x}PO_4 \cdot 2H_2O$ (gel) and (d) $(UO_2)_3(PO_4)_2 \cdot 4H_2O$[26].

Materials

17.2.1 Dense sintered bodies

HUP precipitate is obtained by mixing phosphoric acid and uranyl nitrate solutions. The precipitation has an induction period of a few minutes; filtration gives a yellow-green thixotropic slurry which is washed with a pH2 water and 'dried' at 50–60% relative humidity. The resulting crystalline powder is pressed under $10-50 \, \text{MN m}^{-2}$, yielding optically clear discs which have an almost theoretical density of $3.43 \, \text{g cm}^{-3}$. Scanning electron micrographs show well-densified microstructures exhibiting 120° grain boundaries. The microstructure and grain growth indicate that there is diffusion of matter and thus a true sintering. Liquid phase *sintering* also happens but the solubility is not high and the phenomenon is slow and not pressure-activated. The densification mechanism was discussed by Childs *et al.*[11e] who suggested that viscous flow and even a fracture mechanism can equally occur. This shows that the sintering of HUP (and more generally, hydrates and gels) and ceramics is similar, although the sintering temperature is much lower in HUP, because of hydrogen bonding which is considerably weaker than the usual ionic or covalent bonding. The empirical law concerning the sintering temperature $T_s = 0.8 T_{\text{melt}}$ is roughly followed and T_s is lowered by pressure *sintering*.

17.2.2 Single crystals

Crystals larger than about 100 μm are difficult to grow because of the low solubility of HUP. However, single crystals of a few mms could be prepared by using the gel method[31]. Similar to HUAs, HUP crystals show a perfect cleavage on (001). Sometimes domains are observed in between the phase transitions[10,11,31].

17.3   Thermal and chemical stability

The stability of protonic conductors and hydrates in particular, is not a trivial problem and in spite of numerous studies a clear understanding was obtained only recently. The investigations of the thermal evolution of HUP by Kobets & Umreiko[5], Howe & Shilton[11b] and Pham-Thi & Colomban[26] have shown that the first two water molecules disappear at 320 K and the last one at about 420 K. However, unlike the potassium and sodium analogues which can be reversibly dehydrated at 600 K,

Hydrogen uranyl phosphate

irreversible transformations above 370 K are observed for HUP. It should also be pointed out that the dehydration temperature depends on atmospheric humidity[32]. The loss of three water molecules is accompanied by a shortening of the distance between (UO_2PO_4) slabs from 0.875 to about 0.93 nm, a decrease in crystallinity and a proton transfer to the phosphate group, forming HPO_4^{2-} or even $H_2PO_4^-$ species. In the last case, the transformation is irreversible. Nearly reversible dehydration–hydration equilibrium exists when less than 1/2 water molecule is removed but there are still changes in grain size and shape. Dehydrated samples are likely to be non-stoichiometric and can be represented by $H_3OUO_2PO_4 . 3 - xH_2O$ with $x < 0.1$–0.5. Finally, the condensation of phosphate groups can occur at temperatures above ~ 425 K.

17.3.1 Chemical drying and intercalation

A milder method to dehydrate HUP consists of dipping it into an appropriate solvent. Ethanol, for instance, dehydrates HUP but the material becomes amorphous. Acetone, on the other hand, allows a progressive and reversible dihydration as illustrated in Fig. 17.4. There is a gradual transformation of HUP to $(UO_2HPO_4)(CH_3COCH_3)H_2O$ (α phase) with proton transfer from H_3O^+ to PO_4^{2-}. Three more or less crystalline intermediate phases: β, $(UO_2H_xPO_4)(H_3O)_{1-x} . H_2O$; γ, $(UO_2H_xPO_4)(H_3O)_{1-x}$; and δ, $UO_2H_3OPO_4 . 3 - \varepsilon H_2O$, can be obtained. The δ phase is strongly hydrophilic, which is probably due to surface defects[33]. The water content of acetone influences the intercalation process and when it increases above 5% there is no CH_3COCH_3 intercalation any more. Many other entities such as *hydrazine, pyridine, alkyl ammonium ions* (Fig. 17.2), *cobaltocene, methyl viologen* and *Creux–Tauber complex* can be intercalated or exchanged in HUP[13]. This property is related to the lamellar structure of HUP, to the lability of protonic species and to the easiness of the proton transfer.

17.3.2 Electric field

The application of a low-voltage (1.5 V) electric field to a HUP crystal may reduce or oxidize water, producing H_2 and O_2, respectively. This is similar to dehydration since a proton transfer occurs between the protonated water layer and phosphate tetrahedra, forming HPO_4^{2-} (quasi-reversible step) and $H_2PO_4^-$ (irreversible step), depending on the

259

applied *electric field*[33]. Such irreversible transformations are observed at much higher voltage ($\gg 1.5$ V) than that for a liquid, probably because of slower kinetics in the solid state and because the reaction products remain in the area where the reaction takes place. This explains the very long lifetime of the HUP components.

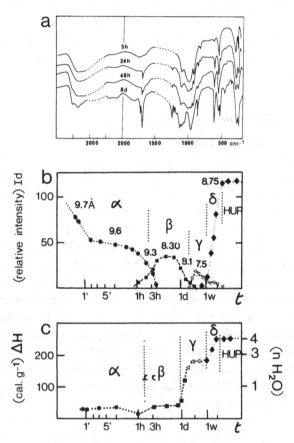

Fig. 17.4. Infrared (a), X-ray (b) and DSC (c) evidence of the gradual transformation of $H_3OUO_2PO_4 \cdot 3H_2O$ material in $(UO_2HPO_4)(CH_3COCH_3)H_2O$ (for α' phase, the interslab distance is 1.08 nm) by dipping for a few days in acetone with 1% H_2O (reproduced with permission of Chapman & Hall)[33]. A logarithmic time scale is used. If dry acetone is used the α phase $(UO_2HPO_4)(CH_3COH_3)$ with an interslab distance of 0.97 nm is obtained. The kinetics of the acetone de-intercalation and rehydration process as a function of exposure to moisture is illustrated by the X-ray intensity (height) plot of the interslab distances and by the plot of total enthalpy measured on DSC traces between 350 and 450 K[33].

17.4   Structure and phase transitions

The crystal structure of $H_3OUO_2PO_4 . 3H_2O$ has been investigated by various authors[1,6,34]. Different crystalline modifications have been reported. However, the tetragonal P4/ncc (D_{4h}^8) space group with four formula units per unit cell has been generally accepted. The structure consists of infinite UO_2PO_4 sheets separated by a two-level water layer (Fig. 17.5). The former contain dumb-bell shaped uranyl ions linked to four different phosphate tetrahedra in octahedral coordination with an intermolecular O . . . O distance of 0.235 nm. The UO_2^{2+} ion is linear with different intramolecular U–O distances of about 0.175 nm[29,34]. Oxygen atoms of the water layer lie on squares and each water molecule may participate in four hydrogen bonds. There are thus more hydrogen bond sites than available hydrogen atoms and a statistical distribution of the protons is required. In this description, oxonium ions are statistically located in the place of water molecules. Oxygen–oxygen distances of 0.256, 0.281 and 0.283 nm correspond to hydrogen bonds between water squares at different levels (a-distance in Fig. 17.5), between water molecules of the same square (c and d distances) and between water and phosphate tetrahedra (b-distance), respectively. The infrared spectra of isotopically diluted HUP crystal at 130 K with a H/D ratio of about 8% show OH stretching frequencies at 3390, 3380, 3260, 3170 and 2920 cm^{-1} which can

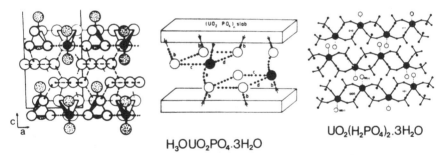

$H_3OUO_2PO_4 . 3H_2O$

$UO_2(H_2PO_4)_2 . 3H_2O$

Fig. 17.5. Structure of HUP-type materials. Black circles correspond to uranium atoms, dashed circles to the oxygen of UO_2 ions and white circles to other oxygen atoms; in the a, b projection, oxygen atoms of tetrahedra are not given. Solid lines correspond to O–U–O bonds and dashed lines to U–O'PO$_3$ bonds. A schematic representation of the water layers in HUP is also given; reprinted with permission from *Solid State Communication* **53** (Ph. Colomban, M. Pham-Thi & A. Novak, *Vibrational study of phase transition and conductivity mechanisms in HUP*, (1985) Pergamon Press PLC)[15]. The projection of $UO_2(H_2PO_4)_2 . 3H_2O$ on the (101) plane shows the ribbon of the structure; UO_2 ions are orthogonal to this plane; water molecules W2 and W3 have been omitted for clarity; reproduced with permission[14].

be assigned, according to the empirical relationship between frequencies and distances[35], to the O...O distances of 0.277 and 0.276 nm ($H_2O...H_2O$), 0.271 and 0.269 nm ($H_2O...PO_4^{2-}$) and 0.263 nm ($H_3O^+...H_2O$), respectively. The spectra of $KUO_2PO_4.3H_2O$, on the other hand, yield higher νOH frequencies corresponding to water–water interaction. This indicates longer (about 0.285 nm) O...O distances, which is expected since the $H_3O^+...H_2O$ hydrogen bond appears to be the strongest[29]. The O...O distances between H_2O and PO_4^{2-} remain similar to those in HUP.

17.4.1 Phase transitions

There are three phase transitions in HUP as shown by DSC[15], vibrational spectroscopy[29], conductivity[8], optical birefringence[31] and X-ray diffraction[11]. One near 270 K is well-defined ($\Delta H = 2.5 \text{ kJ mol}^{-1}$) while those near 200 and 150 K are diffuse[15,29]. Four phases designed by I (above 270 K), II (270–200 K), III (200–150 K) and IV (below 150 K) can thus be distinguished (Figs 17.6 and 17.7). The high temperature phase I is superionic and the low temperature phase II appears to be ferroelectric[13,36,37] Comparison of infrared and Raman spectra of phases VI–I gives some indications about these essentially order–disorder transitions and the following interpretation is proposed. For arsenate analogues, the transitions are displaced to slightly higher temperatures.

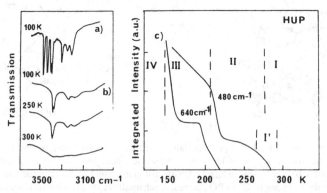

Fig. 17.6. Infrared spectra of an isotopically diluted compound of (a) KUP and (b) HUP in the 4000–2700 cm^{-1} region. Intensity variation of some bands as a function of temperature (c): integrated IR intensity of the H_2O vibrational mode ($\nu = 640 \text{ cm}^{-1}$) and U–OPO$_3$ stretching metal–ligand band ($\nu = 480 \text{ cm}^{-1}$)[29]; reprinted with permission from *J. Phys. Chem. Sol.* **46** (M. Pham-Thi, Ph. Colomban & A. Novak, *Vibrational study of HUP* (1985), Pergamon Press PLC).

Hydrogen uranyl phosphate

In phase I, there is rotational and translational disorder of H_2O and H_3O^+ species, i.e. a quasi-liquid state of the water layer, with considerable orientational disorder of phosphate tetrahedra. The spectroscopic manifestation of such a state is the presence of a structureless broad absorption in the OH stretching and H_2O and H_3O^+ librational region, i.e. individual contributions of protonic species are smeared out while the distinction between H_2O and H_3O^+ is observed only in the OH bending region. In the ferroelectric phase II, the translational disorder is drastically reduced, rotation around the $H_3O^+ \dots H_2O$ axis becomes uniaxial and the phosphate groups become ordered. The OH stretching bands due to H_2O and H_3O^+ can be clearly distinguished. The structure becomes gradually orthorhombic through an intermediate I' phase (Fig. 17.7). The protonic species can also be described in terms of $H_5O_2^{+}$[38] and $H_9O_4^{+}$ entities[5]. In phase II, the rotation of water molecules hydrogen bonded to PO_4 groups becomes uniaxial and the long range order in HUP sheets appears to be established. Finally in phase IV, the last degree of rotational freedom of the water molecules around the intra-square O . . . O directions may be frozen in and the sites of the square are no longer equivalent, as shown by splitting of the OH stretching band at 3500 cm^{-1} due to $H_2O \dots H_2O$ intra-square hydrogen bonded water (Fig. 17.6). The system appears completely ordered. The transitions can be considerably softened by substitution of a small quantity of H_3O^+ by M^+ ions (K^+, Na^+)[30] or by 50% D substitution[39].

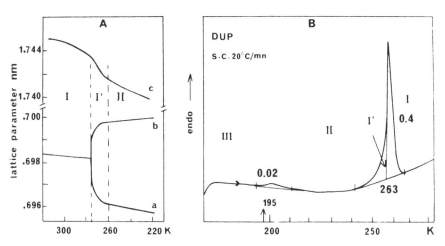

Fig. 17.7. (A) Lattice parameter variation for HUP as a function of temperature[11], with permission. (B) DSC trace of deuterated analogue (DUP)[15] (enthalpies in kcal mol^{-1}).

263

The arsenate analogue, $H_3OUO_2AsO_4 \cdot 3H_2O$ shows similar features as shown by a NMR study[17]: below 170 K the total second moment is characteristic of a rigid network, between 170 and 200 K a reorientation of water and oxonium molecules takes place and above 270 K diffusion of protonic species is observed. The transition to the superionic tetragonal phase occurs at 300 K[17] (see Chapter 31 for a more complete description). *Microwave dielectric relaxation* study of HUP gives similar conclusions[37] and the phase transition temperatures are almost unchanged by H/D substitution (see p. 398).

The strongest OH . . . O hydrogen bond between oxonium and water molecules, characterized by an OH stretching frequency near 2920 cm^{-1} and O . . . O distance of about 0.263 nm, appears to play an important role in the ordering process in HUP as shown by deuterium and potassium substitution[29]. In the case of DUP where the positive isotope effect is expected to lengthen the shortest O . . . O distance much more than the long ones, the long-range order is established sooner than in HUP. When the oxonium ion is exchanged for potassium, there are no phase transitions and the ordering is progressive and short range[31].

17.5 Electrical properties

The room temperature conductivity of an optically clear, sintered HUP disc varies between 1 and $5 \times 10^{-3} \, \Omega^{-1} \, cm^{-1}$ (Fig. 17.8) as a function of the crystallite orientation, the activation energy being close to 0.34 eV[16]; the conductivity along the conduction plane of a single crystal amounts to $6.3 \times 10^{-3} \, \Omega^{-1} \, cm^{-1}$. The anisotropy of proton conductivity is measured as equal to $\sim 10^3$ [13] and confirmed from quasi-elastic neutron scattering[40]. The conductivity versus temperature plot of HUP (*HUAs*) (Fig. 17.8) shows a sudden decrease in conductivity of about two orders of magnitude and an increase of the activation energy to about 0.65 eV below 273 K (303 K). Non-Arrhenius behaviour is observed during the phase transition. Under pressure (>2 GPa), the conductivity plot does not show a discontinuity and the low temperature behaviour ($E_a \sim 0.5$–0.7 eV) is observed over all the 200–350 K temperature range. The structure of HUP contains a water layer and the first papers tried to interpret the remarkable electrical properties by the *Grotthuss mechanism*, a mechanism supposed characteristic of aqueous conduction[13]. Later studies showed that the vehicle mechanism; i.e. hopping of H_3O^+ and H_2O molecules in the water layer was more realistic (cf. Chapters 30 &

Hydrogen uranyl phosphate

31) in agreement with the high conductivity of $NaUP^{16,41}$. In fact, the conductivity of the sodium derivative, $NaUO_2PO_4 \cdot 3H_2O$ and of HUP are similar while that of other salts such as Ag^+, K^+ and Li^+ is more than 100 times lower[13,15]. The relatively small difference in conductivity of HUP and NaUP can be understood by recognizing that the main conductivity mechanism consists of hopping of sodium or oxonium ions in the quasi-liquid water layer ('vehicle' mechanism); the fact that the conductivity of HUP is a little higher is doubtless due to the possibility of a complementary Grotthuss mechanism[16]. The conductivity remains unchanged within a large range of humidity (5–99% at 300 K), where the water stoichiometry remains close to $n = 3^{42}$. This is characteristic of a true intrinsic (bulk) proton conductor[43]. Finally, the defects induced by washing HUP can decrease the conductivity about a hundred fold[37].

Microwave dielectric measurements confirm the ferroelectric nature of order phases, the dipoles being related to the orientation of the protonic species and not to the proton ordering, according to the small temperature shift on H/D substitution (Table 17.1). The extrapolated Curie–Weiss temperature of the first order I → II transition is located below the actual

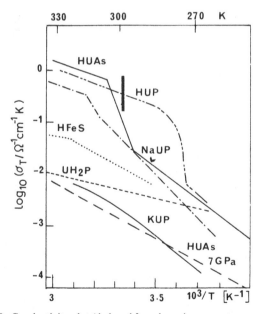

Fig. 17.8. Conductivity plot (deduced from impedance spectroscopy) as a function of inverse temperature for MUP pressure-sintered discs (heating cycle). Bar indicates variations as a function of crystallite orientation[16]. (For abbreviations, see Table 17.1).

Table 17.1. Comparison of conductivity parameters

	Activation energy (eV)						
	HUP	HUAs	NUP	NaUP	KUP	HFeS	UH$_2$P$_2$
Motions							
H$_3$O$^+$ Reorientation	0.13[a]						
H$_2$O Reorientation	0.25[a]	0.27[c]		0.25[a]			
H$_3$O$^+$/H$_2$O Jump							
Phase I	0.34[a]						
Phase II	0.40[a]	0.63[c]					
Conductivity							
Phase I	0.35[b]	0.27[d]		0.3[b]	0.35[b]	0.19[e]	
Phase II	0.7[b]	0.74[d]		0.8	0.6[b]	0.42[e]	0.20[f]
$\sigma_{300\,K}$ (Ω^{-1} cm^{-1})	5×10^{-3}	5×10^{-3}	2×10^{-6}	10^{-4}[b]	7×10^{-6}	10^{-4}[f]	5×10^{-5}

[a] From microwave dielectric relaxation[37], [b] from impedance spectroscopy[16], [c] from NMR[17], [d] from a.c. conductivity[17], [e] reference 51 and [f] reference 14.

HUP: H$_3$OUO$_2$PO$_4$.3H$_2$O; HUAs: H$_3$OUO$_2$AsO$_4$.3H$_2$O; NUP: NH$_4$UO$_2$PO$_4$.3H$_2$O; NaUP: NaUO$_2$PO$_4$.3H$_2$O; HFeS: FeH(SO$_4$)$_2$.4H$_2$O, UH$_2$P$_2$: UO$_2$(H$_2$PO$_4$)$_2$.3H$_2$O.

transition, close to 210 and 170 K for H_2O and H_3O^+ entities, respectively[37] (see also Chapter 25). The electronic conductivity at 300 K is about $10^{-9}\,\Omega^{-1}\,cm^{-1}$ with an activation energy of 0.5 eV[11].

17.6   Related materials

HUP and many other natural or synthetic compounds belong to the torbernite mineral group of general formula $M_{1-y}(UO_2XO_4)(H_3O)_y$. $3H_2O$ where X is P or As and M^+ is a cation such as K^+, NH_4^+, Na^+, Li^+, Ca^{2+}, Pb^{2+}, Ag^{2+} or Al^{3+}[1]. The role of interlayer cations and water molecules in torbernite is similar to that in zeolites, clay minerals, some feldspars and micas, jarosites, autunites and clathrate compounds. In all these materials solid solutions between protonic and ionic derivatives are readily observed: ion exchange and protonic conductivity can thus be expected.

17.6.1   Chain materials

An example is given by $UO_2(H_2PO_4)_2 \cdot 3H_2O$ which has a room temperature conductivity of $10^{-4}\,\Omega^{-1}\,cm^{-1}$ with a low activation energy of 0.2 eV[14]. The structure consists of infinite chains along the (101) axis with uranium atoms bridged by two H_2PO_4 groups (Fig. 17.5). The uranium atom is surrounded by a pentagonal bipyramid of oxygen atoms, one of them being an equatorial water molecule. This water molecule is weakly hydrogen bonded, as shown by its high OH stretching frequency at $3520\,cm^{-1}$, while the bonding between the chains is ensured by stronger hydrogen bonds involving both water molecules and P–OH groups. Protonic species, H_2O and $H_2PO_4^-$, appear highly orientationally disordered which can help proton jumps, as seen by the low activation energy. The chain surface seems to be the most conductive part of this crystal. Intercalation is not observed for this material, the protonic transfer being already complete[20]. Intermediate materials such as $(H_3O)Fe_3(HPO_4)_2(H_2PO_4)_6 \cdot 4H_2O$ are also potential proton conductors[49].

Phosphate and arsenate groups can be substituted by silicate or vanadate groups as in uranophane, for instance[45]. This compound, represented by $Ca(H_3O)(UO_2SiO_4)_2 \cdot 3H_2O$[2,9], is one of the earliest known uranium minerals and is found in the form of yellow to yellow-orange fibres. Its structure, similar to that of HUP, consists of UO_2SiO_4

sheets. However, a recent structure refinement indicates that protons are attached to SiO_4 tetrahedra and not to water[44]; the correct formula is thus $Ca(UO_2SiO_3OH)_2 . 3H_2O$ but the calcium mineral must be expected to be an ionic-exchanger and thus a true oxonium form can also be expected. This last compound may be a proton conductor since similar structures show protonic conductivity.

17.6.2 Alunite *group*

Minerals of alunite–jarosite groups have the general formula $MA_3(SO_4)_2(OH)_6$ where M is H_3O^+, NH_4^+, Na^+, K^+, Rb^+, Ag^+, Tl^+, $0.5Pb^{2+}$ and A is Al^{3+} (alunite) or Fe^{3+} (jarosite). Related synthetic materials $MFe(SO_4)_2(OH)_6$ (M = H_3O^+, NH_4^+), $MGa_3(SO_4)_2(OH)_6$ also form solid solutions[1]. Hydronium alunite can be prepared by hydrothermal reaction from $Al_2(SO_4)_3 . 16H_2O$ at 250 °C while only the mixed K^+/H_3O^+ derivative is obtained by refluxing at 105 °C. Sulphate groups can be substituted partly or completely by HPO_4^- or $HAsO_4^-$ ions[46,47]. Members of the alunite group are isostructural, with R3m space group and solid solutions between Fe and Al derivatives are possible. The structure contains $A(OH)_4(SO_4)_2$ octahedra linked through the basal corners via OH^- species. Cations M^+ occupy cavities surrounded by six OH^- ions and six sulphate oxygens. It should be pointed out that these crystals may contain both OH^- and H_3O^+ ions when a cation or OH^- ion is substituted by water molecules, which seems surprising. However, high resolution NMR investigations confirm the presence simultaneously of H_3O^+, H_2O and OH^- entities in these materials[48] as already observed in antimonic acid[49]. The presence of the hydroxyl group is due to the deficiency in Al^{3+} ions and a more accurate formula is $MAl_{3-x}(SO_4)_2(OH)_{6-3x}(H_2O)_{3n}$.

The room temperature conductivity of hydrated pellets of hydronium alunite, $H_3OAl_3(SO_4)_2(OH)_6$ is about $3 \times 10^{-4} \Omega^{-1} cm^{-1}$[47] and that of dry pellets much lower, $\sim 10^{-9} \Omega^{-1} cm^{-1}$, which indicates a major contribution of the surface conductivity similar to that in $\alpha Zr(HPO_4)_2 . nH_2O$ and in zeolites[43]. The presence of OH^- groups may favour the building up of a conducting water layer.

17.6.3 Silicates

The substitution of K^+ by NH_4^+ ion occurs in many silicates, in some micas, amphiboles and feldspars, for instance: NH_4 sanidine,

Hydrogen uranyl phosphate

$NH_4AlSi_3O_4 . 1/2H_2O$, H_3O^+ orthoclase and H_3O *micas*[1], NH_4 *wadeite*, $(NH_4)_2ZrSi_3O_9$ (amorphous or crystalline form) and NH_4 khibinskite $(NH_4)_2ZrSi_2O_7$, have been synthesized by ion exchange from the corresponding potassium compounds[50]. Their conductivity is low ($\sigma = 10^{-5}\,\Omega^{-1}\,cm^{-1}$ at 300 K with $E_a = 0.5\,eV$) but the materials are quite stable up to 500 K. Oxonium derivatives are more conductive with a larger contribution of the surface conductivity, especially if porous thick films are used. Generally, this class of material does not appear to be well studied.

17.6.4 $FeHSO_4 . 4H_2O$ and rhomboclase *group*

Solid solutions can be formed by $NH_4Al(SO_4)_2$ and $H_3OAl(SO_4)_2$ on the one hand and by $NH_4Fe(SO_4)_2$ and $H_3OFe(SO_4)_2$ on the other[1]. The hydrated form of the latter, $HFe(SO_4)_2 . 4H_2O$ (or $H_5O_2Fe(SO_4)_2 . 2H_2O$), has a high conductivity (Fig. 17.8), $\sigma + 5 \times 10^{-4}\,\Omega^{-1}\,cm^{-1}$ at 300 K and $E_a = 0.2\,eV^{51}$. The structure of the natural form, rhomboclase, has been determined by Mereiter[52] and that of the synthetic parent compound, $InH(SO_4)_2 . 4H_2O$, by Tudo *et al.*[53]. In both cases, the space group is Pnma with $Z = 4$ and the structure is similar to that of HUP: in the $Fe(SO_4)_2$ slabs, iron atoms are equatorially coordinated with four oxygen atoms of different SO_4 ions and on the top and the bottom with the water oxygens. The water molecules are hydrogen bonded to sulphate groups with O . . . O distances of 0.283 and 0.285 nm. The water layer between slabs contains structurally disordered cations, $H_5O_2^+$, with corresponding short O . . . O distances of 0.244 and 0.241 nm for Fe and In sulphate, respectively. Phase transitions are observed in the 320–330 K temperature range and dehydration occurs above 360 K[51].

17.7 References

1. M. Ross and H. T. Evans, *Am. Mineral.* **49** (1964) 1578–1602; **50** (1965) 1–12.
2. R. W. T. Wilkins, A. Mateen and G. W. West, *Am. Mineral.* **59** (1974) 811–19.
3. I. Kh. Moroz, A. A. Valuera, E. A. Sidorenko, L. G. Zhiltosava and L. N. Karpova, *Geokhimiya* (1973) 210–23.
4. F. Weigel and C. Hoffmann, *J. Less-Common Metal* **44** (1976) 99–123.
5. L. V. Kobets and D. S. Umreiko, *Russ. Chemical Reviews* **52** (1983) 509–23.
6. J. Beintema, *Rec. Trav. Chim. Pays-Bas Belg.* **57** (1938) 155–75.
7. Y. Sugitami, H. Kauga, K. Nagashima and S. Fujiwara, *Nippon. Kogoku Zashi* **90** (1969) 52–6.
8. A. T. Howe and M. G. Shilton, *Mat. Res. Bull.* **12** (1977) 701–6.

Materials

9. L. V. Kobets, T. A. Kolevich and D. S. Umreiko, *Koord. Khim.* **4** (1978) 1856–60.
10. M. A. R. de Benyacar and H. L. De Dussel, *Ferroelectrics* **9** (1975) 241–5; M. A. R. de Benyacar, H. L. De Dussel and M. E. J. de Abeledo, *Am. Mineral.* 59 (1974) 763–5.
11. A. T. Howe and M. Shilton, *J. Solid State Chem.* (a) **28** (1979) 345–61; (b) **31** (1980) 393–9; (c) **34** (1980) 137–47; (d) **34** (1980) 149–55; (e) **34** (1980) 341–6; (f) **37** (1981) 37–43; S. B. Lyon and D. J. Fray, *Solid State Ionics* **15** (1985) 21–31.
12. V. Pekarek and M. Benesova, *J. Inorg. Nucl. Chem.* **26** (1969) 1793–1804; A. T. Howe in *Inorganic Ion Exchange Materials*, A. Clearfield (ed) (CRC Press Inc., Boca-Raton (1981)) 134–9.
13. A. V. Weiss, K. Harlt and U. Hofmann, *Z. Naturforsch.* **12b** (1957) 351–5; C. M. Vershoor and A. B. Ellis, *Solid State Ionics* **22** (1986) 65–8; G. Lagaly, *Solid State Ionics* 22 (1986) 43–51; L. Moreno Real, R. Pozas Tormo, M. Martinez Lara and S. Bruque, *Mat. Res. Bull.* **22** (1987) 19–27.
14. R. Mercier, M. Pham-Thi and Ph. Colomban, *Solid State Ionics* **15** (1985) 113–26.
15. Ph. Colomban, M. Pham-Thi and A. Novak, *Solid State Commun.* **53** (1985) 747–51.
16. M. Pham-Thi and Ph. Colomban, *Solid State Ionics* **17** (1985) 295–306.
17. K.-D. Kreuer, A. Rabenau and R. Messer, *Appl. Phys.* A32 (1983) 45–53; 155–8.
18. G. Velasco, *Solid State Ionics* **9** & **10** (1983) 783–92.
19. A. T. Howe, S. H. Sheffield, P. E. Childs and M. G. Shilton, *Thin Solid Films* **67** (1980) 365–70.
20. M. Pham-Thi and G. Velasco, *Rev. Chim. Mineral.* **22** (1985) 195.
21. H. Kahil, M. Forestier and J. Guitton, in *Solid State Protonic Conductors III*, J. B. Goodenough, J. Jensen and A. Potier (eds) (Odense University Press (1985)) 84–99; P. de Lamberterie, M. Forestier and J. Guitton, *Surface Techn.* **17** (1982) 357–67.
22. M. Pham-Thi, Ph. Adet, G. Velasco and Ph. Colomban, *Appl. Phys. Lett.* **48** (1986) 1348–50.
23. J. M. Schreyer and C. F. Baes Jr., *J. Am. Chem. Soc.* **76** (1954) 354–6.
24. A. B. Ellis, M. M. Olken and R. N. Biaginioni, *Inorg. Chim. Acta* **94** (1984) 89–90.
25. F. Weigel and G. Hoffmann, *J. Less-Common Metal.* **44** (1976) 99–123.
26. M. Pham-Thi and Ph. Colomban, *J. Less-Common Metal.* **108** (1985) 189–216.
27. Ph. Colomban and M. Pham-Thi, *Rev. Chimie Minerale* **22** (1985) 143–60.
28. J. Metcalf-Johansen and H. Boye, *Mater. Letts* **3** (1985) 513–21.
29. M. Pham-Thi, A. Novak and Ph. Colomban, in *Solid State Protonic Conductors III*, J. B. Goodenough, J. Jensen and A. Potier (eds) (Odense University Press (1985)) 143–62; M. Pham-Thi, Ph. Colomban and A. Novak, *J. Phys. Chem. Solids* **46** (1985) 493–504; 565–78.

30. G. Velasco, Ph. Adet, Ph. Colomban and M. Pham-Thi, French Patent 8509653; Eur Pat. Appl. EP 208 589 (14 Janv. 1987).
31. E. Manghi and G. Polla, *J. Crystal Growth* **63** (1983) 606–14; **74** (1986) 380–4; E. Manghi and G. Polla, *J. Mater. Sci. Letts* **6** (1987) 577–9.
32. L. V. Kobets, T. A. Kolevich, D. S. Umreiko and V. N. Yaglov, *Zh. Neorgan. Khim.* **22** (1977) 45–8.
33. M. Pham-Thi and Ph. Colomban, *J. Mater. Sci.* **21** (1986) 1591–600.
34. B. Morosin, *Acta Cryst.* **B34** (1978) 3732–4.
35. A. Novak, *Structure and Bonding* **18** (1974) 177–216.
36. M. A. R. De Benyacar and H. L. De Dussel, *Ferroelectrics* **17** (1978) 479–2.
37. J. C. Badot, N. Baffier, A. Fourier-Lamer and Ph. Colomban, *J. Physique (Paris)* **48** (1987) 1325–36.
38. A. N. Fitch, L. Bernard, A. T. Howe, A. F. Wright and B. E. F. Fender, *Acta Cryst.* C39 (1983) 156–62; A. N. Fitch, A. F. Wright and B. E. F. Fender, *Acta Cryst.* **B38** (1982) 2546–54; G. J. Kearley, A. N. Fitch and B. E. F. Fender, *J. Mol. Struct.* **125** (1984) 229–41.
39. Ph. Colomban, J. C. Badot, M. Pham-Thi and A. Novak, *Phase Transitions* **14** (1989) 55–68.
40. C. Poinsignon, A. Fitch and B. E. F. Fender, *Solid State Ionics* **9 & 10** (1983) 1049–64.
41. K.-D. Kreuer, *J. Mol. Struct.* **177** (1988) 265–76.
42. Ph. Barboux, Thesis, University P.&M. Curie, Paris, 1987; Ph. Barboux, R. Morineau and J. Livage, *Solid State Ionics* **27** (1988) 221–5.
43. Ph. Colomban and A. Novak, *J. Mol. Struct.* **177** (1988) 277–308.
44. D. Gingerow, *Acta Crust.* C44 (1988) 421–4.
45. W. P. Bosman, P. T. Beurskens, J. M. M. Smits, H. Behn, J. Mintgens, W. Meisel and J. C. Fuggle, *Acta Cryst.* C42 (1986) 525–8.
46. D. K. Smith, J. W. Gruner and W. N. Lipscombs, *Am. Mineral.* **42** (1957) 594–618.
47. Y. Wing, M. Lal and A. T. Howe, *Mat. Res. Bull.* **15** (1980) 1649–54; S. B. Hendricks, *Am. Mineral* **22** (1937) 773–80.
48. J. A. Ripmester, C. I. Ratcliffe, J. E. Dutrizac and J. L. Jambor, *Can. Mineral.* **24** (1986) 435–47.
49. Ph. Colomban, C. Dorémieux-Morin, Y. Piffard, M. H. Limage and A. Novak, *J. Mol. Struct.* **213** (1989) 83–96.
50. Ph. Colomban, H. Perthuis and G. Velasco, in *Solid State Protonic Conductors II*, J. B. Goodenough, J. Jensen and M. Kleitz (eds) (Odense University Press (1983)) 375–90.
51. I. Brach, Thesis, University of Montpellier, 1986, France; D. Jones, J. Rozière, J. Penfold and J. Tomkinson, *J. Mol. Struct.* **195** (1989) 283–91; **197** (1989) 113–21; I. Brach and J. B. Goodenough, *Solid State Ionics* **27** (1988) 243–9.
52. K. Mereiter, *Tschermaks Min. Petr. Mitt.* **21** (1974) 216–32.
53. J. Tudo, B. Jolibois, G. Laplace and G. Nowogrocki, *Acta Cryst.* **B35** (1979) 1580–3.

18 From crystalline to amorphous (particle) hydrates: inorganic polymers, glasses, clays, gels and porous media

PHILIPPE COLOMBAN AND ALEXANDRE NOVAK

Protonic conduction is encountered in a series of 'composite' compounds ranging from clays to 'porous' polymer structures via gels. In all these materials the distinction between the host lattice and protonic species becomes difficult because of the 'softening' of the rigid lattice. The rigid framework can be composed of big ions such as *Keggin salts* ($d \sim 2.5$ nm), amorphous inorganic polymers ($d \sim 2.5$–10 nm) or very small crystallites or particles such as the well-known clay minerals ($d \sim 50$–100 nm) (Fig. 18.1). The latter can also be prepared in the gel state if interaction with water is sufficiently strong. Some other materials, zirconium phosphate hydrate, $Zr(HPO_4)_2 \cdot nH_2O$, for instance, can be obtained in the crystalline, amorphous and gel states[1,2]. The formation of a gel depends on the existence of defects involving both the framework and associated protonic species and can be described in the following way: 'dissolution' of the host lattice leads to the formation of an associated water layer between solid and liquid, the increase of the solid–liquid interface being correlated. Here can also be cited the alteration phenomena of well-known silico-aluminates (mica, feldspar) in clays by proton exchange of first and second group cations. Fig. 18.2 summarizes the relationships existing between these materials and those discussed in the preceding chapters, (β-alumina, NASICON and HUP). The differentiation between the structure and properties of 'surface' and 'bulk' remains valid if a chemically and thermally stable framework, i.e. of low 'water' content, is considered. At the other extreme, there are microporous structures, frequently called *fractals*[3-5] where the distinction between bulk and surface is not clear.

The particular feature of a surface fractal is a dimensionality which does not correspond to a whole number, for example 1.75 instead of 2. This is due to the extreme 'division' of the surface. Surface fractals have the self-similarity property in the sense that their geometric features do not change if the surface is magnified. Mathematically, surface self-similarity

Crystalline to amorphous hydrates

is represented by the equation $S \sim R^{Ds}$ where S is the surface and Ds is the surface fractal dimension. For a smooth object, $Ds = 2$ implying that a smooth surface is two-dimensional (2D). For fractally rough surfaces, on the other hand, Ds varies between 2 and 3 and its value is a measure of roughness. In the case of linear polymers the characteristic fractal dimension is related to the mass: $M = R^D$. D is equal to 1, 2, and 3 for rods, discs and spheres, respectively, while for fractal 'linear' polymers, D varies between 1.7 and 2.

Such fractal materials with a large specific surface area are usually prepared by hydrolysis of chlorides[6, 7] or alkoxides[8–10]. Fig. 18.3 shows that porosity distribution covers a very broad domain from a few angstroms to several micrometres, i.e. from the dimensions of a cationic site in a structure like β-alumina or HUP (0.5 nm) or a zeolite cavity (0.5–2 nm) to the pores of membranes and organic polymers (5 nm–10 μm). Clays that are gels exist in the form of apparently dry powders in spite of their water content – typically 2 to 20 water molecules per mole of oxide – and are transformed by pressure sintering to objects

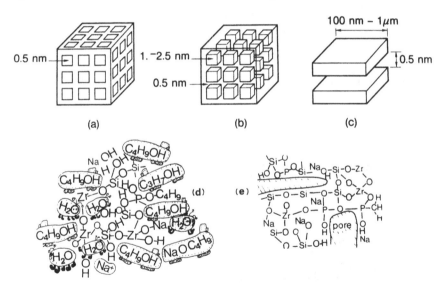

Fig. 18.1. Schematic representation of 'rigid' skeletons in hydrous oxides: (a) a three-dimensional lattice in which small cation sites are interconnected (e.g. Nasicon, $H_2Sb_4O_{11} \cdot nH_2O$, hydrogen β-alumina); (b) crystalline heteropolyacids with Keggin anions and interconnected liquid containing protonic species: particle hydrate structure is assumed to be of the same type but disordered; (c) layer structure (as in e.g. clays, V_2O_5); (d) skeleton of a gel obtained by alkoxide hydrolysis of a complex composition (solvent in the pores is a water–ethanol mixture); (e) porous glass obtained by thermal treatment of the above gel.

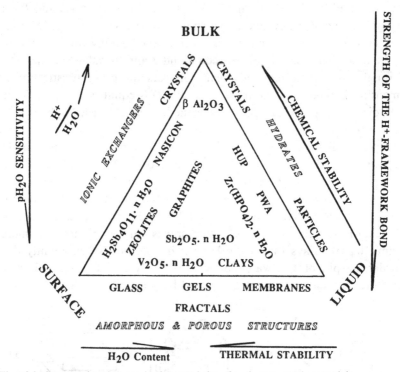

Fig. 18.2. Structural and dynamic characteristics of various protonic materials.

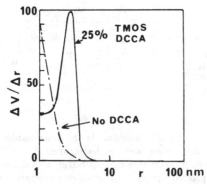

Fig. 18.3. Porosity distribution in a silica gel obtained by alkoxide hydrolysis (TMOS: tetramethoxysilane). Drying control chemical agent (DCCA) allows a modification of porosity distribution[10].

similar to ceramics except for their lack of thermal stability. Their mechanical strength is due to a sintering phenomenon which takes place at room temperature for hydrates (see Chapter 17) and which can be compared to high temperature sintering (800–1800 °C) in ceramics. Several mechanisms leading to densification, such as viscous flow, dissolution–recrystallisation and hydrogen bonding, are possible.

In this chapter, a few typical examples are presented in order to illustrate the main characteristics of hydrous oxides. They include heteropolyacids derived from Keggin's salts, high water content materials with a two-dimensional gel-forming framework, such as $Zr(HPO_4)_2 . nH_2O$, clays, $V_2O_5 . nH_2O$, inorganic polymers prepared by alkoxide hydrolysis and glasses.

18.1 Hydrous heteropolytungstic (molybdic, silicic) acids

18.1.1 Structures and properties

Heteropolyacids exist in a series of hydrated phases, the stable forms of which depend strongly on temperature and relative humidity (Fig. 18.4). They show very high proton conductivity[11–13], $\sigma_{300K} = 8 \times 10^{-2} \Omega^{-1}$ cm^{-1}. The basic structural unit of these acids is the Keggin anion cluster $(PM_{12}O_{40})^{3-}$ where $M = W$ $(H_3PW_{12}O_{40} . nH_2O)$, Mo $(H_3MoW_{12}O_{40} . nH_2O)$ or Si $(H_4SiW_{12}O_{40} . nH_2O)$[14] hereafter referred to as *PWA*, *PMoA* or *PSiA*. Other types of anion are known[15] and the proton may be exchanged by other monovalent cations[16]. The water content (*n*) may be 28–30, 18–21, 12–14, 6 and 0 and depends on M and humidity[11]. The compounds are generally soft and have a tendency to decompose under pressure. The tungsten-containing polyanion can be reduced electrochemically and changes from white to blue, which makes it suitable for use in displays or as inorganic resists[17,18]. A photochemical reduction in the presence of an organic reducing agent can also occur[19]. These materials can thus be used in catalysis[20] and the production of hydrogen[21].

The polyacid structure (Fig. 18.5) consists of an array of globular Keggin ions linked by protonated water layers. The structure of this layer of high water content is not known accurately and only oxygen atoms have been localized[22]. They show a large 'thermal' motion which is likely to be associated with a high dynamic and/or static disorder. The O . . . O distances are close to 0.25–0.26 nm[23]. The structure of the heteropolytungstic acid hexahydrate can be represented by the formula

275

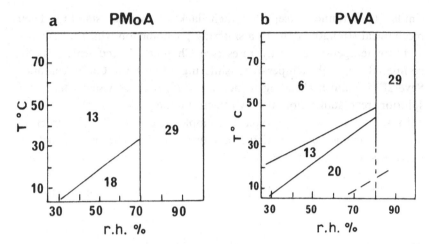

Fig. 18.4. Temperature and relative humidity (r.h.) ranges of $H_3W_{12}PO_{40} \cdot nH_2O$ under an atmosphere of either oxygen or air; with permission[11].

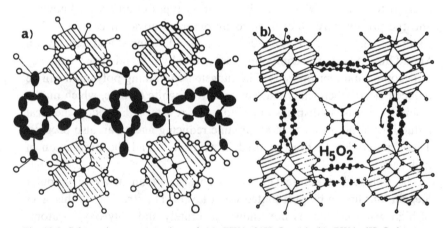

Fig. 18.5. Schematic representation of (a) $PWA \cdot 21H_2O$ and (b) $PWA \cdot 6H_2O$ hetero-polyacids. $PW_{12}O_{40}$ groups are connected by a protonated water layer (a) or disordered $H_5O_2^+$ ions (b)[22,23] (with permission). Hatched polyhedra, Keggin ions. Black ellipses, ellipsoids of protonic species (oxygen and hydrogen) taking into account the thermal motion and the disorder.

Crystalline to amorphous hydrates

$(H_5O_2)_3PW_{12}O_{40}$ and contains a symmetric planar dioxonium ion, $H_5O_2^+$, with an O ... O distance of 0.241 nm. An alternative description consists of saying that H_2O ... H groups are flat pyramids[24]. Neighbouring dioxonium ions are not associated, the corresponding intermolecular O ... O distance being about 0.4 nm[23].

18.1.2 Conductivity and proton dynamics

Kreuer *et al.* have studied the proton transport in single crystals which yield slightly lower conductivity values than those of compacted powders[13,25]. There was no significant difference for hydrogenated and deuterated compounds. This indicates that the *Grotthuss mechanism* is not the main one and that a vehicle mechanism can be involved, as in other proton conductors, such as oxonium and ammonium β-alumina, HUP and antimonic acid. This appears to be consistent with a high structural disorder of oxygen atoms of the protonic species and with non-stoichiometry of the water content.

Heteropolyacids $H_3PW_{12}O_{40} \cdot 14H_2O$ [26,27], $H_3PW_{12}O_{40} \cdot 21H_2O$ [28] and $H_3PMo_{12}O_{40} \cdot 21H_2O$ [28] have been studied by *quasielastic neutron scattering (QNS)* and *nuclear magnetic resonance (NMR)*. QNS results on PWA . $14H_2O$ show a slow, 180° reorientation of the water molecule in the dioxonium ion, with a radius of gyration of 0.076 nm, a mean residence time of 3×10^{-11} s at 295 K and an activation energy of 0.27 eV. The NMR data give a proton diffusion coefficient of about 10^{-7} cm^2 s^{-1} at 295 K which corresponds to a slightly lower conductivity than obtained by electrical measurements[26]. The activation energy values of different hydrates vary as a function of the method used (pulsed field gradient NMR, T_2 relaxation time NMR and conductivity); values vary by as much as a factor of three, the highest values usually being obtained by conductivity measurements.

Fig. 18.6 shows the correlation time, (τ), distribution for protons in PWA . $21H_2O$ water in zeolite 13-X and charcoal[27]. In charcoal, water is adsorbed by the essentially non-polar environment and the τ value is similar to that of layered water. In the zeolite, the pores are smaller (0.9–1.3 nm), the motion is more restricted and the τ value is two orders of magnitude lower. Still lower values have been determined for Nafion^R, while the lowest ones have been observed in PWA . $21H_2O$ due to the small (0.7 nm) dimensions of the region available for proton motion. The gradual decrease of correlation time can also be related to the

277

increasing interactions between a framework of higher acidity and water, in going from charcoal to the *Keggin salts* (Chapter 1).

18.2   Water layers in 2D frameworks

Two-dimensional lattices are particularly favourable for the formation of protonated water layers which can interact to varying extents with the supporting lattice. Clays are among the best known examples. Numerous compounds which can be prepared in the colloidal state show similar properties. Some materials are real ionic exchangers and can accommodate between their layers large quantities of various species such as hydrated or non-hydrated ions, polyfunctional organic molecules, solvents and other molecules.

18.2.1   Clays

18.2.1.1 Framework structure

The hydrous layer silicates commonly known as clay minerals belong to the large family of phyllosilicates[29]. The layer silicates contain continuous two-dimensional tetrahedral sheets in which tetrahedra are linked to form

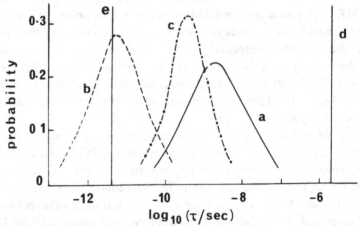

Fig. 18.6. Correlation time distributions at 0 °C for protons in (a) PWA . 21H$_2$O, (b) water in charcoal and (c) water in zeolite 13-X. The correlation times have been calculated from T$_1$ and T$_2$ measurements[27]; with permission. The correlation times for ice (d) and water (e) are also shown.

Crystalline to amorphous hydrates

a hexagonal mesh pattern. The fourth tetrahedral corner points perpendicularly to the layer and at the same time forms part of an adjacent octahedral sheet. The common plane at the interface between tetrahedral and octahedral layers consists of shared oxygen atoms and unshared OH groups. The octahedral cations are usually Mg^{2+}, Al^{3+}, Fe^{3+} or Fe^{2+} whereas Al/Si substitution takes place in the tetrahedra. Electroneutrality of the layers is achieved by means of hydroxide groups and various interlayer entities including individual cations which may be hydrated to varying extents. Clay particles are usually smaller than 0.5 μm and only those of some kaolinites may be larger (several micrometres). The thickness of the layers is usually about ten times smaller than the shortest basal dimension.

18.2.1.2 Water layer

Water, the most common interlayer entity, is usually present in *montmorillonites* (smectites), vermiculites and hydrated halloysites[29, 30]. These three families differ by the sheet thickness and composition. The 'natural' cations may be substituted by other mono- or polyvalent inorganic cations, including ammonium and oxonium ions or organic (polyfunctional) cations[29-31]. One, two, three or four water layers are possible in montmorillonite and the interlayer spacing may be as large as 13 nm, while the usual layer thickness in clays varies between 0.7 and 1.4 nm[29, 32]. The layer stacking is often turbostratic. The specific surface area of montmorillonite particles is typically 700 $m^2 g^{-1}$ and a *gel* state can be achieved with only 0.02–0.04 g clay cm^{-3}. The average distance between particles increases thus to 100 nm with a water layer 10–100 times larger than the particle.

When clay minerals are treated with dilute acids ('activation' process), protons may attack the silicate layers via the interlayer region and exposed edges. Octahedral ions such as Al^{3+} and Mg^{2+} are extracted into the interlayer which promotes a rapid decomposition. However, many montmorillonites resist such a treatment, even when using concentrated acids[29]. Formation of layer defects allows the number of anchoring points for new cations at the surface layer to increase.

18.2.1.3 Electrical properties and proton dynamics

Room temperature conductivity of Na, Ca, Li and 'protonated' clays varies from 10^{-2} to $10^{-4} \Omega^{-1} cm^{-1}$ as a function of water content[33, 34]. Attempts to distinguish the contribution of 'solid clay particles' and of

Materials

'interparticle liquid water' have been made only for Na–Ca montmorillonite gels and it was concluded that the mobility of exchangeable ions increased with increasing concentration[35].

Water dynamics in clays has been studied for Na and Li montmorillonites, vermiculites and hectorites[36,37]. Quasi-elastic neutron scattering and nuclear magnetic resonance indicate the existence of slow and fast motions of water molecules. The fastest motion ($\tau_{RT} \sim 10^{-12}$ s) has the same value as in normal water and is almost temperature independent. The slow motion is anisotropic and associated with the hydrated cation shell. In materials with high water content (interparticular liquid-like phase) the distinction between rotational and translational motions appears possible[36,38].

18.2.2 Zirconium phosphate hydrate

Clay particles allow gel formation by formation of a thick, liquid-like water layer around the microparticles resulting from cations compensating the lack of charge on the framework. The particle size, however, remains limited. Hydrated zirconium phosphate, $Zr(HPO_4)_2 . nH_2O$, on the other hand, can be obtained as a transparent monolithic gel as well as in a powdered amorphous state (particle size <0.1 μm) or in a crystalline state (~ 1 μm–1 mm)[1,2,39]. This compound has been known for a long time for its ion exchange properties, particularly for treatment of irradiated materials and in artificial kidneys. The crystal structure has been investigated by numerous authors[40–43] (see Chapter 16): the arrangement of zirconium and phosphate ions has been determined while the distribution of protonic species is not well-known, especially in amorphous and poorly crystallized samples. Degree of crystallinity and water content are interdependent.

Well-crystallized compounds can be represented by the formula $Zr(HPO_4)_{2-x}(H_3OPO_4)_x(n-x)H_2O$, with $n \sim 1$–1.5 and $x \sim 0.5$ for a sample at room temperature and normal humidity[2]. The compositions of optically clear samples and amorphous powder, on the other hand, are roughly given by $Zr(HPO_4)_2 . nH_2O$ with $n = 6$ and 12, respectively. Loss of crystallinity is associated with defect formation in $Zr(PO_4)_2$ slabs, the appearance of HP_2O_7 and perhaps OH groups directly bonded to zirconium ions. All these defects favour the formation of a water shell around the particles. Modification of the hydrogenated network certainly occurs. There is a large variety of hydrogen bonding, involving H_3O^+, PO_4, H_2O, P–O and P–OH groups.

The conductivity of amorphous, semi-crystalline and well-crystallized $Zr(HPO_4)_2 . nH_2O$ has been determined by different authors. Room temperature values vary between $6 \times 10^{-3}\,\Omega^{-1}\,cm^{-1}$ for amorphous compound ($n \sim 4H_2O$) and $5 \times 10^{-6}\,\Omega^{-1}\,cm^{-1}$ for crystalline material ($n \sim 1$)[44–46]. Higher values ($2 \times 10^{-4}\,\Omega^{-1}\,cm^{-1}$) are observed for oriented polycrystalline samples[47]. The above values have been obtained at 90% relative humidity, while in a dehydrated sample ($n < 1$) the conductivity drops to $10^{-9}\,\Omega^{-1}\,cm^{-1}$. The activation energy is about 0.3 eV up to 200 °C and increases to 0.7 eV at higher temperatures when oxonium ions are destroyed. The conductivity depends strongly on the 'capacity for surface adsorption' and the surface conductivity thus plays a dominant role. This assumption is supported by the fact that the conductivity depends on the partial water pressure. The surface water constitutes an important part of the material, particularly so for smaller particles and amorphous material. The activation energy (0.3 eV) appears to be too high for a pure *Grotthuss mechanism* and probably reflects a participation of both H_2O and H_3O^+ hopping. At high temperature, the higher activation energy doubtless corresponds to proton diffusion in a hydrated lattice of low water content as in protonic β-alumina above 400 °C[48].

18.2.3 V_2O_5 'gels'

V_2O_5, WO_3 and Nb_2O_5 hydrous gels can be prepared by polycondensation of *vanadic, tungstic* and *niobic acid*, respectively. $V_2O_5 . nH_2O$ gel, for example, can be obtained by an appropriate treatment of decavanadic acid and corresponds to a polymer of high molecular weight. It forms crystalline fibres (0.5 μm × 10–50 nm) similar to clay minerals[49–51]. Its internal structure is closely related to orthorhombic V_2O_5. Spontaneous dehydration leads to an almost dry solid or xerogel of approximate formula, $V_2O_5 . 1.6H_2O$. The presence of oxonium ions has been detected by IR spectroscopy; a formula which better describes V_2O_5 gel is $(H_3O)_x V_{2-x}^V V_{5-y}^{IV}(OH)_{2y} . nH_2O$ with $x + 2y \leqslant 0.2$–0.4[52]. Water content (n) can vary from 0.5 to 20 and at higher values, the gel is transformed to a colloidal sol. V_2O_5 gels have semi-conductor properties and their protonic conductivity has been ignored for a long time. In fact, in samples with $n \sim 0.5$ the conductivity ($E_a \sim 0.3$ eV) is electronic and consists of a polaron corresponding to an electron jump between V^{IV} and V^V ions[53-55]*. Compounds with $n \sim 1.6$ contain water and oxonium molecules between layers and protonic conductivity is expected. The activation energy is 0.35 and 0.42 eV above and below 263 K, respectively, and implies an important participation of H_3O^+ hopping (the V^{4+} content is lower than 1%). Water and oxonium ions disappear above 455 and 500–600 K, respectively. The room temperature conductivity of V_2O_5 gel is about $10^{-3}\,\Omega^{-1}\,cm^{-1}$ with a maximum value of $10^{-2}\,\Omega^{-1}\,cm^{-1}$ for $n = 20$[54].

Collapse of the structure with water freezing below 265 K is observed for samples with $n = 5.6$ as occurs similarly in montmorillonite and bentonite clays[56,57]: part of the water freezes on cooling while the remaining water is not frozen. NMR data have been interpreted in terms of a chemical exchange between almost liquid-like and high viscosity regions[58].

18.3   Hydrous oxides

18.3.1   Chemical and topological considerations

Hydrous oxides are a class of inorganic (mostly cation) exchangers and thus (potentially) ionic and protonic conductors. They have been described as particle hydrates consisting of irregular, amphoteric charged particles separated by aqueous solution. It was assumed that the structure within the particle is that of the anhydrous (amorphous) oxide and that (colloidal) particles of about 2.5–5 nm associate to form larger agglomerates[51,58]. The acidity of the interparticle solution depends on the extent of protonation of the surface oxygen. Amphoteric substances represented by M–OH can dissociate into $M^+ + OH^-$ or into $MO^- + H^+$ for anion and cation exchanger, respectively. The first type of dissociation is favoured in acidic and the second in basic medium while both can occur near the isoelectric point. The basicity of M and the strength of the M–O bond with respect to that of the O–H bond play an important role in determining the dissociation type.

Forced hydrolysis of metal salts can be used to prepare colloidal metal

---

* Two-terminal switching devices based on 'semi-conducting' $V_2O_5 \cdot 1.6H_2O$ films were patented by Bullot & Livage[88]. The switching seems to occur when the input joule heating in the channels exceeds the heat loss from the channels, driving a dehydration and transformation into VO_{2-x}[89].

Crystalline to amorphous hydrates

oxides (Cr, Al, Th, Fe, Cu, Ti)[7, 59]. Matijevic showed that metal ions known for their tendency to polymerize give 'amorphous' particles, whereas metal ions that hydrolyze to well-defined species yield crystalline solids[7]. This description, however, appears much too simple since alkoxide hydrolysis, for instance, leads to amorphous hydrous oxides ('gels') of almost any composition[10]. Alkoxide hydrolysis is related to polycondensation reactions as follows

$$M(OR)_n + nH_2O \rightarrow M(OH)_n + nROH$$

$$M(OH)_n \rightarrow MO_{n/2} + n/2H_2O$$

The idea of separation between a non-hydrated solid and a 'liquid' (Fig. 18.7) must be replaced by a structural or defect gradient related to the presence of protons in the bulk, e.g. 2M–OH instead of M–O–M bonds, inducing microporosity (0.1 nm scale) which is linked to macroporosity (10 μm scale); thus water invades the pores until it becomes part of the lattice and develops particular properties due to hydrogen bonding. Additional water molecules can occupy inter-particle space producing a connected viscous liquid region that permeates the 'composite solid'. A protonation equilibrium exists between the outer particle surface and the interparticle 'liquid'.

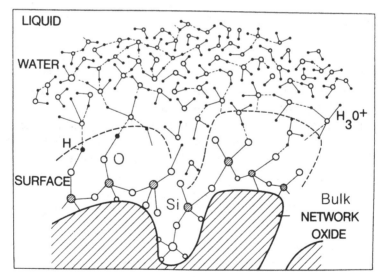

Fig. 18.7. Schematic illustration of water layers which can be formed on an oxide or hydroxide surface or when oxonium ions are present.

Materials

The adsorption and exchange properties of hydrous oxides have been extensively studied; the attempts to understand the local structure on the 0.1–1 nm scale, however, are quite recent. Hydrous oxides are known for multivalent metals (Be, Mg, Zn, Fe, Al, Mn, La, Bi, Cr, In, Sb, Si, Sn, Th, Zr), most of them being amorphous[54,44,60–63].

18.3.2 Antimonic acid

Antimonic acid can be obtained, similarly to $Zr(HPO_4)_2 \cdot nH_2O$, with different water content and degree of crystallinity. Almost amorphous $Sb_2O_5 \cdot nH_2O$ ($n = 5$–6.5) can be prepared by $SbCl_5$ hydrolysis[58,60,64] or by pressing aqueous $KSb(OH)_6$ solution through an ion exchange column. Aging of the gel leads to crystalline material with cubic pyrochlore structure ($A_2M_2O_6O'$) of approximate formula, $Sb_2O_5 \cdot 4H_2O$. It has a 3D array of corner-sharing Sb_2O_6 octahedra with water molecules distributed over the A_2O' array, protons being in the vicinity of A[58,65].

Crystalline antimonic acid, $H_2Sb_4O_{11} \cdot nH_2O$ ($0 \leqslant n \leqslant 3$), can be prepared by ionic exchange from $K_2Sb_4O_{11}$[66] and has the same $(Sb_4O_{11})^{2-}$ framework structure built of edge- and corner-sharing SbO_6 octahedra[67]. This framework possesses large channels containing potassium ions or water. Three kinds of protonic species, H_3O^+, H_2O and OH, exist in antimonic acid and the formula, $(H_3O)_{2-x}Sb_4O_{11-x}(OH)_x \cdot nH_2O$, in which both n and x vary with temperature, appears more realistic[68].

The conductivity curves, $\sigma = \sigma_0 \exp(-E_a/kT)$, of the crystalline forms with $n = 2$ and 3, show two slopes corresponding to $E_a = 0.67$–0.69 and 0.42–0.47 for low and high temperatures, respectively[69]. The change of slope occurs at 160 and 333 K, respectively, the higher water content giving rise to the lower transition temperature. This may indicate that the filling up of channels by water molecules leads to higher conductivity, not only via the σ_0 factor which is proportional to the (higher) number of mobile species but also by increasing the dynamical disorder and thereby reducing E_a. The room temperature conductivity is 2×10^{-3} and $10^{-6} \, \Omega^{-1} \, cm^{-1}$ for $n = 3$ and 2, respectively[69].

Infrared spectra show that the main kinds of protonic species exist as chain fragments of the $H_3O^+(H_2O)_n$ type and isolated H_3O^+/H_2O species which are hydrogen bonded to the rigid framework. A smaller contribution consists of protons trapped in the $Sb_4O_{11}^{2-}$ network in equilibrium

Crystalline to amorphous hydrates

following the reaction

$$H_3O^+ \leftrightharpoons H_2O + H^+/Sb_4O_{11}$$

The proton transfer requires simultaneous hopping of H_3O^+ ions between different sites of the channel and a proton transfer along the chain fragments. The latter can be realized following the *Grotthuss mechanism*. The same applies to proton transfer to the rigid framework. The activation energy corresponding to the jumps of H_3O^+ or H_2O in an open structure can be estimated – by analogy with other protonic conductors – to be about 0.3–0.5 eV while higher values are necessary for jumping through the bottleneck in the channel. In a 'quasi liquid' state, on the other hand, the activation energy of a rotation is close to 0.1 eV and that of the proton jump lower than 0.2 eV. The energy corresponding to proton diffusion in a rigid framework can be much higher: in the case of the spinel structure obtained by decomposition of β- and β″-alumina, it is ~0.6 eV[48].

Phosphate antimonic acid $H_xSb_xP_2O_{3x+5} \cdot nH_2O$ ($x = 1$, 3, 5 and $n = 2$–10) has also been prepared by ionic exchange and exhibits room temperature conductivity in the range 10^{-4} to $10^{-3} \, \Omega^{-1} \, cm^{-1}$ with activation energies of 0.2–0.5 eV[70].

18.3.3   Other hydrous oxides

Many other hydrous oxides are rather good proton conductors (Table 18.1) but for most of them their structures have not been determined. NMR data concerning $SnO_2 \cdot nH_2O$ and $TiO_2 \cdot nH_2O$ indicate proton transport, including the exchange between surface hydroxyl groups and 'acid solution' in both micropores ($d < 10 \, nm$) and macropores ($d > 100 \, nm$)[71]. Many NMR features are similar to those of adsorbed water but different from zeolites where 'OH' groups are motionless on NMR time scales. No effect of water freezing in pores is observed in SnO_2 and TiO_2 hydrous oxides, contrary to framework structures such as HUP and antimonic acid or organic polymer ion-exchange membranes[68, 71].

$In_2O_3 \cdot nH_2O$ (or $In(OH)_3 \cdot nH_2O$) is obtained by reaction of In metal with nitric acid and translucent pellets can be prepared by pressing. X-ray powder patterns give some reflexions corresponding to the cubic structure $In(OH)_3$[60]. Ferric oxide seems also to be a proton conductor[72].

Many gels of the type $M_2O \cdot xZrO_2 \cdot yP_2O_5 \cdot zSiO_2 \cdot nH_2O$ prepared by alkoxide hydrolysis show a conductivity of the order of $10^{-5} \, \Omega^{-1} \, cm^{-1}$ ascribed in part to proton conduction[63]. The structure of the polymeric

Table 18.1. *Activation energy and conductivity of some hydrous oxides* $M_xO_y \cdot nH_2O$ *and* $HM_xO_y \cdot nH_2O$

Compounds	H$_2$O content n		σRT^a (Ω^{-1} cm^{-1})	E_a (eV) HT	E_a (eV) LT	T_c (K)	References
ThO$_2$	3–5	a	5×10^{-6}–4×10^{-4}	0.21	0.48	247	60
Sb$_2$O$_5$	3–6.5	a–p	10^{-4}–10^{-3}	0.16–0.23	0.29–0.31	253	60
In$_2$O$_3$	5–6	p	1–3×10^3	0.25			60
Cr$_2$O$_3$	8.5	p	10^{-5}	0.17			60
Ga$_2$O$_3$	4		10^{-4}	0.35			60
SnO$_2$	2–3		2–4×10^{-4}	0.24	0.4	288	58
ZrO$_2$	1.75–5.6		10^{-4}–10^{-5}	0.35			58
V$_2$O$_5$	0.5		10^{-4}	0.3		220	54, 55
	1.8		10^{-4}–10^{-2}	0.35	0.44	260	54, 55, 62
	12		10^{-2}	0.19			54
	75		10^{-2}	0.16			54

CeO$_2$	2	10^{-5}–10^{-6}	0.18–0.38		62
Nb$_2$O$_5$	3 a	10^{-4}–10^{-5}	0.18–0.34		62
Ta$_2$O$_5$	4 a	10^{-4}	0.25–0.39		62
HSbO$_3$	1–2.5 (ilmenite)	2×10^{-3}	0.19		61
	0.6 (cubic)	9×10^{-4}	0.39		38
	1–2 (pyrochlore)	3×10^{-3}	0.20		61, 64
HTaO$_3$	0.5–1 (pyrochlore)	10^{-4}	0.25		61
HNbO$_3$	0.3–0.8 (pyrochlore)	10^{-4}	0.22		61
NASIGEL	6–12 a	10^{-6}–10^{-5}			63
PSiA	28	2×10^{-2}	0.4		11.25
(H$_4$PSi$_{12}$O$_{40}$)					
PWA	6		0.28		11, 25
(H$_3$PW$_{12}$O$_{40}$)			0.33		11, 25
	14	7×10^{-3}	0.4		11, 25
	21	8×10^{-2}			11, 25
	28–29		0.15–0.25	281	11, 25

T_c, transition temperature; HT, above T_c; LT, below T_c; a, amorphous; p, poor crystallized.
[a] At water pressure $\sim 60\%$ (see references). NASIGEL: M$_2$O.xZrO$_2$.yP$_2$O$_5$.zSiO$_2$.nH$_2$O.

gel is very open with micro- and macroporosity which can support a 'surface conductivity'.

Finally, materials where organic functional branches have been grafted onto an inorganic backbone (*ORMOSIL*, ORganically MOdified SILicates), as shown below

$$\begin{array}{ccc}
| & & | \\
\text{O—Si—O} & & \text{—O—Si—CH}_2\text{—}\langle\bigcirc\rangle\text{—SO}_3\text{H} \quad + \text{acids} \\
| \quad + \text{HCF}_3\text{SO}_3 & & | \\
(\text{CH}_2) & & \\
| & & \\
\text{NH}_2 & &
\end{array}$$

correspond to a transition between organic polymers (Chapter 20) and silicate glasses. Proton conductivity between 10^{-3} and $10^{-7}\,\Omega^{-1}\,cm^{-1}$ was measured in the 25–100 °C temperature range. These materials are thermally stable up to 180–200 °C but are water sensitive[73]. A new family (*AMINOSIL*) $SiO_{3/2}$ R–(HX)$_x$ where R is an amine radical and HX an acid was recently prepared[74].

18.3.4 Silicate glasses

Protonic conduction can be introduced into a $(SiO_2)_n$ lattice by grafting certain groups, as shown above, or by using water layers adsorbed onto this lattice. However, 'lone' protons can also diffuse in silicates. It is well known that usual alkali silica glasses contain 0.1–1% of hydrogen and that certain glasses contain several percent hydrogen[75]. Glasses prepared by dealkalization and hydration in an autoclave are highly porous, similar to those obtained by alkoxide hydrolysis, low-temperature (300–600 °C) thermal treatment and rehydration.

Proton migration in a glass was demonstrated as long ago as 1927 using ammonium hydrogen sulphate melt as a proton source[76]. Hydrogen is present in these glasses in the form of water and OH$^-$ ions, as shown by characteristic infrared bands near 3300 and 3600 cm^{-1}, respectively[77,78].

Proton diffusion in glass induces cracks in the same way as occurs on substitution of Na$^+$ and K$^+$ by Li$^+$ ions which implies that large H$_3$O$^+$ ions are not formed. In fact, ^{18}O and ^{3}H diffusion measurements show that proton diffuses 100 times faster than oxygen[79]. This supports the assumption that the diffusion mechanism consists also of proton jumps and not only of OH$^-$ jumps. The corresponding activation energy has been estimated to be about 0.6–0.8 eV while the oxygen migration energy is close to 2.6 eV[80] which yields *diffusion coefficients* of 10^{-12} and 10^{-10} cm^2 s^{-1} for oxygen and proton, respectively[79–81].

As far as pH glass electrodes are concerned, the proton transfer is also complex: the conductivity mechanism is probably indirect, involving charge transport by alkali ions in the centre of the glass membrane. On the surface, a protonated water layer is formed, leading to surface conductivity of silica glass[82–84]. In the case of porous materials, this phenomenon predominates, as has been observed for many gels. Incidentally, the differentiation between a glass and a gel appears somehow arbitrary. Proton diffusion has been observed in silica gels, the surface of which was protonated by ammonialysis[85]. The *diffusion coefficient* at room temperature was 0.5×10^{-4} cm^{2}s^{-1} and the activation energy about 0.2 eV. One can assume that the silicate glasses undergo the following reactions[86]

$$H_2O + Na^+ \text{ (glass)} \leftrightarrows H^+ \text{ (glass)} + NaOH$$

and

$$2H_2O + Na^+ \text{ (glass)} \leftrightarrows H_3O^+ \text{ (glass)} + NaOH$$

The exact mechanism, however, remains to be determined. The facility to develop conducting hydrated phases has been related to the existence of $H_2Si_{14}O_{29} \cdot n H_2O$ type of compounds[87].

18.4   References

1. A. Clearfield (ed) *Inorganic Ion Exchange Materials* (CRC Press Inc., Boca Raton, Florida (1980)).
2. Ph. Colomban and A. Novak, *J. Mol. Struct.* **177** (1989) 277–308; **198** (1989) 277–96.
3. B. B. Mandelbrot, *Fractals, Form and Dimension* (San Francisco, Freeman (1977)).
4. D. W. Schaeffer, *Science* **243** (1989) 1023–7.
5. J. Teixeira, *J. Appl. Cryst.* **21** (1988) 781–5.
6. R. M. Dell, in *Proc. 7th Int. Reactivity of Solids*, J. S. Anderson, M. N. Roberts and F. S. Stone (eds) (Chapman and Hall, London (1972)) 553–66.
7. E. Matijevic, *Langmuir* **2** (1986) 12–20; *Pure Appl. Chem.* **50** (1978) 1193–210.
8. B. E. Yoldas, *J. Am. Ceram. Soc.* **65** (1977) 387–93.
9. S. Sakka and K. Kamiya, *J. Non-Crystalline Solids* **42** (1980) 403–22.
10. Ph. Colomban, *Ceramics Int.* **15** (1989) 23–50; T. W. Zerda, I. Artaki and J. Jonas, *J. Non-Crystalline Solids* **81** (1986) 365–79.
11. O. Nakamura, T. Kodama, I. Ogino and Y. Miyake, *Chem. Letts.* (1979)

17–18; (1980) 231–5; O. Nakamura, I. Ogino and T. Kodama, *Solid State Ionics* **3/4** (1981) 347–51.

12. A. Hardwick, P. G. Dickens and R. C. T. Slade, *Solid State Ionics* **13** (1984) 345–51.

13. K.-D. Kreuer, M. Hampele, K. Dolde and A. Rabenau, *Solid State Ionics* **28/30** (1988) 589–93.

14. J. F. Kegin, *Proc. R. Soc. Lond.* **A144** (1934) 75–100.

15. L. Chen, Y. Liu and V. Chen, *J. Solid State Chem.* **68** (1987) 132–6; H. T. Evans Jr, in *Perspective in Structural Chemistry*, Vol. 4, J. D. Dunitz and J. A. Iberseed (eds) (J. Wiley, New York (1971)); R. Contant, *Can. J. Chem.* **65** (1987) 568–73; H. Okamoto, K. Yamanoka and T. Kudo, *Mat. Res. Bull.* **21** (1986) 551–7.

16. E. A. Ukshe, L. S. Leonova and A. I. Korosteleva, *Solid State Ionics* **36** (1989) 219–23.

17. B. Tell and F. Wagner, *J. Appl. Phys.* **50** (1979) 5944–9; *Appl. Phys. Lett.* **33** (1978) 837–8.

18. W. C. Dautremont-Smith, Display, **1** (1982) 3–35; M. Tatsumisago and T. Minami, *J. Am. Ceram. Soc.* **72** (1989) 484–6; A. Ishikawa, H. Okamoto, K. Miyauchi and T. Kudo, *Denki Kagaku* **56** (1988) 538–42.

19. E. Papaconstantinou, *J. Chem. Soc. Chem. Commun.* (1982) 903; E. Papaconstantinou, D. Dimotikali and A. Politou, *Inorg. Chim. Acta* **46** (1980) 155–61.

20. G. A. Tsigdinos, Heteropolycompounds of Molybdenum and Tungsten, *Top. Curr. Chem.* **1** (1978) 76–112.

21. E. N. Savinov, S. S. Saldkhanov, V. N. Parmon, K. I. Parmon and K. I. Zamarev, *React. Kinet. Catal. Lett.* **17** (1981) 407–11.

22. M. R. Spirley and W. R. Busing, *Acta. Cryst.* **B34** (1978) 907–10.

23. G. M. Brown, M. R. Noe-Spirlet, W. R. Busing and H. A. Levy, *Acta. Cryst.* **B33** (1977) 1038–46; H. d'Amour and R. Allman, *Z. Krist.* **143** (1976) 1–13.

24. C. J. Kearley, H. A. Pressman and R. C. T. Slade, *J. Chem. Soc. Chem. Commun.* (1986) 1801–2.

25. K.-D. Kreuer, *J. Mol. Struct.* **177** (1988) 265–76.

26. R. C. T. Slade, I. M. Thompson, R. C. Ward and C. Poinsignon, *J.Chem. Soc. Chem. Commun.* (1987) 726–7.

27. A. Hardwick, P. G. Dickens and R. C. T. Slade, *Solid State Ionics* **13** (1984) 345–50.

28. R. C. T. Slade, J. Barker, H. A. Bressman and J. H. Strange, *Solid State Ionics* **28–30** (1988) 594–600.

29. G. W. Brindley and G. Brown (eds) *Crystal Structures of Clay Minerals and their X-Ray Identification*. Mineralogical Soc. Monograph. **5** (London (1980)).

30. W. F. Bradley, R. E. Grim and G. F. Clark, *Z. Kristallogr. Kristallgeom.* **97** (1937) 260–70.

31. M. M. Mortland, *Adv. Agron.* **23** (1970) 75–117; V. C. Farmer and M. M. Mortland, *J. Chem. Soc.* A(1966) 344–51.

32. K. Norrish, *Trans. Faraday Soc.* **18** (1954) 120–34.

Crystalline to amorphous hydrates

33. S. H. Sheffield and A. T. Howe, *Mat. Res. Bull.* **14** (1979) 929–37.
34. A. Cremers, J. Van Loon and H. Landelout, *Clays and Clay Minerals, Proc. 14th Natl. Conf. Berkeley, California 1965*, S. W. Bailey (ed.) (Pergamon Press, New York) 149–92.
35. I. Shainberg, J. O. Oster and J. D. Wood, *Clays and Clay Minerals* **30** (1982) 55–62.
36. D. Cebula, R. K. Thomas and J. N. White, *Clays and Clay Minerals* **28** (1980) 19–26.
37. P. L. Hall, D. K. Ross, J. J. Tuch and M. H. B. Hays, *Proc. Int. Conf. Oxford* **1978** (1979) 121–30.
38. J. Connard, H. Estrade-Szwarkopf, A. J. Dianoux and C. Poinsignon, *J. Physique (Paris)* **45** (1984) 1361–71.
39. G. Alberti, U. Costantino and R. Giuletti, *J. Inorg. Nucl. Chem.* **42** (1980) 1062–3.
40. G. Alberti, M. Casciola and U. Costantino, in *Solid State Protonic Conductors III*, J. B. Goodenough, J. Jensen and A. Potier (eds) (Odense University Press (1985)) 215–46.
41. A. Clearfield and D. Smith, *Inorg. Chem.* **8** (1969) 431–6.
42. J. M. Troup and A. Clearfield, *Inorg. Chem.* **16** (1977) 3311–14.
43. J. Albertsson, A. Oskarsson, R. Tellgreen and J. O. Thomas, *J. Phys. Chem.* **81** (1977) 1574–8.
44. G. Alberti, M. Casciola, U. Costantino and R. Redi, *Gazz. Chim. Ital.* **109** (1975) 921–2.
45. E. Krogh-Andersen, I. G. Krogh-Andersen, C. Knakkegard-Moller, E. D. Simonsen and E. Skou, *Solid State Ionics* **7** (1982) 301–7.
46. G. Alberti, *Acc. Chem. Res.* **11** (1978) 163–98.
47. G. Alberti, M. Casciola, U. Costantino and M. Leonardi, *Solid State Ionics* **14** (1984) 289–95.
48. N. Baffier, J. C. Badot and Ph. Colomban, *Solid State Ionics* **2** (1981) 107–13; **13** (1984) 233–6.
49. P. Aldebert, N. Baffier, N. Gharbi and J. Livage, *Mat. Res. Bull.* **16** (1981) 669–74; J. Livage and J. Lemerle, *Ann. Rev. Mater. Sci.* **12** (1982) 103–22.
50. J. J. Legendre, P. Aldebert, N. Baffier and J. Livage, *J. Colloid Interface Sci.* **94** (1983) 84–91.
51. P. J. Wiseman in *Progress in Solid Electrolytes*, T. A. Wheat, A. Ahmad and A. K. Kuriakose (eds) (CANMET, Energy Mines and Resources Canada, Ottawa (1983)).
52. L. Znaidi, Thesis, University of Paris VI (1989). L. Znaidi, N. Baffier and D. Lemordant, *Solid State Ionics* **28/30** (1988) 1750–5.
53. A. Nareska and W. Streich, *Colloid Polymer Sci.* **258** (1980) 379–85.
54. Ph. Barboux Thesis (1987), University Paris VI.
55. Ph. Barboux, N. Baffier, R. Morineau and J. Livage, *Solid State Ionics* **9/10** (1983) 1073–6.
56. P. Aldebert, H. W. Haesslin, N. Baffier and J. Livage, *J. Colloid Interface Sci.* **98** (1984) 484–8.
57. D. M. Anderson and P. Hoekstra, *Soil Sci. Soc. Amer. Proc.* **29** (1965) 498–504.

Materials

58. W. A. England, M. G. Cross, A. Hamnett, P. J. Wiseman and J. B. Goodenough, *Solid State Ionics* 1 (1980) 231–49.
59. V. Vesely and V. Pekarek, *Talanta* 19 (1972) 219–62.
60. D. J. Dzimitrowicz, J. B. Goodenough and P. J. Wiseman, *Mat. Res. Bull.* 17 (1982) 971–9.
61. U. Chowdry, J. R. Barkley, A. D. English and A. W. Sleight, *Mat. Res. Bull.* 17 (1982) 917–33.
62. R. C. T. Slade, J. Barker and T. K. Halstead, *Solid State Ionics* 24 (1987) 147–53.
63. Ph. Colomban and J. P. Boilot, *Rev. Chim. Minérale* 22 (1985) 235–55.
64. X. Turillas, G. Delabouglisse, J. C. Joubert, T. Fournier and H. Muller, *Solid State Ionics* 17 (1985) 169–73.
65. C. M. Mari, A. Anghileri, M. Catti and G. Chiodelli, *Solid State Ionics* 28–30 (1988) 642–6.
66. Y. Piffard, M. Dion and M. Tournoux, *C.R. Acad. Sci.* C290 (1980) 437–41.
67. H. Y. P. Hong, *Acta Cryst.* B30 (1974) 945–50.
68. Ph. Colomban, C. Dorémieux-Morin, Y. Piffard, M. H. Limage and A. Novak, *J. Mol. Struct.* 213 (1989) 83–96.
69. H. Arribart, Y. Piffard and M. Dorémieux-Morin, *Solid State Ionics* 7 (1982) 91–9.
70. S. Deniard-Courant, Y. Piffard, P. Barboux and J. Livage, *Solid State Ionics* 27 (1988) 189–94.
71. R. C. T. Slade, M. G. Cross and W. A. England, *Solid State Ionics* 6 (1982) 225–30; A. K. Nikumbh, K. S. Rane and A. J. Mukhedkar, *J. Mater. Sci.* 17 (1982) 2503–13.
72. Ph. Colomban, M. Pham-Thi and A. Novak, *Solid State Commun.* 53 (1985) 747–51.
73. Ph. Colomban and J. P. Boilot, *Rev. Chimie Minérale* 22 (1985) 235–55.
74. Y. Charbouillot, D. Ravaine, M. Armand and C. Poinsignon, *J. Non-Crystalline Solids* 103 (1988) 325–30.
75. R. Bartholemew, W. Haynes and R. Shoop, *Soluble Silicates, ACS Symposium Series 194*, J. S. Falcone (ed) (1982) Ch. 17; R. F. Bartholomew, *J. Non-Crystalline Solids* 56 (1983) 331–42; E. Stolper, *Contr. Mineral. Petrol.* 81 (1982) 1–17.
76. G. B. Hurd, E. W. Engel and A. Vernul, *J. Am. Chem. Soc.* 49 (1927) 1447–53.
77. P. Ehrman, M. de Billy and J. Zarzycki, *Verres et Refract.* 2 (1961) 63–72.
78. H. Scholze, *Glastech. Ber.* 32 (1959) 81–8; 142–52; *Naturwissensch.* 47 (1960) 226–7; *Glass Ind.* 47 (1966) 622–8.
79. G. J. Roberts and J. P. Roberts, *Phys. Chim. Glass* 7 (1966) 82–9; F. M. Ernsberger, *Phys. Chim. Glass* 21 (1980) 146–9.
80. R. W. Douglas and J. O. Isard, *J. Soc. Glass Techn.* 33 (1949) 289–335.
81. A. J. Moulson and J. P. Roberts, *Trans. Faraday Soc.* 57 (1961) 1208–16.
82. G. Eisenmann, *Glass Electrodes for Hydrogen and Other Cations* (M. Dekker, New York (1967)).
83. P. W. Schindler and W. Stumm, The Surface chemistry of oxides, hydroxides and oxide minerals, in *Aquatic Surface Chemistry – Chemical*

Processes at the Particle–Water Interface, W. Stumm (ed) (J. Wiley and Sons (1987)) 83–1110.

84. E. N. Boulos, A. V. Lesikar and C. T. Moynihan, *J. Non-Crystalline Solids* **45** (1981) 419–35.

85. J. J. Fripiat, C. Van der Meersche, R. Touvillaux and A. Jelli, *J. Phys. Chem.* **74** (1970) 382–93.

86. I. S. T. Tsung, C. A. Houser, W. B. White, A. L. Wintenberg, P. D. Miller and C. D. Moak, *Appl. Phys. Lett.* **39** (1981) 669–70.

87. T. J. Pinnavola, I. D. Johnson and A. J. Lipsicus, *J. Solid State Chem.* **63** (1986) 118–21.

88. J. Bullot and J. Livage, French Patent 81-13665 (1981). J. Livage, *Mater. Res. Soc. Symp. Proc.* **32** (1984) 125–31.

89. J. G. Zhang and P. C. Eklund, *J. Appl. Phys.* **64** (1988) 729–33.

19 Perfluorinated membranes

GERALD POURCELLY and CLAUDE GAVACH

19.1 Historical background and development

Electrochemical cells (electrolysers, batteries and fuel cells) require *separators*, which allow a flow of specific ionic charges but prevent the transfer of chemical species which remain located either in the cathodic or in the anodic compartment. Among the various separators of electrochemical cells, the ion permeable *organic membranes* are also used for separation processes such as dialysis and electrodialysis.

Ion permeable organic membranes are *polymer films* made from ion exchange material. Their thickness is between 50 and 200 μm. The polymeric matrix contains fixed ionic groups (the functional sites) the charges of which are compensated by mobile ions (the counter-ions) which can be exchanged with adjoining media. In contact with an aqueous electrolyte solution, ion permeable membranes contain also water and sorbed electrolytes. Monofunctional membranes are ion permselective. The fixed sites of cation permeable membranes are sulphonic or carboxylic groups; those of anion permeable membranes are amine or quaternary ammonium groups.

Among the cation permeable membranes, the perfluorinated membranes which have been developed as separators for *fuel cells* and *chlor-alkali electrolysers* show the characteristic features of superselectivity, very high thermal stability and chemical resistance, which are not obtained by the other classes of polymeric ion permeable membranes. Three commercial forms of cation permeable perfluorinated membranes have been proposed:

the *Nafion*® monofunctional perfluorosulphonic membrane produced by Du Pont de Nemours,
the *Flemion*® perfluorocarboxylic membrane produced by Asahi Glass,
the Tokuyama Soda bifunctional membrane (both perfluorosulphonic and carboxylic).

Perfluorinated membranes

For *chlor-alkali electrolysers*, a bilayer *perfluorinated membrane* is produced by Du Pont de Nemours. One layer is perfluorosulphonic, the other is *perfluorocarboxylic*.

In the presence of protons, perfluoroacetic membranes are not ionized. Therefore, only *perfluorosulphonic membranes* are proton conductors having high electrical conductivity combined with a good mechanical behaviour, and a strong chemical resistance even in contact with oxidizing agents.

Perfluorosulphonic Nafion® membrane separators are used in direct contact with electrodes as solid polymer electrolytes (SPE) in fuel cells[1-4]. In this case, the membrane is both the electrolyte and the separator. The use of perfluorosulphonic membranes as SPE started 30 years ago with the US space program Gemini and the realization of low temperature H_2/O_2 SPE fuel cells. Since then, the feasibility of operating the SPE fuel cells on hydrogen/halogen couples has been demonstrated[5]. In addition, the introduction of perfluorinated membranes for use in water and brine electrolysis[6,7] and more recently in organic synthesis has taken place[8].

The interest in perfluorinated ion exchange membranes has been increasing in recent years because of their industrial importance. Much of the research work into their structures and properties has been gathered in a relatively recent monograph[9].

19.2 Synthesis of perfluorinated membranes

Among the cation permeable perfluorinated membranes previously mentioned, only the perfluorosulphonic membranes are proton conductors. Their synthesis is described in this chapter[10,11].

These membranes are produced by Du Pont de Nemours (USA) under the registered trademark Nafion. To synthesize Nafion®, tetrafluoroethylene is reacted with SO_3 to form a cyclic sulphone. After rearrangement, the sulphone can then be reacted with hexafluoropropylene epoxide to produce sulphonylfluoride adducts, with $m > 1$

$$CF_2\!\!=\!\!CF_2 + SO_3 \rightarrow \underset{\underset{\displaystyle O\!-\!\!-\!\!-\!\!SO_2}{|\quad\quad|}}{CF_2\!-\!CF_2} \rightarrow FSO_2CF_2C\!\!=\!\!O$$

$$FSO_2CF_2C\!\!=\!\!O + (m+1)\underset{\underset{\displaystyle CF_3}{|}}{CF_2}\overset{\displaystyle O}{\diagup\diagdown}CF \rightarrow FSO_2CF_2CF_2(\underset{\underset{\displaystyle CF_3}{|}}{OCFCF_2})_m\underset{\underset{\displaystyle CF_3}{|}}{OCFC}\!\!=\!\!O$$

Materials

When these adducts are heated with sodium carbonate, a sulphonylfluoride vinylether is formed. This vinylether is then copolymerized with tetrafluoroethylene to form XR resin

$$FSO_2CF_2CF_2(OCCFCF_2)_mOCF{=}CF_2 + CF_2{=}CF_2 \rightarrow$$
$$\underset{\displaystyle CF_3}{|}$$

$$-(CF_2CF_2)_n{-}\underset{\displaystyle \underset{CF_3}{|}}{C}FO(CF_2{-}\underset{\displaystyle \underset{CF_3}{|}}{C}FO)_mCF_2CF_2SO_2F \quad \text{(XR resin)}$$

This high molecular weight polymer is melt fabricable and can be processed into various forms, such as sheets or tubes, by standard methods. By hydrolysis, this resin is converted into Nafion® perfluorosulphonic polymer

XR resin + NaOH →

$$-(CF_2CF_2)_n{-}\underset{\displaystyle \underset{\displaystyle \underset{|}{CF_2}}{|}}{C}{-}O(CF_2{-}\underset{\displaystyle \underset{CF_3}{|}}{C}FO)_mCF_2CF_2SO_3Na$$
$$\qquad\qquad\qquad\qquad\qquad\qquad\qquad\qquad \text{Nafion®}$$

In this ion exchange membrane, the sodium counter-ion can be easily exchanged with other cations by soaking the polymer in an appropriate aqueous electrolyte solution. For commercial materials, m is usually equal to one unit and n varies from about 5 to 11. This generates an equivalent weight ranging from about 1000 to 1500 g of polymer in its dry hydrogen ion form per mole of exchange site.

19.3   *Structure of perfluorinated membranes*

Perfluorinated ion exchange membranes differ from conventional ion exchange membranes in that they are not cross-linked polyelectrolytes but thermoplastic polymers with pendant acid groups (sulphonic or carboxylic) partially or completely neutralized to form salts. These membranes reveal a high perselectivity to cations and their water content or electrolyte uptake can vary over a large range. It is widely accepted that sizeable ionic clusters scattered in a surrounding organic hydrophobic medium can exist in the polymer. *Ion clustering* in Nafion® has been suggested by a number of techniques including mechanical and dielectric relaxation, NMR, IR, transport experiments, electron microscopy and

Perfluorinated membranes

X-ray studies. Gierke[12] proposed a cluster network model assuming that both the ions and the sorbed solutions are all in clusters. Assuming the clusters to be spherical, their size has been calculated from solvent adsorption data obtained on various polymers of different equivalent weight. The calculated cluster diameter range was between 3 and 5 nm; for a 1200 equivalent weight polymer, the average cluster contains approximately 70 ion exchange sites and 1000 water molecules[13]. In this model, the counter-ions, the fixed sites and the swelling water phase separate from the fluorocarbon matrix into approximately spherical domains connected by short narrow channels. The fixed sites are embedded in the water phase very near to the water fluorocarbon interface. This structure is essentially that of an inverted micelle. Falk[14], using *IR* techniques, which provide information on the microscopic structure of hydrated polymers, has suggested that the hydrated ion clusters are smaller than given by Gierke's calculations and are highly non-spherical in shape, with local intrusions of fluorocarbon phase. By fluorescence spectroscopy, Lee & Meisel[15] have followed the diffusion of heavy cations in the membranes and have confirmed the inverted micellar structure. Gierke & Hsu[13] considered that the clusters are interconnected by short narrow channels in the fluorocarbon backbone network, as illustrated in Fig. 19.1. The diameter of the channels estimated from hydraulic permeability data is about 1 nm. Assuming that the Bragg spacing (near 5 nm from SAXS data) represents the distance between clusters, they correlate the variations of cluster diameter and the number of ion exchange sites per

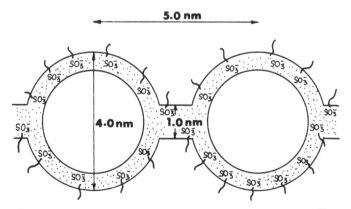

Fig. 19.1. Cluster network model for Nafion® perfluorosulphonic membrane[13]; reprinted by permission of Elsevier Science Publishers, B.V.

297

cluster with the water content as shown in Fig. 19.2, and establish that the pore diameter decreases when the equivalent weight increases. Gierke *et al.*[16] established the existence of clusters with a diameter less than 2 nm and with about 25 ion exchange sites per cluster even in the dry polymer. The increase in the number of exchange sites per cluster with increasing water content led them to suppose that a reorganization of the cluster network occurs during dehydration as shown in Fig. 19.3. Mauritz & Hopfinger[17] assumed that ionic clustering does not exist in the dry polymer. Fujirama *et al.*[18] established that this assumption is applicable to the perfluorinated carboxylic acid polymer but not to the perfluoro-sulphonate polymer. By small angle and wide angle X-ray studies (SAXS and WAXD), Hashimoto *et al.*[19] established that ionic clusters affect the physical properties of the membranes. The size of the ionic clusters was shown to depend on the equivalent weight, the nature of the cation and

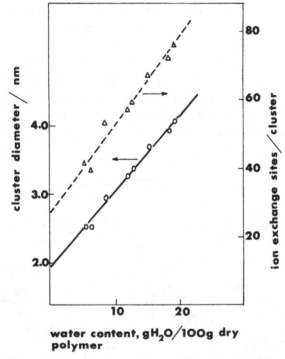

Fig. 19.2. The variation of the cluster diameter (○) and number of ion exchange sites (△) per cluster with water content in Nafion® 1200 EW polymer[13]; reprinted by permission of Elsevier Science Publishers, B.V.

of the fixed sites, and also on temperature. By Mössbauer spectroscopy and *neutron* studies, Pineri *et al.*[20-22] have derived a three-phase model in which fluorocarbon crystallites, ionic hydrophilic clusters and an amorphous hydrophobic region of lower ionic water content coexist. Yeager & Steck[23] correlating various spectroscopic and ionic diffusion results proposed a three-region model as illustrated in Fig. 19.4. Datye *et al.*[24], by computer simulations, have shown that the effect of electrostatic and elastic forces on the pendant ionic groups and the neutralizing counter-ions is such as to form a dipole layer at the surface of an ionic cluster. They proposed a model for selective transport based on the electrostatic potential created by the dipole layer. Their calculations do not imply the existence of channels between the ionic clusters.

The perfluorinated ionomer membranes, with no crosslinking, rely upon their crystalline domains to inhibit the dissolution. The crystalline domains originating from the tetrafluoroethylene material act as cross linked

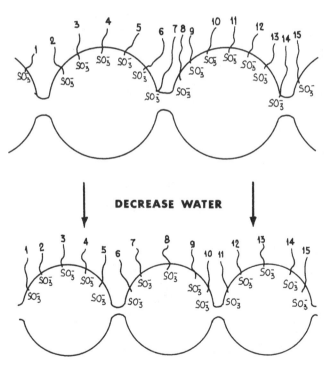

Fig. 19.3. Representation of redistribution of ion exchange sites occurring during the dehydration of polymer[16]; reprinted by permission of John Wiley and Sons Inc.

points. The degree of crystallinity decreases with increasing equivalent weight (EW)[25]. Perfluorosulphonate polymers with EW of less than 1000 are soluble[26], primarily due to the absence of crystallinity.

19.4 Proton transport in perfluorosulphonic membranes

19.4.1 The importance of water in transport properties

The transport properties of perfluorosulphonic membranes are largely influenced by the water content of the membrane, particularly when the membrane is in the acid form. In the dry state, the Nafion® membrane behaves like an insulator but, when hydrated, the membrane becomes conductive as a function of the water content. Yeo[27] established that the minimum threshold corresponds to about six molecules of water per sulphonic site, whereas Pourcelly et al.[28] estimated about seven molecules. Randin[29] has shown that membranes with six molecules of water per sulphonic site have sufficient conductivity for use as a semi-solid proton

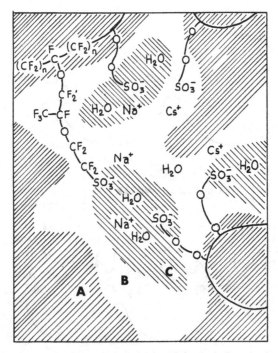

Fig. 19.4. Three-region structural model for Nafion®; A, fluorocarbon; B, interfacial zone; C, ionic clusters[23]; reprinted by permission of The Electrochemical Society.

conductor in WO_3-based electrochromic displays. Because of the strong acidity of the sulphonic groups, the membrane is highly conductive to protons. Yeo *et al.*[5, 30] have reported that the conductivity of Nafion® in concentrated acid electrolytes is one order of magnitude smaller than the electrolyte conductivity. In the case of very dilute electrolytes, the membrane conductivity is higher than that of the electrolyte because the intrinsic conductivity becomes important. Among the polymer-containing acid groups which all have high but similar intrinsic conductivities, the activation energy of proton conduction of Nafion® is low compared with that in other polymers[27]. This low activation energy of proton conduction may be due to the state of water in the membrane phase. In electrolytic processes, the water content of the membrane is monitored by direct contact with the aqueous solution. In a fuel-cell using proton exchange membranes, the water content is determined by steady state contact of humidified gas streams at each membrane interface. Under these conditions, Rieke & Vanderborgh[31] have shown that the proton conductivity reaches a maximum over a temperature range of 55–70 °C even though water content is minimum in this domain. They have established that, in the temperature range 25–50 °C, a variation in water content is less significant than a temperature increase. Consequently, temperature plays also a main role in the kinetics of proton motion in the polymeric membrane. Outwith these last two temperature ranges, the conductivity decreases resulting from a deionization of the sulphonic acid groups and perhaps a change of the hopping distance between cluster zones.

19.4.2 Basic processes in proton transport

The mobility of protons in Nafion® perfluorosulphonic membranes is strongly associated with the water content of the membrane phase. This mobility is directly related to the molality of fixed charged sulphonic sites. By IR measurements, Jenard[32] has observed the protonated form of sulphonic sites for low water contents in these membranes. Excepting these extreme conditions, the proton transfer process in perfluorosulphonic membranes must take account of the ionic composition of the membrane phase. Two cases must be examined: on the one hand, the membrane contains counter-ions, lone protons or protons with other cations, and on the other hand, the membrane contains sorbed acids. In the first case, the conduction is due only to the counter-ions and proton motion is considered to involve the jump from one site located in the vicinity of a sulphonic group to another; in the second case, the anions resulting from the dissociation of sorbed acids may also participate in the conduction process. For a range of low water contents, it has been shown that percolation theory can be applied[33-35].

The Nafion® membrane may be considered as heterogeneous with conductive ion clusters scattered randomly in an insulating fluorocarbon matrix. Aqueous microdomains composed of hydrated fixed groups, counter-ions, co-ions and non-solvating water molecules phase-separate from the polymeric matrix. In order to solve the membrane conductivity problem with percolation theory, an appropriate percolation model should be selected. Assuming the cluster network model to be appropriate, the polymer may be regarded as a set of conductive spherical particles scattered in a continuous medium. This model is called a 'dense packed hard sphere model'. To recall the main features of percolation theory as applied to conduction phenomena in ionomer membranes, the membrane phase is considered as a three-dimensional lattice in which the ionic clusters correspond to the lattice nodes. In such a system, the conductivity $\kappa(x)$ is proportional to $(x - x_c)^t$ where x denotes the concentration of the conducting phase, t is a universal critical exponent and x_c is the concentration of conducting phase at the 'insulator-to-conductor' transition. In a site percolation model, a theoretical estimate for t is 1.5 ± 0.2[33] and Scher & Zallen[36] proposed the critical volume fraction of the conducting phase as a characteristic constant. On this basis, the conductivity law takes the form

$$\kappa = \kappa_0 (\tau - \tau_c)^t$$

where κ_0 is the specific conductivity of the conducting phase, τ its volume fraction and τ_c the critical volume fraction, assumed to be theoretically equal to 0.15. For a Nafion® 117 membrane (EW = 1100), Gavach *et al.*[35] obtained results in relatively good agreement with the theoretical equation over only the low-swelling range. The values of the exponent t are 1.53 for H^+, 1.51 for K^+, 1.57 for Li^+ and 1.43 for Na^+. By applying the same theory, Hsu & Gierke[13] obtained $t = 1.5$ for a Nafion® 125 membrane (EW = 1200) equilibrated with Na^+ ions, while, in the same conditions, Wodski *et al.*[37] obtained $t = 1.61$ for a Nafion® 120 membrane. For higher values of water content, it has been observed[35] that the variation of membrane conductivity does not follow the theoretical law. This fact suggests that, in this case, there are no more isolated domains and the percolation mode cannot be applied. The extrapolated values of the threshold volume fraction τ_c range are between 0.2 and 0.3. They depend on both the dimensionality and the manner in which the clusters are dispersed. These values, larger than the theoretical value of 0.15, imply that ion clusters flocculate into several well-isolated regions[34].

19.4.3 Proton transport when only counter-ions are present in the membrane

The conductance variations with water content have two different origins: the first is a change in the rate constant of the elementary ion transfer step due to interactions between the mobile ions and the fixed sites; the second is related to a change in microstructure of the membrane material. The analysis of electrical conductivity data must be carried out for membranes having the same ionic composition but variable water contents. The *a.c. impedance technique,* using a mercury cell, enables one to follow these conditions[35, 38–40], but other techniques have been used[41–43]. The Nyquist plots of impedance diagrams drawn over the 1 Hz–1 MHz frequency range with H^+ as the counter-ion, are composed of only a fairly linear part for a high degree of swelling. When the water content decreases, a semi-circle appears in the high frequency range. These shapes are the same as obtained with other cations as the counter-ions[28], or when *sorbed acids* are present in the membrane[44]. Pourcelly *et al.*[28] have proposed an analysis of the membrane conductance on the basis of absolute rate theory. Thus, the membrane can be considered as a series of barriers that the mobile counter-ion has to overcome[45–47]. For high degrees of swelling, the high frequency impedance diagram of the membrane phase corresponds to that of a pure resistance. In this state, the membrane phase is sufficiently homogeneous to allow a free communication between two clusters. The height of the energy barrier for the elementary jumps is the same for every pair of adjacent fixed sites, as shown in Fig. 19.5a. For lower degrees of swelling, the high frequency impedance diagram includes part of a semi-circle. In this state of swelling, the conducting phase is composed of aqueous clusters connected to each other by narrow channels. The energy barrier to be overcome for hopping between two adjacent sites in a cluster is lower than that for two sites on either side of the narrow channel (Fig. 19.5b). An accumulation of ions on each side of the channel gives rise to capacitive currents. The overall impedance of the membrane then becomes a pure resistance associated with the motion of mobile counter-ions inside the cluster, in series with the impedance (resistance in parallel with a capacitance) of the narrow channel. The fact that the impedance diagram of the membrane phase has a non-zero value at infinite frequency confirms this model. The shape of the diagram reveals heterogeneity of the membrane phase. The depression of the centre of semi-circles below the abscissa reveals a dispersion of the RC circuits associated with the channels. These considerations and those relative to percolation theory, are consistent with the schematic ionic cluster model.

For a given water content and by applying the Stokes–Einstein equation and the rate process to ionic migration[48], it is possible to calculate, for two different ions, the difference between their activation free energies for the elementary ion transfer reactions, the proton being chosen as the reference ion

$$\Delta(\Delta G^{\circ \neq}) = \Delta G_i^{\circ \neq} - \Delta G^{\circ}{}_{H^+} = RT \ln(\kappa_{H^+}/\kappa_i)$$

in which κ_i is the specific conductivity of the membrane containing ion 'i'. This equation means that, for a given membrane, the rate determining step for the counter-ion diffusion–migration process is the elementary ion transfer reaction between two adjoining sites, all the sites being considered as energetically equivalent; this situation is expected for high swelling degrees as we have seen previously. The values of the difference in free

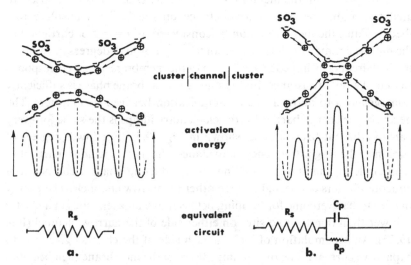

Fig. 19.5a, b. Counter-ion transport model in a cationic exchange membrane; activation energy of the elementary process and equivalent electrical circuit; a, in a high water content range; b, in a low water content range[28] (with permission).

energy, $\Delta(\Delta G^{\circ \neq})$ corresponding to a swelling degree of 16 molecules of water per sulphonic group are, respectively, 0.018 for Li^+, 0.024 for Na^+ and 0.03 eV per atom for K^+. The results confirm that, for the alkali cation series, the rate of diffusion–migration in Nafion® 117 membrane is determined by the electrostatic interactions between the fixed charged groups and the mobile ions, the more hydrated ions having a lower attraction energy for the fixed anionic groups[35].

An attempt has been made to analyse the results for membrane conductivity obtained in the presence of two counter-ions, the proton and an alkali ion. From the results of self-diffusion flux measurements by radiotracer techniques, Gavach *et al.*[35] obtained a value for the individual mobility of Na^+ ions in the membrane containing both Na^+ and H^+ counter-ions. By measuring the conductivity of the membrane containing only H^+ as the mobile ion and applying the Nernst–Einstein relationship, they determined the relative mobilities of these two counter-ions and concluded that the individual mobility of the proton increases by up to 150% as the number of Na^+ ions increases in the membrane. This increase with mole fraction of Na^+ ions has been attributed to an increase of the proton hydration, Na^+ being less hydrated than H^+.

19.4.4 Proton transport in the presence of sorbed acids

Nafion® perfluorosulphonic membranes exhibit a high *permselectivity* to cations especially to protons. E.m.f. measurements carried out on asymmetric 0.1 N, x N acid solutions have shown that deviations from Nernst's law occur from $x = 2$ N for HCl and $x = 3$ N for H_2SO_4[49] while the internal concentrations in acid were, respectively, 0.16 mol of HCl and 0.06 mol of H_2SO_4 per sulphonic fixed site. The electrical transport properties of hydrohalogenic acids in a Nafion® 117 membrane have been studied[44]. The swelling and the amount of sorbed acid species depend on the external acid concentration. For high swelling, the presence of sorbed HF does not affect the membrane conductivity whereas the presence of HCl or HBr increases it strongly until the fraction of sorbed acid molecules over fixed sulphonic groups reaches 5% for HCl and 20% for HBr, as shown in Fig. 19.6. In the case of HCl, this increase cannot be ascribed to a significant increase in the number of charge carriers, protons and chloride ions. The transport number of chloride ions, which has been measured using radiotracer measurements, remains lower than 0.02. The experimental data suggest that, in this case, according to the Gouy–Chapmann theory of the electrochemical diffuse layer, the increase of ionic strength inside clusters resulting from the sorption of acid reduces the mean distance between mobile protons and the perfluorinated surface cavity that bears the fixed sulphonic groups and consequently, the rate of the elementary proton jump is increased, probably due to a more organized structure of water molecules in contact with the perfluorinated chains. For larger amounts of sorbed HCl or HBr, the membrane conductivity remains almost constant. For H_2SO_4, the membrane conductivity increases initially with the amount of sorbed acid, then decreases slightly and finally remains constant. As with HCl, the transport number of HSO_4^- or SO_4^- anions remains lower than 0.01 whatever the external concentration of sulphuric acid. The assumption established previously for HCl may be applied to H_2SO_4, which should confirm the importance of water to ionic transport in perfluorosulphonic membranes. Moreover, the ionic dissociation of sorbed acid must be considered further.

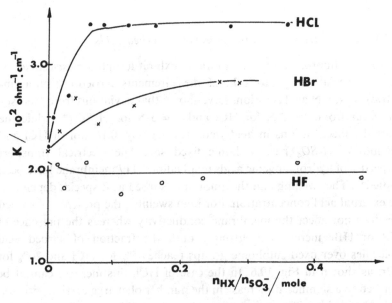

Fig. 19.6. Variations of specific conductivity of a Nafion® 117 membrane at equilibrium with acid solutions *vs* the amount of sorbed acid (maximum swelling)[44] (with permission).

Perfluorinated membranes

Table 19.1. *Conductivity of Nafion® membranes in the protonic form in various electrolytes at 25 °C*

Membrane Nafion®	Form[a]	External electrolyte (wt %)	Conductivity ($10^2 \, \Omega^{-1} \, cm^{-1}$)	Reference
1200 EW	E[b]	H_2O	8.2	5
1200 EW	E[b]	20% HBr	7.8	5
1200 EW	E[b]	30% HBr	6.8	5
1200 EW ⎱	E[b]	24% HBr	9.8	5
1200 EW ⎰	N	24% HBr	7.6	5
1200 EW	E[b]	5% HCl	5.5	30
1200 EW	E[b]	10% HCl	6.1	30
1200 EW	E[b]	20% HCl	2.4	30
1100 EW	E[c]	H_2O	6.1–7.8	31
1100 EW	E[b]	5% HCl	3.8	44

[a] Form depending on the pretreatment; [b] expanded form (boiled in distilled water); [c] expanded form (boiled in nitric acid); N, normal form or as-received form.

19.4.5 Conductivity data for Nafion® membranes

The conductivity of Nafion® membranes depends on the equivalent weight of the ion exchange polymer and its water content, these two parameters being closely related. On the other hand, the conductivity has been found to be affected strongly by the pretreatment of the membrane [5,27]. At room temperature, the maximum in conductivity reaches 0.078 $\Omega^{-1} \, cm^{-1}$ for a Nafion® 1100 EW without sorbed acid and 0.098 $\Omega^{-1} \, cm^{-1}$ for a Nafion® 1200 EW when HBr is present in the membrane phase[30]. Some experimental values of conductivities are summarized in Table 19.1.

19.5 Perfluorinated membranes and related materials

Nafion® materials, and more generally perfluorinated ionomers, are particularly suitable for water and brine electrolysis and, to date, no viable alternative has been found for SPE applications. The *dissolution* of Nafion® membranes allows the preparation of material with high porosity and high electroactive area. Such structures are required for the development of high power density SPE fuel cells. In recent work, Aldebert *et al.*[50] have presented different methods for the preparation of SPE

Materials

electrocatalyst composites, one of the remaining problems being the optimization of the permeability of perfluorinated ionomer to H_2 and O_2 gas.

Polymers having both electronic and ionic conductivities are of interest due to the applications of composite materials in polymeric electrodes or in electrochemical catalysis. Electronic and ionic conductive polymer composites have been synthesized by electropolymerization using pyrrole, bithiophene or aniline trifluoromethane with an hydrophobic ionomer gel of Nafion® type[51,52]. With *pyrrole*, such materials reach conductivities of $1000 \, \Omega^{-1} \, cm^{-1}$ and have good mechanical properties.

19.6   Conclusions

Because of the great industrial importance of the perfluorinated cation exchange membrane, the research effort devoted to this membrane during the past 20 years has greatly exceeded that devoted to any other single ion exchange membrane. Their remarkable chemical stability, their high permselectivity to cations and particularly to protons, associated with a low electrical resistance and very good mechanical properties make them materials particularly suitable for electrolysis in aggressive media. Concerning their structure, in addition to a small amount of crystallinity leading to cohesion of the material, two distinct non-crystalline regions exist, the hydrophobic fluorocarbon phase and the hydrophilic ionic area which is composed of aqueous clusters forming an inverted micellar-like structure. As a result of spectroscopic studies, it has been shown that interactions between the mobile cations and the anionic fixed sites are highly sensitive to the degree of hydration of the membrane. The relatively high cost of these perfluorinated ionomer membranes requires us to develop a better knowledge of the relationship between the unusual ion cluster morphology and the transport properties, in order to produce materials having similar properties but being less expensive. In the specific domain of solid polymer electrolyte technology, methods for the preparation of electrocatalyst solid electrolyte composites have been developed[50-52] and seem to be promising for the development of energetically efficient electrolysers and power generators.

19.7   References

1. W. T. Grubb, *J. Electrochem. Soc.* **106** (1959) 295–8.
2. W. T. Grubb and L. W. Niedrach, *J. Electrochem. Soc.* **107** (1960) 131–4.

3. H. J. R. Maget, in *Handbook of Fuel Cell Technology*, W. Mitchell Jr (ed) (Academic Press, New York (1963)) 253.
4. L. W. Niedrach and W. T. Grubb, in *Fuel Cell Technology*, C. Berger (ed) (Prentice-Hall, Englewood Cliffs, N.J. (1967)) 425.
5. R. S. Yeo and D. T. Chin, *J. Electrochem. Soc.* **127** (1980) 549–55.
6. R. J. Laurence and L. D. Wood, US Patent no. 4272353 (1981).
7. H. Takenaka, E. Torikai, Y. Kawami and N. Wakabayashi, *Int. J. Hydrogen Energy* **7** (1982) 397.
8. Z. Ogumi, M. Inaba, S. I. Ohashi, M. Uchida and Z. I. Takehara, *Electrochem. Acta* **33** (1988) 365–9.
9. Perfluorinated Ionomer Membranes, A. Eisenberg and H. L. Yeager (eds) *A.C.S. Symposium Series no. 180* (1982).
10. O. D. Bonner and L. L. Smith, *J. Phys. Chem.* **61** (1957) 326–9.
11. O. D. Bonner, W. L. Argensinger and A. W. Davidson, *J. Am. Chem. Soc.* **74** (1952) 1044–7.
12. T. D. Gierke, *J. Electrochem. Soc.* **124** (1977) 319c.
13. W. Y. Hsu and T. D. Gierke, *J. Membrane Sci.* **13** (1983) 307–26.
14. M. Falk in *Perfluorinated Ionomer Membranes*, A. Eisenberg and H. L. Yeager (eds) *A.C.S. Symposium Series, no. 180* (1982).
15. P. C. Lee and D. Meisel, *J. Am. Chem. Soc.* **102** (1980) 5477–81.
16. T. D. Gierke, G. E. Munn and F. C. Wilson, *J. Polym. Sci. Polym. Phys. Ed.* **19** (1981) 1687–704.
17. K. A. Mauritz and A. J. Hopfinger, in *Modern Aspects of Electrochemistry*, J. O'M. Bockris, B. E. Conway and F. E. White (eds) Vol. 14 (Plenum Press, New York (1982)) 425.
18. M. Fujimara, T. Hashimoto and H. Hawai, *Macromolecules* **14** (1981) 1309–15.
19. T. Hashimoto, M. Fujimara and H. Hawai, in *Perfluorinated Ionomer Membranes*. A. Eisenberg and H. L. Yeager (eds.) *A.C.S. Symposium Series, no. 180* (1982) 217.
20. B. Rodmacq, J. M. D. Coey and M. Pineri, in Reference 19, p. 170.
21. M. Pineri, R. Duplessix and F. Volino, in Reference 19, p. 249.
22. B. Rodmacq, J. M. Coey, M. Escoubes, R. Duplessix, A. Eisenberg and M. Pineri, in *Water in Polymers*, S. P. Powland (ed), *A.C.S. Symposium Series no. 127* (1980) 487.
23. H. L. Yeager and A. Steck, *J. Electrochem. Soc.* **128** (1981) 1880–4.
24. V. K. Datye, P. L. Taylor and A. J. Hopfinger, *Macromolecules* **17** (1984) 1704–8.
25. H. W. Starkweather Jr, *Macromolecules* **15** (1982) 320.
26. W. T. Grot, US Patent, 4433082 (1984).
27. R. S. Yeo, *J. Electrochem. Soc.* **130** (1983) 533.
28. G. Pourcelly, A. Oikonomou, H. D. Hurwitz and C. Gavach, *J. Electroanal. Chem.* **287** (1990) 43–59.
29. J. Randin, *J. Electrochem. Soc.* **129** (1982) 1215–20.
30. R. S. Yeo and J. C. McBreen, *J. Electrochem. Soc.* **126** (10) (1979) 1682–6.
31. P. C. Rieke and N. E. Vanderborgh, *J. Membr. Sci.* **32** (1987) 313–28.

Materials

32. A. Jenard, Laboratory of Thermodynamic Electrochemistry, U.L.B., Brussels, private communication.
33. S. Kirkpatrick, *Rev. Mod. Phys.* **45** (1973) 574–88.
34. W. Y. Hsu, J. R. Barkley and P. Meakin, *Macromolecules* **13** (1980) 198–200.
35. C. Gavach, G. Pamboutzoglou, M. Nedyalkov and G. Pourcelly, *J. Membrane Sci.* **45** (1989) 37–53.
36. H. Scher and R. Zallen, *J. Chem. Phys.* **5** (1970) 3759–61.
37. R. Wodski, A. Narebska and W. Kwas, *J. Appl. Polym. Sci.* **30** (1985) 769–80.
38. A. I. Meshechkov, O. A. Demina and N. P. Gnusin, *Electrokhimiya* **23** (10) (1987) 1452–4.
39. N. Nedyalkov, D. Schuhmann and C. Gavach, *Spring Meeting Electrochem. Soc., Boston, 1986* (The Electrochemical Society, Pennington, N.J. (1986)) Extended abstracts, no. 1216.
40. M. Nedyalkov and C. Gavach, *J. Electroanal. Chem.* **234** (1987) 341–6.
41. C. S. Fadley and R. A. Wallace, *J. Electrochem. Soc.* **115** (1968) 1264–70.
42. R. A. Wallace, *J. Appl. Polym. Sci.* **18** (1974) 2855–9.
43. R. A. Wallace and J. P. Ampaya, *Desalination* **14** (1974) 121–34.
44. G. Pourcelly, A. Lindheimer, G. Pamboutzoglou and C. Gavach, *J. Electroanal. Chem.* **259** (1989) 113–25.
45. S. Glasstone, K. L. Laider and H. Eyring, in *The Theory of Rate Process* (McGraw-Hill Ed., New York, N.Y. (1941)).
46. F. H. Johnson, H. Eyring and M. J. Polissar, in *The Kinetics Basis of Molecular Biology* (J. Wiley, New York, N.Y. (1963)).
47. H. Eyring and E. M. Eyring, in *Modern Chemical Kinetics* (Van Nostrand Reinhold, New York, N.Y. (1963)).
48. J. O'M. Bockris and A. K. Ready, in *Modern Electrochemistry*, Vol. 1 (Plenum Press, New York, N.Y. (1970)) 387.
49. G. Pourcelly, J. Seta and C. Gavach, Laboratory of Physical Chemistry of Polyphasic Systems, CNRS, UA330, 34033-Montpellier, France, unpublished data.
50. P. Aldebert, F. Novel Castin, M. Pineri, P. Millet, C. Doumain and R. Durand, *Solid State Ionics* **35** (1985) 3–9.
51. G. Gebel, P. Aldebert and M. Pineri, *Macromolecules* **20** (1987) 1425–8.
52. P. Audebert, P. Aldebert and M. Pineri, *Synthetic Metals* **32** (1989) 1–14.

20 Mixed inorganic–organic systems: the acid/polymer blends

JEAN-CLAUDE LASSEGUES

20.1 Introduction

In the past few years there has been a strong demand for thin solid proton conducting films to build 'all-solid state' electrochemical devices[1]. One obvious way to obtain thin films is to use polymers; generally, these polymers have to be modified or mixed with inorganic compounds in order to achieve the required proton conductivity, a similar strategy to that successfully applied to obtain alkali ion conductivity in the polyether/salt complexes[2,3].

Perfluorinated sulphonic or carboxylic membranes have been a first important step in this direction and they already find many practical applications[4–6]. However, they do not conduct in the dry state and the usual limitations caused by the presence of water or of organic solvents may occur.

Thus, recent works have aimed at developing anhydrous proton conducting polymers. A new concept has been put forward simultaneously in several laboratories[7–15]. It consists of mixing *strong acids* such as H_2SO_4 or H_3PO_4 with polymers bearing basic sites (Table 20.1). Let us recall that the pure H_2SO_4 and H_3PO_4 acids are themselves good protonic conductors because of extensive self-ionization and self-dehydration[16–19]. The overall result of these processes is summarized below, with equilibrium concentrations expressed in $mol\,l^{-1}$.

$$6H_2SO_4 \rightleftarrows H_3SO_4{}^+ + H_3O^+ + HSO_4{}^- + HS_2O_7{}^- + H_2S_2O_7 + H_2O$$

| 18.547 | 0.021 | 0.014 | 0.027 | 0.008 | 0.0065 | 0.0002 |

$$\text{(20.1)}$$

$$5H_3PO_4 \rightleftarrows 2H_4PO_4{}^+ + H_3O^+ + H_2PO_4{}^+ + H_2P_2O_7{}^{2-} \quad \text{(20.2)}$$

| 16.815 | 0.890 | 0.461 | 0.429 | 0.461 |

Therefore, even if a polymer is less efficient than water in displacing

Table 20.1. *Abbreviation, chemical formula and T_g of different polymers used in mixtures with strong acids; the latter are indicated together with the explored concentration range*

Polymer	Formula	T_g (°C)	Acid concentration		References
Poly(ethyleneoxide) PEO	$(-CH_2CH_2O-)_n$	-67	H_3PO_4	$0 < x < 2$	9, 10
Poly(vinylalcohol) PVA	$(-CH_2CH-)_n$ OH	58–85	H_3PO_4/H_2O	?	7, 8
Poly(acrylamide) Paam	$(-CH_2-CH-)_n$ $C=O$ H_2N	165–188	H_3PO_4 H_2SO_4	$0.6 < x < 2$ $0.6 < x < 2$	15, 17
Poly(vinylpyrrolidone) PVP	(structure)	175	H_3PO_4 H_2SO_4	$0.5 < x < 8$ $1 < x < 3$	10
Poly(2-vinylpyridine) P$_2$VP	(structure)	104	H_3PO_4 H_2SO_4	$0.5 < x < 3$ $0.5 < x < 3$	27

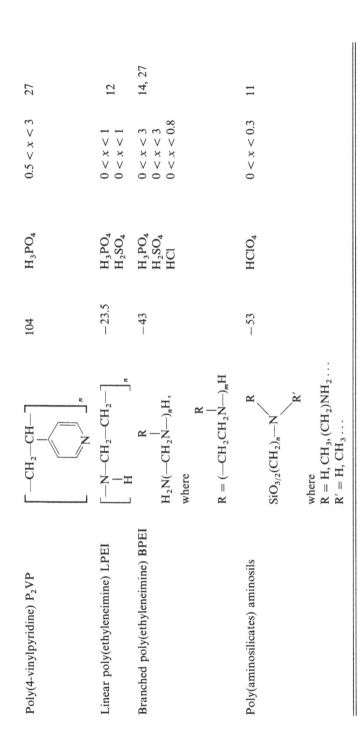

Name	Structure		Acid	x range	Ref.
Poly(4-vinylpyridine) P$_2$VP		104	H$_3$PO$_4$	$0.5 < x < 3$	27
Linear poly(ethyleneimine) LPEI		-23.5	H$_3$PO$_4$ H$_2$SO$_4$	$0 < x < 1$ $0 < x < 1$	12
Branched poly(ethyleneimine) BPEI		-43	H$_3$PO$_4$ H$_2$SO$_4$ HCl	$0 < x < 3$ $0 < x < 3$ $0 < x < 0.8$	14, 27
Poly(aminosilicates) aminosils		-53	HClO$_4$	$0 < x < 0.3$	11

these equilibria towards ionization, one can expect an increased acid dissociation by hydrogen bonding or protonation of the basic sites. In terms of conductivity, an acid/polymer blend is expected to behave somewhere between acidic aqueous solutions and pure acids.

20.2   Preparation

One of the first conditions an acid/polymer blend must satisfy is chemical stability. It is well known, for example, that the C–O bond of ethers and alcohols is broken by strong acids and that this degradation is especially fast in the presence of traces of water[20]. Similarly, Paam is easily hydrolysed in acidic solution to give acids or imides[21] and P_2VP or P_4VP are oxidized by H_2SO_4. Therefore, although most of the polymers considered (Table 20.1) are hydrosoluble, two kinds of preparation have been used depending on their chemical stability.

> When chemical degradation is negligible or very slow, the required quantity of an acid solution is added under stirring to an aqueous solution of the polymer. Then water is eliminated under a stream of dry gas at room temperature[7,8] or by heating at 60/70 °C under vacuum[13–15].
>
> If water has to be avoided, dry organic solutions of the polymer and acid are mixed under an argon atmosphere and the solvent is then evaporated. Convenient solvents for H_2SO_4 and H_3PO_4 are acetonitrile and tetrahydrofuran respectively[9,10].

20.3   Classification of the acid/polymer blends

The main results obtained so far can be presented according to several classifications. We have chosen to consider the kind of interaction established between the acid, X–H, and the basic sites of the polymer, Y. Since we are finally concerned with proton conductivity, it is indeed essential to specify the location of the proton between the two basic atoms, X and Y, in hydrogen-bonded complexes X–H . . . Y or X⁻ . . . ⁺H–Y. Important parameters are of course the relative acidities and basicities of X–H and Y, as measured for example by their pK_a values in aqueous solution.

In the mixtures to be described below, the acids X–H are mainly H_2SO_4 and H_3PO_4 of well-known pK_as. The nature of the *basic polymers* is variable. In addition, it must be kept in mind that their pK decreases when the degree of ionization α increases, as described for example by the following series expansion in α[22].

$$pK = pK_0 + a_1\alpha + a_2\alpha^2 + \cdots \tag{20.3}$$

Indeed, if the macromolecular chain is already carrying positively charged groups, additional work is required to protonate the remaining sites.

In some cases, the pK curves exhibit also a non-monotonous dependence on α because of discontinuous changes in the polymer conformation. This is illustrated in Fig. 20.1 where the acidities of H_2SO_4 and H_3PO_4 are compared with the basicities of some polymers.

Hereafter, we will denote the number of acid moles per polymer repeat unit as x.

20.3.1   Weakly basic polymers

With a pK_0 of the order of -3, *PEO* can be considered as the prototype of a weak polybase. Hydrogen bonding rather than protonation seems to occur with H_3PO_4[9,10]. A phase diagram for the system PEO/H_3PO_4 has been established by calorimetry, X-ray diffraction and NMR[10]. It exhibits two striking features (Fig. 20.2a). A definite compound PEO, $0.75H_3PO_4$ exists which corresponds to a minimum in conductivity, presumably because of its crystallinity and a pronounced eutectic composition PEO, $0.48H_3PO_4$ exists which is associated with a maximum in conductivity. The following model of complexation has been proposed for the eutectic[10].

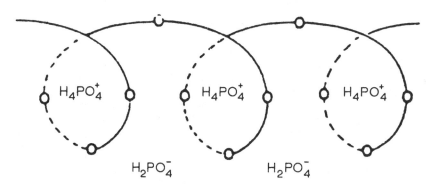

Materials

A maximum of conductivity for eutectic compositions has already been observed in PEO/alkali salt systems[23,24] and also for aqueous solutions of strong acids[25,26], as illustrated in Fig. 20.2b for the H_3PO_4/H_2O system. There is no doubt that this important and apparently rather general feature deserves further theoretical studies.

NMR and quasielastic neutron scattering (QNS) studies have been performed on PEO, $0.42H_3PO_4$ and PEO, $0.66H_3PO_4$ to investigate the phosphorus and proton dynamics in these systems[9]. The temperature dependence of the conductivity is shown to be governed by the segmental

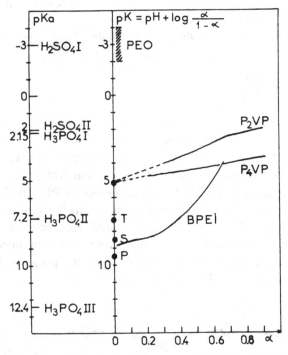

Fig. 20.1. Comparison of the ionization constants of H_2SO_4 and H_3PO_4 (left scale), with those of several polymers measured in water at room temperature without added salt (right scale). For the polymers, the α dependence, when available, is determined by potentiometric titration using the $pK = pH + \log \alpha/1 - \alpha$ empirical equation[22]. For PEO, pK_0 is not precisely known and a possible range of values deduced from those of ethers is indicated. The black circles are pK_0 values for P_2VP and P_4VP ($\simeq 5$) according to M. Satoh *et al.*, *Macromolecules* **22** (1989) 1808–12 and for BPEI according to C. J. Bloys Van Treslong *et al.*, *Recl. Trav. Chim. Pays-Bas* **93** (1974) 171–8. The letters P, S and T refer to the pK_0 values of the primary, secondary and tertiary amino groups in BPEI respectively, estimated by C. J. Bloys Van Treslong, *Recl. Trav. Chim. Pays-Bas* **97** (1978) 9–21.

chain motion in agreement with the previous, more qualitative descriptions derived from the free-volume theory[2, 3].

The PVA/H_3PO_4 system studied by Petty-Weeks *et al.*[7, 8] seems to belong to the same family in the sense that solid solutions are obtained apparently without protonation of the weakly basic oxygens of PVA. However, the exact nature of the interaction is not yet established and the situation is complicated by the existence of a separate H_3PO_4/H_2O phase.

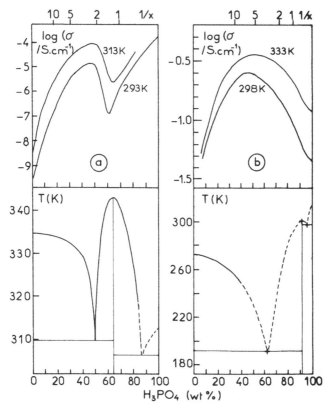

Fig. 20.2. (a) The PEO/H_3PO_4 system. Conductivity (top) and phase diagram (bottom) as a function of H_3PO_4 concentration[10]. The top concentration scale gives the number of PEO repeat units per H_3PO_4 mole. The first eutectic composition corresponds roughly to $1/x = 2$. (b) H_3PO_4/H_2O solution. Conductivity (top)[25] and freezing curve (bottom) according to the *Handbook of Chemistry and Physics*, 61st edn., p. D-249 (continuous line) and reference 26 (crosses). The top concentration scale, again in $1/x$, indicates that the main eutectic composition corresponds roughly to three water molecules per acid mole.

20.3.2 Polymers of intermediate basicity

Polyamides such as Paam and PVP and polyamines such as P_2VP and P_4VP have pK_0 values within the range 3–6. Infrared (IR), Raman[13–15,27] and NMR[28,29] spectroscopies have shown that they are protonated by hydrogen halides and by the first dissociation of H_2SO_4. The situation is less clear with H_3PO_4, which seems to give an equilibrium, varying with x, between protonation $X^- \ldots {}^+H$–Y and hydrogen bonding X–H $\ldots$ Y[27].

When protonation occurs, there is no doubt that P_2VP, P_4VP and Paam are N-protonated whereas PVP is O-protonated. However, the polyamides present a specific feature. The well-known electron delocalization along the O=C—N amide group is highly perturbed as revealed by large variations of frequency and intensity of the amide I ($\nu C{=}O$) and amide III (νC–N) IR absorptions.

It is not excluded that protonated Paam involves a small amount of resonance forms

$$
-C\underset{NH_3{}^+}{\overset{O}{\big\langle}} \quad \longleftarrow\!\!\!\rightarrow \quad -C\underset{\overset{+}{NH_2}}{\overset{OH}{\big\langle}} \tag{20.4}
$$

and the following equilibrium has effectively been observed for protonated PVP[27]:

$$
\text{C}{=}\overset{+}{\text{OH}} \quad \longleftarrow\!\!\!\rightarrow \quad \text{C}{=}\text{O} \tag{20.5}
$$

where the magnitudes of the arrows indicate actual equilibrium displacement.

In all these systems, the specific conductivity is very low at small acid concentration but increases suddenly at $x \simeq 1$ up to values as high as 6×10^{-3} S cm^{-1} for Paam, $1.5H_2SO_4$[15], 10^{-4} for PVP, $2H_2SO_4$[10] and 4×10^{-3} for P_2VP, $2H_2SO_4$[27] at 300K (Table 20.2).

A plateau in the specific conductivity is reached at high acid concentration except for the case of PVP. However, there are not enough experimental points with PVP to decide whether the conductivity decreases further. If this is the case, it should increase again to reach the limiting

Mixed inorganic–organic systems

Table 20.2. *Specific conductivity* ($S\ cm^{-1}$) *at 300 K and 373 K of polymer/ acid mixtures. The conductivity of the pure sulphuric and phosphoric acids is reported for comparison*

Polymer	x	Acid	σ_{300K}	E_{a300K}	σ_{373K}	E_{a373K}	References
		H_2SO_4	1.1×10^{-2}	0.30			16–19
		$H_3PO_4(l)$	5×10^{-2}	0.30			
		$H_3PO_4(s)$	$\sim 2 \times 10^{-4}$	—			
PEO	0.42	H_3PO_4	$\sim 3 \times 10^{-5}$	0.90			9, 10
PVA	0.26	H_3PO_4	$\sim 10^{-5}$	0.95			7, 8
Paam	0.8	H_2SO_4	2×10^{-6}	0.89	8.8×10^{-4}	0.58	15, 27
	1.0		10^{-5}	0.73	1.7×10^{-3}	0.47	
	1.5		6.3×10^{-3}	0.43	1.4×10^{-2}	0.08	
Paam	0.8	H_3PO_4	10^{-5}	1.12	2.4×10^{-3}	0.51	15, 27
	1.0		3.8×10^{-5}	0.89	4.4×10^{-3}	0.41	
	1.5		4.2×10^{-4}	0.64	1.1×10^{-2}	0.29	
P_2VP	1.0	H_2SO_4	3.2×10^{-5}	0.59	9.3×10^{-4}	0.33	27
	1.5		6.5×10^{-4}	0.40	5.4×10^{-3}	0.21	
	2.0		3.9×10^{-3}	0.28			
P_2VP	1.0	H_3PO_4	6.0×10^{-8}	0.93	1.2×10^{-5}	0.33	27
	1.5		2.1×10^{-6}	0.75			
	2.0		1.0×10^{-4}	0.73			
P_4VP	1.0	H_3PO_4	1.0×10^{-6}	0.75	2.4×10^{-4}	0.55	27
	2.0		1.7×10^{-4}	0.76	5.8×10^{-3}	0.26	
PVP	1.33	H_2SO_4	8.0×10^{-6}	0.62			10
	1.5		1.6×10^{-4}	0.62			
	2.0		1.2×10^{-4}	0.51			
PVP	1.33	H_3PO_4	6.0×10^{-6}	0.93			10
	1.5		5.5×10^{-5}	0.72			
	2.0		3.0×10^{-6}	1.5			
BPEI	0.2	H_2SO_4	2.0×10^{-5}	0.77	2.5×10^{-3}	0.43	14, 27
	0.35		7.7×10^{-7}	1.05	2.8×10^{-4}	0.58	
	0.5		2.7×10^{-4}	0.51	6.0×10^{-3}	0.35	
	1.0		3.0×10^{-4}	0.50	6.4×10^{-3}	0.36	
	1.5		8.0×10^{-4}	0.48			
BPEI	0.23	H_3PO_4	2.5×10^{-6}	1.2	1.6×10^{-3}	0.58	14, 27
	0.35		2.7×10^{-7}	1.3	7.0×10^{-4}	0.94	
	0.52		1.4×10^{-5}	0.94	2.2×10^{-3}	0.50	
	1.0		5.0×10^{-4}	0.69			
	1.5		3.0×10^{-4}	0.7	1.3×10^{-2}	0.48	

Table 20.2 (*cont.*)

Polymer	x	Acid	σ_{300K}	E_{a300K}	σ_{373K}	E_{a373K}	References
LPEI	0.1	H_2SO_4	3.5×10^{-9}	1.13	3.3×10^{-6}	0.51	12
	0.46				8.7×10^{-9}	1.1	
	0.73				5.2×10^{-7}	0.77	
LPEI	0.4	H_3PO_4			1.9×10^{-6}	1.4	12
	0.7		7.6×10^{-9}	1.36	3.4×10^{-5}	0.81	
	0.9		1.9×10^{-8}	1.32			
Aminosil	0.02	$HClO_4$	8.0×10^{-7}	0.42	1.0×10^{-5}	0.31	11
	0.1		7.0×10^{-6}	0.58	2.2×10^{-4}	0.32	
	0.3		1.0×10^{-6}	1.0	2.2×10^{-4}	0.48	

At 300 K, H_2SO_4 is liquid and H_3PO_4 either in the supercooled liquid state (l), or in the solid state (s). The activation energy values E_a (eV) are evaluated from the slope of $\log(\sigma T)$ vs $(1000/T)$ plots at 300 and 373 K.

conductivity of the pure acid. Part of these data are reported in Table 20.2 and Fig. 20.3.

Unfortunately, no glass transition temperature (T_g) measurements are available on these systems. Visual observation indicates that, starting from polymers with T_g values above 300K (Table 20.1), addition of acid leads to a progressive plastification until, finally, soft materials that can be processed as thin films are obtained at high acid concentrations.

The sharp increase of conductivity observed with mixtures involving H_2SO_4 at $x \simeq 1$ indicates that the protonation needs to be nearly complete and the number of HSO_4^- anions high enough to create a conduction path for the protons. Then, a system of the type $Y-H^+ \ldots SO_4H^-$ is present. The conduction is supposed to proceed essentially through the hydrogen-bonded anionic chains by analogy with the superionic phases of $MHSO_4$ compounds (M = Cs, $NH_4 \ldots$)[15]. The dynamical disorder of the M^+ cations in these superionic phases would be paralleled by a high chain flexibility of the $Y-H^+$ polycation in the sulphuric acid/polymer mixtures. In addition, we have seen that proton exchange may occur within the resonance forms of protonated amide groups. This could explain the very high conductivities observed with Paam and PVP. Proton exchange in amides is an outstanding problem for biochemists[30].

Mixed inorganic–organic systems

20.3.3 Highly basic polymers

Owing to their high basicity (Fig. 20.1), alkylenimine polymers are readily protonated by acids[31]. The conduction properties of LPEI[12] and BPEI[13–15,27] mixed with H_2SO_4 or H_3PO_4 have recently been investigated.

A new family of compounds called *aminosils* (Table 20.1) has also been presented[11].

The concentration dependence of their conductivity is illustrated in Fig. 20.4. Three conduction regimes can be distinguished.

From $x = 0$ to $\simeq 0.35$ (region I of Fig. 20.4), the LPEI and BPEI polymers mixed with H_2SO_4 or H_3PO_4 are progressively protonated to reach a stage of maximum protonation of about 70% that cannot be overcome because of strong electrostatic repulsions between neighbouring ammonium groups[14]. The associated anions are $SO_4{}^{2-}$ or $HPO_4{}^{2-}$, as shown by IR spectroscopy[13–15]. The maximum in conductivity observed for $x = 0.2$, corresponding to $\simeq 40\%$ protonation, is attributed to the greatest probability of proton exchange between protonated and non-protonated amine sites. The conductivity then decreases towards a minimum for

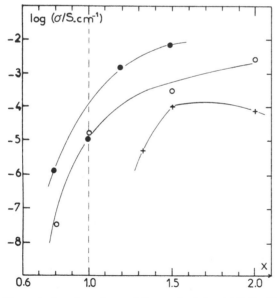

Fig. 20.3. Concentration dependence of the conductivity at 300 K of Paam, xH_2SO_4 (●), P_2VP, xH_2SO_4 (○)[15,27], PVP, xH_2SO_4 (+)[10].

321

$x \simeq 0.35$, i.e. when the polymer has reached its maximum degree of protonation. The conductivity of BPEI, xHCl shows a similar behaviour and supports the hypothesis of conduction based on the exchange process

$$N-H^+ \dots N \rightleftarrows N \dots {}^+H-N \qquad (20.6)$$

This kind of interaction is effectively observed by IR in the N–H stretching region[27]. It must be pointed out, however, that the nature of the anion influences strongly the level of conductivity in the sequence $SO_4{}^{2-} > HPO_4{}^{2-} > Cl^-$ and it is, at present, difficult to decide whether the anion plays an active relay role in the proton transport or simply a passive role of conjugated base affecting the polymer dynamics. The T_g increases systematically from $-43\,°C$ at $x = 0$ up to, respectively, 0, 30 and 60 °C for H_3PO_4, H_2SO_4 and HCl at the maximum of protonation. There is no clear correlation between these T_g increases and the

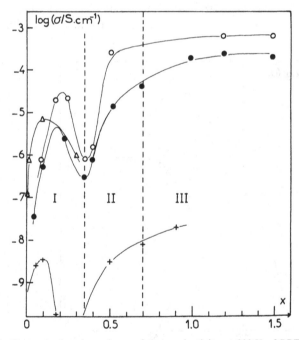

Fig. 20.4. Concentration dependence of the conductivity at 300 K of BPEI, xH$_2$SO$_4$ (○), BPEI, xH$_3$PO$_4$ (●) refs 13–15, 27, Aminosil, xHClO$_4$ (△) ref. 11 and LPEI, xH$_2$SO$_4$ (+) ref. 12.

conductivity differences. One can only conclude that, with all acids, the polymer expands and is progressively rigidified or cross-linked by the onset of new hydrogen bonds of the $N-H^+ \ldots N$ and $N-H^+ \ldots {}^-X$ types.

A new conduction mechanism starts above $x \simeq 0.35$ using H_2SO_4 and H_3PO_4 (region II). It corresponds to the appearance of HSO_4^- or $H_2PO_4^-$ anions. The observed increase of conductivity (Fig. 20.4) is attributed to proton migration along the mixed HSO_4^-/SO_4^{2-} or $H_2PO_4^-/HPO_4^{2-}$ anionic chains by successive proton transfer and anion reorientation steps.

The T_g of these mixtures decreases as a result of the new $O-H \ldots O$ interactions between anions which compete with the previous $N-H \ldots N$ and $N-H \ldots O$ interactions and leave more freedom to the chain motions.

Above $x \simeq 0.7$ (region III), HSO_4^- and $H_2PO_4^-$ anions exist only with excess acid. The conductivity reaches a plateau (Fig. 20.4) which is of the same order of magnitude as the conductivity of the pure acids. In this concentration range, the T_gs are no longer measurable as the mixtures become viscous pastes with increasingly lower viscosity.

Let us note that regions II and III could not be studied with BPEI/HCl, due to their very low conductivity and with Aminosils[11], because the material became white and granular. Although the LPEI and BPEI mixtures with acids present similar variations in conductivity versus acid concentration, their level of conductivity differs by several orders of magnitude (Fig. 20.4 and Table 20.2). This has been ascribed to the crystallinity of LPEI compared to the completely amorphous character of BPEI[14].

20.4 Temperature dependence of the conductivity

The temperature dependence of the conductivity in polymer electrolytes has often been taken as indicative of a particular type of conduction mechanism[2,3]. In particular, a distinction is generally made between systems which show an Arrhenius type of behaviour and those which present a curvature in log σ or log σT vs inverse temperature plots. In the latter case, empirical equations derived from the free-volume theory have been used, as for example the *Vogel–Tamman–Fulcher (VTF)* equation

$$\sigma = AT^{-1/2}\exp[-E_a/k(T - T_0)] \qquad (20.7)$$

where A is a pre-exponential factor related to the number of charge carriers, T_0 is the ideal glass transition temperature, E_a an activation energy and k the Boltzmann constant.

Similarities between the PEO/H_3PO_4 and PEO/alkali salt systems have been pointed out[9,10]. The data for the amorphous PEO, $0.42H_3PO_4$ composition, which corresponds to the maximum of conductivity (Fig. 20.2), have been reproduced by the VTF equation with $A = 2$, $T = 224$ K and $E_a = 0.05$ eV whereas the $x = 0.66$ semi-crystalline composition exhibits Arrhenius behaviour[9].

The PVA, $0.42H_3PO_4$ composition[7] also shows Arrhenius-like behaviour and this is interpreted as due to the presence of a separate H_3PO_4/H_2O conducting phase rather than to a transport mechanism linked to polymer motions.

In all the other systems (Table 20.2), curvature has been observed in the log σT vs ($1000/T$) plots, as illustrated in Fig. 20.5 and reported in

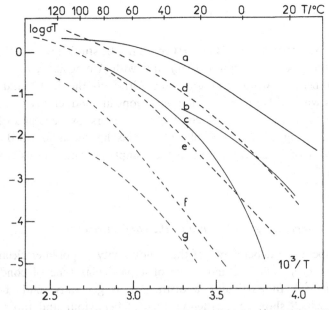

Fig. 20.5. Arrhenius plots of: continuous lines (a) Paam, $1.2H_2SO_4$[15], (b) PVP, $1.5H_2SO_4$[10], (c) PEO, $0.42H_3PO_4$[9]. Dashed lines: BPEI, xH_2SO_4 (d) $x = 0.5$, (e) $x = 0.25$, (f) $x = 0.35$, (g) $x = 0.05$[15,27].

the literature[10,15]. However, if the VTF equation has been successfully applied to the Aminosil/HClO$_4$ mixtures of low acid concentration[11], it was difficult to reproduce the Paam/H$_2$SO$_4$ and BPEI/H$_2$SO$_4$ data[27]. Actually, Fig. 20.5 indicates that for these mixtures a marked curvature exists at high temperature but that the low-temperature regime is nearly linear. For such highly associated and viscous materials, the free-volume concept is questionable.

NMR and *QNS* studies of the polymer and anion dynamics are likely to provide a more direct approach to the understanding of the conduction mechanisms in these systems[9,15].

20.5   Conclusion

In the large family of proton conductors, a new class has now to be included: the mixed inorganic–organic systems involving strong acids X–H and polymers bearing basic sites, Y. The conductivity depends strongly on the nature of the two components and on the acid concentration, as reported in Table 20.2. At low acid concentrations ($x \lesssim 0.5$), only very basic polymers exhibit appreciable conductivities, even with mono-acids, in the order H$_2$SO$_4$ > H$_3$PO$_4$ > HX (X = Cl, Br, I, ClO$_4$ . . .). In the partially protonated polymer, the dominant interactions are intra- and/or interchain hydrogen bonds, Y–H$^+$ . . . Y, between protonated and non-protonated basic sites. Conduction is supposed to occur primarily by proton exchange between these sites.

In contrast, when the polymer has reached its maximum degree of protonation, at higher acid concentration, the conduction is supposed to proceed via the anionic hydrogen-bonded chain. Indeed, mixtures with H$_2$SO$_4$ or H$_3$PO$_4$ present increasing conductivities whereas monoacids give non-conducting salts. Analogies certainly exist with proton-conducting hydrogensulphate salts[15].

For concentrations where excess acid begins to appear, the conductivity often reaches a plateau. The protonated polymer may then be considered simply as a plastified matrix having little influence on the level of conductivity which seems to be determined by the acid self-ionization. However, some particular organic groups, such as *amides*, could well contribute to the conductivity by proton exchange between protonated resonance forms[27,30].

In the presentation of the results, we have stressed the importance of the relative acidities and basicities of the acids and polymers. A first classification has been proposed in terms of pK differences (Fig. 20.1). This approach is very qualitative since pK values measured in aqueous solution are applied to the discussion of anhydrous mixtures. A more direct and appropriate method would be to consider the proton affinity of the basic sites, evaluated for example from vibrational correlation diagrams[32, 33].

Very few electrochemical studies have been performed on these systems to determine the proton transport number and the redox stability. These studies now become necessary.

Finally, it is clear that a better understanding of the proton conduction mechanisms requires also fundamental studies of the polymer and acid proton dynamics at a microscopic level. Preliminary NMR and QNS investigations indicate the way to be followed[9].

Although many fundamental problems remain to be elucidated, practical applications such as *electrochromic devices*[33, 34], *fuel-cells and sensors*[35, 36] have already been proposed. These applications require the production of *electrolytic membranes* of good mechanical stability. To achieve this goal, new interesting directions can be taken. They tend to exploit the very rich polymer chemistry, either to synthesize new systems, as illustrated by the promising *Ormosil* family of compounds[11], or to use interpenetrating polymer networks, as proposed by Petty-Weeks *et al.*[8, 37]. These authors have made a PVA/poly(methacrylic acid–methylenebisacrylamide)/H_3PO_4 membrane which they claim to be water insoluble and of increased bulk modulus while keeping the primary conduction properties of the PVA/H_3PO_4 blend.

More exciting developments could occur in the near-future with systems more closely related to biological processes[38].

20.6   References

1. *Solid State Protonic Conductors – for fuel cells and sensors I, II, III,* J. Jensen, M. Kleitz, A. Potier and J. B. Goodenough (eds) (Odense University Press (1982, 1983, 1985)); and *IV*, R. T. C. Slade (ed.), *Solid State Ionics* **35** (1989) 1–197.
2. M. Armand, *Solid State Ionics* **9 & 10** (1983) 745–54.
3. *Polymer Electrolyte Reviews – 1,* J. R. MacCallum and C. A. Vincent (eds) (Elsevier Applied Science (1987)).
4. *Developments in Ionic Polymers,* A. Wilson and H. J. Prosser (eds) (Applied Science Publishers (1983)).

Mixed inorganic–organic systems

5. P. J. Smith, in *Electrochemical Science and Technology of Polymers – 1*, R. G. Linford (ed.) (Elsevier (1987)) 293–329.
6. P. Aldebert, F. Novel-Cattin, M. Pineri, P. Millet, C. Doumain and R. Durand, *Solid State Ionics* **35** (1989) 3–9.
7. A. J. Polak, S. Petty-Weeks and A. J. Beuhler, *Sensors and Actuators* **9** (1986) 1–7; S. Petty-Weeks and A. J. Polak, *Sensors and Actuators* **11** (1987) 377–86.
8. S. Petty-Weeks, J. J. Zupancic and J. R. Swedo, *Solid State Ionics* **31** (1988) 117–25.
9. P. Donoso, W. Gorecki, C. Berthier, F. Defendini, C. Poinsignon and M. B. Armand, *Solid State Ionics* **28–30** (1988) 969–74.
10. F. Defendini, Thesis, Grenoble (1987).
11. Y. Charbouillot, D. Ravaine, M. B. Armand and C. Poinsignon, *J. Non-Cryst. Solids* **103** (1988) 325–30.
12. R. Tanaka, T. Iwase, T. Hori and S. Saito, in *Proc. Intern. Congr. Polym. Electrolytes*, St Andrews, Scotland (1987) 31–3.
13. M. F. Daniel, Thesis, Bordeaux (1986).
14. M. F. Daniel, B. Desbat, F. Cruege, O. Trinquet and J. C. Lassègues, *Solid State Ionics* **28–30** (1988) 637–41.
15. J. C. Lassègues, B. Desbat, O. Trinquet, F. Cruege and C. Poinsignon, *Solid State Ionics* **35** (1989) 17–25.
16. N. N. Greenwood and A. Thompson, *J. Chem. Soc.* (1959) 3485–92.
17. R. J. Gillespie and E. A. Robinson, in *Non-Aqueous Solvent Systems*, T. C. Waddington (ed.) (Academic Press **4** (1965)) 117–210.
18. R. A. Munson, *J. Phys. Chem.* **68** (1964) 3374–7.
19. *Chemistry of the Elements*, N. N. Greenwood and A. Earnshaw (eds) (Pergamon Press (1984)).
20. S. Searles and M. Tamres in *The Chemistry of the Ether Linkage*, Ch. 6, S. Patai (ed.) (Interscience (1967)).
21. W. M. Kulicke, R. Kniewske and J. Klein, *Prog. Polym. Sci.* **8** (1982) 373–468.
22. M. Mandel, *Polyelectrolytes* in *Encyclopedia of Polymer Science and Technology* **11** (1988) 739–829.
23. C. D. Robitaille and D. Fauteux, *J. Electrochem. Soc.* **133** (1986) 315–24.
24. W. Gorecki, R. Andreani, C. Berthier, M. B. Armand, M. Mali, J. Roos and J. Brinkmann, *Solid State Ionics* **18–9** (1986) 295–9.
25. D. T. Chin and H. H. Chang, *J. Appl. Electrochem.* **19** (1989) 95–9.
26. *Nouveau Traité de Chimie Minérale*, P. Pascal (ed.), Masson, Tome X (1956) 816–27.
27. O. Trinquet, Thesis, Bordeaux (1990).
28. M. Putterman, J. L. Koenig and J. B. Lando, *J. Macromol. Sci. Phys.* **B16** (1979) 89–116.
29. Yu. E. Kirsch, O. P. Komarova and G. M. Lukovkin, *Europ. Polym. J.* **9** (1973) 1405–15.
30. C. L. Perrin, *Acc. Chem. Res.* **22** (1989) 268–75.
31. P. Ferruti and R. Barbucci, *Adv. Polym. Sci.* **58** (1984) 57–92.
32. B. S. Ault, E. Steinback and G. C. Pimentel, *J. Phys. Chem.* **79** (1975) 615–20.

Materials

33. T. Zeegers-Huyskens, *J. Mol. Struct.* **177** (1988) 125–41.
34. M. B. Armand, F. Defendini and B. Desbat, Matériaux à conduction protonique, French Patent No 86.00717 (1986).
35. H. Arribart, C. Padoy, A. Dugast, M. B. Armand, F. Defendini and B. Desbat, Vitrage à transmission variable du type électrochrome, Eur. Patent No 025713 (1987).
36. A. J. Polak, J. A. Wrezel and A. J. Beuhler, Electrochromic Devices, US Patent No 4,664,768 (1986).
37. J. J. Zupancic, R. J. Swedo and S. Petty-Weeks, Water insoluble proton conducting polymers, US Patent No 4,708,981 (1987); Method and apparatus for gas detection using proton-conducting polymers, US Patent No 4,664,757 (1987); Electrochemical method and apparatus using proton-conducting polymers, US Patent No 4,664,761 (1987).
38. D. Hadzi, *J. Mol. Struct.* **177** (1988) 1–21.

21 Incoherent neutron scattering studies of proton conductors: from the anhydrous solid state to aqueous solutions

JEAN-CLAUDE LASSEGUES

21.1 Introduction

Among the experiments probing hydrogen on an atomistic scale, it is well known that *neutron scattering* plays a special role[1-8]. Indeed, the proton has a much higher incoherent scattering cross-section than any other common atom and its motions can be probed on appropriate space and time-scales, ~ 0.5–10 Å and $\sim 10^{-9}$–10^{-13} s, respectively.

Since we are concerned with proton conductors, we will restrict discussion to incoherent scattering. In addition, since the samples considered are mostly liquids and powders, the basic formalism involves simply the modulus, Q of the momentum transfer in

$$\hbar \bar{Q} = \hbar(\bar{k}_0 - \bar{k}) \tag{21.1}$$

where $\bar{k}_0$ and $\bar{k}$ are the incident and scattered neutron wavevectors. The energy transfer is defined as

$$\hbar\omega = (\hbar^2/2m)(k_0^2 - k^2) \tag{21.2}$$

where m is the neutron mass.

Broadly speaking, the aim of an *incoherent neutron scattering* study is to characterize the individual translational and rotational diffusive motions of the scatterer (in this case, the proton held by an ion or molecule) at small energy transfers in the so-called quasielastic neutron scattering (QNS) region and the quantized vibrations at higher energy transfers in

Proton dynamics and charge transport

the inelastic region (*INS*). In the case of proton conductors, there is of course a special interest in relating the above dynamical information to the conduction mechanism and in finding eventually some specific feature of the proton transport itself.

In the following, we will consider the incoherent scattering law $S^{inc}(Q, \omega)$ obtained from the experimental data after the usual standard corrections and the elimination of Bragg diffraction in certain scattering directions.

21.1.1 QNS: translational and rotational motions

$S^{inc}(Q, \omega)$ is generally analysed under the simplifying, but sometimes rather severe, approximation of uncoupled *translational*, *rotational* and *vibrational motions*. In the QNS region, this approximation leads to

$$S^{inc}(Q, \omega) = \exp(-\langle u^2 \rangle Q^2) \cdot S^T(Q, \omega) \otimes S^R(Q, \omega) \qquad (21.3)$$

The vibrational contribution reduces to a Debye–Waller Factor where $\langle u^2 \rangle$ is the mean-square vibrational amplitude of the scatterer resulting from all the vibrations of the system. This factor multiplies the convolution product (symbol $\otimes$) of the translational and rotational scattering laws.

21.1.1.1 Translational motions

In the case of Fick's diffusion, $S^T(Q, \omega)$ takes the simple lorentzian form:

$$S^T(Q, \omega) = \frac{1}{\pi} \frac{D_t Q^2}{\omega^2 + (D_t Q^2)^2} \qquad (21.4)$$

where D_t is the macroscopic *self-diffusion coefficient*. QNS experiments on liquids have indeed shown that in the limit of small Q values, the lorentzian width is proportional to Q^2 [8–11]. The slope gives a D_t value which is in agreement with the one measured by NMR or tracers.

The main interest of the QNS technique lies, however, in the $S^T(Q, \omega)$ behaviour at higher Q values, i.e. when the translational process is followed on molecular distances. A departure from *Fick's law* is generally observed [4–8]. This is indicative of the microscopic aspect of the motion and its interpretation needs the elaboration of more sophisticated models involving characteristic jump lengths, jump times, etc.

In the *random jump diffusion* (RJD) *model*[12] used, for example, in the case of water and aqueous solutions[11–13], a statistical distribution of *jump lengths* is assumed and the following expression is derived

$$S^T(Q, \omega) = \frac{1}{\pi} \frac{\dfrac{D_t Q^2}{1 + D_t Q^2 \tau_0}}{\omega^2 + \left(\dfrac{D_t Q^2}{1 + D_t Q^2 \tau_0}\right)^2} \tag{21.5}$$

where τ_0 represents the *residence time* between instantaneous jumps of mean-square length $\langle l^2 \rangle = 6 D_t \tau_0$.

A residence time τ_0 and a non-negligible jump time τ_1 are also introduced to describe the proton translational motion in solids[14,15] but in contrast with liquids, the jump direction and length are determined by quasi-equilibrium sites which form a periodic interstitial lattice. In this case, as exemplified by studies of hydrogen diffusion in metals[4,14,15], it becomes very interesting to look at the anisotropy of the motion by studying selected crystal orientations relative to $\bar{Q}$, particularly when the conductivity itself is anisotropic.

It is not possible here to go into the details of the various proposed models[4–8,12]. Let us briefly note that in the simplifying assumption of a cubic lattice of parameter L the *Chudley–Elliott (CE) model*[14] gives

$$S^T(Q, \omega) = \frac{1}{\pi} \frac{\dfrac{\hbar}{\tau_0} \left(1 - \dfrac{\sin QL}{QL}\right)}{\omega^2 + \left[\dfrac{\hbar}{\tau_0} \left(1 - \dfrac{\sin QL}{QL}\right)\right]^2} \tag{21.6}$$

where the *jump time*, τ_1, is considered as negligible in comparison with the residence time $\tau_0 = L^2/6D$. In all these models, $S^T(Q, \omega)$ is a lorentzian when τ_1 is neglected and its expression tends to the limiting case of Eqn (21.4) at very low Q values.

21.1.1.2 Rotational motions

As the *rotational motions* are bound to a fixed point in the molecule or ion, $S^R(Q, \omega)$ takes the following general expression

$$S^R(Q, \omega) = A_0(Q) \, \delta(\omega) + \frac{1}{\pi} \sum_{i=1}^{n} A_i(Q) L_i(\omega) \tag{21.7}$$

Proton dynamics and charge transport

where the so-called *Elastic Incoherent Structure Factor* (*EISF*), $A_0(Q)$, multiplies a delta function. $A_0(Q)$ corresponds to the fraction of elastically scattered intensity, since

$$\sum_{i=0}^{n} A_i(Q) = 1$$

and characterizes the spatial distribution of the scatterer at equilibrium. The second term of Eqn (21.7) involves a weighted sum of lorentzians $L_i(\omega)$. The width of these lorentzians is proportional to the rotation rate.

In the case of molecular liquids, the *Sears model* of rotational diffusion on a sphere of radius a has often been adopted[16]

$$S^R(Q, \omega) = j_0{}^2(Qa)\,\delta(\omega) = \frac{1}{\pi}\sum_{l=1}^{\infty}(2l+1)j_l{}^2(Qa)\frac{l(l+1)D_r}{\omega^2 + [l(l+1)D_r]^2} \quad (21.8)$$

where D_r is a rotational diffusion coefficient and the $j_l(Qa)$ terms are spherical Bessel functions.

For solids, better-defined positions are generally explored during the reorientational motions. If, for example, the scattering particle performs stochastic jumps among S sites equally distributed on a circle of radius a, the result after powder averaging is

$$S^R(Q, \omega) = B_0(Q)\,\delta(\omega) + \frac{1}{\pi}\sum_{l=1}^{S-1}B_l(Q)\frac{\dfrac{\hbar}{\tau_l}}{\omega^2 + \left(\dfrac{\hbar}{\tau_l}\right)^2} \quad (21.9)$$

with

$$B_l(Q) = \frac{1}{S}\sum_{j=1}^{S}j_0\left(2Qa\sin\frac{\pi j}{S}\right)\cos\left(2l\frac{\pi j}{S}\right) \quad (21.10)$$

and

$$\tau_l = \tau_1\frac{\sin^2\dfrac{\pi}{S}}{\sin^2\dfrac{\pi l}{S}}, \qquad \tau_1 = \frac{\tau_{res}}{1 - \cos\dfrac{2\pi}{S}} \quad (21.11)$$

where τ_{res} is the mean residence time on a site.

More sophisticated models have been proposed but an important

332

property of Eqn (21.7) remains. It is in principle always possible to determine experimentally the EISF and therefore to obtain a good idea of the geometry of the motion. If the scatterer can occupy S sites of coordinate $\bar{r}_l$ with a probability p_l, the EISF can indeed be calculated numerically[17] according to

$$B_0(Q) = \left| \sum_{l=1}^{S} p_l \exp(i\bar{Q}\bar{r}_l) \right|^2 \tag{21.12}$$

with

$$\sum_{l=1}^{S} p_l = 1$$

21.1.2 INS: vibrational motions

For systems presenting discrete inelastic features due to internal vibrational modes of molecules or ions, a quantum-mechanical formalism is convenient

$$S^{V}(Q, \omega) \propto \sum_{i} \sum_{f} P_i \sum_{j} \langle f | \exp(i\bar{Q}\bar{u}_j)|i \rangle^2 \, \delta(E_i - E_f - \hbar\omega) \tag{21.13}$$

where P_i is the thermal occupancy of the initial state i of energy E_i, j is the final state of energy E_f and $\bar{u}_j$ is the nuclear displacement vector of scatterer (proton) j. Eqn (21.13) has been developed for the harmonic oscillator and other simple cases[6]. In contrast with optical spectroscopy where energy transfers are measured with a higher precision but intensities are difficult to exploit because the dipole moment or polarizability operators are often unknown, INS intensities can be directly calculated since the neutron scattering amplitude simply involves the $\bar{Q} \cdot \bar{u}_j$ product. A quantitative approach to the INS spectra may thus be developed from normal coordinate analysis methods[18–20].

For broad spectral densities occurring at lower frequency in liquids or dynamically disordered solids such as proton conductors, the time dependent formalism is better adapted[8,12,21]. It leads for example to the frequency distribution function

$$g(\omega) \sim \omega^2 \lim_{Q \to 0} \frac{S^{inc}(Q\omega)}{Q^2} = \frac{1}{2\pi} \int dt \, \exp(-i\omega t)\langle v(0)v(t) \rangle \tag{21.14}$$

As indicated above, $g(\omega)$ (usually called $P(\omega)$, see Chapters 25 & 30), is the Fourier transform of the velocity autocorrelation function, $\langle v(0)v(t) \rangle$, of great importance in molecular dynamics calculations. In addition, $g(\omega)$ is homogeneous with the $\omega^2 I(\omega)$ formalism now currently used to analyse the low-frequency infrared and Raman spectra[8, 22, 23]. Finally, $g(\omega)$ is proportional to the frequency dependent ionic conductivity[24] (see Chapter 25).

21.2   Anhydrous solid protonic conductors

We have included in this category hydronium salts such as $H_3O^+ClO_4^-$ or H_3O^+ β-alumina in which all water molecules are protonated.

21.2.1   QNS studies

Conductivity and diffusivity values are reported in Table 21.1 for some systems. As far as we know, only NH_4^+ β-alumina[25] and $CsHSO_4$ have been studied at high resolution to look at the proton translational motion. The limiting D_t values deduced from the QNS broadening at low Q are in agreement with those measured by NMR or tracers (Table 21.1). $CsHSO_4$ is the only case where a departure from Fick's law could be detected with reasonable accuracy at higher Q values[26, 27]. Typical spectra are given in Figs 21.1a and b. Fig. 21.1c indicates the Q dependence of the translational lorentzian width. Although the RJD model Eqn (21.5) or the CE model Eqn (21.6) are crude approximations for the description of the proton diffusion in this system, they have been fitted to the data for a first estimate of mean jump lengths and residence times. It can be observed that in the investigated Q range there is little difference between the two kinds of theoretical curves. The jump lengths are situated between 1 and 1.8 Å and the residence times τ_0 are in the range $1-2.5 \times 10^{-10}$ s.

From these preliminary results it is clear that future studies have to overcome several problems.

> Precise structural work is needed to correlate the jump lengths deduced from QNS with possible crystallographic sites for the proton. Two different sets of data have, for example, been published on the $CsHSO_4$ superionic phase[28, 29] and a third, more precise neutron diffraction study on $CsDSO_4$ has appeared[30]. The modelling of the oxygen and proton positions is indeed a difficult task in these highly disordered materials.

Table 21.1. *Specific conductivity* σ, *proton self-diffusion coefficient* D_t *and residence time between reorientations* τ_{res} *for some anhydrous solid state protonic conductors*

System	Conductivity			Diffusivity			Reorientations		
	$\sigma/\text{S cm}^{-1}$ E_a/eV	(T/K) $(\Delta T/\text{K})$	Ref.	$D_t/\text{cm}^2\text{ s}^{-1}$ E_a/eV	(T/K) $(\Delta T/\text{K})$	Ref.	τ_{res}/s E_a/eV	(T/K) $(\Delta T/\text{K})$	Ref.
$H_3O^+ClO_4^-$	3.4×10^{-4} 0.37	(298) (263–321)	65	1.6×10^{-9} 0.36	(298) (290–304)	65	$(2 \pm 1) \times 10^{-12}$ 0.077	$(294)^a$ (253–323)	37
H_3O^+, β-alumina	$\sim 10^{-10}$ 0.8	(300)	39				2.5×10^{-12} 0.056	$(395)^b$ (395–485)	25
NH_4^+, β-alumina	1.5×10^{-6} 0.5	(300)	39	$\sim 10^{-8}$	(351)	25	0.7×10^{-12} 0.035	$(373)^c$ (298–473)	25, 36
$CsHSO_4$	2.7×10^{-3} 0.33	(420) (414–480)	66	1.2×10^{-7} 0.23	(420) (414–460)	68	~ 1 and 10×10^{-12}	$(420)^d$	26, 27
Py, H_2SO_4	2.0×10^{-6} 1.13	(298) (269–315)	67				11.5	$(300)^e$	35

[a] H_3O^+ reorientations determined by NMR to be either three-fold or isotropic.
[b] 44% rotating H_3O^+ about their three-fold axis.
[c] 75% rotating NH_4^+ about their three-fold or two-fold axis.
[d] HSO_4^- local oscillations and slower reorientations of unknown geometry.
[e] Coupled oscillations of the pyridinium and HSO_4^- ions by angles of about $\pm 20°$ and $\pm 10°$ respectively.

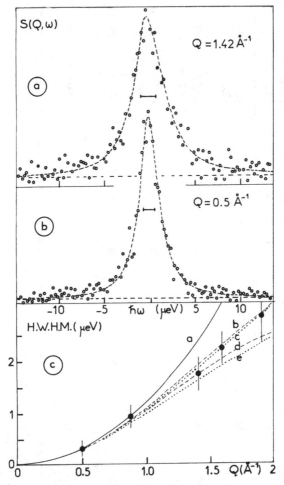

Fig. 21.1. Typical QNS spectra of CsHSO$_4$ at 440 K recorded with the back-scattering IN10 spectrometer at I.L.L. (Grenoble). The experimental points have been fitted by the convolution product of a translational lorentzian by the resolution function (FWHH indicated on the figure). A flat background has been added to simulate the broad rotational contribution.

The halfwidth of the fitted lorentzians is reported in Fig. 21.1c (black circles) as a function of momentum transfer. The solid line a corresponds to a D_tQ^2 law fitted at small Q values. The dashed curves correspond to the Chudley–Elliott model with $L = 1.5$ (line b) and 1.8 Å (line d) and the dotted curve to the Random Jump Diffusion model with $\langle l^2 \rangle^{\frac{1}{2}} = 0.93$ (line c) and 1.22 Å (line e).

Unless the present optimal resolution of $\sim 1\,\mu eV$ is improved, only D_t values larger than $\sim 10^{-7}\,cm^2\,s^{-1}$ can be accurately measured. This is clearly illustrated by successful studies of ionic conductors involving less favourable scatterers (Li^+, Na^+, Ag^+, Cl^- etc . . .) but faster diffusion[31–34].
It is crucial to explore larger Q ranges in order to discriminate between the various models.

QNS studies at medium resolution (10–100 μeV) have shown that all the compounds of Table 21.1 present rather fast oscillatory and reorientational motions of the hydrogenated ions[25, 35–38]. In jump models given by Eqns (21.9–21.11), the residence times are situated in the 10^{-11}– 10^{-12} s range (Table 21.1). Analysis of the EISFs has brought interesting information not only on the geometry of the motion but also, as for the β-alumina derivatives, on the percentage of rotating protons when the ions do not occupy identical crystallographic sites[25]. In the hydrogen-sulphates of cesium, triethylenediamine (TED) or pyridine (Py), the HSO_4^- anions seem to perform large amplitude oscillatory motions and less frequent reorientations[35, 38]. Although at a preliminary stage, the QNS results on $CsHSO_4$ clearly show that a 'nearly-free' rotation of the HSO_4^- ions, previously invoked in the literature, has to be excluded[27].

21.2.2 INS studies

Several compounds of Table 21.1 exhibit a phase transition between poorly and highly conducting phases. The changes of the INS spectra through this transition are often striking, as illustrated for $CsHSO_4$ in Fig. 21.2. The low temperature spectrum exhibits rather well-defined peaks mainly due to translational (T) and rotational (R) oscillations of the hydrogen-bonded anions. In the superionic phase, these vibrations are strongly damped and give a broad continuous profile[27, 39, 40]. The $g(\omega)$ spectra of $H_3O^+ClO_4^-$ present similar behaviour and an assignment of the ClO_4^- and H_3O^+ R and T motions has been given[41] (see p. 167).
In both cases, it has been remarked that at small energy transfers, ($\hbar\omega \leqslant 5$ meV), $g(\omega)$ changes from a Debye to a liquid-like shape. This is the short-time dynamics region, intermediate between QNS and INS. It has been little considered until now and would certainly deserve more quantitative studies in parallel with those performed by optical spectroscopy[42].

However, when discrete INS excitations are observed, their assignment, even at a qualitative level, complements very usefully the analysis of the IR and Raman spectra as illustrated by a recent determination of the H_3O^+ internal and external modes in $H_3OM(SO_4)_2$ (M = Fe, In and Al)[43,44]. A representative spectrum is given in Fig. 21.3a.

21.3 Hydrated solid protonic conductors

The more highly conducting solid protonic conductors such as *Nafion*[®45,46], $HUO_2PO_4 \cdot 4H_2O(HUP)$[39,47,48], heteropoly acid hydrates[49,51], etc. contain water generally associated with H_3O^+ or $H_5O_2^+$ ions. Therefore, several kinds of protons having different dynamical behaviour are present.

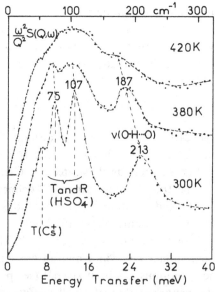

Fig. 21.2. INS spectra of $CsHSO_4$ recorded with the IN6 spectrometer (I.L.L., Grenoble, France) and presented under a form proportional to Eqn (21.14) but without extrapolation to $Q = 0$. As the observed modes present no detectable Q dependence, several spectra have been added together in the Q range $1-1.2 Å^{-1}$ (at $\hbar\omega = 0$) in order to improve the statistics. The temperatures of 300, 380 and 420 K correspond to three stable phases of $CsHSO_4$ including the superionic phase ($T_{tr} \sim 415$ K). The wavenumber and assignment of some peaks are indicated.

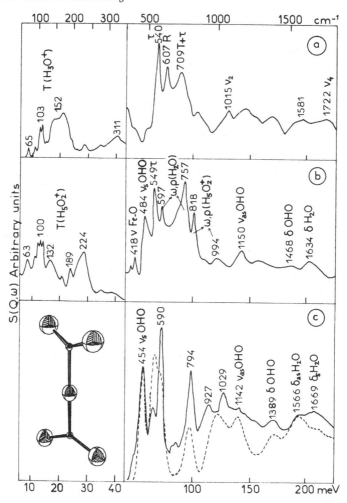

Fig. 21.3. INS spectra of (a) [H₃O][Fe(SO₄)₂] at 20 K recorded with the TFXA spectrometer (Rutherford Appleton Laboratory, UK)[43], (b) [H₅O₂][Fe(SO₄)₂]·2H₂O, same conditions[57], and (c) (H₅O₂)₃(PW₁₂O₄₀) (TPA, 6H₂O) at 5 K recorded with the IN1BeF spectrometer (I.L.L., Grenoble, France)[56]. The experimental points have been smoothed by a continuous solid line. In (c), the dashed line corresponds to a spectrum fitted to the experimental one from a force field of the $H_5O_2^+$ internal modes. The total vibrational ellipsoids of these internal vibrations are schematized in the insert[56]. Part of the assignment is reported under Jones *et al.*'s nomenclature: ω, wagging; τ, torsion; ρ, rocking; δ, in-plane bending; R, rotational libration; T, translational oscillation; ν_s, symmetric stretching; ν_{as}, antisymmetric stretching; ν_2, symmetric H_3O^+ bending; ν_4, doubly degenerate H_3O^+ bending.

Proton dynamics and charge transport

21.3.1 QNS studies

QNS experiments using different instrumental resolution and energy transfer ranges have been performed in order to disentangle the diffusive proton motions of the various species. One of the first attempts has been made on an acid Nafion® membrane containing c. 15% water by weight[46]. It has been found that on a scale of about 10 Å the water molecules move practically as freely as in bulk water ($D_t = 1.8 \times 10^{-5}$ cm² s⁻¹ at RT) but that long range proton diffusion between clusters is slower ($D_t \sim 1.6 \times 10^{-6}$ cm² s⁻¹) in agreement with macroscopic NMR[45] and tracer[46] measurements. It is clear from this example that QNS brings spatial and temporal information which cannot be attained by any other technique. The connection between these diffusive processes and proton conductivity is still a matter of discussion[45,52] especially as the Nafion® conducting properties seem to depend rather strongly on storage, pretreatment and water content. It is the author's opinion that the fast local motion might be due, in part, to proton exchange within acidic clusters. Comparative QNS experiments on H^+ and Na^+ or K^+ exchanged membranes could provide further insight into this problem, provided conductivity is measured on the same membranes. This kind of approach has been adopted with *HUP* ($H_3O^+:UO_2PO_4^-$, $3H_2O$) and NaUP ($Na^+:UO_2PO_4^-$, $4H_2O$) and with V_2O_5 gels, HV_2O_5, $1.6H_2O$ and NaV_2O_5, $1.6H_2O$[48]. High resolution QNS experiments indicate that the acidic compounds not only have higher D_t values than the Na ones but present different Q dependencies of the translational width. The faster departure from Fick's law in the Na^+ compounds compared with the H^+ ones is attributed to longer jump lengths in the former. It had been previously shown on an oriented HUP sample that the translational motion was highly anisotropic[47]. Furthermore, medium resolution QNS experiments on HUP have been analysed by assuming that the H_3O^+ ions reorient about 10 times faster than the water molecules (Table 21.2).

Slade *et al.* have investigated the QNS spectra of 12 *tungstophosphoric acid* (TPA) hydrates $H_3PW_{12}O_{40} \cdot xH_2O$ with $x = 6, 14, 21$[49,50]. They also deduce the existence of 180° jump rotations of water molecules with $\tau_{res} \sim 10^{-10}$ s at room temperature and faster H_3O^+ reorientations ($\sim 10^{-11}$ s). Actually in HUP as in TPA, xH_2O, the major argument used to attribute more rapid rotations to H_3O^+ than to H_2O is based on structural considerations. It is stated that H_3O^+ rotations can occur without a change in the hydrogen bond distributions whereas H_2O rotations are almost certainly coupled[47,49–51].

Table 21.2. *Specific conductivity σ, proton self-diffusion coefficient D_t and residence time between reorientations τ_{res} for some hydrated solid state protonic conductors*

	Conductivity			Diffusivity			Reorientations		
System	$\sigma/S\,cm^{-1}$ E_a/eV	(T/K) $(\Delta T/K)$	Ref.	$D_t/cm^{-2}\,s^{-1}$ E_a/eV	(T/K) $(\Delta T/K)$	Ref.	τ_{res}/s	(T/K)	Ref.
Nafion	$\sim 4 \times 10^{-3a}$ 0.22	(298) (240–300)	45	7×10^{-6a} or 1.6×10^{-6b} 0.42	(298) (288–308)	45 46			
HUP	3×10^{-3c} 0.35	(300) (275–320)	39	3×10^{-7}	(292)	48	(H_2O) $\sim 10^{-10}$ (H_3O^+) $\sim 4 \times 10^{-12}$	(303) (318)	47
TPA, $6H_2O$	$\sim 6 \times 10^{-5}$ 0.28	(290) (255–303)	71	$\sim 10^{-11}$	(350)	49 71	(H_2O) 1.15×10^{-10} (H_3O^+)	(314)	49 51
TPA, $14H_2O$	$\sim 1.1 \times 10^{-3}$ 0.33	(290) (255–303)	71	1×10^{-6} 0.19	(300) (280–320)	71	(H_2O) $\sim 10^{-10}$ (H_3O^+) 2.1×10^{-11}	(290)	50 51
TPA, $21H_2O$	$\sim 3.4 \times 10^{-3}$ 0.44	(290) (278–310)	71	3×10^{-6d} 0.19	(298) (275–325)	71	(H_2O) $\sim 10^{-10}$ (H_3O^+) $\leqslant 10^{-11}$	(290)	51

[a]Values for Nafion 117 boiled in aqueous nitric acid and then air dried[45].

[b]Values for acid Nafion containing 15% water by weight. D_t has been measured by the tracer method. An additional short-range diffusion with $D \simeq 1.8 \times 10^{-5}\,cm^2\,s^{-1}$ at 298 K is revealed by QNS (see text).

[c]Different values of the conductivity have been reported for HUP depending on its method of preparation and on its two-dimensional character (see for example references 69 and 70).

[d]A value of $7.8 \times 10^{-6}\,cm^2\,s^{-1}$ has been deduced for D_t at 295 K by QNS[51].

Some characteristic values of the proton dynamics in hydrated protonic conductors are reported in Table 21.2.

21.3.2 INS studies

The characterization of the vibrational modes involving the hydrated proton is of crucial interest with respect to proton conduction. A huge amount of spectroscopic work has been performed in this field, as nicely summarized in a detailed review extending up to 1988 by Ratcliffe & Irish[53,54]. Since this date, several interesting papers dealing with the *INS* spectra of H_3O^+[41,43,44] and $H_5O_2^+$[55-57] have appeared. We have already noted the work on trivalent metal acid sulphate monohydrates[43,44] in which INS characterization of H_3O^+ was achieved in a spectral region which is obscured in the IR and Raman spectra by intense counter ion vibrations.

The vibrational assignment of $H_5O_2^+$ has always been a subject of debate[53-56] because, in contrast to H_3O^+, $H_5O_2^+$ can adopt a variety of conformations. It seems that on well-defined compounds involving the $H_5O_2^+$ ion, consistent INS vibrational analysis is now available. Jones *et al.*[57] have proceeded by comparison between the low temperature INS spectra of $HClO_4, 2H_2O$ and $HM(SO_4)_2 . 4H_2O$ (M = Fe, In). Kearley *et al.*[56] have adjusted a force field to the observed IR, Raman and INS spectra of $H_4SiW_{12}O_{40} . 6H_2O$ (TSA . 6H_2O) and $H_3PW_{12}O_{40} . 6H_2O$ (TPA . 6H_2O) and they arrive at a similar assignment to Jones & co-workers (Fig. 21.3).

In the more complex $HUO_2PO_4 . 4H_2O$ compound, the INS spectra were found to be consistent with the presence of both H_3O^+ and $H_5O_2^+$ ions[55].

21.4 Acidic aqueous solutions

Several years ago, J. W. White *et al.*[58,59] raised the very fundamental question as to whether density fluctuations associated with the fast proton kinetics in acidic solutions could be distinguished in the QNS spectra from the normal diffusion processes of the hydrogenated species at equilibrium.

More recently, after a very extensive theoretical study of the diffusion and electrical conductance in *HCl–water* mixtures[60], Hertz *et al.* tried again by QNS to find some evidence for a proton self-diffusion coefficient D^+[61]. It was postulated that the mean self-diffusion coefficient of the proton in aqueous HCl could be written as

$$D_t = D_{H_2O}(1 - x_H^a) + D^+ x_H^a \qquad (21.15)$$

where D_{H_2O} is the 'water molecule' self-diffusion coefficient in the bulk of the solution, including the hydration spheres of the Cl^- ions and x_H^a is the mole fraction of protons present as the H^+ ion. The experiments were performed as a function of acid concentration at room temperature and in a momentum transfer range limited to $Q \leqslant 1 \text{ Å}^{-1}$ in order to neglect the rotational contributions. These authors concluded that there was an absence of fast H^+ ion motion, D^+ being supposed to be of the same order of magnitude as D_{H_2O}. Furthermore NMR, tracer and QNS results agree on the fact that the mean *self-diffusion coefficient*, D_t, decreases markedly when the acid concentration increases[61].

Finally, a new attempt has been made by Lassègues & Cavagnat[62] on aqueous H_2SO_4 to see whether the abnormally high proton mobility induces an abnormal proton diffusivity, as expected according to the Nernst–Einstein law. Pure water and cesium sulphate solutions have been studied for comparison. With the acid solutions, the temperature and concentrations have been varied and the spectra recorded with several resolutions in a rather extended Q range (~ 0–2 Å^{-1}). Thus mean transport coefficients for the translational and rotational motions of water molecules have first been determined. They indicate a marked slowing down of the translational diffusion when salt or acid is added to water, as for the HCl/H_2O system. The rotational motion is much less affected. In H_2SO_4 solutions corresponding to the maximum of specific conductivity, some evidence is obtained for a weak and broad quasielastic component, not present in pure water or in Cs_2SO_4 solutions (Fig. 21.4). It is taken as a signature of the abnormal proton diffusivity. Numerical values are given in Fig. 21.5.

The best conditions for the observation of this additional component are small Q values ($\leqslant 0.5 \text{ Å}^{-1}$) and low temperatures. The first condition allows the rotational contribution to be neglected and the second provides a better separation between the fast and slow proton diffusion processes as they seem to be characterized by activation energies of about 0.087 and 0.19 eV, respectively.

Of course, since the two previous studies[61,62] led to contradictory results, further experiments are needed before a definite conclusion can be drawn. However, it must be pointed out that even if the additional component in H_2SO_4/H_2O is spurious or due to a cause other than abnormal proton diffusivity, the QNS studies of acidic aqueous solutions yield not only mean transport coefficients D_t and D_r but also information at the microscopic level as for example the residence time τ_0 of Eqn (21.5) involved in the translational motion (Fig. 21.5). These quantities constitute essential parameters of the whole conduction process.

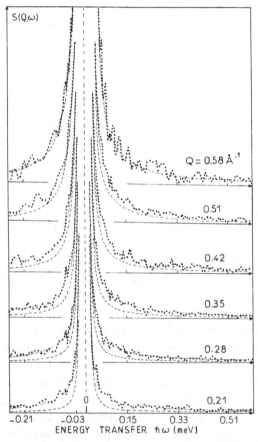

Fig. 21.4. QNS spectra of a 3.2 M H_2SO_4 solution at 263 K taken at various scattering angles with a resolution of 19 μeV (IN5 instrument of the Laue–Langevin Institute, Grenoble)[62]. The momentum transfer value at $\hbar\omega = 0$ is indicated for each spectrum. The experimental points are compared with a theoretical $S(Q, \omega)$ dashed curve corresponding to the convolution of the resolution function by a single translational lorentzian characterizing the mean water self-diffusion $D_t = 0.532 \times 10^{-5}$ cm^2 s^{-1} and with a convolution of the latter by a fast translational motion involving 12% of the protons (solid line). The fast motion is characterized by an apparent self-diffusion coefficient $D^+ = 2.2 \times 10^{-4}$ cm^2 s^{-1} at 263 K.

21.5   Conclusion

The neutron is a very sensitive probe for all kinds of motions, translation, rotation and vibration, experienced by the proton in the 10^{-9}–10^{-13} s time-scale but this sensitivity is paid by $S^{inc}(Q, \omega)$ profiles which are often very complex. QNS and INS, or in other terms, diffusive and quantized motions, are usually separated to provide a simple working hypothesis. The spectra of highly conducting materials indicate that there is not such

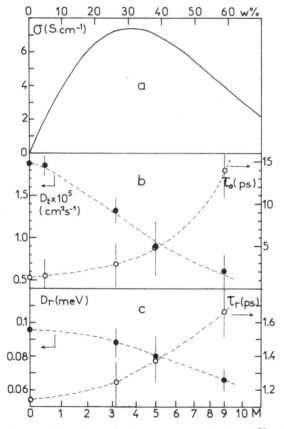

Fig. 21.5. Characteristic values related to proton conduction (a)[72], self-diffusion (b) and reorientation (c) in sulphuric acid aqueous solutions at 290 K as a function of H_2SO_4 concentration expressed in weight % (top scale) or molarity (bottom scale). For pure H_2O, the QNS study[62] yields the same D_t value as measured by NMR spin-echo or tracer methods[73]. In the same graph (b) is reported the residence time τ_0 measured by QNS[62] and defined in Eqn (21.5). The rotational diffusion constant D_r defined in Eqn (21.8) is reported in graph (c) together with the characteristic rotational time $\tau_r = \hbar/6D_r$.

a clear separation between fast diffusive motions and damped low frequency vibrations. Some systems seem to exhibit a continuous spectrum of correlation times in the observation window.

However, progress has undoubtedly been made in the disentanglement of the various contributions on some selected systems. The vibrational assignment of H_3O^+ and $H_5O_2^+$ in low temperature solids is one example. Another reassuring result is the agreement between the D_t values found by NMR or tracer techniques or determined at small momentum transfers by QNS on solids or liquids. From this well established macroscopic value, one can hope that preliminary QNS results obtained at higher Q values will be confirmed and extended. Indeed they are expected to lead either to the determination of characteristic jump-lengths and residence times in solids or to the finding of an abnormal proton diffusivity in solutions. This is a domain where QNS can bring a unique and important contribution to the understanding of the conduction mechanism.

The reorientation rates of hydrogenated species such as H_2O, H_3O^+, NH_4^+, HSO_4^-, etc. ... in proton conductors have in many cases the right order of magnitude to be observed by QNS. In some well-defined systems it has been possible to determine the correlation time and also the geometry of the motion, again a specific advantage of the QNS technique. However, when $(H_3O^+)(H_2O)_n$ species are involved, the distinction between H_3O^+ and H_2O rotations is virtually impossible in aqueous solution and very difficult in hydrated solid conductors.

Finally, it can be concluded that all the dynamical parameters measured up to now by QNS and INS certainly constitute useful information to establish a complete picture of the conduction process. However, such a picture is still not available: some pieces of the puzzle are missing. For example there is very little or no experimental evidence for the decisive proton transfer step supposed to occur in many models between H_3O^+ and H_2O. Recent QNS[63] and INS[64] studies on some particular hydrogen-bonded systems give, however, hope that fast proton exchange could be observable.

21.6   References

1. H. Boutin and S. Yip, *Molecular Spectroscopy with Neutrons* (M.I.T. Press, Cambridge, USA (1968)).
2. W. Marshall and S. W. Lovesey, *Theory of Thermal Neutron Scattering* (Clarendon Press, Oxford (1971)).

Incoherent neutron scattering

3. J. A. Janik, *The Hydrogen Bond – Recent Developments in Theory and Experiments*, P. Schuster *et al.* (eds) (North-Holland Publ. Co., Amsterdam (1976)) 893–935.

4. T. Springer, *Quasielastic Neutron Scattering for the Investigation of Diffusive Motions in Solids and Liquids*, Springer Tracts in Modern Physics (Springer-Verlag 64 (1972)).

5. R. K. Thomas, *Molecular Spectroscopy*, The Chemical Society. Specialist Periodical Reports, London **6** (1979) 232–318.

6. J. Howard and T. C. Waddington, *Advances in Infrared and Raman Spectroscopy*, R. J. M. Clark and R. E. Hester (eds) **7** (1980) 86–211.

7. M. Bee, *Application of Quasielastic Neutron Scattering to Solid State Chemistry, Biology and Material Science* (Adam Hilger, Bristol (1988)).

8. A. J. Dianoux, *The Physics and Chemistry of Aqueous Ionic Solutions*, M. C. Bellissent-Funel and G. W. Neilson (eds) (D. Reidel Publ. Co. (1987)) 147–79.

9. M. Besnard, A. J. Dianoux, P. Lalanne and J. C. Lassègues, *J. Physique* **38** (1977) 1417–22.

10. J. C. Lassègues, M. Fouassier, M. Besnard, H. Jobic and A. J. Dianoux, *J. Physique* **45** (1984) 497–503.

11. J. Texeira, M. C. Bellissent-Funel, S. H. Chen and A. J. Dianoux, *Phys. Rev.* A **31** (1985) 1913–17.

12. P. A. Egelstaff, *An Introduction to the Liquid State* (Academic Press (1967)).

13. M. C. Bellissent-Funel, R. Kahn, A. J. Dianoux, M. P. Fontana, G. Maisano, P. Migliardo and F. Wanderlingh, *Molec. Phys.* **52** (1984) 1479–86.

14. C. T. Chudley and R. J. Elliott, *Proc. Phys. Soc.* **77** (1961) 353–68.

15. W. Gissler and N. Stump, *Physica* **65** (1973) 109–17.

16. V. F. Sears, *Can. J. Phys.* **45** (1966) 1299–311.

17. L. Ricard and J. C. Lassègues, *J. Phys.* C (*Solid State Physics*) **17** (1984) 217–31.

18. H. Jobic, J. Tomkinson and A. Renouprez, *Molec. Phys.* **39** (1980) 989–99.

19. G. J. Kearley, *J. Chem. Soc. Faraday Trans.* **82** (1986) 41–7.

20. J. Tomkinson, M. Warner and A. D. Taylor, *Molec. Phys.* **51** (1984) 381–92.

21. J. C. Lassègues and Ph. Colomban, *Solid State Protonic Conductors II for fuel cells and sensors*, J. B. Goodenough, J. Jensen and M. Kleitz (eds) (Odense University Press (1983)) 201–13.

22. M. Perrot, M. H. Brooker and J. Lascombe, *J. Chem. Phys.* **74** (1981) 2787–94.

23. P. Faurskov Nielsen, P. A. Lund and E. Praestgaard, *J. Chem. Phys.* **75** (1981) 1586–7.

24. H. Sato, *Topics in Applied Physics*, S. Geller (ed) (Springer-Verlag, Berlin 21 (1977)) 3–37.

25. J. C. Lassègues, M. Fouassier, N. Baffier, Ph. Colomban and A. J. Dianoux, *J. Physique* **41** (1980) 273–80.

26. Ph. Colomban, J. C. Lassègues, A. Novak, M. Pham-Thi and C. Poinsignon, *Dynamics of Molecular Crystals*, J. Lascombe (ed) (Elsevier (1987)) 269–75.

27. J. C. Lassègues, C. Poinsignon and A. Chahid (to be published).

Proton dynamics and charge transport

28. B. V. Merinov, A. I. Baranov, L. A. Shuvalov and B. A. Maksimov, *Sov. Phys. Crystallogr.* **32** (1987) 47–50.
29. Z. Jirak, M. Dlouha, S. Vratislav, A. M. Balagurov, A. I. Beskrovnyi, V. I. Gordelii, I. D. Datt and L. A. Shuvalov, *Phys. Stat. Sol.* (a) **100** (1987) K117–22.
30. A. V. Belushkin, W. I. F. David, R. M. Ibberson and L. A. Shuvalov, *Acta Cryst.* **B47** (1991) 161–6.
31. N. H. Andersen, K. N. Clausen and J. K. Kjems, *Methods of Experimental Physics*, D. L. Price and K. Sköld (eds) (Academic Press, 23.B (1987)) 187–241.
32. K. Funke, A. Höch and R. E. Lechner, *J. Physique* C **6-17** (1980) 41–52.
33. G. Lucazeau, J. R. Gavarri and A. J. Dianoux, *J. Phys. Chem. Solids* **48** (1987) 57–77.
34. S. M. Shapiro and F. Reidinger, *Physics of Superionic Conductors: Topics in Current Physics*, M. B. Salamon (ed) (Springer-Verlag (1979)) 45–75.
35. J. C. Lassègues, B. Desbat, O. Trinquet, F. Cruege and C. Poinsignon, *Solid State Ionics* **35** (1989) 17–25.
36. J. D. Axe, L. M. Corliss, J. M. Hastings, W. L. Roth and O. Muller, *J. Phys. Chem. Solids* **39** (1978) 155–9.
37. M. H. Herzog-Cance, M. Pham-Thi and A. Potier, *J. Mol. Struct.* **196** (1989) 291–306.
38. M. F. Daniel, Thesis, Bordeaux, France (1986).
39. Ph. Colomban and A. Novak, *J. Mol. Struct.* **177** (1988) 277–308.
40. A. V. Belushkin, I. Natkaniec, N. M. Pakida, L. A. Shuvalov and J. Wasicki, *J. Phys. C (Solid State Phys.)* **20** (1987) 671–87.
41. M. Pham-Thi, J. F. Herzog, M. H. Herzog-Cance, A. Potier and C. Poinsignon, *J. Mol. Struct.* **195** (1989) 293–310.
42. K. Funke, *Solid Electrolytes, General Principles. Characterization, Materials, Applications*, P. Hagenmuller and W. Van Gool (eds) (Academic Press (1978)) 77–92.
43. D. J. Jones, J. Penfold, J. Tomkinson and J. Roziere, *J. Mol. Struct.* **197** (1989) 113–21.
44. D. J. Jones and J. Roziere, *Solid State Ionics* **35** (1989) 115–22.
45. R. C. T. Slade, J. Barker and J. H. Strange, *Solid State Ionics* **35** (1989) 11–15.
46. F. Volino, M. Pineri, A. J. Dianoux and A. De Geyer, *J. Polym. Sci.: Polym. Phys. Ed.* **20** (1982) 481–96.
47. C. Poinsignon, A. Fitch and B. E. F. Fender, *Solid State Ionics* **9 & 10** (1983) 1049–54.
48. C. Poinsignon, *Solid State Ionics* **35** (1989) 107–13.
49. H. A. Pressman and R. C. T. Slade, *Chem. Phys. Lett.* **151** (1988) 354–61.
50. R. C. T. Slade, I. M. Thomson, R. C. Ward and C. Poinsignon, *J. Chem. Soc. Chem. Commun.* (1987) 726–7.
51. H. A. Pressman, Thesis, Exeter (1988).
52. F. W. Poulsen, *High Conductivity Solid Ionic Conductors*, T. Takahashi (ed) (World Publ. Co., Singapore (1988)).
53. C. I. Ratcliffe and D. E. Irish, *Water Science Reviews 2*, F. Franks (ed) (Cambridge University Press (1986)) 149–214.

Incoherent neutron scattering

54. C. I. Ratcliffe and D. E. Irish, *Water Science Reviews 3*, F. Franks (ed) (Cambridge University Press (1988)) 1–77.
55. G. J. Kearley, A. N. Fitch and B. E. F. Fender, *J. Mol. Struct.* **125** (1984) 229–41.
56. G. J. Kearley, R. P. White, C. Forano and R. C. T. Slade, *Spectrochim. Acta* **46A** (1990) 419–24.
57. D. J. Jones, J. Rozière, J. Penfold and J. Tomkinson, *J. Mol. Struct.* **195** (1989) 283–91.
58. J. C. Lassègues and J. W. White, *Molecular Motions in Liquids*, J. Lascombe (ed) (D. Reidel (1974)) 439–59.
59. A. D. Taylor, J. W. White and J. C. Lassègues, *Protons and Ions Involved in Fast Dynamics Phenomena*, P. Lazlo (ed) (Elsevier (1978)) 123–46.
60. See H. G. Hertz, *Chemica Scripta* **27** (1987) 479–93, and literature cited therein.
61. H. Bertagnolli, P. Chieux and H. G. Hertz, *J. Chem. Soc. Faraday Trans. 1* **83** (1987) 687–95.
62. J. C. Lassègues and D. Cavagnat, *Mol. Phys.* **68** (1989) 803–22.
63. B. H. Meier, R. Meyer, R. R. Ernst, A. Stöckli, A. Furrer, W. Halg and I. Anderson, *Chem. Phys. Lett.* **108** (1984) 522–3.
64. F. Fillaux, B. Marchon, A. Novak and J. Tomkinson, *Chem. Phys.* **130** (1989) 257–70.
65. A. Potier and D. Rousselet, *J. Chem. Phys.* **70** (1973) 873–8.
66. A. I. Baranov, L. A. Shuvalov and N. M. Shchagina, *JETP Lett.* **36** (1982) 459–62.
67. T. Takahashi, S. Tanase, O. Yamamoto, S. Yamauchi and H. Kabeya, *Int. J. Hydrogen Energy* **4** (1979) 327–38.
68. R. Blinc, J. Dolinsek, G. Lahajnar, I. Zupancic, L. A. Shuvalov and A. I. Baranov, *Phys. Stat. Sol.* (b) **123** (1984) K83–7.
69. E. Skou, I. G. Krogh Andersen, K. E. Simonsen and E. Krogh Andersen, *Solid State Ionics* **9 & 10** (1983) 1041–8.
70. K. D. Kreuer, I. Stoll and A. Rabenau, *Solid State Ionics* **9 & 10** (1983) 1061–4.
71. R. C. T. Slade, J. Barker, H. A. Pressman and J. H. Strange, *Solid State Ionics* **28–30** (1988) 594–600.
72. P. Pascal, *Nouveau Traité de Chimie Minérale*, Masson et Cie Ed., Tome XIII, 1357–60.
73. R. Mills, *J. Phys. Chem.* **77** (1973) 685–8.

22 NMR studies of local motions in fast protonic conductors

S. V. BHAT

22.1 Introduction

Nuclear magnetic resonance (NMR) is a well-established local probe for the study of dynamics as well as structure in condensed matter systems[1]. The various NMR spin Hamiltonian parameters (see below) accurately reflect the symmetry and the nature of motions at and around the site of the nucleus being studied. When combined with the variations of thermodynamic parameters such as temperature and pressure, NMR can provide a wealth of microscopic, structural and dynamical information, such as lattice spacings, potential barriers and jump/attempt frequencies of the motion. Therefore it is not surprising that NMR has played a major role in the study of fast ionic conductors (FICs). Advantages of the NMR technique stem from its nuclear specificity, it being a true local probe in the sense that inequivalent sites within a unit cell can be distinguished and its non-destructive nature. Fast protonic conductors (FPCs) are eminently suited for investigations by NMR since protons or hydrogen nuclei are about the best nuclei that can be studied by most of the NMR techniques.

There are a number of review articles[2-7] summarizing and critically examining the NMR studies of FICs, and many of them contain some discussion of fast protonic conductors. Many features of the NMR studies are similar for protonic and other fast ionic conductors. However, there are also aspects peculiar to FPCs, an important one being the influence of local dynamics. In this article, we shall be specifically concerned with NMR studies of local motions in FPCs in contrast to the long range diffusion which is being discussed elsewhere in this book (see p. 412).

22.2 NMR as a probe of structure and dynamics

In general, an NMR spectrum is determined by some or all of the terms of the following Hamiltonian[8]

$$H = H_z + H_{r.f.} + H_{cs} + H_j + H_D + H_Q + H_{SR} \qquad (22.1)$$

where $H_z = \gamma \hbar \bar{I} . \bar{H}_o$ is the Zeeman interaction for a nucleus with spin $\bar{I}$ and gyromagnetic ratio γ placed in the external magnetic field H_o,

$H_{r.f.} = H_1 \cos \omega t$ provides the time dependent perturbation of amplitude H_1 causing the NMR transitions such that the frequency $\omega = \gamma H_o$,

$H_{cs} = \gamma \bar{I} . \tilde{\sigma} . \bar{H}_o$ describes the effect of shielding of the nucleus by the surrounding electrons from the applied field,

$H_j = \sum_{i \neq j} \bar{I}_i . \tilde{J} . \bar{I}_j$ accounts for the interaction between nuclear spins via the intervening electrons,

$$H_D = \sum_{i<j} \hbar \gamma_i \gamma_j r_{ij}^{-3} \left(\bar{I}_i . \bar{I}_j - \frac{3(\bar{I}_i . \bar{r}_{ij})(\bar{I}_j . \bar{r}_{ij})}{\gamma_{ij}^2} \right)$$

is the direct magnetic dipole–dipole interaction between the nuclear spins $\bar{I}_i$ and $\bar{I}_j$ separated by a distance $\bar{r}_{ij}$ and probably is the most frequently encountered interaction in the study of fast protonic conductors,

$$H_Q = \frac{eQ}{2I(2I - 1)\hbar} \bar{I} . \tilde{V} . \bar{I}$$

is the interaction between the non-cubic electric field gradient (EFG) V at the nuclear site and the quadrupole moment Q of the nucleus with its spin $I \geqslant 1$ (and thus not relevant for protons ($I = \frac{1}{2}$) except in cases where protons are replaced by deuterons ($I = 1$) and their NMR is studied), and

$$H_{SR} = \sum_i \bar{I}_i . \tilde{C}_i . \bar{J}$$

is the spin rotation interaction; C is the spin rotation tensor. In Eqn (22.1) the first two terms on the right hand side lead to the basic NMR experiment and the remaining terms cause modification of the signals through effects on (i) line position, (ii) line shape, (iii) line width, and (iv) multiplicity of the signals.

If there is motion (either local or long range), depending upon the relative magnitudes of the time scale of the motion and the inverse of the strength of the particular interaction the NMR spectrum will be profoundly influenced. An important consequence is that there will be averaging out of the particular interaction to a more or less degree: (a) the interactions with zero trace (H_D, H_Q) can be averaged out to zero if the motion is isotropic and rapid enough (similar to the situation in liquids); and (b) the interactions with non-vanishing traces (chemical shift H_{CS} and spin–spin coupling H_j) will be averaged to their respective isotropic values. As the temperature is increased and the system transforms from the 'rigid lattice' to the 'diffusive' state, the change is reflected in the 'motional narrowing' of the signal. Such a narrowing, in general, occurs in two steps; the first signifies the local dynamics averaging out the intramolecular interactions and the second is due to the long range translational motion making the intermolecular interactions vanish.

Further, either the local or the long range dynamics in the system lead to fluctuating magnetic moments providing non-resonant pathways for relaxation to dissipate the excess energy of the excited nuclei. The spin–lattice relaxation time in the Zeeman field, T_1, the spin–lattice relaxation time in the r.f. field (or the so-called spin–lattice relaxation time in the rotating frame), $T_{1\rho}$, and the spin–spin relaxation time, T_2, probe the spectral densities of the fluctuations. T_1, T_2 and $T_{1\rho}$ together enable one to study a wide range of time scales of dynamics, from nearly 10^{-10} s to 10^{-3} s. Interpreting the motional narrowing experiments and the relaxation time measurements in terms of an appropriate model of the dynamics, (the most commonly used model being the Bloembergen, Purcell and Pound (BPP) model[9]) the time–time correlation functions τ_c are extracted at various temperatures (T), and fitted to an Arrhenius type equation,

$$\tau_C = \tau_{CO} \, e^{E_a/k_B T} \tag{22.2}$$

where $1/\tau_{CO}$ is the *attempt frequency* and E_a is the potential barrier. The details of the analysis, the problems thereof and the attempts to resolve them are not discussed here for want of space. The interested reader is referred to various articles on this issue[3–7,10].

In protonic conductors, where the protonic transport generally involves groups like NH_4^+, $H^+(H_2O)_n$, H_2O and the like, the local rotational/ hopping motions of these species have a significant influence on the long range translational motion. While it is possible in some cases to have only the translational motion of H^+, more often, the long range motion consists of two steps: a rotation on the local scale of a molecular group followed by a jump of the charge carrying species from one group to another (as in the case of the Grotthuss mechanism). In such cases, the local motion can act as the rate limiting process for the translational diffusion between interconnected sites occupied by groups executing local motion. Further, most of the NMR parameters, except the diffusion coefficient D directly determined by the pulsed field gradient techniques, are affected in a similar and sometimes, unless looked at carefully, indistinguishable manner by both local and long range motions. Therefore it becomes important to separate contributions from local and long range motions to derive the translational parameters. Of course, there are innumerable cases where there is only local reorientational and jump motion of protonic groups without any accompanying translational motion. Indeed, the problem of making this distinction between the two types of motion is not a trivial one. Examples can be cited of the ^{1}H NMR study of $N_2H_6SO_4$[11], where the motional narrowing observed for temperatures above 150 K was incorrectly attributed to translational diffusion whereas, in fact, the narrowing is due only to local *reorientational motions* of the NH_2 and the NH_3 groups. The long range translational motion sets in only at a much higher temperature of 475 K as shown by a careful NMR study[12] and also a ^{2}D quadrupole perturbed NMR study[13].

In the following, we shall illustrate the application of various NMR techniques to understand the different aspects of local dynamics in FPCs. Where relevant, we shall also touch upon the long range motion briefly. We shall try to point out at appropriate places the problems associated with the interpretation of NMR results and attempts to resolve the problems. In Table 22.1, the information obtained on protonic conductors using various NMR techniques is presented.

22.3 High resolution NMR study of ammonium ferrocyanide hydrate (*AFC*)

22.3.1 Potentials of chemical shift spectroscopy

An early wideline NMR study of $(NH_4)_4Fe(CN)_6 \cdot 1.5H_2O$, (*AFC*), by Whittingham *et al.*[14], showed that the ^{1}H NMR signal is narrowed by translational diffusion even at room temperature. However, it was not possible from that study to determine the nature of the diffusing species, i.e. whether it is the ammonium ions or the individual protons. Structurally, there are two non-equivalent NH_4 groups (48 'g' sites and 16 'b' sites of the space group Ia3d) and together with the water protons, one may expect three different NMR signals. However, the narrowing resulting from complete isotropic local motion with some averaging of the

Table 22.1. *Protonic conductors investigated by NMR. The attempt frequency* $v_0 = 1/\tau_{CO}$ *and the potential barrier* E_a *are given where available*

Sr. No	Substance	Type of NMR experiment conducted	$v_0(s^{-1})$	$E_a(eV)$	References
1.	$H_3O(UO_2PO_4)3H_2O(HUP)$ and $H_3O(UO_2AsO_4)3H_2O(HUAs)$				
	HUP	T_1, T_2	10^{13}	0.21	24
		T_1, T_2, PFG		0.26	25
		$T_1, T_{1\rho}, T_2$		0.22	27
		PFG		0.27	26
2.	NH_4 β-alumina	Line narrowing	1.5–4×10^{13}	0.5	28, 30
		T_1, T_2	3.5×10^{10}	0.2	29
3.	$(NH_4)_{1.0}(H_3O)_{0.67}Mg_{0.67}Al_{10.33}O_{17}$ (ammonium hydronium β-alumina)	PFG		0.202	31
		Line narrowing, T_1, T_2		0.07	32
4.	Nafion (perfluoroethylene sulphonic acid)	$T_1, T_{1\rho}, T_2$	5.9×10^8		33
		T_1			34
	S1: 14.3% H_2O by mass	$T_1, T_{1\rho}, T_2$		0.45	35
	S2: 26.2% H_2O by mass			0.62	
5.	$TaS_2(NH_3)_{0.9}$	$T_1, T_{1\rho}$	1.3×10^{13}	0.21	36
	$TaS_2(NH_3)_x$; $x = 0.8, 0.9, 1.0$	$T_1, T_{1\rho}$			37
	$TiS_2(NH_3)_{1.0}$				
6.	TiS_2H_x	Line narrowing		0.34	38
	VS_2H_x	Line narrowing		0.27	38
7.	$TiH_{1.7}$	T_1	9.18×10^{12}	0.51	39
8.	NH_4NO_3	Line narrowing, $T_1, T_{1\rho}$	5×10^{13}	0.52	40

Proton dynamics and charge transport

9. $HNbO_3$	Line narrowing	10^7	0.05	41
10. $H_{1.4}ReO_3$	T_1, $T_{1\rho}$, T_2	9.09×10^6	0.1	42
$\quad\ H_xReO_3$		2.8×10^7	0.1	
11. $H_{0.39}WO_3$	Line narrowing, T_1	2.1×10^7	0.2	43
$\quad\ H_{0.46}WO_3$	$T_{1\rho}$	10^7	0.17	43
$\quad\ H_{0.35}WO_3$	Line narrowing	10^{10}	0.13	44
	T_1	2.7×10^{11}	0.06	45
12. Intercalation compounds of h-WO_3 with H^+, H_2, H_2O, NH_3 and NH_4 as intercalants:				
$\quad$ h $-$ $(H_2)_{0.11}WO_3$	Line narrowing	6.6×10^{10}	0.23	46
$\quad$ h $-$ $(NH_4)_x WO_3$; $x = 0.36$	Line narrowing	9.1×10^{10}	0.25	46
$\qquad\qquad\qquad\ \ x = 0.32$			0.27	46
$\quad$ h $-$ $(NH_3)_{0.16}WO_3$	Line narrowing	7.1×10^9	0.23	46
13. $(NH_4)_{0.25}WO_3$	Line narrowing, T_1, $T_{1\rho}$	10^7	0.14	47
$\quad\ (NH_4)_{0.33}WO_3$	Line narrowing	3×10^6	0.08	48
$\quad\ (NH_4)_{0.23}WO_3$	T_1, T_2	$\approx 10^5$	0.04	70
14. $H_{0.34}MoO_3$	T_1, $T_{1\rho}$, T_2	10^{13}	0.23	49
$\quad\ H_{0.36}MoO_3$	T_1, $T_{1\rho}$, T_2	3.3×10^7	0.11	
$\quad\ H_{1.71}MoO_3$				
15. $LaNi_5H_6$	$T_{1\rho}$, Line narrowing	2×10^{12}	0.25	50
16. $CaF_{2-x}H_x$	Line narrowing		0.45	51
17. PEO $(H_3PO_4)_x$	T_1, $T_{1\rho}$, T_2	6×10^{11}	0.05	52
$\quad$ PEO: poly(ethylene)oxide				

continued

Table 22.1. Continued

Sr. No Substance		Type of NMR experiment conducted	$\nu_0(s^{-1})$	E_a(eV)	References
18. $CsOH \cdot H_2O$		T_1, $T_{1\rho}$, T_2, PFG	8.3×10^{10}	0.14	53
19. $NH_4Sn_2F_5$		T_1		0.13	54
20. $(NH_4)_{0.22}(NH_3)_{0.34}(TiS_2)_{0.22}$ $(NH_4)_{0.21}(Ti_2)_{0.21}$		T_1, Line narrowing		0.138	55
21. $Sb_2O_5 \cdot nH_2O$ $n = 3, 4$		T_1, $T_{1\rho}$, T_{1D}, T_2	1.3×10^{14}	0.37	56, 57
22. $(N^2H_4)Zr_2(PO_4)_3$ $(^2H_3O)Zr_2(PO_4)$		MASS, Chemical shift study			58
23. $V_2O_5 \cdot 1.62H_2O$ $CeO_2 \cdot 2H_2O$ $Nb_2O_5 \cdot 3.2H_2O$ $Ta_2O_5 \cdot 3.24H_2O$		T_1, T_2		0.05 0.04 0.04 0.02	59
24. $Al_4(Si_4O_{10})_2(OH)_4 \cdot xH_2O$ (Montmorillonite clay) with interlayer cation, M;	M Na^+ Li^+ Ni^{2+}	T_1, T_2	7.9×10^9 2.0×10^{10} 1.6×10^{10}	0.08 0.09 0.1	60
25. $SnO_2 \cdot nH_2O$		T_1, T_2	$10^{16}(T_1)$	$0.37(T_1)$ $0.27(T_2)$	61
26. $TiO_2 \cdot nH_2O$		T_1, T_2	10^{16}	0.12	61

27. $(NH_4)_4Fe(CN)_6 \cdot xH_2O$	Line narrowing	10^9	0.20	14, 17
	T_1	7.7×10^{11}	0.12	18
	$T_{1\rho}$	1.4×10^9	0.20	
	T_2	2.2×10^8	0.16	
28. NH_4TaWO_6	T_1	10^{10}	0.25	62
	T_2		0.3	
29. $HTaWO_6 \cdot H_2O$ (pyrochlore)	Line narrowing, T_1, T_2	3×10^{10}	0.25	63, 21
$HTaWO_6 \cdot H_2O$ (pyrochlore)	T_2	2.6×10^6	0.12	21
	T_1	2.6×10^{10}	0.1	
	$T_{1\rho}$	3.1×10^7	0.11	
$HTaWO_6$	Line narrowing, T_1, T_2		0.3	63
30. $HTaWO_6 \cdot 1.5H_2O$ (layered)	T_2	4.5×10^9	0.13	21
	T_1	2.0×10^{11}	0.12	
	$T_{1\rho}$	3.1×10^9	0.3	
31. $HNbWO_6 \cdot H_2O$ (pyrochlore)	T_2	1.3×10^7	0.16	22
	T_1	3.1×10^{10}	0.12	
	$T_{1\rho}$	1.2×10^7	0.11	
$HNbWO_6 \cdot 1.5H_2O$ (layered)	T_2	3.6×10^{10}	0.16	
	T_1	2.0×10^{11}	0.15	
	$T_{1\rho}$	1×10^{11}	0.21	
32. Heteropolyacid hydrates				
$H_3PW_{12}O_{40} \cdot 28H_2O$ (TPA $\cdot$ 28H$_2$O)	PFG		0.25	64
$H_3PW_{12}O_{40} \cdot 21H_2O$ (TPA $\cdot$ 21H$_2$O)				
$H_4Si_{12}O_{40} \cdot 28H_2O$			0.39	

continued

Table 22.1. Continued

Sr. No	Substance	Type of NMR experiment conducted	$v_0(s^{-1})$	$E_a(eV)$	References
33.	TPA . 21 H_2O	PFG, T_2		0.19	65
	TPA . 14 H_2O	PFG, T_2		0.17	66
		T_1, T_2	10^{12}	0.16 for reorientation 0.37 for self diffusion	
34.	$H_3PM_{12}O_{40} . 21H_2O$ $M = W, Mo$	T_1, $T_{1\rho}$, T_2	3.16×10^{18}	0.53	67
35.	$A_x(NH_3)_xMS_2$	Line narrowing, T_1			68
	A = Li, Na, K, Ca, Sr, Ba, NH_4				
	M = Ti, Nb, Ta, Mo				
	$(NH_4)_x(NH_3)_yTiS_2$		2.5×10^{13}	0.24	
	$(NH_4)_x(NH_3)_yNbS_2$		1.67×10^{13}	0.24	
	$(NH_4)_x(NH_3)_yTaS_2$		1.1×10^{13}	0.19	
	$(Na)_x(NH_3)_yMoS_2$		5×10^3	0.46	
	$Sr_x(NH_3)_yMoS_2$		6.3×10^{12}	0.23	
36.	$A_x^+(H_2O)\{MS_2\}^{x-}$	Wide-line, T_1			
	A^+ M				
	Li Nb		1.6×10^{13}	0.31	69
	Na Nb		7.7×10^{13}	0.32	
	K Nb		1.3×10^{13}	0.23	
	Rb Ta		2.1×10^{12}	0.13	
	$Li_{0.33}(H_2O)_2NbS_2$	T_1		0.32(2D) 0.23(3D)	69

NMR studies

intermolecular dipolar interactions by translational motion still leaves residual line widths of about 0.018 mT at room temperature and the different signals are not resolved. However, a '*high resolution NMR experiment*', the high resolution having been obtained by working at 500 MHz, gives rise to two moderately resolved signals (Fig. 22.1), at 12.42 ppm and 8.87 ppm with respect to TMS in $CDCl_3$[15]. On the basis of the relative intensities of the signals, the signal at 12.42 ppm is attributed to the two NH_4 groups and that at 8.87 ppm is assigned to the water protons. A temperature variation experiment carried out at 270 MHz indicates reductions in the widths of both the signals as well as a reduction in the chemical shift difference between the two signals with increasing temperature. The inference could be drawn that both the NH_4 groups and the H_2O groups together take part in translational diffusion. Use of this type of chemical shift spectroscopy obviously has many potential applications in gaining understanding of dynamics of species in inequivalent sites. In fact the high resolution NMR in the solid state as it is normally practised, i.e. magic angle spinning (MAS), multipulse narrowing and the combined rotation and multipulse spectroscopy (CRAMPS) can be very powerful in the study of protonic conductors.

22.4   High pressure NMR studies of AFC

22.4.1   Activation volume and phase transition

While the high resolution NMR experiment was not able to distinguish between the two NH_4 groups, a wideline NMR experiment carried out at high hydrostatic pressures is able to achieve this goal, albeit indirectly[16]. Fig. 22.2 shows the NMR signal at room temperature for various pressures. Around 0.45 GPa there is a sudden change in the lineshape, the single narrow signal at ambient pressure giving rise to a central narrow signal and a broader background signal. On the basis of the ratio of the relative intensities it is concluded that the narrow signal is due to the 'b'

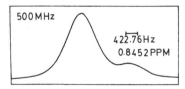

Fig. 22.1. ^1H NMR spectrum of $(NH_4)_4Fe (CN)_6 1.5H_2O$ at 500 MHz.

site NH_4 groups and the protons of the water of crystallization, and the broader signal represents the 'g' site NH_4 groups. The broadening of the signal due to the latter indicates the increase in the local potential barrier on pressurization for these NH_4 groups via the occurrence of a phase transition. The result is also confirmed by a compressibility study[16].

It is interesting to note that even at 77 K, and at a pressure of 1.5 GPa, the broad signal still has a width of only 0.39T, indicating the averaging of the intramolecular dipolar interaction by local dynamics, which, in this case, is most probably quantum mechanical tunnelling. The study[17] also determines, below the transition pressure, the activation volume which turns out to be about 6% of the molar volume and indicates that the mechanism for diffusion is vacancy assisted ionic hopping.

22.5   Relaxation time studies of AFC

22.5.1   Distinction between local and long range motion

The relaxation studies in AFC illustrate how local and long range motions can be distinguished by careful measurements of different relaxation times.

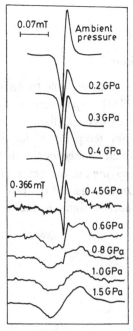

Fig. 22.2. ^1H wide line NMR signals of $(NH_4)_4Fe(CN)_6 1.5H_2O$ at various pressures.

While according to the BPP model, the minimum in the $T_{1\rho}$ vs. temperature plot is expected to occur at a lower temperature than that of the T_1 minimum, in fact in AFC the behaviour is opposite[18] as shown in Fig. 22.3. The minimum in $T_{1\rho}$ appears to occur above room temperature while the minima of T_1 occur well below 300 K. This clearly indicates that the two relaxation processes have different origins. It is concluded that while translational diffusion causes $T_{1\rho}$ relaxation, it is the local reorientational dynamics that determines T_1 relaxation.

22.6   Effects of low dimensionality

22.6.1   Cubic and layered $HMWO_6 . xH_2O$

Dimensionality of ionic transport lower than three can have drastic consequences for NMR spectra as shown by Richards[19,3]. Recently, Bjorkstam *et al.*[20] pointed out the importance of properly modelling the correlation functions in obtaining correct parameters from NMR data. Thus it becomes important to decide on the relative importance of the two effects. With this object in mind, compounds of the type $HMWO_6 . xH_2O$ (M = Nb, Ta) were prepared in both three dimensional (cubic pyrochlore) and two dimensional (layered) structures and were

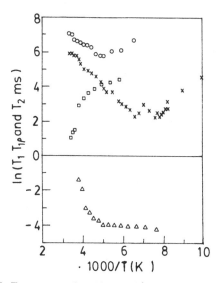

Fig. 22.3. Temperature dependences of ^1H T_1 at 300 MHz (O) and 15 MHz (×), of $T_{1\rho}(\square)$ and $T_2(\triangle)$ for $(NH_4)_4$ Fe $(CN)_6 . 1.5H_2O$.

361

studied by various NMR techniques. Significant differences were observed in the behaviour of the 3-D and the 2-D systems[21,22] (Fig. 22.4). In the 2-D system the motional narrowing occurs at a much lower temperature and is much sharper compared to that in the 3-D system. The T_1 and $T_{1\rho}$ data give meaningful motional parameters only when fitted to a 2-D model. However, τ_{co}s orders of magnitude lower are observed even for the 3-D structure and most probably this indicates the need to employ more appropriate correlation functions. At low temperatures (< 200 K), the signal consists of a doublet due to the water protons and a central signal due to the lattice protons. The decrease in peak separation of the doublet and the width of the central signal as a function of temperature are shown in Fig. 22.5[23]. The reorientational dynamics of the water protons is seen to decrease the peak separation with temperature until about 220 K when both types of protons are together seen to participate in long range diffusion, the local dynamics thus acting as the precursor to the translational motion.

22.7   Conclusion

In this short review, we have tried to illustrate how various NMR techniques can be utilized to gain understanding of the local proton dynamics in FPCs and its relevance to the study of long range translational diffusion.

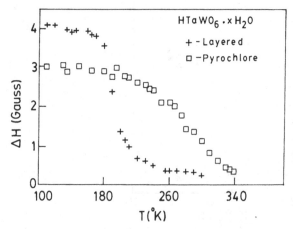

Fig. 22.4. Motional narrowing of the central part of the ^{1}H NMR signal in $HTaWO_6 \cdot xH_2O$ in layered ($+$) and cubic ($\square$) structures.

NMR studies

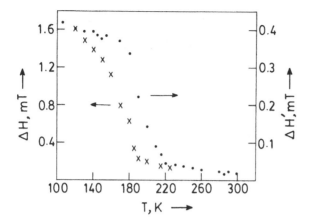

Fig. 22.5. Temperature dependences of the peak separation ΔH ($\times$) of the doublet and the linewidth $\Delta H'$($\cdot$) of the central signal in layered $HNbWO_6 \cdot 1.5H_2O$.

22.8 References

1. See, for example, C. P. Slichter, *Principles of Magnetic Resonance* 2nd edition (Springer-Verlag, Berlin (1978)).
2. M. S. Whittingham and B. G. Silbernagel in *Solid Electrolytes, General Principles, Characterization, Materials and Applications*, P. Hagenmuller and W. van Gool (eds) (Academic Press (1978)) 93–108.
3. P. M. Richards in *Physics of Superionic Conductors*, M. B. Salamon (ed) (Springer-Verlag, Berlin (1979)) 141–74.
4. J. B. Boyce and B. A. Huberman, *Physics Reports* **51** (1979) 189–265. (This is a general review on fast ionic conductors but contains a fairly detailed section on NMR.)
5. J. L. Bjorkstam and M. Villa, *Magnetic Resonance Review* **6** (1980) 1–57.
6. J. L. Bjorkstam, *J. Mol. Struct.* **111** (1983) 135–50.
7. J. H. Strange, *Cryst. Latt. Def. Amorph. Mat.* **14** (1987) 183–99.
8. M. Mehring, *High Resolution NMR in Solids* 2nd edition (Springer-Verlag, Berlin (1983)) 9.
9. N. Bloembergen, N. Purcell and R. V. Pound, *Phys. Rev.* **73** (1948) 679–712.
10. R. C. T. Slade, *Solid State Commun.* **54** (1985) 1035–8.
11. S. Chandra and N. Singh, *J. Phys. C (Solid State Phys.)* **16** (1983) 3099–103.
12. D. R. Balasubramanyam and S. V. Bhat, *Proc. Solid State Phys. Symp. India*, **27C** (1984) 168.
13. J. W. Harrel Jr and E. M. Peterson, *J. Chem. Phys*, **63** (1975) 3609–12.
14. M. S. Whittingham, P. S. Connel and R. A. Huggins, *J. Solid State Chem.* **5** (1972) 321–7.
15. D. R. Balasubramanyam and S. V. Bhat, *Solid State Ionics* **23** (1987) 267–70.
16. D. R. Balasubramanyam, S. V. Bhat, M. Mohan and A. K. Singh, *Solid State Ionics* **28–30** (1988) 664–7.

Proton dynamics and charge transport

17. D. R. Balasubramanyam and S. V. Bhat, *J. Phys. Condensed Matter* **1** (1989) 1495–502.
18. G. Mangamma and S. V. Bhat, *Solid State Ionics* **35** (1989) 123–5.
19. P. M. Richards, *Solid State Commun.* **25** (1978) 1019–21.
20. J. L. Bjorkstam, J. Listerud, M. Villa and C. I. Massara, *J. Magn. Res.* **65** (1985) 383–94.
21. S. V. Bhat and G. Mangamma, *International Conf. on Solid State Ionics, Nov. 5–11, 1989, Japan*, Extended abstracts, p. 176.
22. G. Mangamma, V. Bhat, J. Gopalakrishnan and S. V. Bhat, *International Conf. on Solid State Ionics, Nov. 5–11, 1989, Japan*, Extended abstracts, p. 177.
23. Adapted from G. Mangamma, Ph.D. thesis, Indian Institute of Science, Bangalore (1990), unpublished; and G. Mangamma and S. V. Bhat, to be published.
24. P. E. Childs, T. K. Halstead, A. T. Howe and M. G. Shilton, *Mat. Res. Bull.* **13** (1978) 609–19.
25. R. E. Gordon, J. H. Strange and T. K. Halstead, *Solid State Commun.* **31** (1979) 995–7.
26. Y.-T. Tsai, W. P. Halperin and D. H. Whitmore, *J. Solid State Chem.* **50** (1983) 263–72.
27. H. Metcalfe, T. K. Halstead and R. C. T. Slade, *Solid State Ionics* **26** (1988) 209–15.
28. D. Gourier and B. Sapoval, *J. Phys.* **C12** (1979) 3587–96.
29. R. C. T. Slade, P. F. Fridd, T. K. Halstead and P. McGeehin, *J. Solid State Chem.* **32** (1980) 87–95.
30. H. Arribart, H. Carlos and B. Sapoval, *J. Chem. Phys.* **71** (1982) 2336–43.
31. Y.-T. Tsai, S. Smoot, D. H. Whitmore, J. C. Tarczon and W. P. Halperin, *Solid State Ionics* **9 & 10** (1983) 1033–40.
32. A. R. Ochadlick Jr, W. C. Bailey, R. L. Stamp, H. S. Story, G. C. Farrington and J. L. Briant, in *Fast Ion Transport in Solids*, P. Vashishta, J. N. Mundy and G. K. Shenoy (eds) (North Holland, Amsterdam (1979)) 401–4.
33. N. G. Boyle, J. M. D. Coey and V. C. Brierty, *Chem. Phys. Lett.* **86** (1982) 16–19.
34. N. Shivashinsky and G. B. Tanny, *J. Appl. Polymer Sci.* **26** (1981) 2625–37.
35. R. C. T. Slade, A. Hardwick and P. G. Dickens, *Solid State Ionics*, **9 & 10** (1983) 1093–8.
36. T. K. Halstead, K. Metcalfe and T. C. Jones, *J. Magn. Res.* **47** (1982) 292–306.
37. R. L. Kleinberg and B. G. Silbernagel, *Solid State Commun.* **33** (1980) 867–71.
38. C. Ritter, W. Muller-Warmuth and R. Schollhorn, *Solid State Ionics* **20** (1988) 283–9.
39. C. Korn and D. Zamir, *J. Phys. Chem., Solids* **31** (1970) 489–502.
40. M. T. Riggin, R. R. Knispel and M. M. Pintar, *J. Chem. Phys.* **56** (1972) 2911–18.

NMR studies

41. J. L. Fourquet, M. F. Renou, R. DePape, H. Theveneau, P. P. Man, O. Lucas and J. Pannetier, *Solid State Ionics* 9 & 10 (1983) 1011-14.
42. M. T. Weller and P. G. Dickens, *Solid State Ionics* 9 & 10 (1983) 1081-6.
43. P. G. Dickens, D. J. Murphy and T. K. Halstead, *J. Solid State Chem.* 6 (1973) 370-3.
44. K. Nishimura, *Solid State Commun.* 20 (1976) 523-4.
45. M. A. Vannice, M. Boudart and J. Fripiat, *J. Catalysis* 17 (1970) 359-65.
46. H. Moller, W. Muller-Warmuth, F. Ruschendorf and R. Schollhorn, *Zeit. Phys. Chem.*, *N.F.*, 151 (1987) 121-31.
47. M. S. Whittingham and R. A. Huggins, in *Fast Ion Transport in Solids*, W. Van Gool (ed) (North Holland, Amsterdam (1973)) 645-52.
48. L. D. Clark, M. S. Whittingham and R. A. Huggins, *J. Solid State Chem.* 5 (1972) 487-93.
49. R. C. T. Slade, T. K. Halstead and P. G. Dickens, *J. Solid State Chem.* 34 (1980) 183-92.
50. T. K. Halstead, N. A. Abood and K. H. J. Buschow, *Solid State Commun.* 19 (1976) 425-8.
51. G. Villeneuve, *Mat. Res. Bull.* 14 (1979) 1231-4.
52. P. Donoso, W. Gorecki, C. Berthier, F. Defendin, C. Poinsignon and M. B. Armand, *Solid State Ionics* 28-30 (1988) 969-74.
53. J. Gallier, B. Taudic, M. Stahn, R. E. Lechner and H. Dachs, *J. Phys. France*, 49 (1988) 949-57.
54. J. P. Battut, J. Dupuis, H. Robert and W. Granier, *Solid State Ionics* 8 (1983) 77-81.
55. G. W. O'Bannon, W. S. Glaunsinger and R. F. Marzke, *Solid State Ionics* 26 (1988) 15-23.
56. W. A. England and R. C. T. Slade, *Solid State Ionics* 1 (1980) 231.
57. W. A. England and R. C. T. Slade, *Solid State Commun.* 33 (1980) 997-9.
58. N. J. Clayden, *Solid State Ionics* 24 (1987) 117-20.
59. R. C. T. Slade, J. Barker and T. K. Halstead, *Solid State Ionics* 24 (1987) 147-53.
60. R. C. T. Slade, J. Barker, P. R. Hirst, T. K. Halstead and P. I. Reid, *Solid State Ionics* 24 (1987) 289-95.
61. R. C. T. Slade, M. G. Gross and W. A. England, *Solid State Ionics* 6 (1982) 225-30.
62. M. A. Butler and R. M. Biefeld, *Solid State Commun.* 29 (1979) 5-7.
63. M. A. Butler and R. M. Biefeld, *Phys Rev.* B19 (1979) 5455-62.
64. K.-D. Kreuer, M. Hampele, K. Dolde and A. Rabenau, *Solid State Ionics* 28-30 (1988) 589-93.
65. R. C. T. Slade, J. Barker, H. A. Pressman and J. H. Strange, *Solid State Ionics* 28-30 (1988) 594-600.
66. R. C. T. Slade, I. M. Thompson, R. C. Ward and C. Poinsignon *J. Chem. Soc. Chem. Commun.* (1987) 726-7.

67. A. Hardwick, P. G. Dickens and R. C. T. Slade, *Solid State Ionics* **13** (1984) 345–50.
68. E. Wein, W. Muller-Warmuth and R. Schollhorn, *Solid State Ionics* **22** (1987) 231–40.
69. U. Roder, W. Muller-Warmuth, H. W. Spiess and R. Schollhorn, *J. Chem. Phys.* **77** (1982) 4627–31.
70. R. C. T. Slade, P. G. Dickens, D. A. Claridge, D. J. Murphy, and T. K. Halstead, *Solid State Ionics* **38** (1990) 201–6.

23 Vibrational spectroscopy of proton conductors

PHILIPPE COLOMBAN AND ALEXANDRE NOVAK

23.1 Introduction

Vibrational (infrared, Raman and neutron) spectroscopy can give useful information about proton conductors, i.e. about structural as well as dynamical aspects. Infrared spectroscopy appears particularly suited for such studies since AH stretching modes (A = O, N, halogen) give rise to strong absorption bands in a region (4000–1700 cm^{-1}) where there is not much interference from other groups. Various protonic species can thus be studied, even at very low concentrations.

As far as crystalline structures are concerned, X-ray and/or neutron diffraction methods and vibrational spectroscopy are complementary. The former can determine with accuracy the structure of the rigid framework but have some difficulties in locating protonic species, particularly if they are (statically or dynamically) disordered and if their concentration is low. The latter are well-suited for the identification of protonic entities and their types of association, the investigation of some structural details such as different crystallographic sites, (non)equivalent molecules, protonation sites, A–H and A . . . B distances and the nature and degree of structural disorder.

Spectroscopic data can also be used to study dynamical aspects, either as proton dynamics deciphered from the AH stretching band profile or as the dynamics of phase transitions. They are helpful in determining the order (first or second), the nature (displacive, order–disorder, reconstructive) and particularly the mechanism of the transformation at the molecular level. This can also shed some light on the conductivity mechanism, which can change considerably in going from one phase to the other.

Information about phase transitions can be obtained by studying the frequency, intensity and width of infrared and Raman bands as a function of temperature and pressure. The low-frequency bands associated with

367

lattice modes appear generally more sensitive to phase transitions than are the high frequency intramolecular bands.

23.2   Hydrogen bonding

A few protonic species in solid conductors can exist 'free' but most are hydrogen bonded to some degree. When an A–H . . . B *hydrogen bond* between a *Brönsted acid* AH and a base B is formed, the corresponding potential curve of the proton is modified (Fig. 23.1a): the A–H distance increases while the A . . . B distance decreases with respect to the sum of the Van der Waals radii of A and B atoms. The vibrational levels get closer and the AH stretching frequency is lowered. Here are thus the main criteria of hydrogen bonding as far as distances and frequencies are concerned. Moreover, the AH stretching absorption bands are also characterized by an increase of intensity and band breadth on hydrogen bonding.

In the case of the OH . . . O system, most frequently encountered in protonic conductors, hydrogen bonds are usually divided into weak, medium-strong and strong with O . . . O distances >0.27, 0.27–0.26 and 0.26–0.24 nm, respectively[1-3]. Typical examples of weak bonds are found in hydrates, of medium-strong bonds in self-associated carboxylic acids and of strong bonds in acid salts of many inorganic and organic acids[1-4].

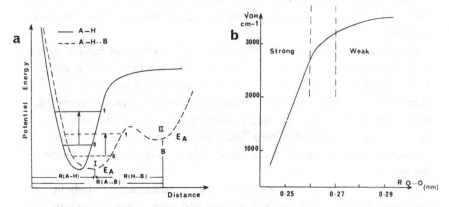

Fig. 23.1. Potential energy of the proton involved in the A–H . . . B hydrogen bond (a), and $\nu_{OH} = f(d_{O-H...O})$ correlation curve after Novak (b)[1].

23.3 Relationship between OH stretching frequencies and O . . . O distances

Empirical relationships between e.g. OH stretching frequencies and O . . . O distances have been established and are useful in predicting the latter from the former[1,4]. Over the years, most of these predictions have been confirmed and the spectroscopic scale of hydrogen bond strength appears to have become more accurate. Fig. 23.1b shows a νOH versus RO . . . O curve in which the νOH frequency can vary over the whole spectrum from 3700 to 500 cm^{-1}. The slope of the curve changes, in going from weak to medium-strong and strong hydrogen bonds, with respective values of 10, 60 and 120 cm^{-1}/pm. The variation of frequency with distance is thus much higher for strong hydrogen bonds. Similar relationships for NH . . . O and NH . . . N hydrogen bonds have also been found[5].

In choosing the frequencies, some caution must be taken since, in the spectra, multiple OH stretching bands due to different factors may be observed. They can be due to different protonic species as, for instance, in antimonic acid (cf. Chapter 11) where H_2O, H_3O^+ and —OH entities occur. In cases where only one species exists, multiple bands may arise because of the presence of crystallographically non-equivalent molecules and/or correlation field splitting as is found, for instance, in H_3O^+ β-alumina. In hydrates, there is also an intramolecular coupling of H_2O vibrations such that in centrosymmetric crystals, infrared and Raman frequencies are different. Finally, the band multiplicity may be due to Fermi resonances such as occur between ν_s and δH_2O modes.

23.4 Isotopic dilution method

In order to get rid of intra and intermolecular coupling as well as Fermi resonances, the *isotopic dilution* technique has been used. A given crystal is deuterated to about 90–95% with H/D ~ 0.05–0.1 (or vice versa, D/H ~ 0.05–0.1); under these conditions practically every OH oscillator is surrounded by OD groups and gives rise to a single OH stretching band, the frequency and band shape of which reflects its immediate environment. This frequency can then be safely used for correlation purposes. If, in a crystal, all protonic entities are equivalent with a single O–H distance only one OH stretching frequency is expected and there will be as many OH frequencies as there are different O–H distances.

This is illustrated by the infrared spectra of HUP where the νOH bands of an isotopically diluted crystal at 100 K could be identified at 3390–3380, 3260 and 3170 and 2920 cm^{-1} and have been assigned to H_2O–H_2O, H_2O–PO_4 and H_3O^+–H_2O interactions[6,7]. On the other hand, HUP infrared spectra at 300 K show only a broad pattern between 2000 and 3500 cm^{-1} and the distinction between H_3O^+ and H_2O does not appear possible.

23.5 Structure determination

Vibrational spectroscopy is useful in identification of protonic entities and their configuration. It can determine the type of association; for instance, whether M–OH entities form infinite chains, cyclic or open dimers, and the position of a proton in strong, symmetric hydrogen bonds where it can distinguish a truly centred proton from a statistically symmetrical case[8].

One of the best examples in the identification of protonic species has been in hydrogen β-alumina (Fig. 23.2). If silver β-alumina is partially reduced by hydrogen, its infrared spectrum shows narrow bands at 3614 and 3543 cm^{-1} assigned to non-hydrogen bonded Al–OH$^+$ groups, i.e. to protons trapped by interstitial oxygens and by lattice oxygens, respectively (proton defects)[9]. A stronger reduction leads, after rehydration in air, to a hydrated proton giving rise to a narrow band at 3498 cm^{-1} which corresponds to a H_3O^+ ion at a Beevers–Ross (BR) site[10]. A similar narrow band, near 3508 cm^{-1}, is observed for stoichiometric β-alumina in which only BR sites are available[9]; in non-stoichiometric samples, oxonium ions are also found in mid-oxygen (mO) sites and are characterized by broader bands at lower frequencies.

The spectrum of H_3O^+ β″-alumina, on the other hand, shows a single broad band at a frequency similar to that of the stoichiometric sample, but the band width is about 20 times larger[11]. The band broadening is ascribed essentially to the static structural disorder of oxonium ions in the conducting plane because these ions occupy many different positions which are displaced to varying extents from the ideal sites. The spectrum of fully hydrated samples of β-alumina shows bands characteristic of $H_5O_2^+$ and $H_7O_3^+$ entities: narrow, high frequency OH stretching bands due to terminal H_2O groups (3585 and 3380 cm^{-1}) and broader, low frequency absorptions corresponding to the terminal H_3O^+ (2900 cm^{-1}) and the OH ... O central hydrogen bond (2500 cm^{-1}) can be identified[10,11]. Similar spectra are observed for β″-aluminas[11]. *Inelastic neutron scattering* (INS) is a particularly sensitive method for detecting protonic species, giving rise to vibrational modes in the regions where other bands may interfere. Neutron vibrational spectroscopy has no selection rules; for hydrogen-containing materials, the incoherent scattering is dominated by the modes carrying protons due to their large incoherent scattering section[9,12]. Some illustrations of INS on metal oxides are given in Figs 7.5, 7.6, 11.2, 21.2, 21.3 and 23.3. Evidence of free and of non-specific bonded proton is given.

23.6   Disordered crystals

Statistically symmetric hydrogen bonds are disordered; in such disordered crystals, the apparent crystalline symmetry can depend on the method of investigation. Diffraction methods give usually higher symmetry than spectroscopic methods, but even in spectroscopy the behaviour of external and internal vibrations is different.

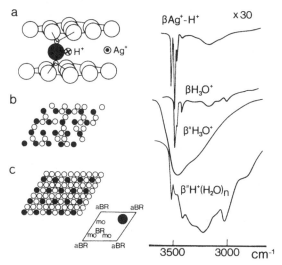

Fig. 23.2. Structure and infrared spectra of hydrogen β/β''-alumina. (a) Silver β-alumina conduction plane with an Al–OH$^+$ defect; the filled circle represents the oxygen of the Al–O–Al bridge between spinel blocks. (b) Conduction plane of H_3O^+ β/β''-alumina. (c) Fully packed conduction layer of $H^+(H_2O)_n$ β/β''-alumina. For (b) and (c) the filled circles indicate oxygen atoms in the Al–O–Al bridge and open circles indicate H_3O^+ and/or H_2O molecules. The different crystallographic sites designated BR (Beevers–Ross), aBR (anti-Beevers–Ross) and mO (mid-oxygen) are indicated. The infrared spectra are of corresponding single crystals with the electric vector of the incident light perpendicular to the *c*-axis (parallel to the conducting plane) (with permission[9]).

371

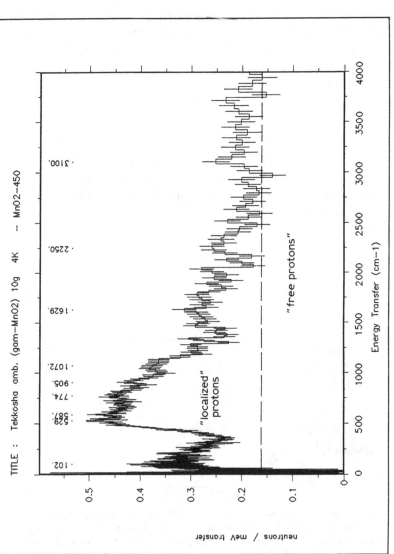

Fig. 23.3. Inelastic neutron scattering spectra of γ-MnO_2 powder. The spectrum of the sample is recorded at 4 K and the spectrum of the same sample heated previously at 450 °C has been subtracted. Bonds are assigned to localized protons (OH^-, H_2O) whereas the diffuse and broad continuum above 100 cm^{-1} seems related to a 'proton gap' (proton lattice gap) (with courtesy[26]). The INS spectrum of $H_{0.96}MnO_2$ prepared by chemical reduction of γ-MnO_2 shows three peaks near 1120, 2200 and 3100 cm^{-1} which are assigned to the $0 \rightarrow 1$, $0 \rightarrow 2$ and $0 \rightarrow 3$ transitions of an isotropic oscillator, respectively. This triplet may correspond to a non-specific bonded proton at the centre of the octahedral cavity[27].

Vibrational spectroscopy

The paraelectric phase of CsH_2PO_4, for instance, contains a disordered OH...O hydrogen bond with statistically distributed protons in two off-centre equivalent sites; its crystalline symmetry is that of $P2_1/m$ space group with $Z = 2$[13]. The selection rules derived from the above symmetry predict nine Raman and six infrared active lattice modes which in fact have been observed. The internal vibrations, on the other hand, do not obey the same selection rules and do not see centres of symmetry. It turns out that in CsH_2PO_4[14], as in several other disordered crystals such as squaric acid[15] or protonic conductors $CsHSO_4$[16] and NH_4HSeO_4[17] (paraelectric phase), the external vibrations see the (average) symmetry of the crystal in the same way as X-ray or neutron diffraction while the internal vibrations see only the (lower) symmetry of a molecule (or ion) and not the statistical (higher) one. It should be pointed out that this type of disorder affects the selection rules for internal vibrations but not the band breadth of either internal or external modes.

The distinction between cyclic *dimers* $(M-OH)_2$ and infinite *chains* $(M-OH)_n$ is based on the mutual exclusion rule for centrosymmetric cyclic dimers, i.e. infrared and Raman frequencies should not coincide. In the case of carboxylic acids R–COOH, the C=O stretching mode is particularly sensitive and the difference between infrared and Raman frequencies can be as much[18] as $60-70$ cm^{-1}.

Hydrogen sulphates and selenates, $MHXO_4$, some of which are superionic conductors, present also these two association types: $KHSO_4$ and $KHSeO_4$ contain both cyclic dimers and infinite chains and the Se–O and Se–OH stretching modes are sensitive to the type of association[8]. It turns out that the difference in these two frequencies is much greater for cyclic dimers ($130-100$ cm^{-1}) than for chains ($80-90$ cm^{-1}), the OH–O hydrogen bond being weaker in the former. It has thus been suggested that $CsHSeO_4$ and high temperature superionic phases of $CsHSO_4$ and NH_4HSeO_4 contain cyclic dimers.

Crystallographically symmetric hydrogen bonds can be either truly or statistically symmetric and diffraction methods cannot distinguish the two cases. In spectroscopy, the OH stretching frequencies of truly symmetric hydrogen bonds are usually much lower, e.g. near 700 cm^{-1} in $NaH(CH_3COO)_2$[19] or 900 cm^{-1} in $KH(CF_3COO)_2$[20] than for statistically symmetric 1300 cm^{-1} in squaric acid[15]. The best criterion, however, is the value of the $\nu OH/\nu OD$ isotopic frequency ratio. This decreases from 1.36 (free OH group) with decreasing νOH frequency and may reach values close to unity (oxalic acid dihydrate)[1] or even below for all asymmetric and statistically symmetric hydrogen bonds; in the case of a truly centred O–H–O bond, the value of $\nu OH/\nu OD$ increases above the value of the harmonic oscillator (1.41). Protonic conductors such as $CsHSO_4$, NH_4HSeO_4, $M_3H(SO_4)_2$ (M$=$K, Rb, NH_4) all contain statistically symmetric hydrogen bonds (Chapter 11).

Another type of structural disorder is orientational; it has been observed for a number of crystals containing ammonium ions, in particular, in the protonic conductors $NH_4\beta\text{-}Al_2O_3$, $(NH_4)_3H(SO_4)_2$ and NH_4HSeO_4. The last two crystals undergo several phase transitions (cf. Chapter 11) from ordered to an increasingly disordered state in which the ammonium ions play an important role. The spectroscopic results concern in particular the low-frequency Raman[9,11] and inelastic[9,10] neutron spectra. Low-temperature ordered phases give rise to well-structured narrow bands (due to external modes of cations and anions) which broaden progressively and lose most of their structure with increased temperature[9,21]. In the spectra of the high temperature superionic phase, all the structure is smeared out but there is a considerable broadening of the Rayleigh wing. The spectrum is characteristic of a *plastic phase* implying a rapid reorientation of ions on given sites. The internal bands also broaden; the νNH band of NH_4HSeO_4 for example, loses all structure indicating a high (at least C_{3v} but probably O_h) site symmetry of the ammonium ion[17]. Band broadening is temperature dependent and due to dynamical disorder, unlike in β''-alumina where the νOH band breadth is caused by static disorder.

23.6.1   OH stretching band breadth

In the case of weak and medium-strong hydrogen bonds, νOH band breadth can be due to intermolecular coupling and structural disorder; it may frequently be reduced to normal values by deuteration and lowering of temperature. Strong hydrogen bonds, on the other hand, can yield very broad (several hundred cm^{-1}) single or multiple bands which are an intrinsic property of such hydrogen bonds and which are due to anharmonicity and the particular shape of the potential curves of the proton in both fundamental and excited states.

23.7   Potential barrier and conductivity

The direct conductivity, σ, is given by the response of the system to a longitudinal electric field parallel to the direction of diffusion of the

Vibrational spectroscopy

conducting ions

$$\sigma = \frac{n^2 e^2 d^2}{kT} v_0 \exp \frac{E_a}{kT}$$

where e is the ion charge, v_0 the fundamental vibrational frequency, d the jump length and E_a the activation energy. In optical vibration spectroscopy, $\sigma(\omega)$ reflects the response of the system to a transverse electric field; v_0 and E_a can be estimated, provided that translational vibrations are well identified and the interactions with the framework are weak[7,21-23]. At low temperatures, conducting cations (or cation clusters) oscillate in their potential wells. At high temperatures, an increasing proportion of mobile cations is an energy state higher than the potential barrier v_0. In a crude model (harmonic oscillator), $v_0 = 2c^2 d^2 m v_0^2$, where m is the mobile species mass; the choice of an appropriate d value is difficult. The resulting v_0 barrier may be of the same order of magnitude as E_a.

Conductivity can be deduced from vibrational spectra: in IR spectroscopy, the absorption coefficient $\alpha(\omega)$ is related to $\sigma(\omega)$: $\alpha(\omega) = 4\pi\sigma(\omega)/nc$, n being the refractive index and c the velocity of light[24]. In Raman spectroscopy, the scattered intensity $I(\omega)$ is related to conductivity by $\sigma(\omega) \propto \omega I(\omega)/n(\omega) + 1$, $n(\omega)$ being the Bose–Einstein population factor[25]. Finally, the inelastic incoherent neutron scattering function $P(\omega)$ is proportional to the Fourier transform of the 'current correlation function' of the mobile ions. $P(\omega)$ is homogeneous with $\omega^2 I(\omega)$ formalism. However, since $P(\omega)$ reflects mainly single particle motions, its comparison with $\sigma(\omega)$ could provide a method for the evaluation of correlation effects. (For further discussion, see also Chapter 9 and p. 333.)

23.8   Phase transitions

There is an extensive literature on the application of vibrational spectroscopy *to phase transition* investigations. The phase transitions in proton conductors can be studied in much the same way as in other crystals as far as the order, nature and mechanism are concerned. Particularly interesting examples are those of HUP, $CsHSO_4$, $(NH_4)_3H(SO_4)_2$ and NH_4HSeO_4 described in Chapters 11 and 17.

23.9   References

1. A. Novak, *Structure and Bonding, Berlin* **18** (1974) 177–216.
2. W. C. Hamilton and J. A. Ibers, *Hydrogen Bonding in Solids* (New York,

Proton dynamics and charge transport

W. A. Benjamin Inc. (1968)); S. N. Vinogradov and R. H. Linnell, *Hydrogen Bonding* (Van Nostrand Reinhold (1971)) 59.

3. D. Hadzi and S. Bratos, in *The Hydrogen Bond II*, P. Schuster, G. Zundel and C. Sandorfy (eds.) (North-Holland Publishing (1979)) 565.

4. G. Ferraris and G. Ivaldi, *Acta Cryst.* **B44** (1900) 341–44; J. Howard, J. Tomkinson, J. Eckert, J. A. Goldstone and A. D. Taylor, *J. Chem. Phys.* **78** (1983) 3150.

5. A. Lautié, F. Froment and A. Novak, *Spectroscopy Lett.* **9** (1976) 289–99.

6. M. Pham-Thi, Ph. Colomban and A. Novak, *J. Phys. Chem. Solids* **46** (1985) 493–504; 565–78.

7. Ph. Colomban, M. Pham-Thi and A. Novak, *Solid State Commun.* **53** (1985) 747–51.

8. Ph. Colomban, M. Pham-Thi and A. Novak, *J. Mol. Struct.* **161** (1987) 1–14.

9. Ph. Colomban and A. Novak, *J. Mol. Struct.* **177** (1988) 277–308; Ph. Colomban, G. Lucazeau and A. Novak, *J. Phys C (Solid State Phys.)* **14** (1981) 4325–33.

10. Ph. Colomban, R. Mercier and G. Lucazeau, *J. Chem. Phys.* **67** (1977) 5244–51.

11. Ph. Colomban and A. Novak, *Solid State Commun.* **32** (1979) 46–71; *Solid State Ionics* **5** (1981) 241–4.

12. R. C. T. Slade, A. Ramanan, P. R. Hirstand and H. A. Pressman, *Mat. Res. Bull.* **23** (1988) 793–8; W. S. Glaunsinger, M. J. McKelvy, F. M. Larson, R. B. Von Dreele, J. Eckert and N. L. Roos, *Solid State Ionics* **34** (1989) 281–6.

13. D. Semmingsen, W. D. Ellenson, B. C. Frazer and G. Shirane, *Phys. Rev. Lett.* **38** (1977) 1299–302.

14. B. Marchon and A. Novak, *J. Chem. Phys.* **78** (1983) 2105–20.

15. D. Bougeard and A. Novak, *Solid State Commun.* **27** (1978) 453–7.

16. M. Pham-Thi, Ph. Colomban, A. Novak and R. Blinc, *J. Raman Spectrosc.* **18** (1987) 185–94.

17. I. P. Aleksandrova, Ph. Colomban, F. Denoyer, N. Le Calvé, A. Novak, B. Pasquier and A. Rozycki, *Phys. Stat. Sol. (a)* **114** (1989) 822–34.

18. J. de Villepin, M. H. Limage, A. Novak, N. Toupry, M. Le Postollec, H. Poulet, S. Ganguly and C. N. R. Rao, *J. Raman Spectrosc.* **15** (1984) 41–6.

19. A. Novak, *J. Chim. Phys.* **11–12** (1972) 1615–25.

20. D. Hadzi, B. Orel and A. Novak, *Spectrochim. Acta* **29**A (1973) 1745–53.

21. Ph. Colomban and G. Lucazeau, *J. Chem. Phys.* **72** (1980) 1213–24.

22. J. B. Bates, T. Kaneda and J. C. Wang, *Solid State Comm.* **25** (1978) 629–32.

23. Ph. Colomban, R. Mercier and G. Lucazeau, *J. Chem. Phys.* **75** (1981) 1388–99.

24. W. Hayes, G. F. Hopper and F. L. Pratt, *J. Phys. C (Solid State Phys.)* **15** (1982) L675–80.

25. M. J. Delaney and S. Ushida, *Solid State Commun.* **19** (1976) 297–301.

26. F. Fillaux, H. Ouboumour, J. Tomkinson and L. T. Yu, *Chem. Phys.* **149** (1991) 459–69.

27. F. Fillaux, H. Ouboumour, C. Cachet, J. Tomkinson and L. T. Yu, *Physica B* (1992) (in press).

24 Raman spectroscopic studies of proton conductors

R. FRECH

24.1   Introduction

Raman scattering spectroscopy is a versatile and powerful technique for studying composition, structure and dynamics in a wide variety of condensed phases. Transitions between vibrational states are measured in the usual Raman scattering experiment, as in an infrared transmission experiment. However, the selection rules governing Raman scattering differ from those of infrared absorption, so the two techniques should be regarded as complementary. Since the scientific literature regarding the application of Raman spectroscopy to studies of proton conductors is quite extensive, this chapter will not attempt to provide a comprehensive review. Rather the different kinds of structural and dynamical information which can be obtained from these studies will be surveyed, with examples given which illustrate the versatility and power of the technique.

In the next section, the theory of the Raman effect will be briefly reviewed, followed by Section 24.3, the application of Raman spectroscopy to studies of proton conductors. The description of structural information will be divided according to data obtained from band frequencies, intensities, bandwidths, and studies of phase transitions. The chapter will then conclude with several illustrations of dynamical information which can result from an appropriate analysis.

24.2   The Raman effect

Raman scattering has been the subject of numerous excellent reviews[1-3] and will be only briefly summarized here. In general terms, an external electromagnetic field interacts with matter, inducing an oscillating electric moment which acts as a source of scattered electromagnetic radiation. Generally the interaction is treated in the semi-classical approximation, in which the electromagnetic field is a continuum, but the energy states

377

are quantized. When the incident radiation of frequency ω_0 is polarized in the ρ direction, the intensity of the scattered light polarized in the σ direction may be written, following Wang[3]

$$I_{\rho\sigma}{}^{jk} = (\omega^4/4c^4)|\alpha_{\rho\sigma}{}^{jk}|^2 I_0 \delta(\omega - \omega_0 - \omega_{jk}) \qquad (24.1)$$

Here I_0 is the incident intensity and $\alpha_{\rho\sigma}{}^{jk}$ is the $\rho\sigma$ component of the Raman polarizability tensor originating in a transition from state j to state k. The tensor component is

$$\alpha_{\rho\sigma}{}^{jk} = \sum_i \frac{\langle j|P_\rho|i\rangle\langle i|P_\sigma|k\rangle}{h[\omega_{ik} + \omega]} + \frac{\langle j|P_\sigma|i\rangle\langle i|P_\rho|k\rangle}{h[\omega_{ij} - \omega]} \qquad (24.2)$$

where P_ρ is the ρ component of the electric moment operator and $\omega_{ik} = \omega_i - \omega_k$. If $\omega_{ij} < 0$, the scattered light will be at a lower frequency than the incident light, which is the case of Stokes scattering. If $\omega_{ij} > 0$, the scattered light will be at a higher frequency than the incident light, corresponding to anti-Stokes scattering.

In vibrational Raman scattering, which is the primary technique of interest in studies of proton conductors, the Born approximation is invoked to write the state $|i\rangle$ as the product of an electronic wave function, $|e\rangle$, and a vibrational wave function, $|v_m\rangle$, i.e. $|i\rangle = |e\rangle|v_m\rangle$. The subscript m on the vibrational wave function designates the mth normal vibrational mode. Usually the transition $j \to k$ occurs between the vibrational level v_m in the ground electronic state g and the vibrational level v_m' also in the ground electronic state, so the transition $|j\rangle \to |k\rangle$ may be written $|g\rangle|v_m\rangle \to |g\rangle|v_m + 1\rangle$. Here the harmonic oscillator selection rule $v_m' = v_m \pm 1$ has been invoked, and the value $v_m + 1$ has been chosen, corresponding to Stokes scattering. In this case, Eqn 24.2 can be considerably simplified by writing the Raman tensor as

$$\alpha_{\rho\sigma}{}^m = \langle v_m + 1|\alpha_{\rho\sigma}{}^{gg}|v_m\rangle \qquad (24.3)$$

where the elements of the polarizability tensor $\alpha_{\rho\sigma}$ are

$$\alpha_{\rho\sigma}{}^{gg} = \sum_e \frac{\langle g|P_\rho|e\rangle\langle e|P_\sigma|g\rangle}{h[\omega_{eg} + \omega]} + \frac{\langle g|P_\sigma|e\rangle\langle e|P_\rho|g\rangle}{h[\omega_{eg} - \omega]} \qquad (24.4)$$

In the Placzek approximation, which is appropriate for a system with a nondegenerate electronic ground state and in which resonance does not occur, the Raman tensor is real and symmetric.

24.3 Applications of Raman spectroscopy to the study of proton conductors

24.3.1 Structural information

24.3.1.1 Frequencies

Perhaps the widest use of Raman spectroscopy is in the identification of molecular or ionic species. A measurement of the Raman spectrum provides a number of Stokes-shifted bands, each occurring at a particular frequency ω_m corresponding to the energy spacing of the set of vibrational energy levels associated with the normal mode m. The frequencies of the modes depend on both the atomic masses and the bonding of the individual atoms. Often the same subset of atoms covalently bound in the same molecular architecture, e.g. the NH_4^+ cation, will result in approximately the same normal mode frequency pattern, providing a precise molecular 'fingerprint' for identification purposes. These characteristic or group frequencies have been studied in a variety of systems and a wealth of literature exists which describes spectra–structure correlations[4]. Since Raman scattering and infrared transmission spectra yield complementary information about the spectral frequency pattern due to the differences in the selection rules, it is important that both kinds of data be examined before structural conclusions can be drawn. In this context, isotopic substitution of the compound being studied can be quite helpful, since the normal mode frequencies shift in a way which can be calculated, often providing confirmation of a tentative assignment. The isotopic shift associated with the substitution of deuterium for hydrogen is large, consequently proton-conducting solids are especially amenable to this technique.

There have been numerous spectroscopic studies of proton-conducting β- and β″-alumina in which the identification of the species is of concern. Usually these studies involve a mixture of infrared transmission and Raman scattering measurements. The basic structure of both of these aluminas is well-understood from diffraction studies[5, 6], consisting of closely packed spinel blocks of aluminium and oxygen atoms, separated by loosely packed conduction planes perpendicular to the *c*-axis containing bridging oxygen ions and the mobile cations. In the β-alumina structure the bridging oxygen atoms in the conduction plane form a hexagonal network. In between these oxygen atoms are the Beevers–Ross (BR) sites, the anti Beevers–Ross (aBR) sites, and the mid-oxygen (mO)

Proton dynamics and charge transport

sites, with site symmetries D_{3h}, D_{3h} and C_{2v}, respectively. However, the identification of the protonated ionic and molecular species present in the conduction plane relies heavily on spectroscopic measurements, as illustrated by the study by Bates *et al.*[7] of lithium β-alumina hydrated with H_2O. Samples were also prepared hydrated with HDO and D_2O to verify the spectral assignments. Vibrational bands in the Raman spectrum at 3622, 3571, 3472 and 3409 cm^{-1} were identified as originating in various OH$^-$ species. The authors have speculated that there are two kinds of OH$^-$ species, interstitial OH$^-$ resulting from dissociation of an H_2O molecule and OH$^-$ resulting from the bonding of a proton to one of the bridging oxygens in the conduction plane. Bands at 3212, 3116 and 2947 cm^{-1} were assigned as intramolecular OH stretching motions of water molecules located in the conduction plane. A band at 2740 cm^{-1} was assigned as H_3O^+, while a band at 2572 cm^{-1} was attributed to $H_5O_2^+$. However, these assignments (and in some cases, the reported band frequencies) have been questioned by Lucazeau[8] in an infrared study which also included deuterated compounds. He assigned bands at 3508 and 3520 cm^{-1} to H_3O^+ ions on BR sites, bands at 3433 and 3395 cm^{-1} to H_3O^+ ion pairs doubly occupying mO sites, and bands at 3500 and 3495 cm^{-1} to single H_3O^+ ions occupying mO sites. Three bands were then assigned to the asymmetric $H_3O^+H_2O$ species (to be distinguished from the symmetric $H_5O_2^+$ species); a band at 3380 cm^{-1} was attributed to a terminal H_2O, a band at 2900 cm^{-1} to a terminal H_3O^+, and a band at 2540 cm^{-1} to the central hydrogen in the O–H $\cdots$ O structure. Finally a series of bands at 3235, 2880, 2710, 3585 and 3545 cm^{-1} was assigned to the $H_3O^+(H_2O)_2$ species.

In addition to providing a (not always unambiguous!) identification of the species present, the frequency shift of a normal mode upon deuterium substitution can often be used to quickly identify the particular type of atomic displacements constituting the normal mode. An excellent example is a study of NH_4^+ β-alumina by Bates *et al.*[9] in which two bands at 168 and 145 cm^{-1} were observed in the Raman spectrum. In a similar measurement of a β-alumina crystal containing ND_4^+ ions, the bandcentre of the 168 cm^{-1} mode was shifted to 152 cm^{-1}. If this band were due to a translational mode, the calculated isotopic frequency shift ratio would be

$$v(NH_4^+)/v(ND_4^+) = [M(ND_4^+)/M(NH_4^+)]^{1/2} = 1.106 \quad (24.5)$$

However, if the band were due to a librational mode, the calculated

Raman spectroscopic studies

isotopic frequency shift ratio would be

$$v(NH_4^+)/v(ND_4^+) = [I(ND_4^+)/I(NH_4^+)]^{1/2} = 1.414 \quad (24.6)$$

where I is the principal moment of inertia. Experimentally, the frequency shift ratio is found to be 1.105, providing a compelling argument for identifying the mode as originating in the translational motion of the ammonium ion. Further, the modes were seen in $a(a'a')c$ and $a(a'a)c$ scattering geometries, but not in $a(ca')c$ and $a(ca)c$ scattering geometries, which indicates that the displacements of the ammonium ions lie in the conduction plane.

Frequency data can also provide important information about the potential energy environment of the mobile ionic species. In the previously mentioned study Colomban *et al.*[11] which included NH_4^+ β-alumina, the motion of the cations in the conduction plane was assumed to occur in the periodic potential

$$V(x) = 1/2 V_0 [1 - \cos(2\pi x/\delta)] \quad (24.7)$$

where V_0 is the barrier height for motion from one site to an adjacent site separated by the period d. The barrier height is then given by

$$V_0 = 2mc^2 v^2 d^2 \quad (24.8)$$

where v is the translational mode frequency and m is the mass of the cation. (This calculation is quite dependent on the value chosen for the jump distance d, and an alternative method of estimating V_0 from the temperature dependence of the bandwidth will be described in a later section.) An NH_4^+ translational mode frequency of 150 cm^{-1} combined with BR → aBR distance of 3.2 Å gives a value for V_0 of 78 kJ mol^{-1}, which is the activation energy associated with the cation jump. A similar analysis of H_3O^+ β-alumina using the value of 140 cm^{-1} for the translational mode frequency gives a value of 65 kJ mol^{-1} for the activation energy.

24.3.1.2 Intensities

Absolute Raman scattering intensities are quite difficult to measure accurately and consequently are usually not used to calculate absolute concentrations of species. However relative Raman band intensities can provide a great deal of information about the occupancy and orientation of the scattering species on the sites in a condensed phase system through

381

the properties of the polarizability tensor (Eqn 24.4). Again the β-alumina system offers several examples of this application. Since there are significantly more sites in the conduction plane than there are mobile cations, this necessarily introduces disorder into the cation sublattice. There has been some debate as to the distribution of the cations among the possible sites, and Raman spectroscopy has often addressed these questions with a degree of success. The interpretation of the resulting Raman spectra is not trivial, since disorder in the cation sublattice means that the usual factor group or site group analysis cannot be applied without taking account of the disorder in some manner.

In a Raman study of NH_4^+ β-alumina[10], the disorder in the conduction plane was accounted for explicitly by assuming that the ammonium ions are distributed among the various sites in several possible configurations with each ion acting as an independent scattering source. The choice of sites was based on a model which showed that when excess ammonium ions were added to stoichiometric β-alumina in which the ammonium ions occupied BR sites, an added ammonium ion displaced an original ammonium ion from the BR site so that both ions occupied neighbouring mO sites, essentially sharing the BR site as an ion pair. In this model, the splitting of the degenerate intramolecular ammonium ion modes depends on the effective site symmetry, while the intensity depends on the orientation of each ion with respect to the laboratory fixed axes. The individual scattering contributions from each set of equivalent ions were added incoherently to give the total scattering intensity. Comparison of the calculated spectra with experiment indicated that the ammonium ions occupied three sets of non-equivalent sites with the effective symmetry at one set being $C_s(\perp)$ or C_1 and the effective symmetry at the other two sets either C_{2v} or $C_s(\parallel)$. Here the designations $\perp$ and $\parallel$ refer to the conduction plane.

24.3.1.3 Bandwidths

Although the band intensity in Eqn 24.1 is written with an infinitesimally narrow bandwidth (via the Dirac delta function), in practice the delta function is replaced by a more realistic band shape factor with a finite bandwidth. There are several contributions to the bandwidth including vibrational relaxation and dephasing, inhomogeneous broadening, and lifetime broadening. Since each of these separately is difficult to calculate exactly, the bandwidth is often represented by a phenomenological function based on an appropriate model. Bandwidth analyses are most relevant when one particular mechanism provides the dominant contribution to the bandwidth.

Bandwidth data have been utilized by Colomban & Lucazeau[11] in studies of the β-alumina system to follow the transformation of the non-stoichiometric material $(1 + x)M_2O \cdot 11Al_2O_3$ (where M is a monovalent cation) to the corresponding stoichiometric compound with $x = 0$. According to the authors, when non-stoichiometric NH_4^+ β-alumina is heated, NH_3 is eliminated and the free protons react with charge-compensating interstitial oxygen ions and are subsequently solvated by H_2O. This is seen in the infrared spectra by the replacement of the ammonium ion vibrational modes by the oxonium vibrational modes. The loss of non-stoichiometry is monitored in the Raman spectrum by the reduction in the bandwidths (from 20 to $2 \, cm^{-1}$) of spinel block modes. The degree of non-stoichiometry can be estimated from the bands associated with low frequency cation oscillations and from the intensity of a shoulder at $230 \, cm^{-1}$ which originates in the vibration of an aluminium–interstitial oxygen–aluminium bridge (Frenkel defect).

In some proton conductors, the high values of ionic conductivity are found only in a particular phase. Raman spectroscopy in conjunction with other techniques has been utilized extensively in such systems to deduce information about the structural changes accompanying phase transitions. One of the more carefully studied systems is $CsHSO_4$ (and the related compound $CsHSeO_4$). The room temperature phase is monoclinic[12] with zig-zag chains of HSO_4^- anions perpendicular to a plane of loosely packed Cs^+ cations. The hydrogen bonding is evident in the Raman spectra[13,14] from the v(S–O) acceptor band at $998 \, cm^{-1}$ and the v(S–OH) donor band at $860 \, cm^{-1}$ which can be seen in the 300 K spectrum of Fig. 11.5B. At 318 K there is a transition from Phase I into Phase II, accompanied by a shift in the v(S–O) acceptor band from 998 to $1024 \, cm^{-1}$ and a corresponding shift in the v(S–OH) donor band from 860 to $855 \, cm^{-1}$, as noted in the 350 K spectrum of Fig. 11.5B. These shifts originate in a weakening of the hydrogen bonding, as the S–OH bond becomes longer and the S–O bond becomes shorter. Further, the abrupt $26 \, cm^{-1}$ shift in v(S–O) in conjunction with shifts in the skeletal bending region suggests significant rearrangement of the SO_4^{-2} tetrahedra. This rearrangement has been interpreted as a structural change from zig-zag chains of HSO_4^- to cyclic dimers, consistent with structural phase changes observed in similar systems.

At 417 K there is another transition from Phase II to Phase III in which the conductivity is of the order of $10^{-2} \, \Omega^{-1} \, cm^{-1}$[15], although there are both ionic and protonic contributions to the conductivity[16]. The further decrease of the strength of the hydrogen bonding in the new phase is evident in the 420 K spectrum of Fig. 11.5B by the frequency shifts of the $v(S-O)$ band (to 1035 cm^{-1}) and the $v(S-OH)$ band (to 830 cm^{-1}). Even more dramatic is the disappearance of the lattice modes into a broad laser wing. This, in conjunction with the collapse of the factor group splitting in the internal optic mode region of the HSO_4^- anions, is compelling evidence that Phase III is a plastic phase. A plastic phase is characterized by a high degree of dynamic disorder in the sublattice of a polyatomic species, here the HSO_4^- ions, in which the disorder is due to the rapid reorientational motion of the species. It has been postulated in other superionic plastic phase systems that the high conductivity of the cations is dynamically enhanced by coupling with the reorientational motion of the anions[17]. It is interesting to note that upon deuteration of $CsHSO_4$, the Raman spectrum of the room temperature phase (I) of $CsDSO_4$ is more similar to that of Phase II than Phase I of $CsHSO_4$.

24.3.2   Dynamical information

Raman scattering spectroscopy can also provide dynamical information in addition to structural information in a number of proton-conducting systems. One of the more important parameters in fast ion conductor studies is the activation energy required for thermally activated movement of the charge carrier from one site to another. The 'attempt' mode of the mobile cation is the translatory vibrational motion which evolves into a translational degree of freedom with increasing temperature. The bandwidth of such a mode often exhibits a striking temperature dependence, dramatically broadening with increasing temperature as the degree of disorder in the cation sublattice necessarily increases. Analysis of this bandwidth data yields the activation energy for charge carrier motion. The bandwidth Γ of such a mode has been modelled by Andrade & Porto[18], who wrote

$$\Gamma = a + bT + c\tau_c / [1 + \omega^2 \tau_c^2] \qquad (24.9)$$

where a, b and c are constants, ω is the frequency, and the correlation time τ_c is

$$\tau_c = \tau_0 \exp(\Delta U / kT) \qquad (24.10)$$

Raman spectroscopic studies

The activation energy ΔU has been calculated for lithium fast ion conductors using this formalism[19], and a modified form of this equation was used by Colomban & Lucazeau in a study of in-plane cation translatory modes in β-alumina[11].

Proton-conducting systems containing molecular or polyatomic ionic species which undergo reorientational motion or orientational changes within a lattice can also be studied with temperature-dependent Raman spectroscopy. The significant molecular parameter is the activation energy barrier for thermally activated orientational structural changes. This parameter is important in systems in which the Grotthuss mechanism operates, for the crucial step in such a mechanism is often the reorientational motion of a polyatomic species which has given up a proton to an adjacent species and must reorient itself to receive a proton from another nearby species. The bandwidth of a librational or torsional mode whose vibrational motion becomes a thermally activated reorientational motion has been divided into a vibrational and an orientational contribution by Rakov[20], who wrote

$$\Gamma(T) = \Gamma_{\text{vib}} + A \exp(-E_a/RT) \qquad (24.11)$$

Here Γ_{vib} is the vibrational contribution to the bandwidth, A is a constant, and E_a is the activation energy barrier which must be overcome in order for the polyatomic species to reorient itself within the lattice.

An analysis based on this formalism has been given[21] for lithium hydrazinium sulphate, $LiN_2H_5SO_4$. This compound crystallizes in an orthorhombic space group[22], with the NH_2 of the hydrazinium cation hydrogen bonded into an extended chain parallel to the crystallographic c-axis and the NH_3^+ hydrogen bonded to the LiO_4 and SO_4 tetrahedra. Proton conductivity is highest along the c-axis[23] ($\sigma = 2 \times 10^{-8}$ S cm^{-1} at 25 °C) and both the vehicle mechanism and the Grotthuss mechanism have been invoked to explain the conductivity[24]. The hydrazinium torsional, librational and translational modes were identified using the deuterium-substituted compound[25], and bandwidths were measured from 12 to 300 K. Fig. 24.1 shows the temperature dependent bandshape of the hydrazinium internal torsional mode at 572 cm^{-1}. Analysis of the bandwidth data according to Eqn 24.9 gave an activation energy of 0.072 eV. The three hydrazinium librational modes at 381, 289 and 201 cm^{-1} (11 K frequencies) were similarly analysed and yielded reorientational activation energies of 0.06, 0.031 and 0.027 eV, respectively. These data, combined with moment of inertia calculations for each mode, suggest that the first librational mode be assigned to motion about the internuclear N–N axis of the hydrazinium ion, while the other two modes be assigned to motion perpendicular to the internuclear axis. It should be noted that none of the hydrazinium translatory modes showed any unusual temperature-dependent behaviour, while all librational and torsional modes exhibited significant band broadening with increasing temperature. Although these observations do not establish the Grotthuss mechanism as being correct, they do argue strongly against a vehicle mechanism as contributing significantly to the protonic transport.

A rather novel application of Raman spectroscopy to the study of fast proton transport has been given by Chandra *et al.*[26], who placed a pressed microcrystalline pellet of molybdic acid, $MoO_3 . 2H_2O$, under a 20 V d.c. bias and measured the spectrum at different positions between the cathode and the anode. A number of spectral changes were noted when the d.c. bias was turned on for a period of several hours. The appearance of bands at 795 and 414 cm^{-1} was tentatively assigned to the librational motion of H_3O^+ which was formed as an intermediate in the electrolysis of $MoO_3 . 2H_2O$ to $MoO_3 . H_2O$. An increase in the intensity of bands at

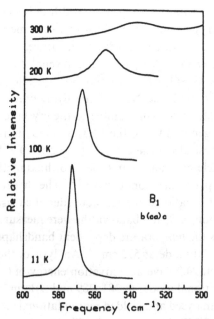

Fig. 24.1. Raman spectra of the hydrazinium ion internal torsional mode in single crystal $LiN_2H_5SO_4$ at various temperatures, with permission[21].

Raman spectroscopic studies

888 and 846 cm^{-1} was also noted and attributed to the increased concentration of H_3O^+, as was the increasing intensity of lattice modes at 77 and 88 cm^{-1}.

The peak intensities of the bands at 795 and 414 cm^{-1} were measured as a function of distance x from the cathode. By assuming that the peak intensity of each band was proportional to the concentration of the diffusing species responsible for the bands, the band intensity I can be written in terms of a solution of the diffusion equation appropriate for the sample geometry (see Eqn 2 of Reference 26),

$$I(x, t) = [I_0/(\pi Dt)^{1/2}] \exp(-x^2/4Dt) \qquad (24.12)$$

Analysis of the intensity data then results in a value of the diffusion coefficient D. The values calculated by the authors from the intensity profiles of the two bands at 414 and 795 cm^{-1} are 1.7×10^{-7} and 1.3×10^{-7} cm^2 s^{-1}, respectively, which can be compared to the value for the self-diffusion coefficient of water[27], 2.34×10^{-5} cm^2 s^{-1}.

24.4 References

1. J. A. Konigstein, *Introduction to the Theory of the Raman Effect* (D. Reidel, Dordrecht, Holland (1972)).
2. W. Hayes and R. Loudon, *Scattering of Light in Crystals* (Wiley, New York (1978)).
3. C. H. Wang, *Spectroscopy of Condensed Media* (Academic Press, Orlando (1985)).
4. (a) L. J. Bellamy, *The Infrared Spectra of Complex Molecules*, Vol. II (Chapman and Hall, London and New York (1980)). (b) G. Socrates, *Infrared Characteristic Group Frequencies* (Wiley, Chichester (1980)).
5. C. R. Peters, M. Bettman, J. W. Moore and M. D. Glick, *Acta Cryst.* B27 (1971) 1826–34.
6. M. Bettman and C. R. Peters, *J. Phys. Chem.* 73 (1969) 1774–80.
7. J. B. Bates, N. J. Dudney, G. M. Brown, J. C. Wang and R. Frech, *J. Chem. Phys.* 77 (1982) 4838–56.
8. G. Lucazeau, *Solid State Ionics* 8 (1983) 1–25.
9. J. B. Bates, T. Kaneda and J. C. Wang, *Solid State Commun.* 25 (1978) 629–32.
10. J. B. Bates, T. Kaneda, J. C. Wang and H. Engstrom, *J. Chem. Phys.* 73 (1980) 1503–13.
11. Ph. Colomban and G. Lucazeau, *J. Chem. Phys.* 72 (1980) 1213–24.
12. S. Yokota, *J. Phys. Soc. Japan* 51 (1982) 1884–91.
13. M. Pham-Thi, Ph. Colomban, A. Novak and R. Blinc, *Solid State Commun.* 55 (1985) 265–70.

Proton dynamics and charge transport

14. Ph. Colomban, M. Pham-Thi and A. Novak, *Solid State Ionics* **24** (1987) 193–203.
15. N. G. Hainovsky, Y. T. Pavlukchin and E. F. Hairetdinov, *Solid State Ionics* **20** (1986) 249–53.
16. Ph. Colomban, M. Pham-Thi and A. Novak, *Solid State Ionics* **20** (1986) 125–34.
17. A. Kvist and A. Bengtzelius, in *Fast Ion Transport in Solids*, W. van Gool (ed) (North-Holland, Amsterdam (1973)) 193.
18. P. d. R. Andrade and S. P. S. Porto, *Solid State Commun.* **13** (1973) 1249–54.
19. D. Teeters and R. Frech, *Phys. Rev.* **B26** (1982) 4132–9.
20. A. V. Rakov, *Tr. Fiz. Inst., Akad. Nauk SSSR* **27** (1964) 111–48.
21. R. Frech and S. H. Brown, *Solid State Ionics* **35** (1989) 127–32.
22. I. D. Brown, *Acta Cryst.* **17** (1964) 654–60.
23. K.-D. Kreuer, W. Weppner and A. Rabenau, *Solid State Ionics* **3/4** (1981) 353–8.
24. K.-D. Kreuer, A. Rabenau and W. Weppner, *Angew. Chem. Int. Ed. Eng.* **21** (1982) 208–9.
25. S. H. Brown and R. Frech, *Spectrochim. Acta* **44A** (1988) 1–15.
26. S. Chandra, N. Singh, B. Singh, A. L. Verma, S. S. Khatri and T. Chakrabarty, *J. Phys. Chem. Solids* **48** (1987) 1165–71.
27. D. Eisenberg and W. Kauzmann, *The Structure and Properties of Water* (Clarendon Press, Oxford (1969)).

25 Frequency dependent conductivity, microwave dielectric relaxation and proton dynamics

PHILIPPE COLOMBAN AND JEAN-CLAUDE BADOT

25.1 Definitions

Electrical properties can be determined at various frequencies. The interaction between an electromagnetic wave and condensed matter can be described using permittivity and conductivity concepts. Fig. 25.1 shows the permittivity, ε, versus frequency, v (or more usually $\omega = 2\pi v$). In the high frequency region, there are various resonances arising from ionic (molecular) and electronic motions. The latter are usually described in terms of optical spectroscopy. In the low frequency region, on the other hand, dipolar and space charge relaxations, usually called dielectric relaxation, are expected. The space charge region, often observed in usual a.c. conductivity measurements, may be described in terms of a (complex) impedance formalism[1] (see Chapters 26 & 27). These relaxations are directly related to the bulk conductivity and to electrode/electrolyte interfacial phenomena. They depend strongly on the microstructure on a $0.1-100\ \mu m$ scale (porosity, surface topology) and on chemical reactions at the interface (polarization, diffusion)[1-3]. This low frequency domain corresponds to 'free charges', which can move in association with an alternative electric field, while the vibrational region at higher frequencies corresponds to 'atom bonded fixed charges' ('dipoles').

How are ions able to move in a solid? The standard answer to this question states that two different kinds of ionic motions can be discerned, namely oscillatory motion and jump diffusion (see Chapter 30). In fact, the motion is not only limited to oscillations and to statistical hopping from site to site. Polyatomic ions (NH_4^+, H_3O^+) may undergo more or less complex rotations and other non-periodic local, non-hopping translational and non-statistical hopping motions are also possible. Such phenomena can be studied experimentally by neutron scattering and dynamic conductivity spectra (see Chapters 21 & 30).

389

Proton dynamics and charge transport

The interaction between an electromagnetic wave and condensed matter can be expressed by the equation

$$J_{(\omega)} = \sigma_{(\omega)}^* . E_{(\omega)}$$

where J is the current density, E the electric field and σ^* ($\sigma^* = \sigma' + i\sigma''$) the complex conductivity (as an additional complication the material can be anisotropic). The imaginary part, σ'', gives the component of the displacement current (at $\pi/2$ with the applied field) and thus does not contribute to power loss while the real part gives the component of the current in phase with the applied field and thus contributes to power loss.

The equation can then be written

$$J = i\omega\varepsilon_0\varepsilon^* E = i\omega\varepsilon_0(\varepsilon' - i\varepsilon'')E$$

ε^* is the complex (relative permittivity) with ε' and ε'' being the real and imaginary permittivity, respectively. The complex conductivity, σ^*, is given by $\sigma^* = \omega\varepsilon_0\varepsilon'' + i\omega\varepsilon_0\varepsilon'$, where $\varepsilon_0 = 1/36\pi . 10^{-9}$ F m^{-1} = 8.854 × 10^{-14} F cm^{-1}. The dielectric constant is a measure of the electrical polarizability of the material, i.e. the ability of charges and dipoles in the material to respond to the electric field. The loss factor, $\varepsilon''/\varepsilon'$, is a measure of the extent to which this motion of charges and dipoles irreversibly extracts energy from the electromagnetic field and dissipates it as heat in the material.

In the low frequency region, the equation

$$I = j\omega\varepsilon^* \frac{s}{t} V$$

where I is the current, V the potential, s and t the surface and thickness of the pellet, respectively, is usually used.

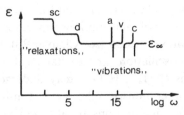

Fig. 25.1. Schematic representation of the permittivity ε' as a function of frequency ($\nu = \omega/2\pi$): sc, space charge relaxation; d, dipolar relaxation; a, atomic or molecular absorption; v and c, valence and core electronic absorption, respectively.

Frequency dependent conductivity

In the domain of optical frequencies, we could define the complex *refractive index* $n^*(\omega)$ from the complex permittivity by the relation

$$n^*(\omega) = [\varepsilon^*(\omega)]^{1/2}$$

for non-magnetic materials.

25.1.1 Languages

In theoretical as well as in applied studies, the concept of dielectric relaxation is fragmented and different methods of evaluation, analysis and measurement are used. Samples transmit and reflect electromagnetic waves and the amplitudes and phases of the transmitted and reflected waves can be measured. If at least two of these scalar quantities are known from experiment, they can be used to derive the complex electric conductivity, $\sigma^*(\omega)$. This is done with the help of Maxwell's equations and the boundary conditions which guarantee the continuity of the electric and magnetic field components at the surfaces of the sample. In the electromagnetic spectrum, the frequency range from 0.01 Hz to 10^{13} Hz is now easily covered. Coaxial waveguides are practicable up to about 20 GHz and 'free space' measurements are practicable up to 150 GHz (~ 5 cm^{-1}), i.e. in the low frequency range of the Fourier transform interferometer used in infrared spectroscopy. The absorption coefficient, $\alpha(\omega)$, is related to the conductivity by $\alpha(\omega) = 4\pi\sigma(\omega)/nc$, where n is the optical refractive index and c the velocity of light in free space. Raman scattered intensity, $I(\omega)$, is also related to the conductivity: $I(\omega)$ is proportional to $(1/\omega)(n^{\mathrm{B}}(\omega) + 1)\sigma(\omega)$ where $n^{\mathrm{B}}(\omega)$ is the Bose–Einstein occupation number. Of course, quasi- and inelastic neutron scattered intensity leads also to knowledge of the frequency dependent conductivity via the $P(\omega)$ function ($P(\omega) = \omega\sigma(\omega)$, see Chapters 21 and 30).

Chemists often reason in terms of dielectric permittivity (ε) or infrared spectra (α) while electrical engineers and physicists consider *tangent loss* (tg$\delta = \varepsilon''/\varepsilon'$) and alternative current conductivity (σ), respectively. It turns out that in most materials, free charges (where the complex impedance formalism is well-suited) and bound charges (where the description of the permittivity $\varepsilon^* = \varepsilon'(\omega) - i\varepsilon''(\omega)$ is more suitable) can be distinguished. Such a distinction is not straightforward in superionic conductors, however, in which some conducting species are in a quasi-liquid state and in particular, for proton conductors. In the latter, the proton (attached to the conducting species or itself the mobile species) represents a good probe for studying the corresponding motion and its properties can be used not only by NMR (see Chapters 22 and 28) and neutron scattering (see Chapter 21) but also in dielectric relaxation measurements. The proton, because of its small size, creates a larger dipole than other conducting ions and dielectric relaxation is easily observed using classical a.c. bridges ($\leqslant 10$ MHz) for liquids and solids containing protonic species. Since the motions studied correspond to molecular reorientations which are strongly temperature dependent, temperature can be used as a 'frequency' scale in techniques such as Deep Level Transient Spectroscopy (*DLTS*) or Thermally Stimulated Capacitance (*TSCap*) applied to semiconductors and some polymers or glasses[4] (see Chapter 9).

25.1.2 Measuring instruments

The complex impedance method was widely used up to 1970 in order to determine conductivity in the 0.01–10 MHz frequency region. Only a few superionic conductors were investigated in the microwave region, i.e. in the region between low frequency infrared (Raman) spectra and the lower frequency conductivity measurements, using mainly fixed-frequency apparatus and the procedure was always time-consuming. Recently, network analysers have been developed for measuring passive components and electronic devices in the new area of microwave communications (TV, satellite) and are commercially available. These instruments allow continuous measurement of dielectric properties up to a few tens of GHz. Furthermore, their high sensitivity and the large dipole associated with protonic species allow precise measurements; well-defined relaxations can therefore be observed, as shown in our previous work[3,5,6]. A better description of charge motions in solids can thus be given than that of the 'classical universal dielectric response' of A. K. Jonscher[7].

25.2 Dielectric relaxation

25.2.1 Ionic relaxation and frequency dependent conductivity

Charge transfer can be studied from a macroscopic point of view, but emphasis can also be given to local phenomena.

Frequency dependent conductivity

(i) The macroscopic approach used by electrochemists, for instance, considers the solid electrolyte as a homogeneous material and employs low frequency or d.c. conductivity measurements and/or tracer diffusion.

(ii) The emphasis on local phenomena, adopted by solid state scientists, studies the local structure and the dynamical properties by NMR and/or vibrational spectroscopy.

The experimental results obtained by the first groups are normally interpreted as if the mobile, charged defects perform a random jump diffusion (simple hopping models, see Chapters 4 and 30). The second group provides abundant evidence of non-random hopping. Correspondence between the activation energy deduced from NMR or vibrational spectroscopy on the one hand and from d.c. measurements on the other does not appear straightforward. There have been attempts to obtain a better description of the actual processes occurring using neutron scattering* and dielectric relaxation over a very wide frequency range in order to determine the motion of ions in both the space and time domains.

Jonscher[7] tried, from studies of polymer dielectrics which showed numerous relaxations in the intermediate frequency region, to unify the different approaches with the 'universal dielectric response concept'. This description remains phenomenological and corresponds essentially to the first group. At the other extreme is the work of Funke who, from neutron scattering results on AgI, has tried to pass from the local to the macroscopic scale.[8, 9] More recent work of his deals with ionic liquids (molten salts)[10]. Conductivity measurements at frequencies above 10^6 Hz, in which the behaviour of an ionic conductor may deviate from its low frequency limiting conductivity, σ_{dc}, are rare (CuBr[11], Na β-Al$_2$O$_3$[12, 13], Ag$_2$HgI$_4$O$_2$, RbAg$_4$I$_5$[8, 14], NASICON[15]) and few of these are using the complete representation σ'' vs. $f(\sigma')$ or ε'' vs. $f(\varepsilon')$. For some compounds, the link between the conductivity obtained from electrical measurements and lattice vibrations determined by infrared, Raman and neutron scattering spectroscopy has been established. A simple mathematical transformation allows the transition from the spectra to the conductivity.

---

* In neutron scattering, the elastic part reflects the static correlations (Bragg peaks reveal the unit cell structure and the diffuse pattern shows the non-periodic, local disorder). The inelastic scattering is due to the periodic motions of atoms or ions (phonons) and the quasi-elastic scattering is caused by any kind of non-periodic motion (or magnetic disorder). The transition between low frequency periodic motions (translational and rotational oscillation) and diffusive motion (non-periodic) is not well understood.

Fig. 25.2 gives a representation of the curves $\sigma(\omega)$ vs. $\log(\omega)$ and ε'' vs. ε' for the main models. First, non-interacting or 'free' charges: we observe a constant value for σ and a semicircle for ε'' vs. ε'. $\sigma(\omega)$ is constant up to about 10^{11} Hz; at higher frequencies, the model is not valid and far infrared (phonon) spectra are observed. The constant conductivity is related to the fact that a randomly hopping defect has no memory; hence

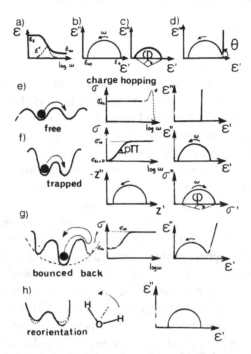

Fig. 25.2. Typical plots obtained for the main models and realistic examples. (a) Frequency dispersion of ε' and adsorption ε'' for a pure Debye model on (b) the corresponding Cole–Cole plot (ε'' vs. ε'). (c) Cole–Cole plot commonly encountered with a real dielectric solid showing an arc of circle centred out of the ε' abscissa and (d) with conduction loss. (e) Charge hopping model: diagrammatic representation of the potential well and corresponding plots of $\sigma(\omega)$ vs. $\log \omega$ and ε'' vs. ε' for non-interacting 'free' charges; (f) similar plots for system with trapped charges; examples of σ'' vs. σ' and Z'' vs. Z' plots are also given. Note the relation between the slope of $\sigma(\omega)$ and the depression of the semicircle in the σ'' vs. σ' plane; also note that ω increases in opposite directions in the ε'' vs. ε' and σ'' vs. σ' plots; (g) plots for a system with interacting charges (bounced back effect).

The molecular reorientation case and ferroelectricity additional effect are illustrated in (h). * Dashed line indicates the additional features due to phonons (far infrared region).

Frequency dependent conductivity

any hop should be correlated only to itself. Second, trapped charges associated with e.g. molecular reorientation or ring diffusion lead to formation of a dipole and the conductivity varies from $\sigma_{dc} = 0$ to a value $\sigma(\infty)$ at high frequencies. The third type of model takes into account the interactions between jumps with a 'bounced back effect' linked to the modification of the crystalline field seen by the ion after jump. This idea was discussed by Richards in 1977[16]; we used it in our HUP study in 1987[3] and Funke has applied it to other compounds. These individual models must be completed in the case of ferroelectric materials by taking into account collective atomic motions[17,18]. The transformation from vibration to translational motion has received attention by Volkov *et al.*[19]. It was shown that the bounced back effect occurs when the potential well is no longer parabolic; the scattering processes are separated into elastic and inelastic components, with a wide well-shaped response, $\sigma(\omega)$, which can be represented as the difference of two *Drude* forms. The superionic conductors have large, anharmonic, multiminimum wells and their non-zero conductivity is related to the highest barrier transition. Such *multiminimum well* structures lead to ordering at low temperature, which is effectively observed in $RbAg_4I_5$ 5 K $\varepsilon''(\omega)$ spectra[19] or in $CsHSO_4$ 77 K $P(\omega)$ spectra[20].

All the above descriptions use the *Debye model*, characterized by an arc of a circle in the plot ε'' vs. ε', and a unique relaxation time. In most cases (polymers, glasses, liquids[7,21]), however, the spectrum does not correspond to an arc of a circle and is frequently interpreted in terms of a relaxation time distribution. The latter broadens with increasing temperature. Such distributions can be either intrinsic (disordered compounds) or due to lack of accuracy in the measurements: fixed frequency measurements with too-widely spaced intervals, or insensitive apparatus. As we shall see later, protonic conductors give rise to better defined but more complex spectra because of the existence of various protonic and polyatomic species corresponding to fixed or mobile charges; strong dipoles lead frequently to ferroelectric phenomena.

25.2.2 Experimental methods

Under the action of an alternating electric field, the electrical response of a system having dipolar interaction may be characterized by the complex permittivity $\varepsilon^* = \varepsilon'(\omega) - i\varepsilon''(\omega)$ as discussed above. Methods to measure the frequency-dependent permittivity use coaxial lines. The cell of our measuring apparatus is a circular coaxial line whose inner conductor is interrupted by the sample[3] and is loaded with a well-characterized impedance Z_0 or a short-circuit. The sample and the inner conductor are identical (the same diameter $2a$), which simplifies the calculation. The impedance, Z_s, of the sample is connected in series with the load Z_0 and at the interface between sample and the load we have $Z' = Z_0 + Z_s$[3]. In this axial geometry, the size of the sample is lower than half the wavelength; only radial propagation Hϕ of the magnetic field and the axial component E_z of the electric field exist. For this type of propagation, we define the impedance $Z_s(\gamma_r)$ of the sample at $r = a$, i.e. $Z_s(\gamma_a)$ as

$$Z_s(\gamma a) = -\frac{i\omega_0\mu_0}{2\pi\gamma a} d \frac{J_0(\gamma a)}{J_1(\gamma a)}$$

where $\gamma = k(\varepsilon' - i\varepsilon'')^{1/2}$ is the radial propagation constant, d is the sample thickness, $\mu_0 = 4\pi . 10^{-7}$ Hm^{-1}, $k = \omega/c$ ($c = 3 \times 10^8$ m s^{-1}, J_0 and J_1 are respectively zero-order and first-order Bessel functions. Let P be the reference plane at the coaxial line/sample interface and Z' the impedance measured in this plane; $Z_s(\gamma a)$ can be calculated from a knowledge of Z', i.e.

$$Z' = \frac{Z_s + iZ_0 \text{tg}(\beta d)}{Z_0 + iZ_s \text{tg}(\beta d)}$$

for a short circuit or

$$Z' = \frac{(Z_s + Z_0) + iZ_0 \text{tg}(\beta d)}{Z_0 + i(Z_s + Z_0)\text{tg}(\beta d)}$$

for a sample in series with Z_0, where $\beta = 2\pi$. For frequencies below 1 GHz, the quasi-static approximation can be used: $Z(\gamma a) = 1/\sigma^*$. $d/\pi a^2$ which is similar to the usual plate capacitor formula $I = \sigma^* s/dV$.

25.2.3 Dielectric relaxation and the Debye model

The *Debye model* is usually applied to the gas phase at low pressure without dipole–dipole interactions[21]. In the case of condensed matter, both short range specific and long range electrical dipole–dipole interactions can occur and deviations from the Debye model may be observed.

Frequency dependent conductivity

Nevertheless, *dielectric relaxation* is usually described in Debye formalism terms.

The frequency dependent complex permittivity is expressed by

$$\frac{\varepsilon^*(\omega) - \varepsilon_\infty}{\varepsilon_s - \varepsilon_\infty} = \frac{1}{1 + i\omega\tau}$$

where ε_∞ and ε_s are limiting values of $\varepsilon^*(\omega)$ when $\omega \to \infty$ (optic domain) and $\omega \to 0$ (static limit) respectively, i.e. the dielectric constant ε deduced from vibrational spectrometry and low frequency conductivity (Fig. 25.2). τ is known as the Debye relaxation time and $fc = 1/2\pi\tau$ is the loss peak frequency. The relaxation time, τ, is a measure of the nominal time scale on which molecular reorientation or ion jumps can occur. If the dipoles interact strongly with each other the equation becomes slightly different

$$\frac{\varepsilon^*(\omega) - \varepsilon_\infty}{\varepsilon_s - \varepsilon_\infty} = \frac{1}{1 + (i\omega\tau)^{1-\alpha}}$$

where α is an empirical parameter ($0 \leqslant \alpha \leqslant 1$) proportional to the degree of deviation from the Debye model[7,20,21]. Fig. 25.2 shows *Cole–Cole plots* (ε'' versus ε') for an ideal Debye solid, a real case with $\alpha \neq 0$ and a case with a high conduction loss (i.e. with both fixed and free charges). In the last case the free charges give rise to a divergence in ε''.

As discussed above, all the representations of Fig. 25.2 (ε, σ or Z) give rise to a circular arc whose centre is usually displaced below the real axis giving rise to ϕ angle $= \alpha\pi$ which slightly differs from π. This deviation is also directly related to the slope of $\sigma(\omega)$ ($\sim \pi/2$), $\sigma(\omega) \propto \omega^{2-\alpha}$ if $\omega Z < 1$ and $\sigma(\omega) \propto \omega^\alpha$ if $\omega Z > 1$.

25.3   Relaxation assignment in protonic conductors

Fig. 25.3 shows typical Cole–Cole plots (ε'' vs. ε') of various protonic materials[3,5,6,22,23]. A low frequency divergence in ε'' due to high conductivity is observed. Various relaxations have been assigned to those of different protonic species. Crystalline samples usually show well-shaped semi-circles while only an envelope is observed for 'amorphous' materials such as $V_2O_5 . nH_2O$. This represents static disorder of protonic entities (or in other words, a distribution of relaxation times). However, it is possible to distinguish various structured off-centre relaxations after deconvolution[24].

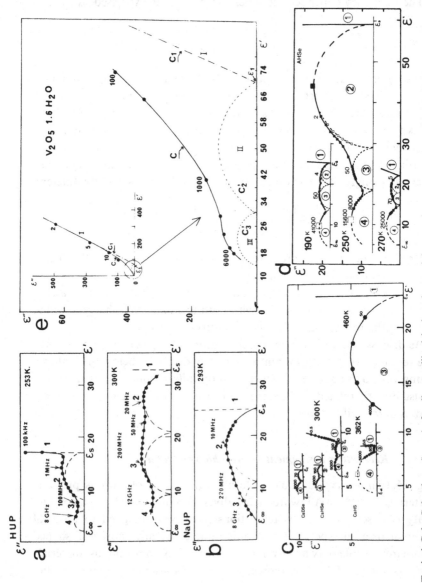

Fig. 25.3. Typical Cole–Cole plots ε'' vs. ε' of well-crystallized hydrated materials at various temperatures: (a) ($H_3OUO_2PO_4 \cdot 3H_2O$: HUP), (b) (NaUO$_2$PO$_4 \cdot 3H_2O$: NaUP[3]), (c) CsHSO$_4$ (CsHS), CsHSeO$_4$ (CsHSe) and CsDSeO$_4$ (CsDSe)[6], (d) NH$_4$HSeO$_4$ and of ill-crystallized material, (e) ($V_2O_5 1.6H_2O$ 'gel'[5]). Frequencies are given in MHz. Dashed lines correspond to deconvolution traces (with permission).

Frequency dependent conductivity

The first problem is to exclude 'geometrical' relaxation. When an electric field is applied, charges are localized on the sample under the influence of both the field and the diffusion gradient. On reversal of the field, charges find a new equilibrium so that the macroscopic dipole, the dimensions of which are those of the sample, and the new polarization are opposed to the previous one. The resulting Debye-like relaxation frequency depends on the sample thickness.

An assignment of intrinsic relaxation phenomena can be proposed by comparing materials having similar structure and conductivity but different protonic species. A comparison of $H_3OUO_2PO_4 . 3H_2O$ (*HUP*) and $NaUO_2PO_4 . 3H_2O$ (*NaUP*) Cole–Cole plots, for instance, helped to assign the H_3O^+ reorientation by defects[3]. Other relaxations are due to H_2O and ion jumps. In $MHXO_4$ ($MDXO_4$) compounds, the substitution of selenium by sulphur decreases strongly the HXO_4^- reorientation while substitution of hydrogen by deuterium reduces the proton jump relaxation to half its initial value due to the mass effect (Fig. 25.3)[5,6]. When cesium ions are substituted by ammonium ions, a new very slow relaxation in the 3–10 MHz region[6] is introduced. A regular NH_4^+ tetrahedron has no permanent dipole but a distorted NH_4^+ ion has a dipole and the dipole intensity is strongly correlated to ferroelectric behaviour.

Another criterion is to compare the relaxation frequency range with the characteristic relaxation frequency. In a previous review paper, Colomban & Novak[20] gave characteristic relaxation frequencies on the basis of conductivity, NMR and neutron scattering literature. In Fig. 25.4, we present a diagrammatic representation of $\log_{10} \tau(s)$ as a function of inverse temperature for a number of well-crystallized compounds. Typical regions for polyatomic reorientation, H_2O reorientation, and ion jumps are shown. Reorientation of dipolar polyatomic ions such as H_3O^+ or HSO_4^- is observed in the 10^{-11}–10^{-12} s region. Fig. 25.4*b* shows the correlation plots obtained with V_2O_5 gels containing only protonic species (H_3O^+, H_2O, OH) or after exchange of the H_3O^+ ions by Li^+, Na^+, K^+ or Ba^{2+}. We observe that the general picture is not modified: at the top slow M^+ jump relaxations are seen whereas rapid H_3O^+ reorientation takes place near the bottom. However, the water within the cation hydration shell has specific behaviour and the region for H_2O reorientation is split into areas X and Y.

Fig. 25.5 shows that activation energies are characteristic of the type of motion: dipolar polyatomic ion reorientation exhibits the lowest activation energies (~ 0.1 eV) according to NMR or neutron scattering measurements[20]. Reorientation of water molecules in the solid state (ionic conductor) is currently found near 10^{-10} s, close to 10^{-11} s for free water and to 10^{-9} s for bound water. Activation energies are slightly higher (~ 0.3 eV). Finally, a charge jump (proton or ion jump) occurs at about 10^{-8} s. The corresponding activation energy varies from 0.1–0.2 eV for proton jump to 0.3 eV for H_3O^+ jumps. Comparison of *activation energy* versus relaxation time for various protonic materials (Fig. 25.5a) with that of one type of material (V_2O_5 gels) in which the charged cations

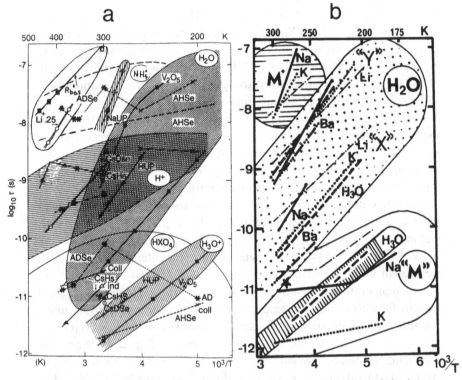

Fig. 25.4. Correlation plot of relaxation time $\tau = 1/2\pi f c$, as a function of the inverse temperature. (a) For crystalline material from the bottom (rapid motions) to the top (slow motions) we have delimited the relaxation time area of HXO_4 and H_3O^+ reorientation, H^+ (M^+) jump relaxation and NH_4^+ reorientations. The H_2O relaxation area crosses the figure due to the wide τ range. $Rb_{0.5} = Rb_{0.5}Cs_{0.5}HSO_4$. For other symbols, see caption of Fig. 25.3. Collective (coll) and individual (ind) relaxation times are given for ferroelectric materials. The stars represent the τ value of liquid H_2O ($\sim 10^{-11}$ s). Discontinuities and slope changes are related to phase transitions. (b) Plot for $V_2O_5 \cdot 1.6H_2O$ 'gels' containing various cations. Note the splitting of the H_2O reorientation area into two domains (X and Y), according to the presence of two kinds of water ('free' water and water of the cation hydration shell)[24].

400

Frequency dependent conductivity

have been substituted indicates that local field effects (cation size, charge) are more important and that motions specifically related to superionic conduction are slow and exhibit the lowest activation energy.

25.3.1 Conductivity

Fig. 25.6 shows the curves of $\log_{10}\sigma$ as a function of frequency (ω) for some typical materials. Contrary to the plots of non-protonic materials, steps are observed between 'σ_{dc}' and 'σ_∞', according to the presence of well-defined relaxations. This shows that the monotonic 'universal dielectric response'[7] is an oversimplification. In the case of $CsHSO_4$ at room

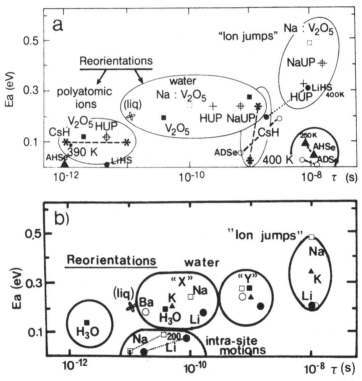

Fig. 25.5. Activation energy (E_a) vs. relaxation time τ for various materials at 300 K (for symbols see Figs 25.3 and 25.4). Dashed lines join values corresponding to higher temperature. Polyatomic ion reorientation, water reorientation proton and ion jump domains are shown. The value for liquid H_2O (white cross) is shown. The same picture is observed for $V_2O_5 \cdot 1.6H_2O$ gels containing various types of cation (H_3O^+, Li^+, Na^+, K^+, Ba^{2+}). In this case intra-site motions give rise to low activated relaxations[24].

401

temperature where the dynamic disorder is rather low, two relaxations assigned to HSO_4^- reorientation and proton jump are clearly observed. In the high temperature superionic phase, on the other hand, the frequency dependence becomes almost non-existent. Similar behaviour is observed for HUP between 250 and 300 K. Introduction of new, dipole-forming species allows disorder and low frequency conductivity to increase: for example, NH_4HSeO_4. Note that relaxation steps are not observed on poorly crystallized samples (e.g. V_2O_5 gel).

The difference between $\sigma_{(\infty)}$ and σ_{dc} is very small for the true superionic conductor phase, according to the description of conduction in terms of a quasi-liquid (free charge model). We have thus a transition from a 'bound charge' configuration to a 'free charge' state at the superionic phase transition.

25.4 *Phase transitions, ferroelectricity and collective motions*

Microwave relaxation can be very useful for the study of phase transitions in protonic materials and their existence (and origin) can be evidenced by considering the following parameters.

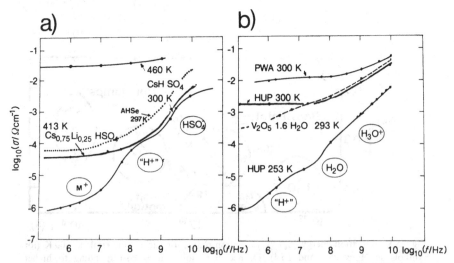

Fig. 25.6. Dynamic conductivity ($\log_{10}\sigma(\omega)$ vs. frequency) (a) for anhydrous $CsHSO_4$ (at 300 K and 460 K), NH_4HSeO_4 (AHSe) at 297 K and $Cs_{0.75}Li_{0.25}HSO_4$ (413 K) and (b) for hydrated materials: HUP (253 and 300 K), $V_2O_5 \cdot 1.6H_2O$ gel (293 K) and $H_3PW_{12}O_{40} \cdot 29H_2O$ (PWA) at 300 K. The main relaxations are indicated.

Frequency dependent conductivity

(i) ε_s, which represents the contribution of all dipolar species and is very sensitive to structural modifications.

(ii) For each relaxation, $\Delta\varepsilon = \varepsilon_1 - \varepsilon_2$ ($\varepsilon_s - \varepsilon_\infty$, if there is only one relaxation). $\Delta\varepsilon$ is proportional to the number of dipoles and to the dipole moment of each relaxation ($\Delta\varepsilon = N\mu^2$)[21]. (The usual dipole moment of H_2O is 1.84 D and that of the OH^- group is 1.51 D.)

(iii) The relaxation time τ, which is shown in Fig. 25.7, is also very sensitive to phase transition and thus information on dynamic parameters can be obtained.

(iv) Unlike in the fixed frequency technique, no artefacts associated with e.g. relaxation peak shifts are possible if the conductivity is measured over a large frequency range.

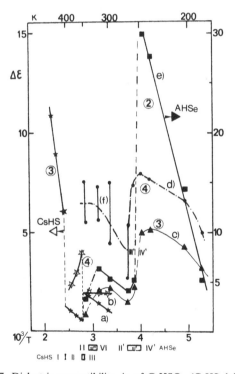

Fig. 25.7. Dielectric susceptibility $\Delta\varepsilon$ of $CsHSO_4$ (CsHS, left) and of NH_4HSeO_4 (AHSe, right) vs. temperature for NH_4^+ reorientation (2-type), charge jump relaxation (3-type) and HXO_4^- reorientation (4-type), respectively. Phase stability domains are schematized below[20, 28]. The data labelled (f) are very dependent on the sample history due to incomplete phase transitions (phase-mixing), with permission[6].

(v) Deviations from Debye law (given by the parameters α or ϕ) can also be a source of information, but the reasons for such deviations are presently not well understood.

Fig. 25.7 shows two interesting examples, $CsHSO_4$[6] and NH_4HSeO_4[25]. The former undergoes two phase transitions at 318 (I $\rightarrow$ II) and 417 K (II $\rightarrow$ III) (see Chapter 11): the first corresponds to a conversion of infinite $(HSO_4^-)_n$ chains into cyclic dimers with considerable orientational disorder[26]. The relaxation time of the HSO_4^- reorientation becomes more and more rapid (Fig. 25.4) whereas the proton jump frequency becomes slower. The second transition gives rise to a plastic superionic phase III[26,27] with a rapid reorientation of HSO_4^- anions (the relaxation times reach the far infrared and low frequency Raman domain; a strong broadening of the Rayleigh line is observed, see Chapter 11), a translational local disorder of Cs^+ cations, and protons diffusing over long distances. For the I $\rightarrow$ II transition, an increase in $\Delta\varepsilon_4$ (reorientational relaxation of HSO_4^-) and a decrease in $\Delta\varepsilon_3$ (charge jump relaxation) have been observed. The increase in $\Delta\varepsilon_4$ is consistent with the appearance of orientational disorder of the anion (the dipole increases) while the decrease in $\Delta\varepsilon_3$ shows that the proton dipole and/or its number N diminishes, associated with lower conductivity and dimer formation. $\Delta\varepsilon_4$ behaviour is characteristic of a ferroelectric/paraelectric transition: the $1/\Delta\varepsilon = f(T)$ plot follows a Curie–Weiss law.

25.4.1 Ferroelectricity

A plot of $1/\Delta\varepsilon$ vs. temperature T allows the extrapolated *Curie–Weiss temperature* and the corresponding C constant in the expression $(\Delta\varepsilon)^{-1} = (T - T_0)/C$ (Table 25.1) to be measured. The Curie constant C is defined by the relation $C = N\mu^2/k$. The constant T_0 is related to the 'exchange integral' which takes into account the dipole–dipole interaction. In the case of a first order transition, T_0 must be lower the T_c.

Previous low frequency studies have failed to detect the ferro-paraelectric nature of the I $\rightarrow$ II transition and $CsHSO_4$ was considered as the only non-ferroelectric material of the $MHXO_4$ family of compounds with an infinite chain structure. This is probably due to the low dipole moment value and to the fact that $\Delta\varepsilon_3$ and $\Delta\varepsilon_4$ susceptibilities vary in opposite directions.

At the II $\rightarrow$ III superionic transition, $\Delta\varepsilon_3$ increases significantly, indicating an increase of the proton jump distance and/or of the number of mobile charges at a given frequency. The number of mobile species in superionic conductors remains an important but unsolved question. Estimations between 0.5 and 30% have been given on the basis of various models for non-protonic superconductors. It should be pointed out that below the transition point, the proton jump is probably intradimer while above 417 K an interdimer jump is needed for conduction to occur. This increase continues with increasing temperature in phase III, indicating that the conductivity parameters are also modified.

Ammonium hydrogen selenate undergoes several phase transitions[28, 29], in particular in the region between 250 and 300 K. There is a transition from the ferroelectric (IV') to paraelectric (II) phase via an *incommensurate (III') phase* (252–262 K). $\Delta\varepsilon$ plots show very large dielectric susceptibility changes. On cooling, a 10 K gap is observed between the increase of $\Delta\varepsilon_2$ (NH_4^+ reorientational relaxation) and $\Delta\varepsilon_4$ (HSO_4^- reorientational relaxation). This gap corresponds to the incommensurate phases which can be thus related to the shift between the appearance of orientational order of ammonium (ferroelectric ordered sublattice) and of hydrogen selenate ions (crystal sublattice, associated with the large size of the anions).

Analysis of $\Delta\varepsilon$ plots recorded on $H_3OUO_2PO_4 \cdot 3H_2O$ also gives a better description of the phase transition mechanism[3]. In particular, the dimensionality effect concerning the ferroelectric sublattice can be deduced from $\Delta\varepsilon$ analysis. The extrapolated Curie–Weiss temperatures are about 210 K (Table 25.1) for the H_2O sublattice (ferroelectric order, in the layer) and 170 K for H_3O^+ orientation (dipole perpendicular to the layer), the C constant being approximately the same for the two dipole types (Table 25.1). It is thus possible to estimate the value of the H_3O^+ dipole to be about 3.2 D in HUP.

25.4.2 Individual and collective relaxation times

The *relaxation time* can be deduced from the Debye-like circular arc. A plot of τ values versus the inverse of temperature ($10^3/T$) (Fig. 25.4) allows a measure of activation energy (Fig. 25.5). The separation of domains already discussed above is clearly visible, from fast reorientational motions of dipolar polyatomic ions such as HXO_4^- and H_3O^+ to slow reorientation of NH_4^+ ions. Motions of water molecules cover a broad region: they are slow in gel, medium in superionic materials (e.g. HUP) and fast in liquid water.

Table 25.1. *Curie–Weiss temperature (T_0) and constant C of some protonic conductors*

Compounds	Relaxation	T_0(K)	C(K)	Ref.
HUP	H_3O^+	170	1000	3
	H_2O	210	1000	3
$CsHSO_4$	HSO_4^-	280	300	6
$Cs_{0.9}Li_{0.1}HSO_4$	HSO_4^-	290	167	22
NH_4HSeO_4	$HSeO_4$	235	370	25
	'H$^+$'	240	100	25
	ND_4^+	250	350	25
	NH_4^+	160	500	25

If a material is ferroelectric, local field effects must be taken into account in order to determine the intrinsic individual relaxation time. The collective relaxation times are characterized by a critical slowing down, as observed for the reorientations of H_2O and H_3O^+ in HUP[3] or of $HseO_4^-$ and NH_4^+ in NH_4HSeO_4[6,27,29]. The Ising model and molecular field approximation lead to the definition of a new relaxation time[17,18]

$$\tau' = \frac{C}{T} \frac{\tau}{\Delta\varepsilon}$$

which characterizes the individual reorientation of each molecule, whereas the relaxation time τ is related to a 'collective' mechanism of molecular reorientation (long range dipole–dipole interactions (Fig. 25.2)). The individual relaxation time is often more rapid and no drastic change is usually observed at the ferro/paraelectric transition, contrary to the characteristic slowing down of collective relaxation times.

25.5   References

1. J. Ross MacDonald (ed), *Impedance Spectroscopy* (John Wiley and Sons, New York (1988)).
2. M. G. S. Thomas, P. G. Bruce and J. B. Goodenough, *J. Electrochem. Soc.* **132** (1985) 1521–8.

Frequency dependent conductivity

3. J. C. Badot, A. Fourier-Lamer, N. Baffier and Ph. Colomban, *J. Physique (France)* **48** (1987) 1325–36.
4. D. V. Lang and R. A. Logan, *Phys. Rev. Letts* **39** (1977) 635–8; C. Hurtes, M. Boulou, A. Mitonneau and D. Bois, *Appl. Phys. Letts* **32** (1978) 821–4; D. Bois and A. Chantre, *Rev. Phys. Appl.* **15** (1980) 631–46.
5. J. C. Badot, N. Baffier, A. Fourier-Lamer and Ph. Colomban, *Solid State Ionics* **28–30** (1988) 1617–22.
6. J. C. Badot and Ph. Colomban, *Solid State Ionics* **35** (1989) 143–9.
7. A. K. Jonscher, *Dielectric Relaxation in Solids* (Chelsea Dielectric Press, London (1983)).
8. K. Funke, *Mat. Res. Soc. Symp. Proc.* **165** (1989) 43–56.
9. K. Funke in *Superionic Solids and Solid Electrolytes: Recent Trends*, A. Laskar and S. Chandra (eds) (Academic Press, New York (1989)) 569–629.
10. K. Funke, J. Hermeling and J. Kümpers, *Z. Naturforsch.* **43a** (1988) 1094–102.
11. C. Clemens and K. Funke, *Ber. Bunsenges Phys. Chem.* **79** (1975) 1119–24.
12. D. P. Almond, A. R. West and R. J. Grant, *Solid State Commun.* **44** (1982) 1277–80.
13. U. Strom and K. L. Ngai, *Solid State Ionics* **5** (1981) 167–70; **9/10** (1983) 283–6; K. L. Ngai and U. Strom, *Phys. Rev.* **B38** (1988) 10350–6.
14. K. Funke and H. J. Schneider, *Solid State Ionics* **13** (1984) 335–44; T. Wong, M. Brodwin, D. F. Shriver and J. I. McOmber, *Solid State Ionics* **3/4** (1981) 53–6.
15. J. R. Dygas and M. E. Brodwin, *Solid State Ionics* **18–19** (1986) 981–6; R. Vaitkus, A. Orliukas, N. Bukun and E. Ukshe, *Solid State Ionics* **36** (1989) 231–3.
16. P. M. Richards, *Phys. Rev.* **B16** (1973) 1393–401; P. M. Richards, in *Fast Ion Transport in Solids*, P. Vashishta, J. N. Mundy and G. K. Shenoy (eds) (North Holland (1979)) 349–52; T. Wong and M. Brodwin, *Solid State Commun.* **36** (1980) 503–9.
17. H. Kolodziej, *Dielectric and Related Molecular Processes*, Vol. 2, M. Davies (ed) (London, Chemical Society (1975)) 249.
18. Y. Makita and I. Seo, *J. Chem. Phys.* **51** (1969) 3058–67.
19. A. A. Volkov, G. V. Kozlov and J. Petzelt, *Ferroelectrics* **95** (1989). 23–7; A. A. Volkov, G. V. Kozlov, J. Petzelt and A. S. Rakitin, *Ferroelectrics* **81** (1988) 211–14.
20. Ph. Colomban and A. Novak, *J. Mol. Struct.* **177** (1988) 277–308.
21. K. S. Cole and R. H. Cole, *J. Chem. Phys.* **2** (1941) 341–69; I. M. Hodge and C. A. Angell, *J. Chem. Phys.* **67** (1977) 1647–58; P. Debye, *Polar Molecules* (Dover, New York (1945)).
22. J. C. Badot, T. Mhiri and Ph. Colomban, *Solid State Ionics* **46** (1991) 151–7.
23. U. Mioc, J. C. Badot and Ph. Colomban, *Solid State Ionics* (submitted).
24. J. C. Badot, A. Fourier-Lamer and N. Baffier, *J. Physique (Paris)* **46** (1985) 1325–33.
25. Ph. Colomban and J. C. Badot, *J. Phys. Condensed Matter* (1992) (in press).
26. Ph. Colomban, J. C. Lassègues, A. Novak, M. Pham-Thi and C. Poinsignon,

in *Dynamics of Molecular Crystals*, J. Lascombe (ed) (Elsevier, Amsterdam (1987)) 269–74.

27. A. I. Baranov, L. A. Shuvalov and N. M. Shchagina, *J.E.T.P. Letts.* **36** (1982) 459–62.

28. I. P. Aleksandrova, Ph. Colomban, F. Dénoyer, N. Le Calvé, A. Novak, B. Pasquier and A. Rozycki, *Phys. Stat. Sol.* (a) **114** (1989) 531–43.

29. R. Mizeris, J. Grigas, V. Samulionis, V. Skritski, A. I. Baranov and L. A. Shuvalov, *Phys. Stat. Sol.* (a) **110** (1988) 429–36.

26 Measuring the true proton conductivity

KLAUS-DIETER KREUER

A knowledge of the true, bulk proton conductivity has been both desirable and difficult to achieve in the characterization of most known solid proton conductors. It was almost 100 years after the first electrical measurements on *ice*[1] that von Hippel demonstrated that single crystals of pure ice must have a proton conductivity far below the values reported previously[2].

Quantitative data on bulk proton transport are required especially for the understanding of proton transport mechanisms (see Chapters 29 & 31) including the implications for the use of solid proton conductors in operational electrochemical cells (see Chapters 32 & 39).

In this Chapter, proton conductivity refers to the displacement of protonic species (e.g. H^+, H_3O^+, OH^-) in small electric fields across a sample close to thermodynamic equilibrium. This conductance relates directly to the self diffusion coefficient of the corresponding species, which may be significantly smaller than the chemical diffusion coefficient, in particular in the presence of another highly mobile species (e.g. conduction electrons in hydrogen bronzes; see Chapter 7).

There is no standard procedure for the measurement of proton conductivity and experimental techniques as well as structural and chemical considerations have to be adapted to the material under investigation. There are some peculiar features common to most solid proton conductors which make it difficult to identify the proton as the charge carrier, to determine its accurate conductance (self diffusion) and to separate bulk proton transport from artefacts (such as transport along surfaces, grain boundaries, domain walls, dislocations, second phases). How to cope with these will be discussed in the following section.

26.1   The sample

A prior condition is to have well-defined samples. Of great advantage is the use of rather perfect single crystals, but these are available for few solid proton conductors only (e.g. for heteropolyacidhydrates, $LiN_2H_5SO_4$, β-aluminas, HUP and $CsHXO_4$ (X = S, Se)). Most compounds are available as fine powders only and many are inherently structurally distorted (e.g. hydrogen bronzes, layer hydrates). Therefore, conductivity contributions from grain boundaries, dislocations etc., have to be considered. Those, as well as bulk proton transport itself, may be affected by the surrounding atmosphere which sometimes is prevented from chemical equilibration with the sample. Proton conductivity of $SrCeO_3$-related materials, for instance, is a result of hydrogen incorporation from the gas phase (see Chapter 8). Hydrates especially tend to form tightly adsorbed aqueous layers which may contribute to the observed conductivity and prevent equilibration of the solid with the water partial pressure of the atmosphere. Even phase transitions and related conductivity changes may be affected drastically by the water partial pressure, as has been observed for $CsHXO_4$ (X = S, Se) (see Chapter 11).

The atmosphere should therefore either be controlled or the sample should be encapsulated in a small, gas-tight volume thus allowing at least isochemical measurements for samples which do not equilibrate with the gas phase.

26.2   *H$^+$-conductivity measurement by a.c.-impedance spectroscopy*

Formerly, conductivity was frequently measured by d.c. techniques using either reversible electrodes (e.g. H_xPd, H_xMoO_3, H_xTi) and low voltages or just by electrolysing the sample at voltages above the decomposition potential. The former often suffers from electrode polarization effects, whereas the latter yields reliable results only if proton conductivity is the rate determining step for the overall current.

A more advantageous method is *a.c.-impedance spectroscopy*, which has become a standard method for the measurement of ionic conductivities in general. The basic principles have been described many times and the interested reader may refer to the excellent review by Gabrielli[3]. A small applied potential difference allows measurements close to thermodynamic equilibrium. The accessibility of an extended frequency range (typically $1-10^7 s^{-1}$) allows the separation of impedance contributions from the sample itself and from the electrode/electrolyte interface using equivalent circuits to assist the interpretation of the data obtained. Unfortunately interpretation is unambiguous only for simple circuits[4] and the different

Measuring the true proton conductivity

contributions to the sample impedance may often hardly be separated. Fig. 26.1 shows a simple scenario for a powder sample including bulk conductance (1), grain boundary conductance both in parallel (2) and in series with the grains (3) (typical for a proton conducting hydrate). As can be seen from the corresponding simulated impedance spectra the bulk resistance can be extracted only for a high grain boundary resistance compared to the bulk resistance ($R_3 > R_1$, spectrum (b)). Otherwise the bulk conductance may be short circuited by the grain boundary conductance ($R_3 \leqslant R_1$) leading to a concealing of the first semicircle in the impedance spectrum (spectrum (a)). For heterogeneous particle hydrates, $1/R_3$ is actually considered to be the 'true' proton conductivity (see Chapters 16–20).

A.c.-impedance measurements are usually carried out with blocking electrodes (e.g. gold for protons); the comparison of the results with those of d.c. experiments (e.g. Hebb–Wagner type measurements using one

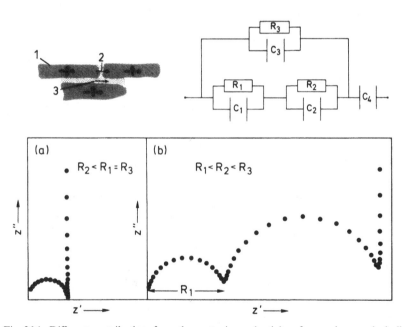

Fig. 26.1. Different contributions from the protonic conductivity of a powder sample: bulk conductivity (1), grain boundary conductivity in series (2) and parallel (3) to the grains. A simple equivalent circuit including simulated impedance spectra for two different cases is shown. The bulk resistance (R_1) can be extracted only from spectrum (b).

blocking and one reversible electrode with a defined chemical potential for hydrogen[5]) allows the separation of the conductance of electrons (holes) and of ions. But whether or not the ionic species are protonic must be decided from other experiments or considerations.

In some cases structural and chemical considerations may plausibly exclude diffusion of any species except protonic ones (e.g. for HUP, $Sb_2O_3 \cdot nH_2O$, H_3O^+ β-alumina). Combined a.c. and d.c. measurements involving the transfer of hydrogen (see Chapters 27 & 36) may also give valuable indication about the nature of the charge carrier.

26.3 $^1H^+$-*diffusion coefficient measurement by PFG-NMR*

Any charge transport by protons must directly correspond to an adequate transport of protonic mass. A straightforward way to determine the true proton conductivity, therefore, is to record directly the proton self-diffusion coefficient. The traditional way to do this is through tracer techniques. They may be very accurate but require difficult experimental work and measurement periods of the order of days or weeks for a single component and temperature. A further disadvantage is the inherent system perturbation by isotope substitution, which is particularly critical for the substitutions $^1H/^2H$ and $^1H/^3H$.

A very elegant alternative, which provides proton self-diffusion coefficients with good precision, within a shorter time, without isotopic labelling is the *pulsed magnetic field gradient spin echo NMR technique* (PGSE- or PFG-NMR). Measurement of self-diffusion coefficients by NMR have been reported in numerous studies since the discovery of spin echoes by Hahn[6]. In his pioneering study, several effects on spin echoes were described, one of which was the diffusional effect on echo amplitudes in an inhomogeneous magnetic field. This stimulated the basic idea of the PFG-NMR technique, which was put forward in a paper by McCall *et al.*[7]; first experiments including a detailed analysis of the results were later presented by E. O. Stejkal & J. E. Tanner[8]. The application of this method, however, was restricted to the measurement of relatively high diffusion coefficients for a long time (e.g. diffusion in liquids, gels, liquid crystals, solvent/polymer systems, surfactant/water systems).

It was only recently that improvement in experimental techniques allowed extension of the method to the study of solid electrolytes and in particular to solid proton conductors. Despite its great potential it is, however, still rarely used. The method should therefore be described in some detail including its specific advantages and limitations.

It involves NMR in a linear magnetic field gradient superposed on the homogeneous magnetic field, thus defining space in one direction. This allows observation of the nucleus distribution along that vector (imaging) or the general probability distribution for lateral displacement of the nuclei during a given time in the direction of the gradient. The latter is directly related to the self-diffusion coefficient.

As already mentioned above, PFG-NMR is based on simple spin echo (SE) experiments. The most common SE-sequence used today utilizes a 90° pulse followed by a 180° pulse after time τ. The dephasing of the spins occurring during that time in the plane perpendicular to the magnetic field is refocused after a time 2τ leading to a so-called 'spin echo' (Fig. 26.2). The pulse sequence does not refocus the normal random spin dispersion which is due to transversal spin relaxation processes (T_2). The possible refocusing is complete only if the precession frequency of each nucleus is constant during the experiment. In the presence of a magnetic field gradient, however, this is only the case when the nuclei experience the same magnetic field for the time of the experiment (2τ), i.e. when there is no diffusion within the magnetic field gradient. Otherwise, the echo is attenuated to some extent depending on the field gradient (G) and the diffusion coefficient (D). For an incoherent random motion, such as self-diffusion, the echo amplitude is then given by[8]:

$$M_{(2\tau)} = M_0 \exp\{-2\tau T_2^{-1} - \gamma^2 D[G_0^2 2/3\tau^3 + G^2\delta^2(\Delta - \delta/3)$$
$$- GG_0\delta(t_1^2 + t_2^2 + \delta(t_1 + t_2) + 2/3\delta^2 - 2\tau^2)]\} \quad (26.1)$$

where γ is the gyromagnetic ratio and G_0, G, τ, δ, Δ, t_1 and t_2 are defined

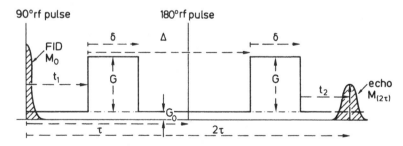

Fig. 26.2. The most common PFG-NMR pulse sequence (see text).

413

Proton dynamics and charge transport

in Fig. 26.2. The magnetic field gradients are actually pulsed in real experiments in order to provide gaps for the rf-pulses which are applied in the presence of the homogenous magnetic field B_0 only. This is necessary to ascertain an identical turn of all nuclear spins by rf-pulses of limited band width. The two gradient pulses have to be precisely balanced in order to prevent an extra echo attenuation and a displacement of the echo in the time domain. Such effects can be suppressed partially by the application of a small background field gradient G_0[9,10].

The proton is the nucleus with the highest gyromagnetic ratio γ and it has a dipolar moment only ($s = \frac{1}{2}$). The observed echo attenuations are therefore bigger than for other nuclei (Eqn 26.1) and the absence of quadrupolar interactions leads to commonly long T_2, thus allowing sufficiently long observation windows (2τ) of several milliseconds for most fast solid proton conductors (favourably it should be $\tau \leqslant T_2$). Recent progress in power electronics and numerical optimization of the field gradient coil design allows the creation of magnetic field gradients up to 5000 G cm^{-1}[11]. Assuming T_2 to be of the order of milliseconds, this enables one to measure proton self-diffusion coefficients as low as 10^{-9} cm^2 s^{-1} (corresponding to a proton conductivity of the order of 10^{-5} Ω^{-1} cm^{-1} for the proton as a majority charge carrier ($n \sim 10^{21}$ cm^{-3})).

It should be noted here that the application of PFG-NMR to other species different from the proton is also of particular interest. The migration of protons in condensed matter is not independent of the dynamics of its next environment. For example, it may also be associated with the diffusion of other species, although their determination appears to be less advantageous. Smaller gyromagnetic ratios lead to lower echo attenuations (Eqn 26.1) and, especially for nuclei with higher spin quantum numbers, quadrupolar interaction may drastically reduce T_2 and therefore the available observation window (2τ). Another limitation may be a long T_1 (spin-lattice relaxation time) which extends the recovery time between individual measurements. More experimental details, including a comprehensive list of references, may be found in reference 12. The use of the PFG-NMR technique for the measurement of proton transport provides several significant advantages. The method directly yields the self-diffusion coefficient without relying on model frameworks. Like any other NMR-method it is uniquely selective for the observation of one kind of nucleus (isotope) which is especially an advantage in the case of multi-component self-diffusion (e.g. e^- and H^+ in hydrogen bronzes, D^+

Measuring the true proton conductivity

in partially deuterated protonic samples, Cs^+ and H^+ in $CsHSO_4$). In contrast to other NMR techniques PFG-NMR is exclusively sensitive to diffusion. Depending on D and 2τ, the nuclei have to migrate of the order of 0.00001–0.1 cm within the magnetic field gradient; local modes (e.g. librations, reorientations, diffusion of other species) do not contribute to the echo attenuation. Different diffusion processes (e.g. bulk, grain

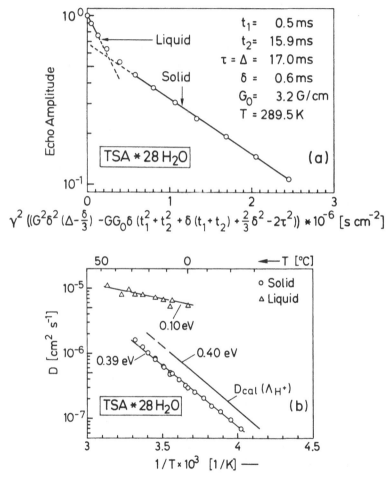

Fig. 26.3. H-diffusion coefficient measured on $TSA \cdot 28H_2O$ by the PFG-NMR technique[18]. (a) Echo attenuation as a function of applied magnetic field gradient showing separation of the contributions from the mother liquor and the solid. (b) The resultant diffusion coefficient as a function of the reciprocal temperature; the diffusion coefficient calculated from the proton conductivity measured by a.c.-impedance spectroscopy is given for comparison (see Chapters 29–31).

Proton dynamics and charge transport

boundary, surface diffusion), including the number of nuclei involved in each process, can be resolved as long as they do not mix during the time 2τ, which can be varied to some extent. Variation of 2τ and therewith of the diffusion length also allows one to investigate hindered diffusion (e.g. diffusion barriers in heterogeneous particle hydrates).

It is clear that PFG-NMR sees any protonic diffusion regardless of whether the proton is part of a neutral (e.g. H_2O) or charged species (e.g. H^+, H_3O^+, OH^-) whereas a.c.-impedance spectroscopy is sensitive only to displacement of ionic charge. The application of both techniques therefore frequently allows one to identify the proton (or protonic species) as the charge carrier, to quantitatively determine its diffusion coefficient (conductance) and sometimes even to deduce the underlying conduction mechanism (see Chapter 31).

As pointed out above, PFG-NMR has, as yet, been applied to very few solid protonic conductors[13-18]. As an example, let us consider the results obtained on the heteropolyacidhydrate $H_4SiW_{12}O_{40} \cdot 28H_2O$ (TSA $\cdot 28H_2O$)[18]. Fig. 26.3a shows the normalized echo amplitude versus the applied magnetic field gradient of TSA $\cdot 28H_2O$ in its respective mother liquor (this guarantees a defined state of hydration of the solid). The contributions from both are clearly separated and shown as a function of temperature (Fig. 26.3b). The diffusion coefficient calculated from the ionic conductivity of a single crystal is given for comparison. Practically identical activation enthalpies indicate the same elementary process for diffusion and conductivity; the discrepancy in the absolute numbers gives valuable information about the conduction mechanism (see Chapter 31).

26.4 References

1. M. E. Ayrton and M. M. Perry, *Proc. Phys. Soc. (Lond.)* **2** (1887) 178.
2. A. v. Hippel, *J. Chem. Phys.* **54** (1971) 134–44; 145–60.
3. C. Gabrielli, *Solartron Instruments*, Technical Report Number 004/83 (1984).
4. *Impedance Spectroscopy*, J. Ross Macdonald (ed) (J. Wiley, N.Y. (1987)) 96.
5. C. Wagner, *Z. Elektrochem.* **60** (1956) 4–7.
6. E. L. Hahn, *Phys. Rev.* **80** (1950) 580–94.
7. D. W. McCall, D. C. Douglass and E. W. Anderson, *Ber. Bunsenges. Phys. Chem.* **67** (1963) 336–40.
8. E. O. Stejskal and J. E. Tanner, *J. Chem. Phys.* **42** (1965) 288–92.
9. M. I. Hrovat and C. G. Wade, *J. Magn. Reson.* **44** (1981) 62–75.
10. M. I. Hrovat and C. G. Wade, *J. Magn. Reson.* **45** (1981) 67–80.
11. T. Dippel, K.-D. Kreuer, M. Hampele and A. Rabenau, *Proc. XXV Congress Ampère*, Stuttgart (1990).

Measuring the true proton conductivity

12. P. Stilbs, Progress in NMR Spectroscopy **19** (1987) 1–45.
13. R. E. Gordon, J. H. Strange and T. K. Halstead, *J. Solid State Commun.* **31** (1979) 995–7.
14. Y.-T. Tsai, W. P. Halperlin and D. H. Whitemore, *J. Solid State Chem.* **50** (1983) 263–72.
15. Y.-T. Tsai, S. Smoot and D. H. Whitemore, *Solid State Ionics* **9/10** (1983) 1033–40.
16. R. Blinc, J. Dolinsek, G. Lahajnar, I. Zupancic, L. A. Shuvalov, A. I. Baranov, *Phys. Stat. Sol.* (b) **123** (1984) K83–7.
17. R. C. T. Slade, J. Barker, H. A. Pressman and J. H. Strange, *Solid State Ionics* **28–30** (1988) 594–600.
18. K.-D. Kreuer, M. Hampele, K. Dolde and A. Rabenau, *Solid State Ionics* **28–30** (1988) 589–93.

27 D.c. techniques and a.c./d.c. combination techniques

ERIC SKOU, INGE G. KROGH ANDERSEN AND
ERIC KROGH ANDERSEN

27.1 Introduction

The a.c.-*impedance method* described in the previous chapter is often employed to characterize the electrical properties of solid proton conductors. The method gives a rapid answer to the question 'does the material under investigation conduct electricity' and the method is not critical with respect to size, shape and quality of the sample used.

If the a.c.-impedance method shows that the material is a conductor there still remain the following questions.

(1) Is the material an electronic or an ionic conductor or a mixture of both?

(2) Which kind of ion is responsible for the conduction in the case of several possible moving ions, e.g. is $CsHSO_4$ a cesium ion conductor, a protonic conductor, a sulphate/hydrogensulphate conductor or a mixture of some or all of these possibilities?

(3) Are the materials in the case of polycrystalline materials true *bulk conductors* or do they possess *surface conductivity*, where the conduction takes place in the boundary phase between the individual grains?

(4) If proton conduction is verified, is the conduction mechanism then proton hopping between partially occupied proton sites accompanied by a reorganization of the anion lattice or is the mechanism proton transport by a carrier (e.g. water, hydrazine or ammonia) which possibly counter-diffuses when the proton is discharged at the cathode (vehicle mechanism)?

Only the third question can be answered directly on the basis of simple a.c. measurements. When two semicircles are observed in the impedance plots (imaginary part vs. real part), they can often be assigned to bulk and grain boundary conduction paths. If only one semicircle is present additional experiments are necessary to determine whether the conducting phase is the grain boundary or the bulk. Single crystals can be used or several samples with very different particle sizes may be prepared[1, 2].

27.2   E.m.f. methods

For materials with mixed electronic and ionic conduction *e.m.f. measurements* on a concentration cell have often been used to separate the different contributions. In this method a concentration gradient is established across the sample, which is equipped with electrodes reversible to only one ion, e.g. platinum black electrodes and different hydrogen partial pressures. The potential difference is measured, preferably at several partial pressure differences. The *transference number* of the ion can then be deduced from the deviation from the *Nernst law* behaviour by use of the equation[3, 4]

$$E = t \frac{RT}{2F} \ln \left(\frac{p''}{p'} \right)$$

where t is the transference number and p', p'' the partial pressures.

The method has been used by several workers to demonstrate protonic conduction (Takahashi *et al.* (1976)[5], amine/sulphuric acid systems; Sheffield *et al.* (1979)[6], clays; Spaziante (1980)[4], different hydroxides and salts; Kreuer *et al.* (1981)[1], lithium hydrazinium salts, HUP and HUAs; Lundsgaard *et al.* (1982)[7], HUP; and Iwahara *et al.* (1986)[8], strontium cerates). The method can, however, be experimentally difficult because no leakage between the anode and cathode compartments can be tolerated. It also needs selective electrodes to avoid interference from impurities. In the above mentioned investigation by Lundsgaard *et al.*, platinum electrodes worked well, while palladium electrodes gave poor and irreproducible results. An additional difficulty is that most proton conductors are either permeable to hydrogen gas due to a high water content[9] or are difficult to compress to completely dense samples, because sintering cannot be used due to limited thermal stability. Some of these difficulties were overcome by Kreuer *et al.*[1], who in measurements on dense pellets of HUP and HUAs used hydrogen tungsten bronze electrodes with different hydrogen activities. The method did not, however, give information about the magnitude of the conductivity and will not be treated further here.

In the following, methods will be reviewed, where the conductivity is measured during an electrolysis (d.c.) experiment. Simultaneous use of a.c.-impedance measurements may be used to identify the contributions from the electrode reactions and grain boundaries. A.c. measurements in combination with a relaxation of the d.c. current may also be used to find out whether concentration gradients have been built up as a result of the passage of current.

27.3   D.c. methods

27.3.1   Wagner polarization experiments

A popular method used to separate different ionic and electronic contributions to the total conductivity is an electrolysis experiment with the use of electrodes blocking for all conducting species but one. The method was used by Wagner and coworkers (Wagner (1957)[10] and references therein) to investigate the electron and ion conductivities in cuprous halides and sulphide. In the experiments a sample of the cuprous halide was placed between a copper and a graphite electrode. When the cell was polarized with copper as the positive electrode, cuprous ions started to migrate towards the graphite electrode, where they were discharged and formed metallic copper. At the copper electrode copper was oxidized and supplied cuprous ions to the electrolyte. In this way the current continued to flow and was a measure of both the ionic and *electronic conductivities* as shown in Fig. 27.1a. If, on the other hand, the cell was polarized with graphite as the positive electrode, the cuprous ions would initially migrate towards the copper electrode. The graphite electrode could not, however, supply cuprous ions and a copper deficit was created at this electrode. This caused the current to decay until finally the copper migration in the electrical field was counteracted by a diffusion in the opposite direction. Under these steady state conditions the current was carried exclusively by electrons as shown in Fig. 27.1b.

The current through the cell is usually determined from the potential drop across the cell. In order to avoid interference from the electrode reactions special current sensing electrodes are often employed (Slade *et al.*[11]). By careful placement of ionic and electronic probes on the circumference of the sample, the specific conductivities can even be determined without knowledge of the exact sample geometry. For a thorough discussion of these methods the reader should consult the

D.c. techniques

papers by Bruce, Evans & Vincent (1988)[12] and Dudley & Steele (1980)[13].

When the polarizing voltage is applied to a cell with ionically blocking electrodes, a current decay is often observed before a stable value is reached. The reason for this is, that immediately after application of the voltage, the charging of the double layer at the electrode/electrolyte interfaces can sustain an ionic current. An electrode reaction is therefore not needed until the double layer is fully charged. For this reason it is common practice to determine the *transference number* of a species for which the electrodes are reversible as the ratio between the stationary current and the initial current[14].

Kreuer *et al.* (1981)[1] have used ionically *blocking electrodes* to show negligible electronic conduction in lithium hydrazinium sulphate, HUP and HUAs. Lyon & Frey (1985) and Slade *et al.* (1987) have used a similar method on HUP[15] and on hydrous oxides[11].

A simplification of the technique, applicable for protonic conductors is shown in Fig. 27.2. The cell is here equipped with platinum black electrodes, which can be made blocking/non-blocking for protons by purging the cell with either nitrogen or hydrogen gas. This system has been used, for example, by Maiti & Freund (1981)[16] on hydroxyapatite.

One of the figures from the chapter on zeolites in this book is included here also to show one of the limitations of the method. In the experiment shown in Fig. 27.3, proton conducting lithium chabazite is purged with argon in order to detect a possible electronic conductivity. The treatment is seen to lower the conductivity by almost three orders of magnitude. It

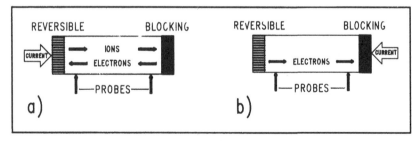

Fig. 27.1. Wagner polarization experiments. (a) Current is sent from the non-blocking electrode to the blocking electrode. Both ionic and electronic currents can be sustained. (b) Current is sent from the blocking electrode to the non-blocking electrode. There is no electrode reaction to sustain the ionic current, so only an electronic current can be sustained.

cannot be concluded, however, that the remaining conductivity is electronic, because even small 'impurity' reactions at the electrodes may sustain protonic conduction. The conversion rate at the electrodes becomes lower when the current is diminished, so a clean-up can take a very long time. The method therefore tends to give the upper limit of the electronic conductivity.

27.3.2 The Hittorf (Tubant) method

All ionic conductors contain ions of opposite signs, which in principle may contribute to the conductivity. If more than one ion can move in a

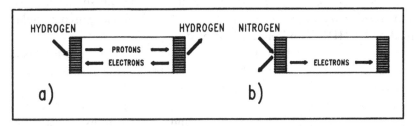

Fig. 27.2. An experiment with protonic conductors, equivalent to the experiment in Fig. 27.1. The electrodes are made non-blocking/blocking by purging with either hydrogen (a) or an inert gas (b).

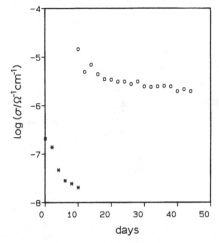

Fig. 27.3. Conductivity of humid lithium chabazite exposed to various gases. Symbols: *, $Ar + H_2O$; O, $H_2 + H_2O$.

common host lattice, concentration gradients may build up, when current is passed through the material, and cause internal losses. This has for example been shown to be the case for polyethylene oxide polymer containing a dissolved lithium salt[17] and may be suspected to play a role in protonic conductors consisting of mixtures of similar polymers and small molecule inorganic acids. Multiple cations may give rise to similar problems in compounds, where migration of the anion can be ruled out for structural reasons (uranyl phosphates, zirconium phosphates, hetero polyacids, zeolites and clays). Uncertainty about which ion is the conducting one will certainly be the case with acid salts. Also the problem whether a 'protonic conductor' actually conducts protons or whether it conducts hydroxyl ions deserves attention.

The classical method of determining transference number is, apart from the e.m.f. method described previously, long time electrolysis with electrodes reversible to only one of the conducting species. Subsequent chemical analysis of the electrolyte at the vicinity of the anode and cathode is then performed. The method was extensively used by Hittorf on liquid electrolytes and some solids[18] and later on by Tubant in numerous studies of pure salts and mixtures of salts (Tubant (1914, 1920, 1921)[19, 20]).

A setup as proposed by Tubant is shown schematically in Fig. 27.4. In the figure a rod of a silver salt is placed between two silver electrodes. The electrolyte is divided into three parts, which fit precisely together so good electrical contact is ensured. Current is passed through the cell from left to right. The total process will then involve transport of silver from the left electrode to the right electrode. In Fig. 27.4a the situation in the electrolyte, when it is a pure silver ion conductor is shown: silver ions will go into the electrolyte at the anode side where they replace the silver ions migrating in the electrolyte. At the end of the experiment no change in the size of the individual electrolyte parts will be observed. In Fig. 27.4b the electrolyte is assumed also to be an anion conductor. The silver will be oxidized to silver ions as before, but as the anions now also migrate in the electrolyte, a charge corresponding to the anion transport has to be neutralized by the cations. The part close to the anode will therefore gain in weight and the part close to the cathode will lose a similar weight. By separating the electrolyte parts after the experiment and by measuring the mass changes of the parts, the nature of the conducting ion and the transference numbers can be determined. With *mixed conductors*, the experiment must not be extended so long that a stationary concentration gradient is established. This gradient has not been established, if the central part of the electrolyte is unchanged after the experiment.

The method can also be used to determine the ratio between the ionic and electronic conductivities by measuring the change of the anode relative to the total charge passed. This was first done by Hittorf (1851)[18], who was able to follow the change of silver sulphide from a predominantly ionic conductor at low temperatures to a predominantly electronic conductor at high temperatures.

Protonic conduction is verified if hydrogen is evolved at the cathode in an electrolysis experiment with a hydrogen source at the anode. This has been used very elegantly by Kreuer[1], who used the blue colouring of tungsten bronzes when hydrogen is intercalated as a proof of proton conduction in lithium hydrazinium sulphate, HUAs and HUP. Several other workers have used the formation of hydrogen as a proof of proton conduction (Murphy (1964)[21], Takahashi *et al.* (1976, 1979)[5,23], Chandra (1985)[22], Iwahara *et al.* (1986)[8]).

In an electrolysis experiment with hydrogen electrodes, Krogh Andersen

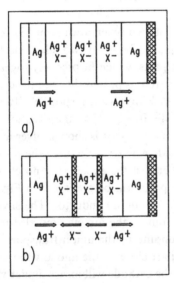

Fig. 27.4. A Hittorf experiment on solid electrolytes as proposed by Tubant. A silver salt is used as an example. The electrolyte is divided into three parts and placed between silver electrodes. Full lines, size before electrolysis. Broken lines, size after electrolysis. Cross hatched area, gain during electrolysis. Only silver ion conduction is assumed in (a) while additional anion conduction is assumed in (b).

et al.[24] used the evolution of ammonia from ammonium zeolites (described in detail in the chapter on zeolites in this book) to show the conduction mechanism to be ammonium ion migration and not proton hopping.

Skou *et al.* (1989)[25] have used electrolysis of $HLiZr(PO_4)_2 . nH_2O$ between platinum electrodes in hydrogen atmosphere to determine the lithium transference number. A sample 10 mm in diameter and *c.* 1 mm in thickness was electrolysed at 0.5 V in humid hydrogen gas until 40% of the ions originally present were exchanged (the experiment lasted 150 days). Then the platinum electrodes were scraped off and the pellet divided into two halves, so the parts close to the anode and to the cathode could be analysed independently for lithium. The analysis showed concentration changes corresponding to a transference number of lithium of *c.* 0.1.

If possible, electrodes blocking for the least conducting species should be used in this kind of experiment, as it gives the highest accuracy in the determination of the transference numbers.

27.4   A.c./d.c. combination techniques

In a recent review on proton conduction in solids (Poulsen (1989)[26]) it is stated: 'It is only when hydrogen evolution can be maintained over long times – with due supply of protons of some form at the anode, that the hydrogen evolution can be taken as a "proof" of protonic conductivity'. This suggests that a d.c. electrolysis experiment should be carried out until at least a significant amount and preferably all of the charge carriers originally present in the sample have been exchanged with protons supplied through the anode reaction. By combining this experiment with a simultaneous measurement of the a.c. conductivity, with measurements of the a.c. conductivity when the d.c. load is relaxed and with purging with inert and/or dry gases, valuable information about the nature of the charge carriers and the conduction mechanism can be gained.

In a typical experiment of this kind a sample with electrodes which are non-blocking to hydrogen (e.g. platinum black), should be purged for some time with an inert gas of the same humidity as the gas used in the rest of the experiment. This will remove oxygen from the sample and thereby avoid water formation and possible damage of the sample, when hydrogen gas is supplied. A.c. measurements taken during this time should reveal the contributions from the bulk and the grain boundary to the conduction and show the electrodes to be blocking.

After purging with nitrogen, the cell should be purged with hydrogen.

The impedance diagram should now show the electrodes to become non-blocking and thereby allow a distinction between grain boundary conduction and electrode polarization. Then the d.c. load should be applied and the a.c. and d.c. conductivities followed as functions of time.

In some cases, the conductivity change with time and/or a difference between the a.c. and d.c. conductivities, which cannot be attributed to electrode polarization, is observed. The *a.c. conductivity* should then be measured with the d.c. load applied and at different times after the load has been removed. This option is available in some commercial impedance analysers, or can be achieved by the use of summation amplifiers to add a small a.c. signal to the polarizing voltage and to subtract the d.c. current from the cell response. This is shown schematically in Fig. 27.5.

During the electrolysis experiment, change of the purging gas from humid to dry will show the sensitivity to the water content in the sample. A replacement of hydrogen gas with nitrogen gas will show if there is any significant electronic conductivity.

At least three different conditions for conduction may be distinguished by use of this combination of a.c. and d.c. measurements and are shown schematically in Fig. 27.6.

27.4.1 Stationary carrier (*proton* hopping)

When the proton is the only conducting ion and when the conduction takes place by proton jumps from site to site (*Grotthuss mechanism*), the situation is simple. In that case there will be no change in conductivity with time and the a.c. and d.c. conductivities will be equal.

Slade *et al.*[27] have investigated the conductivity behaviour of the

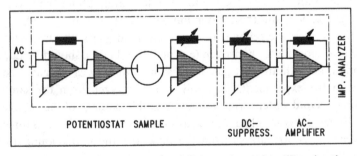

Fig. 27.5. Schematic representation of a setup, which can be used to determine the a.c. conductivity under load.

framework hydrate $Sb_2O_5 . 5H_2O$ by a combined a.c./d.c. method. They passed a charge corresponding to several times the amount of protons in the sample. In a hydrogen atmosphere no change in conductivity was observed. This proves the material to be a protonic conductor. The a.c.-impedance analysis showed both bulk and grain boundary conduction. Sensitivity to the relative humidity of the purging gas showed water to be important for the conduction process, but no change in the conductivity with time after removal of the d.c. load was observed. Removal of hydrogen from the cell showed that no significant electronic conductivity was present.

A similar experiment on 12-tungstosilicic and 12-tungstophosphoric acid hexahydrates by Slade *et al.*[28] showed a similar behaviour, except that the major conduction path was through the grain boundaries.

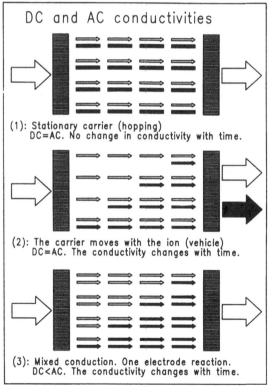

Fig. 27.6. Three different conditions for conduction, which may be separated by the use of a.c./d.c. combination techniques.

27.4.2   *The carrier moves with the ion* (vehicle mechanism)

Kreuer *et al.*[1,29] have proposed a proton conduction mechanism, where the proton is attached to a vehicle (water, ammonia, hydrazine etc.), and the complex moves as a whole. If, in a d.c. experiment the vehicle is not supplied together with protons at the anode, migration of the proton-vehicle complex will generate a vehicle deficit close to the anode. This will cause the conductivity to decrease, if no comparable alternative conduction path exists. The decline will be observed in both the d.c. and the a.c. conductivities.

A clear example of this was seen with the earlier mentioned ammonium zeolites[24], where a gradual change in conductivity took place, when ammonia was liberated at the cathode and replaced by water at the anode.

Krogh Andersen *et al.* have investigated the protonic conduction in $H_2Ti_4O_9 \cdot 1.2H_2O$[30]. Extended d.c. experiments showed protonic conduction. The a.c. conductivities were measured between 0.5 and 2 h after the measuring cell was removed from the d.c. setup and were found to be one order of magnitude higher than the d.c. conductivities. This is shown in Fig. 27.7a. When the a.c. conductivity was measured under d.c. load, the a.c. and d.c. conductivities were identical. After removal of the d.c. load the a.c. conductivity increased with time as shown in Fig. 27.7b. This clearly indicated a vehicle mechanism with water as the vehicle. Water diffuses much slower through the composite platinum anode than hydrogen. Protons can therefore be supplied to the electrolyte easier than water.

A similar behaviour has recently been observed in both the bulk and the grain boundary of zeolite Y which contained tin(IV) oxide[31].

It should be emphasized that a time-independent current cannot be taken as the only proof of a hopping mechanism. If the amount of vehicle is much larger than the amount of protons, the vehicle concentration gradient caused by the current may not be sufficient to influence the conductivity.

27.4.3   *Mixed conduction*

If a compound contains more than one cation as in mixed salts, it is possible that both ions will migrate in the electrical field. If the electrode is non-blocking to only one of the ions (e.g. protons), the other ion will pile up at the vicinity of the cathode. If the anion cannot move, as is the case with a fixed anion lattice, the 'blocking' ion will obstruct the passage of the 'non-blocking' ion. This will cause the d.c. conductivity to decrease with time, but the a.c. conductivity will not decrease to the same extent. This can be seen from the following argument. Assume that the specific conductivities of the two ions are the same and that the sites close to the

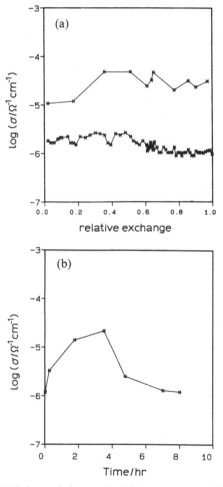

Fig. 27.7. (a) A.c. and d.c. conductivities of $H_2Ti_4O_9$. The cell is purged with hydrogen saturated with water vapour at 20 °C. Measuring temperature 25 °C. The lower curve is the stationary d.c. conductivity of the cell. The upper curve is the a.c. conductivity measured between 0.5 and 2 h after the d.c. load was removed. The relative change was computed as the ratio between moles $H_2Ti_4O_9$ and moles of unit charge passed through the cell (with permission[30]). (b) A.c. conductivity of $H_2Ti_4O_9$ as a function of time after application of a d.c. load of 0.5 V. At time zero the d.c. load was removed and it was applied again after 3.5 h. Other conditions as in (a) (with permission[30]).

cathode are almost completely occupied by the 'blocking' ion as illustrated in Fig. 27.6(3). In this case, there will be almost no d.c. conductivity left. The a.c. conductivity, however, will remain unchanged, as the specific conductivity is the same all over the electrolyte, and the a.c. conductivity is not sensitive to whether the electrodes are blocking or non-blocking.

In a d.c. experiment on monolithium salts of α-zirconium phosphate by Skou *et al.*[25] a difference between the a.c. and d.c. conductivities developed during the experiment and the d.c. conductivity ended one order of magnitude below the a.c. conductivity. The difference did not relax more than 10% within a period of 4 h after the d.c. load was removed. This was taken as an indication of simultaneous proton and lithium ion conduction and was proved by the Hittorf experiment described earlier.

27.5   Conclusion

A technique is described where a.c. measurements are combined with d.c. experiments and the use of several purging gases. The technique makes it possible to distinguish between several conducting ions and gives valuable information about the conduction mechanism.

27.6   References

1. K.-D. Kreuer, W. Weppner and A. Rabenau, *Solid State Ionics* 3/4 (1981) 353–8.
2. E. Krogh Andersen, I. G. Krogh Andersen, C. Knakkergaard Møller, K. E. Simonsen and E. Skou, *Solid State Ionics* 7 (1982) 301–6.
3. T. Norby, *Solid State Ionics* 28–30 (1988) 1586–91.
4. P. M. Spaziante in *Hydrogen as an Energy Vector*, Proc. of Int. Seminar, Brussels, 1980, 390–407.
5. T. Takahashi, S. Tanase, O. Yamamoto and S. Yamauchi, *J. Solid State Chem.* 17 (1976) 353–61.
6. S. H. Sheffield and A. T. Howe, *Mat. Res. Bull.* 14 (1979) 929–35.
7. J. S. Lundsgaard, E. Krogh Andersen, E. Skou and J. Malling, *Solid State Protonic Conductors I*, J. Jensen and M. Kleitz (eds) (Odense University Press (1982)) 222–34.
8. H. Iwahara, T. Esaka, H. Uchida, T. Yamauchi and K. Ogaki, *Solid State Ionics* 18–19 (1986) 1003–7.
9. S. Yde-Andersen, J. S. Lundsgaard, J. Malling and J. Jensen, *Solid State Ionics* 13 (1984) 81–5.
10. J. B. Wagner and C. Wagner, *J. Chem. Phys.* 26 (1957) 1597–606; H. Kobayashi and C. Wagner, *J. Chem. Phys.* 26 (1957) 1609–14.

D.c. techniques

11. R. C. T. Slade, J. Barker and T. K. Halstead, *Solid State Ionics* **24** (1987) 147–53.
12. P. G. Bruce, J. Evans, C. A. Vincent, *Solid State Ionics* **28–30** (1988) 918–22.
13. G. J. Dudley and B. C. H. Steele, *J. Solid State Chem.* **31** (1980) 233–47.
14. P. M. Blonsky, D. F. Shriver, P. Austin and H. R. Allcock, *Solid State Ionics* **18–19** (1986) 258–64; M. Watanabe, S. Nagano, K. Sanui and N. Ogata, *Solid State Ionics* **28–30** (1988) 911–17.
15. S. B. Lyon and D. J. Fray, *Solid State Ionics* **15** (1985) 21–31.
16. G. C. Maiti and F. Freund, *J. Chem. Soc. Dalton Trans.* (1981) 949–55.
17. P. R. Sørensen and T. Jacobsen, *Electrochim. Acta* **27** (1982) 1671–5.
18. J. W. Hittorf, *Ann. Phys. Chem.* **84** (1851) 1–28.
19. C. Tubant and E. Lorenz, *Z. Phys. Chem.* **87** (1914) 513–42.
20. C. Tubant, *Z. f. Elektroch.* **26** (1920) 358–63; C. Tubant, *Z. anorg. u. allg. Chemie* **115** (1921) 105–26.
21. E. J. Murphy, *J. Appl. Phys.* **35** (1964) 2609–14.
22. S. Chandra in *Transport–Structure Relations in Fast Ion and Mixed Conductors*, 6th RISØ International Symposium on Metallurgy and Materials Science 1985, 365–70.
23. T. Takahashi, S. Tanase, O. Yamamoto, S. Yamauchi and H. Kabeya, *Int. J. Hydrogen Energy* **4** (1979) 327–38.
24. E. Krogh Andersen, I. G. Krogh Andersen, E. Skou and S. Yde-Andersen, *Solid State Ionics* **18–19** (1986) 1170–4.
25. E. Skou, I. G. Krogh Andersen, E. Krogh Andersen, M. Casciola and F. Guerrini, *Solid State Ionics* **35** (1989) 59–65.
26. F. W. Poulsen, Proton conduction in solids in *High Conductivity Solid Ionic Conductors*, T. Takahashi (ed) (World Publishing Co (1989)).
27. R. T. C. Slade, G. P. Hall and E. Skou, *Solid State Ionics* **35** (1989) 29–33.
28. R. T. C. Slade, H. A. Pressman and E. Skou, *Solid State Ionics* **38** (1990) 207–11.
29. K.-D. Kreuer, A. Rabenau and W. Weppner, *Angew. Chem.* **94** (1982) 224–5.
30. E. Krogh Andersen, I. G. Krogh Andersen and E. Skou, *Solid State Ionics* **27** (1988) 181–7.
31. I. G. Krogh Andersen, E. Krogh Andersen, N. Knudsen and E. Skou, *Solid State Ionics* **46** (1991) 89–93.

28 NMR in gels and porous media

JEAN-PIERRE KORB AND FRANÇOIS DEVREUX

28.1   Introduction

Some of the most successful methods of *nuclear magnetic resonance* used in the study of *disordered porous media* are presented. This is the case for *nuclear relaxation* and the *pulsed field gradient method* in which a specific solvent is present in various fully or partially saturated, organic or inorganic, porous media. The methods have proven useful for the measurements of *pore size distribution*[1-3], *diffusion coefficients*[3,4] and *permeability*[5] in various fully or partially saturated organic or inorganic porous media. Systems of interest include cements, porous rocks, glasses and resins, sandstones, clays, ceramics and aerogels. Most of these studies are performed on a liquid present in the pore structure. Of course, this supposes that the liquid does not modify the matrix structure. In some cases, it is also possible to use nuclear relaxation without the presence of a solvent. Finally there are some recent applications of *magnetic resonance imaging* in these systems.

28.2   Nuclear relaxation of solvent imbibed in porous materials

28.2.1   A narrow distribution of pores

Though confinement of fluids within random porous media can lead to drastic changes in their physical properties which are not very well known, the observation of an exponential nuclear relaxation of the longitudinal or transverse magnetization

$$R_{z\,\text{or}\,xy}(t) = \exp\left(-\frac{t}{T_{1\,\text{or}\,2}}\right) \tag{28.1}$$

is evidence of a fast exchange process among the solvent molecules in the porous system. In that case one obtains without ambiguity the *spin–lattice*

432

NMR in gels and porous media

$1/T_1$ and *spin–spin* $1/T_2$ *relaxation rates* which give potential information about the microdynamics in the pores. Moreover, the observation of a linear dependence of these rates on the filling ratio V/V_p of the volume occupied by the solvent V over the total open-pore volume V_p shows the homogeneity of the solvent at every coverage[4] (Fig. 28.1a). This supports the two-fraction, fast-exchange model where one of the fluid components preferentially wets the pore surfaces while the other fills the pore domain. As a consequence, the longitudinal or transverse nuclear relaxation rates of a solvent inside a pore are an average of bulk (b) and surface (s)

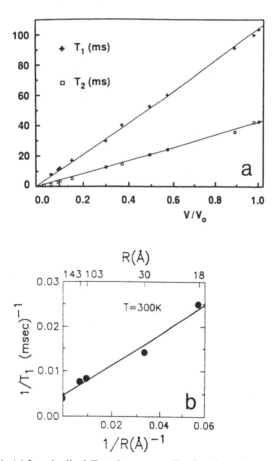

Fig. 28.1. (a) Longitudinal T_1 and transverse T_2 relaxation times for water as a function of filling ratio[4] (with permission). (b) Longitudinal relaxation rate of deuterons as a function of pore radius for nitrobenzene-d_5 at 300 K[7].

433

Proton dynamics and charge transport

contributions

$$\frac{1}{T_{1\,\text{or}\,2}} = \frac{1}{T_{1\,\text{or}\,2\text{b}}} + \frac{\varepsilon S}{V}\frac{1}{T_{1\,\text{or}\,2\text{s}}} \tag{28.2}$$

where S is the total area of the liquid–solid interface and ε refers to the distance from a single molecular coverage at the interface. A BET analysis of nitrogen adsorption isotherm data usually gives values of S. Measurement of weight of solvent after soaking gives the total open pore volume, V_p. Experimental evidence for the validity of Eqn (28.2) has been obtained recently from proton relaxation of water in porous silica glasses with various filling factors[4] (Fig. 28.1a). Another verification of Eqn (28.2) has been obtained by analysing the quadrupolar spin–lattice relaxation rates $1/T_1$ of deuterons of nitrobenzene-d_5 confined within a series of porous glasses with pore sizes ranging from 17 to 143 Å[7]. A linear relationship between $1/T_1$ and the inverse of pore radius justifies also the two-fraction fast-exchange model (Fig. 28.1b).

28.2.2 A large distribution of pores

The problem is different in the presence of a large distribution of pores. Now, the surface/volume ratio in Eqn (28.2) is dependent on the location in the system and consequently, the longitudinal magnetization has a non-exponential relaxation. The pore size dispersion gives rise to a distribution of spin–lattice relaxation rates that can potentially be extracted from the non-exponential, longitudinal magnetization decay[3, 8, 9]. A possible representation of the porous system consists of a superposition of quasi disconnected categories of pores. This is like a network of a distribution of open regions (voids) connected by very narrow channels. The apparent diffusion coefficient through different voids is thus reduced by the high surface ratio (void/channel). As almost all the solvent is concentrated in the voids, the observed relaxation signal comes mainly from these open regions. For a diffusion coefficient of about $D \sim 10^{-6}$–10^{-5} cm^2 s^{-1}, one reaches the condition of bounded diffusion for each category of pore size R_n smaller than 1–10 μm ($n \in \{0, 1, \ldots, n_{\text{max}}\}$). One then has for the normalized longitudinal magnetization $R_z(t)$, a superposition of exponential decays weighted by the volume fraction $W_n = V_n/V_p$ of pores of size R_n in the total volume V_p

$$R_z(t) = \sum_{n=0}^{n_{\text{max}}} W_n\, e^{-t/T_{1n}} \tag{28.3}$$

where

$$\frac{1}{T_{1n}} = \frac{1}{T_{1b}} + \frac{\varepsilon S_n}{V_n} \frac{1}{T_{1s}} \qquad (28.4)$$

Similar equations arise for the transverse relaxation rate $1/T_{2n}$. The problem now is to extract the pore distribution, W_n, from the observed $R_z(t)$. From a mathematical point of view, this could be done by a Laplace inversion of Eqn (28.3). Examples of this method can be found in the review paper of W. P. Halperin *et al.*[10]. Other methods have used a prerequisite distribution that must be verified a posteriori. This has been carried out recently for the nuclear relaxation of ^{13}C and ^{2}H of methanol and nitromethane adsorbed on an organic polymeric resin crosslinked by paramagnetic divalent metal ions[3,9] (Fig. 28.2). The results have been interpreted with a fractal distribution of categories of quasi-disconnected spherical pores, each being composed of N^n spherical pores of radius $R_n = R_0/\alpha^n$ ($\alpha > 1$), with $n_{max} = \log(R_0/R_{min})/\log \alpha$. Introducing the *fractal dimension* D_f through the relation $N \sim \alpha^{D_f}$, leads to

$$W_n = \frac{4\pi R_0^3}{3V} \alpha^{(D_f - 3)n} \qquad (28.5)$$

A continuous approach is given by Chachaty *et al.*[9]. According to Eqns (28.2 and 28.4), one notes that $R_z(t)$ tends to a power law $\sim 1 - t^{3 - D_f}$ at short time, as previously stated by Mendelson[8]. The best fits obtained with these equations, either for longitudinal or transverse relaxation, are displayed in Fig. 28.2 for ^{13}C and ^{2}H of methanol. This has led to the range of values $2.60 < D_f < 2.70$ for the surface fractal dimension, in good agreement with the small angle X-ray scattering ($D_f = 2.6$) and pulsed gradient field ($D_f = 2.6$) results on the same system[9]. There is a good reproducibility in the determination of D_f using both nuclei over a wide frequency range, for two solvents of very different properties. The frequency and temperature dependences of the surface relaxation of methanol have been explained by a reorientation of the adsorbed molecule in the coordination sphere of the paramagnetic ion. The bulk relaxation has been interpreted in terms of a translational diffusion of the solvent ($D \sim 10^{-6}$ cm^2 s^{-1}) relaxed by a distribution of paramagnetic probes, two or three times smaller than in gel of the same composition and one order of magnitude smaller than in the pure solvent[3].

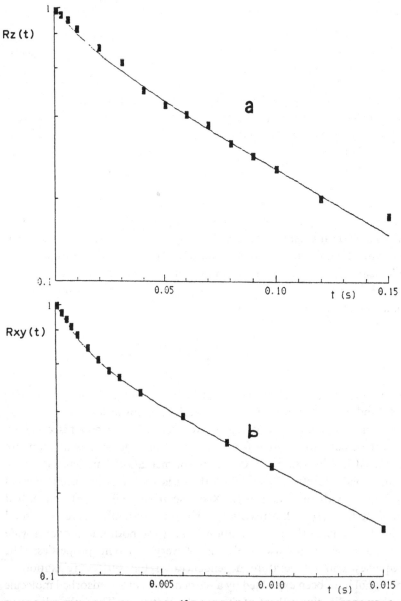

Fig. 28.2. (a) Time dependence of the ^{13}C longitudinal magnetization at 75.5 MHz for methanol adsorbed on organic polymeric resin crosslinked by paramagnetic VO^{2+} ions. The continuous line has been calculated with $T_{1s} = 10$ ms and $T_{1b} = 95$ ms[9]. (b) Time dependence of the ^{2}H transverse magnetization at 46 MHz for deuterated nitromethane adsorbed on organic polymeric resin crosslinked by paramagnetic VO^{2+} and diamagnetic Cd^{2+} ions. The continuous line has been calculated with $T_{2s} = 0.9$ ms and $T_{2b} = 11$ ms[9].

28.3 Pulsed field gradient experiments

The pulsed field gradient spin echo technique (PGSE) is one of the most suitable methods for studying molecular diffusion[4,6,9,12] and pore size distributions[9] in heterogeneous media. It has been applied for measuring the self diffusion of water in porous glasses[4] and the self diffusion of methanol in a resin of an organic polymer crosslinked by diamagnetic divalent metallic ions[9].

In a typical proton PGSE experiment using the usual $\pi/2$, τ, π r.f. sequence, the dephasing of the spins precessing in the transverse plane is monitored by a field gradient pulse G of duration $\delta \sim 2$ ms applied between the two $\pi/2$, π r.f. pulses. A second gradient pulse identical to the first one is applied after a delay $\Delta \sim 100$–200 ms to stimulate refocussing of the spins. In the absence of diffusion, the effects of the two pulses compensate each other and the echo received after 2τ has the same intensity as in a standard spin echo measurement of spin–spin relaxation rate $1/T_2$. In the presence of diffusion there is an imbalance between the defocussing and refocussing processes. This leads to an attenuation of the spin–echo signal $A(G)/A(0)$ given by a sum of the dephasings between the two pulsed gradients weighted over the conditional probability which describes the spin dynamics[6,11]. The method is applicable when T_2 is sufficiently long in order to observe a significant transverse magnetization after the time delay Δ. Fortunately, this is generally the case for liquids.

28.3.1 A narrow distribution of pores

Here there are two limiting regimes where one can calculate such attenuation[11]. The restricted diffusion regime occurs when the free diffusion length $R_F = \sqrt{(2D\Delta)}$ of the imbibed solvent exceeds the average size R of the pores, $R \ll R_F$. Here one has

$$\frac{A(G)}{A(0)} = \exp - \left[\frac{(\gamma \delta G R)^2}{5} \right] \qquad (28.6)$$

This regime is favoured for large values of Δ. The variation of the echo attenuation with the field gradient intensity gives information on the size of the pore R. The free diffusion regime occurs when $R \gg R_F$, and

$$\frac{A(G)}{A(0)} = \exp - [\gamma^2 \delta^2 G^2 D(\Delta - \delta/3)] \qquad (28.7)$$

This gives the diffusion coefficient from the slope at low value of the gradient intensity. This method has been applied to measuring the self diffusion of water in porous glasses[4]. The significant enhancement of diffusion in partially filled porous glasses has been explained by a mechanism involving indirect molecular exchange between the liquid and the vapour phase.

28.3.2 A large distribution of pores

If we consider the fractal distribution W_n of n_{max} categories of pores given in Eqn (28.5), the attenuation of the echo may be written as the following weighted sum over the restricted and free diffusion contributions

$$\left\langle \frac{A(G)}{A(0)} \right\rangle = \sum_{R_{min}}^{R_F} W_n \exp\left[-\frac{\gamma^2 \delta^2 G^2 R_n^2}{5} \right]$$

$$+ \left(\sum_{R_F}^{R_0} W_n \right) \exp\left[-\gamma^2 \delta^2 G^2 D \left(\Delta - \frac{\delta}{3} \right) \right] \tag{28.8}$$

The distinction between the free and bounded diffusion occurs when $R_n = R_F$. The diffusion coefficient D is determined from the slope at low values of $\gamma \delta G \Delta$. This form of attenuation is no longer exponential and presents a marked curvature for high values of $\gamma \delta G \Delta$ due to the contribution of pores of radii R_n lower than R_F. In Fig. 28.3 are shown some typical attenuation curves for resins[9]. The deviation from an exponential is present even at small values of Δ and is especially pronounced for high Δ. The surface fractal dimension $D_f = 2.6$ is found from a fit using Eqn (28.8) and the simplex minimization procedure[9,12].

Another problem of interest comes from the possible existence of very slow diffusion between the pores. Though this effect is very difficult to measure by NMR it has been detected using a dye molecule in solution in porous Vycor glass by the technique of forced Rayleigh scattering[13]. An indirect measurement of permeability k has also been given through the strong correlation $k \propto \Phi^4 T_1^2$ observed with the square of T_1 and the fourth power of the tortuosity Φ (Fig. 28.4)[5]. This suggests that other methods should be applied to characterize more precisely the morphology and transport properties in disordered porous media. Each of these methods probes a different length scale of the pore morphology. So combining different methods could be very fruitful. For instance the use of small angle X-ray scattering, nuclear relaxation and pulse gradient fields have given the same pore size distribution in a polymeric organic resin over four orders of magnitude from 40 Å to 50 μm[9].

28.4   Nuclear relaxation in fractal aerogels

Some porous materials display self-similarity or a fractal structure in a certain length range. This is the case of silica aerogels. Small angle scattering techniques have been so far the most popular method to characterize their geometrical arrangement. In this section, we present an alternative NMR method which has been recently proposed to determine the fractal dimension in these systems. It is based upon the nuclear relaxation induced by paramagnetic impurities diluted in the lattice[14]. Contrary to the diffusion techniques which involve an analysis of scattered intensity in reciprocal space, this method gives directly the mass-to-distance relation in real space: $M(r) \sim r^{D_f}$, with D_f being the fractal dimension. If one applies a saturation comb to kill the spin magnetization

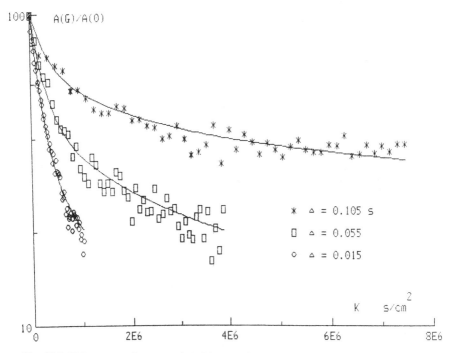

Fig. 28.3. Echo attenuation curves in PGSE experiments on methanol adsorbed on organic polymeric resin crosslinked by diamagnetic Cd^{2+} ions[9].

439

of nuclei belonging to the fractal lattice, the recovery of this magnetization reflects the spatial repartition of the nuclear spins which relax to their equilibrium polarization by dipolar flipping with the dilute magnetic ions. Since the time constant for dipolar coupling relaxation increases with distance as r^6, the recovered magnetization at time t after saturation will be that of the spins which are contained in a sphere of radius $r \sim t^{1/6}$ and therefore will retain a time dependence as: $m(t) \sim t^{D_f/6}$.

The experiments have been performed on different forms of silica obtained by sol–gel condensation of silicon alkoxides in a solution containing chromium nitrate as relaxing agent. The alcogel, which results from the gelation of the polymerization solution, consists of an interconnected network of branched polymers, in which solvent molecules are embedded. The aerogel is a porous structure obtained by hypercritical drying of the alcogel. The amorphous and polycrystalline samples are dense materials resulting from sintering of aerogels at different temperatures. Fig. 28.5 shows the recovery of the ^{29}Si magnetization after a saturation comb in five compounds: an alcogel, two aerogels with different

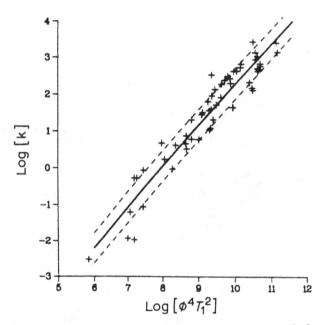

Fig. 28.4. Experimental logarithmic plot of the permeability k vs. $\Phi^4 T_1^{\,2}$ for a set of water-saturated sandstones; k is given in millidarcies, Φ is in percent and T_1 in ms. (From reference 5, with permission; see also references therein.)

paramagnetic impurity concentrations, an amorphous and a crystallized sample. The relaxation is strongly non-exponential for all the samples studied. From the log–log plot in Fig. 28.5, the time dependence of the magnetization recovery follows a power law $m(t) \sim t^{\alpha}$ in an extended time range, before reaching a saturation plateau.

The dimension D_f is obtained from the slopes of the power-law regime: $D_f = 6\alpha$. The resulting values are 2.85 for the alcogel, 2.3 and 2.1 for the aerogels, 3.1 for the amorphous and 2.85 for the crystalline powder. It is not surprising to find values near $D_f = 3$ for the densified materials. For the aerogels, the NMR results are consistent with those of small angle X-ray scattering (SAXS) experiments which give $D_f = 2.3$. However, the SAXS result is $D_f = 2.0$ for the alcogel. This disagreement is easily explained by noticing that the relation $M(r) \sim r^{D_f}$, from which the 'NMR' D_f is deduced, describes the mass distribution around the magnetic impurities. Since the chromium spins can be found everywhere in the

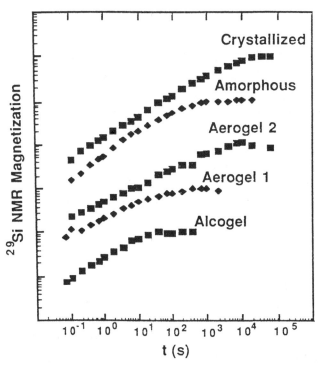

Fig. 28.5. Time dependence of the ^{29}Si magnetization recovery after saturation for different forms of silica[14].

alcogel solution, the NMR relaxation does not reflect the spatial correlations on the silica polymer, but the mass distribution in the gelled solution, which has no reason to be different from a regular tridimensional one.

It is quite unusual to observe a power law over such a long time range, because the direct dipolar coupling between nuclear and electronic spins is generally relayed at long distances by the nuclear spin diffusion, which gives rise to an exponential recovery. However, this short cut is avoided in the present experiments because of different effects. Firstly, in the aerogels, the fractal geometry imposes a tortuous path to the magnetization leading to a slow anomalous diffusion. Secondly, in both the aerogel and amorphous samples, the structural disorder may cause considerable shifts of the NMR lines of neighbouring spins making the dipolar flipping quite inefficient. Finally, in all samples, as the relaxation occurs under magic angle spinning conditions, the averaging of the secular part of the dipolar coupling between nuclear spins inhibits the flip-flop transitions between neighbouring ^{29}Si and quenches the nuclear spin diffusion process.

28.5   NMR imaging and microscopy

Recent applications of nuclear imaging of solvent in porous media include the drying of a limestone block by analysing either one-dimensional projections (profiles) or three-dimensional NMR images, as a function of the amount of water remaining in the system at various stages of drying[15]. The spatial resolution is limited in this disordered solid system. Combined PGSE method and NMR imaging has allowed the observation of an image of solvent flow[16]. For instance, the Poiseuille velocity distribution has been verified on a microscopic scale inside a 0.7 mm capillary at a transverse spatial resolution of 25 μm[16].

28.6   Conclusion

Some of the most successful methods of nuclear magnetic resonance used in the study of disordered porous media have been reviewed and their applications to the measurement of pore size distribution, diffusion coefficients and permeability have been presented. This is the case of nuclear relaxation and pulsed field gradient method of a specific imbibed solvent in various fully or partially saturated organic or inorganic porous media. The cases of narrow and large distributions of pores have been treated separately. Examples of paramagnetic relaxation without solvent have been presented for fractal aerogels. Finally some recent applications of magnetic resonance imaging in these systems have been mentioned.

28.7   References

1. M. Lippsicas, J. R. Banavar and J. Willemsen, *Appl. Phys. Lett.* **48** (1986) 1544.
2. D. P. Gallegos, D. M. Smith and C. J. Brinker, *J. Colloid Interface Sci.* **124** (1988) 186–98.
3. J.-P. Korb, B. Sapoval, C. Chachaty and A. M. Tistchenko, *J. Phys. Chem.* **94** (1990) 953–8.
4. F. d'Orazio, S. Bhattacharja, W. P. Halperin and R. Gerhardt, *Phys. Rev. Lett.* **63** (1989) 43–6.
5. J. R. Banavar and L. M. Schwartz, *Phys. Rev. Lett.* **58** (1987) 1411–14.
6. E. O. Stejskal and J. E. Tanner, *J. Chem. Phys.* **42** (1965) 288.
7. G. Liu, Y. Z. Li and J. Jonas, *J. Chem. Phys.* **90** (1989) 5881–2.
8. K. S. Mendelson, *Phys. Rev.* **B34** (1986) 6503.
9. J.-P. Korb and C. Chachaty (Proc. of the GERM XI, Draveil, France (1990); C. Chachaty, J.-P. Korb, J. R. C. Van der Maarel, W. Brass and P. Quinn, *Phys. Rev.* **B44** (1991) 4771–93.
10. W. P. Halperin, F. d'Orazio, S. Bhattacharja and J. C. Tarczon, *Molecular Dynamics in Restricted Geometries*, J. Klafter and J. M. Drake (eds) Ch. 11 (Wiley, New York (1989)) 311–50.
11. J. Karger, H. Pfeifer and W. Heink, *Adv. Magn. Reson.* **12** (1989) 1.
12. J. A. Nelder and B. Mead, *Comp. J.* **7** (1964) 308.
13. W. D. Dozier, J. M. Drake and J. Klafter, *Phys. Rev. Lett.* **56** (1986) 197–200.
14. F. Devreux, J. P. Boilot, F. Chaput and B. Sapoval, *Phys. Rev. Lett* **65** (1990) 614–17.
15. G. Guillot, A. Trokiner, L. Tarrasse and H. Saint James, *J. Phys. D* (Appl. Phys.) **22** (1989) 1646–9.
16. P. T. Callaghan, C. D. Eccles and Y. Xia, *J. Phys. E* (*Sci. Instrum.*) **21** (1988) 820–2.

29 Mobility in hydrogen-containing oxide bronzes: the atomic-level detail

ROBERT C. T. SLADE

Mobility of H-atoms in hydrogen oxide bronzes and related systems has excited considerable interest in view of their possible applications in e.g. catalysis and electrochromic electrodes for display devices. In practical materials the structure of the host oxide may be ill-defined and inter-crystallite regions may contribute significantly to observed (bulk) mobilities. This article aims to present a consistent picture (derived from the published work of many scientists internationally) of the atomic level phenomena responsible for intracrystallite H-atom mobility in well-characterized single phases. The detailed characterization of such phases is discussed by Dickens & Chippindale (Chapter 7).

29.1 Applicable techniques

The primary techniques used in probing the atomic level detail of motions of protonic species in hydrogenous oxide bronzes are (a) nuclear magnetic resonance and (b) quasielastic scattering of neutrons. The particular applicability of these techniques arises from the high gyromagnetic ratio (and consequent high sensitivity in NMR) and high cross-section for the incoherent scattering of neutrons (emphasizing motions of protonic species when protons are present) characteristic of the ^1H nucleus. There are a number of conditions to be met in order that the interpretation of the results of such measurements should be reliable and unambiguous. Ideally these include (i) rigorous chemical characterization of the materials under investigation, (ii) a self-consistent and chemically sensible interpretation of the results of measurements using each technique, (iii) interrelationship of the results of several techniques on near-identical samples and (iv) testing of the complete multi-technique set of results against the predictions of a range of dynamical models. It is not clear that all these conditions have been met in studies in this area.

29.1.1   Nuclear magnetic resonance

Measurements have included ^{1}H relaxation times, ^{1}H and ^{2}D absorption spectra and (in cases where lines are already narrowed by translation of protonic species) application of combined magic-angle-rotation and multiple-pulse ^{1}H line-narrowing (*CRAMPS ^{1}H NMR*). For the fastest self-diffusions ($D > 10^{-8}$ cm^2 s^{-1}) direct measurement of D is possible using the pulsed field gradient (PFG) technique (see Chapters 26 & 28).

Correct interpretation of variable temperature ^{1}H relaxation time measurements at more than one spectrometer operating frequency can lead to characterization of a dynamical process (e.g. for fast ^{1}H self diffusion, a continuous increase is seen in T_2 at high temperatures and $T_2 \sim T_1$ above the minimum in T_1, i.e. for $D > 10^{-8}$ cm^2 s^{-1}). Results for phases with slower motions can be ambiguous due to sample decomposition at higher temperatures.

In the absence of fast motions of protonic species (in the 'rigid lattice': in practice at temperatures where the characteristic time for motion $\tau > 10^{-5}$ s) broad ^{1}H dipolar absorption spectra related to the (static) crystal structure (e.g. via the second moment M_2 of the lineshape) are observed. Rapid ^{1}H self-diffusion leads to averaging-out of the inter-nuclear dipolar interactions and a much narrower (linewidth ~ 2 kHz) ^{1}H absorption spectrum (observable on a conventional high resolution [liquids] instrument) characteristic of the powder pattern arising from the spatial average of a (much weaker) anisotropic tensor (shift) interaction. CRAMPS ^{1}H NMR spectroscopy averages-out (by magic-angle-spinning) this anisotropy and leads to observation of isotropic shifts (the multiple-pulsing removes what little is left of the ^{1}H–^{1}H dipolar coupling). In the case of ^{2}D NMR absorption spectra, the (strong) quadrupolar interaction leads to broad lines even in the presence of rapid self-diffusion (which nonetheless leads to a spatial average of the electric-field-gradient tensor).

For in-depth treatments of the backgrounds to these techniques the reader is referred to the books by Wolf[1] (relaxation times), Abragam[2] (broadline spectra) and Mehring[3] (high resolution spectra of solids).

445

29.1.2   Quasielastic scattering of neutrons (QNS)

An incident neutron beam (wavelength λ) impinging on a sample is scattered both elastically (coherently [Bragg scattering $\equiv$ diffraction] and incoherently [over 4π]) and inelastically (with changed neutron energies on incoherent scattering). With appropriate instrumentation at a neutron source, it is possible to detect motions (especially reorientations and self-diffusions of protonic species) with characteristic times $\tau < 10^{-9}$ s via Doppler (quasielastic) broadenings of the incoherent 'elastic peak'. Experimentally, the incoherent quasielastic scattering law $S_{inc}(Q, \omega)$ is determined as a function of the magnitude, Q_{el}, of the scattering vector Q at the elastic peak ($Q_{el} = 4\pi \sin(\theta)/\lambda$ for scattering angle θ). Possible dynamic models are then tested by comparison of calculated scattering laws (convoluted with the instrumental resolution function) with the experimental $S_{inc}(Q, \omega)$ (see Chapter 21).

The variation of $S_{inc}(Q, \omega)$ with Q_{el} is dependent on the type(s) of motion being detected and hence can be characteristic of them. For purely reorientational motions, the 'elastic peak' consists of both an elastic (unchanged) and a quasielastic contribution (broadened, with the half-width independent of Q_{el} for the temperature-dependent [approximately Lorentzian] broadening). The variation of the elastic-incoherent-structure-factor (EISF, the ratio of the elastic intensity to the elastic + quasi-elastic intensity) with Q_{el} is then, in principle, characteristic of (and calculable for) the geometry of the reorientational motion. A fast self-diffusion of all the protonic species leads to a (Lorentzian) broadening of the whole of the elastic peak, with (for a three-dimensional motion) a half-width for the broadening varying as DQ_{el}^2 at low Q_{el} values (i.e. at $Q_{el} < 2/a$ for a jump length a). In systems exhibiting more than one motion with similar timescales a more complicated behaviour is observed.

For an in-depth treatment of the application of QNS techniques the reader is referred to the book by Bée[4].

29.2   Results of atomic-level investigations

Investigations of mobility of protonic species have concentrated primarily on tungsten and molybdenum bronzes. In the case of hydrogen oxide bronzes, apparent inconsistencies can arise in comparing studies of materials prepared chemically ($Zn/HCl(aq)$) and materials prepared by hydrogen spillover ($H_2(g)/Pt$-catalyst). In phases in which H is present

Hydrogen-containing oxide bronzes

solely in hydroxyl groups only translational motions are present. In cases where H is present in polyatomic groupings (e.g. coordinated H_2O in higher H-content hydrogen molybdenum bronzes, NH_4^+ ions in *ammonium bronzes*) both reorientational and translational motions are evident. Ritter[5] reviewed NMR studies on some of these systems.

29.2.1   Hydrogen tungsten bronzes and related phases

The majority of studies (see also Chapter 7) have concerned mobility in phases with an ReO_3-related host framework (Fig. 29.1) and inserted H present in hydroxyl groups in the approximately square windows. NMR studies reveal translational motions of H-atoms (activated hopping between sites corresponding to *hydroxyl groups* coordinated to different metal atoms). These motions are too slow to generate observable quasi-elastic broadenings in QNS studies. Hexagonal hydrogen tungsten bronze (denoted h-$H_x WO_3$) has also been studied.

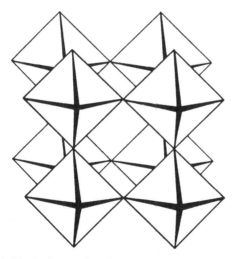

Fig. 29.1. The ReO_3-type host framework: MO_6 octahedra vertex-share to create a three-dimensional network of voids interconnected by *square windows* defined by oxygen atoms. In hydrogen bronzes derived from hosts of this type the H-atoms are present attached to the oxygen atoms (as hydroxyl groups lying in the windows)[9].

447

Proton diffusion mechanisms

29.2.1.1 $H_{0.5}WO_3$

Vannice *et al.*[6] reported a high mobility ($\tau^0 = 4 \times 10^{-12}$ s, $E_a = 5.4$ kJ mol^{-1}) for H-atoms in this phase, this conclusion being based on a possible minimum in T_1 (there was no corresponding increase in T_2, however). Dickens *et al.*[7] and Nishimura[8] determined (from relaxation times and absorption spectrum line-narrowing respectively) that H-atom migration was slow and characterized (when data were treated using the simplest theories appropriate to three-dimensional diffusion) by an anomalously large pre-exponential jump time ($\tau^0 \sim 10^{-8}$ s[7]). The rigid lattice M_2 value is fully consistent with the known structure[7,9]. Ritter *et al.*[10] attributed the pre-factor anomaly to a localized H-atom hopping (of reduced dimensionality) between sites corresponding to structurally identifiable squares of O-atoms (those defining square windows in the host framework; see Fig. 29.1). Application of the semi-empirical Waugh–Fedin relationship then removes the anomaly ($\tau^0 \sim 10^{-12}$ s, $E_a = 44$ kJ mol^{-110}).

29.2.1.2 $H_xW_{1-y}O_3$ and H_xReO_3

On substitution of the host (ReO$_3$-type) framework metal atoms, NMR studies reveal the dynamical picture above to be retained[11-13].

29.2.1.3 h-H_xWO_3

Much interest resulted from an NMR relaxation study[14], which was interpreted in terms of very rapid H-atom self-diffusion in one-dimension in this material (along the direction c of the hexagonal and triangular tunnels characteristic of the h-WO$_3$ host framework; see Fig. 29.2). Restricted dimensionality in self-diffusion is, however, unlikely in view of the small size of H and the interconnection of these tunnel types via square windows akin to those present in ReO$_3$-type hosts (see Fig. 29.2). QNS studies failed to confirm that model[15,16]: a sample prepared following Gerand & Figlarz[17] (hydrothermal synthesis of h-WO$_3$ followed by hydrogen spillover) showed minimal quasielastic broadening from low temperature up to ambient, while a sample prepared by insertion into material prepared by gaseous oxidation of ammonium tungsten bronze (i.e. similarly to reference 14) showed appreciable quasielastic broadening arising from a low activation energy reorientation. Chemical analysis revealed the latter sample to contain constitutional NH$_4{}^+$ ions (see below).

29.2.2   *Hydrogen molybdenum bronzes*

The structure of the parent lamellar MoO_3 host is illustrated in Fig. 29.3. Studies of mobility in derived hydrogen bronzes have been numerous. This is due to the occurrence of rapid H-atom self-diffusion at ambient

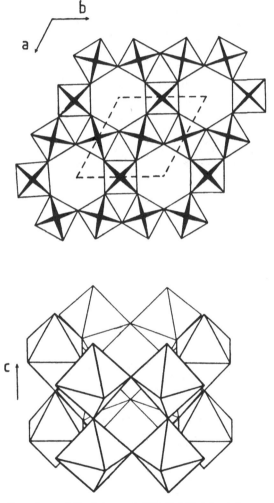

Fig. 29.2. Two views of the hexagonal $WO_3(h\text{-}WO_3)$ host framework: WO_6 octahedra vertex share to produce an open structure with *hexagonal* and *triangular tunnels* in the c direction. These tunnels are interconnected by *square windows* akin to those in the ReO_3-type host oxides. In $h\text{-}H_xWO_3$, the H is present in hydroxyl groups (exact location to be determined).

449

temperature in the higher H-content phases and to the number of single phase materials. Some incompatibility between studies on the red monoclinic phase ($H_x MoO_3$, $x \approx 1.7$) prepared by spillover and by wet chemical methods has been claimed. The former method leads cleanly to that phase. The only high H-content phase which can be prepared pure via

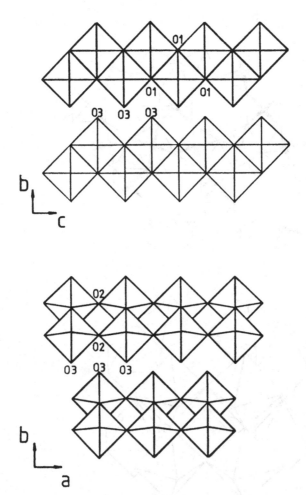

Fig. 29.3. The lamellar MoO_3 structure: MoO_6 octahedra edge and vertex share to produce layers two octahedra thick. H-atom sites in derived hydrogen bronzes vary with stoichiometry. In $H_{0.3}WO_3$ the H-atoms are present in the intralayer ribbon of O2 atoms (extending in the c direction) bonded to O2 in hydroxyl groups[31]. In $H_{1.7}MoO_3$ the H-atoms are attached to terminal (O3) oxygens giving a mixture of hydroxyl groups and coordinated H_2Os in the interlayer region[25,31].

Zn/HCl(aq) reduction is the highest H-content ($x \approx 2.0$) green mono-clinic phase[18]. Lower H-content phases are obtainable chemically using coproportionation reactions of MoO_3 and H_xMoO_3 in the presence of $H_2O(l)$[18].

For the lowest H-content (blue orthorhombic, $x \approx 0.2$–0.4) phase, slow H-atom self-diffusion (hopping between sites corresponding to attach-ment as hydroxyl) along the intra-layer $O2 \ldots O2 \ldots O2 \ldots$ ribbon in the c direction (see Fig. 29.3) has been established by NMR relaxation studies[14,19]. Simple treatment of the relaxation data leads to an anoma-lous prefactor ($\tau^0 \approx 3 \times 10^{-8}$ s, $E_a = 11$ kJ mol^{-1}), interpretable as a consequence of the low-dimensionality of the motion[20]. ^1H NMR M_2 and lineshape studies (for the rigid lattice)[19,21] indicate non-random distribu-tion of H-atoms in sites in the intralayer O2-atom ribbons. Clustering of H-atoms into pairs (neighbouring $O2 \ldots O2$ lines occupied by H) and higher clusters occur.

Numerous NMR studies have been concerned with relaxation and structure in the red monoclinic phase ($x \approx 1.7$)[19–23]. Rapid H-atom self-diffusion ($\tau^0 \sim 10^{-13}$ s, $E_a \approx 22$ kJ mol^{-1}) leads to clearly defined T_1 minima and $T_2 \approx T_1$ at higher temperatures. 1H NMR M_2 and lineshape (rigid lattice) studies[19,21,24] are consistent with a distribution of Mo-coordinated hydroxyl groups and H_2O molecules in the interlayer region (all H attached to O3 in Fig. 29.3, as determined also by neutron powder diffraction[25]). Lineshapes in the line-narrowing region[19,23,24,26] (and also a rotation study of an oriented 'single crystal'[26]) are fully inter-pretable in terms of the powder pattern of an axially symmetric aniso-tropic (spatially averaged) shift tensor, the result of motional averaging over the sites in the interlayer region[19]. Low temperature ^2D lineshape studies[27,28] are also consistent with the presence of two hydrogenic species, with a higher temperature motionally narrowed (spatially averaged) pattern narrower than would correspond to either simple 180°-flips of H_2O molecules about their C_2-axes or diffusion over H-atom sites related by two-fold axes[28]. CRAMPS ^1H spectra[24] led to observation of the isotropic shift (also determinable from the simple ambient temperature ^1H absorption spectrum above) for this phase, with that of a second component (containing *c.* 10% of the H present) also being observed in a material prepared by the wet chemical route.

QNS investigation of dynamics in (red monoclinic) $H_{1.68}MoO_3$[29] allowed characterization of two slowly exchanging dynamic hydrogen populations (with temperature dependent relative sizes) and motional processes associated with them. One population (hydroxyl groups) undergoes rapid self-diffusion (D(295 K) $= 4 \times 10^{-6}$ cm^2 s^{-1}, comparing well with the value obtained by the PFG NMR technique for $H_{1.65}MoO_3$[30]). The second population (coordinated H_2O) has a higher activation barrier to H-atom self-diffusion and takes part in a four-fold reorientational motion ($E_a = 26$ kJ mol^{-1}, pre-exponential residence time $\tau_{res}{}^0 \sim 10^{-14}$ s). A four-fold symmetry about the Mo–OH$_2$ bond is evident in a simple approximate structural model for the interlayer region[27] (Fig. 29.4). This picture becomes compatible with the results of NMR studies (above) when it is realised that chemical exchange renders all H-atoms equivalent on the much longer NMR timescale.

Investigations of other hydrogen molybdenum bronze phases have been few in number. Motion in the blue monoclinic ($x \approx 0.9$) and green monoclinic phases again leads to narrowed ^{2}D NMR absorption spectra[27]. In all but the lowest H-content (blue orthorhombic) phase, H-atoms are found (from both rigid lattice ^{1}H absorption spectra[19,21] and inelastic neutron scattering vibrational spectra[31]) to occur in both hydroxyl groups and coordinated H_2O molecules. NMR relaxation studies suggest that H-atom self-diffusion increases its dimensionality as x increases[21]. A QNS study of $H_{1.98}MoO_3$ (green monoclinic) shows a switch to two-fold reorientation (τ_{res}(304 K) $= 4 \times 10^{-10}$ s) of the coordinated H_2O molecules for this phase[32]. The interlayer network is very nearly fully occupied

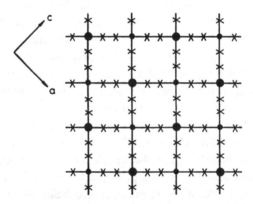

Fig. 29.4. The interlayer region in $H_{1.7}MoO_3$: (●) terminal O3 (upper layer), (●) terminal O3 (lower layer), (×) H-atom sites (no more than one occupied per O3 ... O3 line, no more than two occupied per O3, average occupancy 0.425). The projection onto a plane midway between the layers approximates to a square grid, with H-atom sites on the lines and O3-atoms at intersections.

(very nearly all H present in terminally coordinated H_2O) and the interlayer spacing is smaller than in the red monoclinic phase[18]. NMR relaxation studies do show fast H-atom self-diffusion[21]. However, at the shorter timescale characteristic of QNS studies, quasielastic broadenings due to fast diffusion of hydroxyl hydrogens (the relative population of H in which is very small when $x \approx 2$) are not experimentally discernible. Two-fold reorientation of H_2O coordinated to Mo has also been found (by NMR relaxation studies) to occur in the related lamellar molybdic acids $MoO_3 . 2H_2O$ and $MoO_3 . H_2O$[33].

29.2.3 Ammonium oxide bronzes and related phases

The hexagonal h-WO_3 (Fig. 29.2) and lamellar MoO_3 (Fig. 29.3) host structures provide a very open environment in which to probe the dynamics of the motions of $NH_4{}^+$ and other hydrogenic entities present. Hexagonal *ammonium* tungsten bronzes are obtainable with care[34]. Oxidation of these (elevated temperature, flowing O_2) leads to a fully oxidized oxide framework (X-ray patterns indicate the h-WO_3 framework persists) with residual ammonium ion, e.g. $[(NH_4)_2O]_{0.085}WO_3$[35,36] (a hexagonal polytungstate). Passage of $NH_3(g)$ over blue orthorhombic hydrogen molybdenum bronze results in a bronze containing both $NH_4{}^+$ (very nearly all the hydrogen present) and hydroxyl groups, e.g. $(NH_4)_{0.24}H_{0.03}MoO_3$[37,38].

NMR has been used in studying dynamics in ammonium tungsten bronze, h-$(NH_4)_xWO_3$[38,39]. The ammonium ions are located in the cavities lying along the hexagonal axis of the h-WO_3 structure (Fig. 29.2), the structure-limited maximum ammonium content occurring at $x = 0.33$. A composite picture of three distinct motions involving hydrogenic entities has emerged. At low temperatures relaxation is dominated by low activation energy $(3–5 \text{ kJ mol}^{-1})$ reorientation of ammonium ions (the corresponding T_1 minimum is at temperatures lower than those accessed). At intermediate temperatures a second T_1 minimum is observed. In this temperature range (prior to line-narrowing/increasing T_2 values) M_2 values are lower than is consistent with ammonium ions reorienting on fixed sites. Examination of the frequency dependence of the observed T_1 minimum[40] reveals a behaviour noted elsewhere[41] as being characteristic of low-dimensionality translational motion. Ammonium ions are hopping to-and-fro between cavity sites in a motion akin to the rattling of a slack chain. At the highest temperatures (around ambient) an increase in T_2 is observed, corresponding to a longer range (diffusive) motion involving hydrogens. In view of the ammonium ions themselves blocking long-range translation of NH_4^+, this motion must involve dissociation of some NH_4^+, with H^+ then translating either via sites corresponding to attachment to the framework as hydroxyl (i.e. in a similar manner to that occurring in the hydrogen bronzes) or possibly by hopping directly from an ammonium ion to an ammonia molecule (some H^+ then being localised on the framework) or a combination of both mechanisms.

QNS investigation of ammonium tungsten bronze[36] detected temperature-dependent quasielastic broadenings characteristic of reorientations over a remarkably large temperature range (100–300 K). Such behaviour is indicative of very low activation barrier(s) for the ammonium ion, which is consequential on the size and nature of the cavities occupied and consistent with the low temperature of the reorientational T_1 in NMR relaxation data (above). The *EISF* was temperature-dependent also, being characteristic of reorientation about a three-fold axis at low temperature and changing with temperature toward that characteristic of random tumbling at the highest temperatures. The window of six oxygen atoms linking cavities on the 'hexagonal axis' of hexagonal bronzes (Fig. 29.2) is actually puckered[42], resulting in a three-fold axis which could be the origin of the three-fold reorientation of the NH_4^+ ion (hydrogen-bonded to three atoms of the window). The temperature dependence of the EISF implies that reorientational barriers are low about all axes and tumbling therefore becoming more random at higher temperature. Very similar behaviour was found for the oxidised form (the hexagonal polytungstate above), in which the h-WO_3 framework persists.

A QNS study has also been made of the ammonium containing lamellar molybdenum bronze $(NH_4^+)_{0.085}H_{0.03}MoO_3$ (above)[43], in which the NH_4^+ ions lie in the interlayer region of the MoO_3 host (Fig. 29.3). While the general features observed are similar to those for the ammonium tungsten bronze (low activation barrier reorientations, temperature-dependent EISF), the low temperature behaviour is that for a four-fold reorientation of NH_4^+ about its C_2-axis, consistent with the ammonium ions being located within the 'squares' (actually flattened tetrahedral cavities) of terminal-O3 atoms in Fig. 29.4 (no interlayer H attached to O3 in this case).

Acknowledgement

The author gratefully acknowledges many years of fruitful discussions and continuing collaborations in this area involving Peter Dickens (Oxford) and Tom Halstead (York).

29.3   References

1. D. Wolf, *Spin-Temperature and Nuclear-Spin Relaxation in Matter* (Oxford University Press (1979)).
2. A. Abragam, *The Principles of Nuclear Magnetism* (Oxford University Press (1961)).
3. M. Mehring, *High Resolution N.M.R. Spectroscopy in Solids* (Springer-Verlag, Berlin (1976)).
4. M. Bée, *Quasielastic Neutron Scattering* (Adam Hilger, Bristol (1988)).
5. C. Ritter, *Z. Phys. Chem. N.F.* **151** (1987) S51.
6. M. A. Vannice, M. Boudart and J. J. Fripiat, *J. Catal.* **17** (1969) 359.
7. P. G. Dickens, D. J. Murphy and T. K. Halstead, *J. Solid State Chem.* **6** (1973) 370.
8. K. Nishimura, *Solid State Commun.* **20** (1976) 523.
9. P. J. Wiseman and P. G. Dickens, *J. Solid State Chem.* **6** (1973) 374.
10. C. Ritter, W. Müller-Warmuth and R. Schöllhorn, *Ber. Bunsenges. Phys. Chem.* **90** (1986) 357.
11. M. T. Weller and P. G. Dickens, *Solid State Ionics* **9–10** (1983) 1081.
12. P. G. Dickens and M. T. Weller, *J. Solid State Chem.* **48** (1983) 407.
13. S. J. Hibble and P. G. Dickens, *Mat. Res. Bull.* **20** (1985) 343.
14. C. Ritter, W. Müller-Warmuth, H. W. Spiess and R. Schöllhorn, *Ber. Bunsenges. Phys. Chem.* **86** (1982) 1101.
15. R. Hempelmann, P. R. Hirst, A. Magerl, R. C. T. Slade and H. Wipf, in *ILL Experimental Reports and Theory College Activities – 1984*, p. 336 (Report of Experiment 6-14-62).
16. R. C. T. Slade and P. R. Hirst, in *Transport–Structure Relationships in Fast Ion and Mixed Conductors* (Risø National Laboratory, Denmark, 1985), 377.
17. B. Gerand and M. Figlarz, in *Spillover of Adsorbed Species* (Elsevier, Amsterdam (1983)) 275.
18. J. J. Birtill and P. G. Dickens, *Mat. Res. Bull.* **13** (1978) 311.
19. R. C. T. Slade, T. K. Halstead and P. G. Dickens, *J. Solid State Chem.* **34** (1980) 183.
20. C. Ritter, W. Müller-Warmuth, H. W. Spiess and R. Schöllhorn, *Ber. Bunsenges. Phys. Chem.* **86** (1982) 110.
21. C. Ritter, W. Müller-Warmuth and R. Schöllhorn, *J. Chem. Phys.* **83** (1985) 6130.
22. A. C. Cirillo and J. J. Fripiat, *J. Physique* **39** (1978) 247.
23. A. C. Cirillo, L. Ryan, B. C. Gerstein and J. J. Fripiat, *J. Chem. Phys.* **73** (1980) 3060.

Proton diffusion mechanisms

24. R. E. Taylor, L. M. Ryan, P. Tindall and B. C. Gerstein, *J. Chem. Phys.* **73** (1980) 5500.
25. P. G. Dickens, A. T. Short and S. Crouch-Baker, *Solid State Ionics* **28–30** (1988) 1294.
26. A. T. Nicol, D. Tinet and J. J. Fripiat, *J. Physique* **41** (1980) 423.
27. R. C. T. Slade, T. K. Halstead, P. G. Dickens and R. H. Jarman, *Solid State Commun.* **45** (1983) 459.
28. T. M. Barbara, S. Sinha, J. Jonas, D. Tinet and J. J. Fripiat, *J. Phys. Chem. Solids* **47** (1986) 669.
29. R. C. T. Slade, P. R. Hirst, B. C. West, R. C. Ward and A. Magerl, *Chem. Phys. Lett.* **155** (1989) 305.
30. R. E. Taylor, M. M. Silva-Crawford and B. C. Gerstein, *J. Catal.* **62** (1980) 401.
31. P. G. Dickens, J. J. Birtill and C. J. Wright, *J. Solid State Chem.* **28** (1979) 185.
32. R. C. T. Slade, P. R. Hirst and H. A. Pressman, *J. Mater. Chem.* **1** (1991) 429.
33. R. H. Jarman, P. G. Dickens and R. C. T. Slade, *J. Solid State Chem.* **39** (1981) 387.
34. P. G. Dickens, A. J. Halliwell, D. J. Murphy and M. S. Whittingham, *Trans. Faraday Soc.* **67** (1971) 794.
35. P. R. Hirst, Ph.D. thesis (University of Exeter, 1986).
36. R. C. T. Slade, P. R. Hirst and B. C. West, *J. Mater. Chem.* **1** (1991) 781.
37. P. G. Dickens, S. J. Hibble and G. S. James, *Solid State Ionics* **20** (1986) 213.
38. R. C. T. Slade, A. Ramanan, P. R. Hirst and H. A. Pressman, *Mat. Res. Bull.* **23** (1988) 793.
39. L. D. Clark, M. S. Whittingham and R. A. Huggins, *J. Solid State Chem.* **5** (1972) 487.
40. R. C. T. Slade, P. G. Dickens, D. A. Claridge, D. J. Murphy and T. K. Halstead, *Solid State Ionics* (in press).
41. T. K. Halstead, K. Metcalfe and T. C. Jones, *J. Magn. Res.* **47** (1982) 292.
42. M. F. Pye and P. G. Dickens, *Mat. Res. Bull.* **14** (1979) 1397.

30 Conductivity mechanisms and models in anhydrous protonic conductors

PHILIPPE COLOMBAN AND ALEXANDRE NOVAK

During the past ten years, there have been numerous efforts to give an adequate description of superionic conduction, but they concern exclusively non-protonic conductors. In these compounds, neither tunnelling mechanisms nor oriented hydrogen bonds have been considered. However, in certain anhydrous compounds containing oxonium and ammonium ions, for instance, protonic conduction can also be interpreted in terms of such 'non-protonic'models. We shall start by giving a review of general models before discussing the problems particular to protonic conduction.

30.1 Theoretical interpretations of superionic conduction

Three main types of theoretical description can be distinguished. (i) Models derived from liquid or highly disordered solid electrolytes such as glasses or polymers[1-4]. (ii) Continuous models based on solutions of Langevin's equation[5]. These models can be improved by taking into account cooperative effects corresponding to liquid models (*Brownian* or stochastic particles) fitted to a periodic medium. (iii) *Hopping* models or *lattice gas* models, initially developed for phase transitions[6-9]. They can consider dimensions higher than one and are closely associated with studies on reorientational motion in plastic crystals and on *surface melting* such as occurs with rare gas layers on graphite[10-12].

Fig. 30.1 shows that for the diffusion of a particle, several cases can be distinguished depending on the relative magnitudes of residence time τ_0 on site A and of *time of flight* τ_1 to site B. Existence of (polyatomic) protonic species establishing an oriented bond implies a necessary correlation between translational and rotational motions during a diffusive jump.

Comparison of $P(\omega)$ functions obtained for different limiting cases indicates a quasi-continuity existing between a viscous liquid and a solid sometimes formed by highly overdamped oscillators (Fig. 30.2). The $P(\omega)$ function corresponds to the Fourier transform of the autocorrelation function of particle velocities. It is calculated from $S_{inc}(Q, \omega)$ spectra density determined by incoherent neutron scattering, for instance

$$P(\omega) = \lim_{Q \to 0} \omega^2 \frac{S_{inc}(Q, \omega)}{Q^2}$$

where Q is the *momentum transfer*[13] (see Chapter 21).

There is also $P(\omega) = \omega\sigma(\omega)$ where $\sigma(\omega)$ is the frequency dependent conductivity (see Chapter 25). In the case of a low conductivity ionic solid, $P(\omega)$ is typically of an oscillating behaviour (a peak at $\omega = \omega_0$, more or less damped, see Fig. 11.2) and there are only very rare diffusive events which contribute to $P(\omega = 0)$. In a liquid, the spectrum is centred at $\omega = 0$, since all the particles diffuse. When a liquid becomes (more) viscous a pseudo-oscillating behaviour may be observed (c), while the oscillator damping in a superionic conductor may decrease the difference between the time of flight between two sites and the time of oscillation on a site (Fig. 30.2), leading to a quasi-liquid state[14]. In order to simplify the model, either the diffusive or the oscillatory behaviour is assumed to be predominant. The choice may depend on the supposition of *relaxation time*

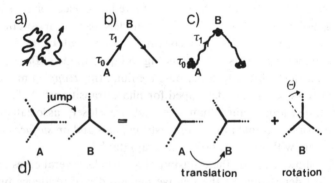

Fig. 30.1. Diagrammatic representation of fast particle diffusion ($D \sim 10^{-5}$–10^{-7} cm^2 s^{-1}): (a) in a non-viscous liquid the particle is diffusing all the time (Brownian motion); (b) in a hydride the jumps are rare and the residence time τ_0 is very long compared to the time of flight τ_1; (c) in a superionic conductor a quasi-liquid state is supposed to be synonymous to $\tau_0 \sim \tau_1$: the particle undergoes large amplitude oscillations during τ_0 in the vicinity of its site and is 'diffusing' during τ_1; (d) in the case of a protonic species which can create a directed hydrogen bond the diffusive jump from site A to site B can be decomposed into translational and rotational motion.

separation or on the method applied: usual conductivity measurements select long times while spectroscopy measurements such as infrared absorption $\alpha(\omega)$ select short times (see Chapter 25). Study of conductivity as a function of frequency up to the GHz range and determination of relaxation times appears thus particularly useful. The transformation from vibration-to-diffusional movement is not so straightforward as illustrated by the picture of $P(\omega) = f(\log \omega)$ (Fig. 30.2) or that of $\sigma(\omega) = f(\log \omega)$ (Fig. 25.2). A nice discussion of the transition between harmonic, damped oscillator models, multipotential well models and the Drude model has been given recently by Volkov *et al.*[15] in the light of submillimetric dielectric measurements.

30.1.1 Liquid-like models for highly disordered electrolytes

Glasses and *polymer* electrolytes are in a certain sense not solid electrolytes but neither are they considered as liquid ones. A glass can be regarded as a supercooled liquid and solvent-free polymer electrolytes are good conductors only above their glass transition temperature (T_g), where the structural disorder is dynamic as well as static. These materials appear macroscopically as solids because of their very high viscosity. A conductivity relation of the *Vogel–Tamman–Fulcher* (*VTF*) type is usually

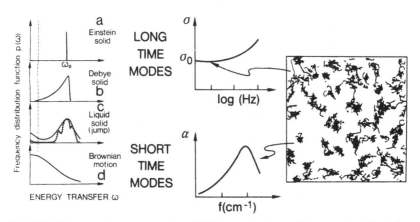

Fig. 30.2. Diagrammatic representation of diffusion paths observed by numerical simulation (with permission[1]) and their correspondence on the conductivity curve $(\sigma(\omega))$ and infrared absorption $(\alpha(\omega))$. Comparison of $P(\omega)$ functions obtained for an ionic solid (a, Einstein model; b, Debye model), a superionic compound (c, dashed line), a viscous liquid (c, solid line) and a Brownian liquid (d)[13].

459

Proton diffusion mechanisms

assumed[8], $\sigma = \sigma_0 \exp(-B/k(T - T_0))$. Whereas in the *Arrhenius* conductivity law $\sigma T = \sigma_0 \exp(-E_0/kT)$, where E_0 is *activation energy*, in the VTF form, T_0 is only an additional parameter and B a constant. In spite of its energy dimensions, B cannot be interpreted as a true activation energy. A more complex description is given by the *Williams–Landel–Ferry (WLF)* relationship initially established for the description of viscosity

$$\log \frac{\eta(T)}{\eta(T_s)} = \log a_T = -\frac{C_1(T - T_s)}{C_2 + (T - T_s)}$$

where T_s is an arbitrary reference temperature, a_T the mechanical shift factor and C_1 and C_2 are constants[16]. Coupling of the above relationship with the empirical *Walden's rule* for liquids[17], $D_n = (1/r_i) C$ (r_i is the radius of the mobile species) or with the *Stokes–Einstein equation*, $\sigma = DCe^2/kT$ (C, concentration; e, charge) leads to the WLF conductivity law

$$\log \frac{\sigma(T)}{\sigma(T_s)} = \frac{C_1(T - T_s)}{C_2 + (T - T_s)}$$

The conductivity then becomes

$$\sigma = \frac{Ce^2}{k} \exp \frac{-a}{C_2 + T - T_0}$$

where T_0 is the equilibrium glass transition temperature. A more microscopic approach based on percolation theories has also been given[8]. However, none of these theories takes into account ion–ion correlations which may be important, particularly for protonic materials. Typical polymer electrolytes for instance, have a ('mobile') ion concentration between 1 and 10 molar, which corresponds to an average ion–ion separation of 0.5–0.8 nm.

30.1.1.1 Glassy materials
The understanding of transport in glassy electrolytes appears to be even less developed than that in polymers. Glassy electrolytes are inhomogeneous on a 1–10 nm scale and the conductivity has been discussed in terms of intra and intercluster jumps[8]. For iono-covalent hard glasses, a

modified hopping model – with the parameters of the usual hopping model (see Chapter 4) for a hard framework reinterpreted as average values in a glass – can be used[18,19]. The main discussion here, as in all solid electrolytes, concerns the percentage of mobile ions. At high concentrations, most carriers are strongly coupled to counter charges and the situation becomes very similar to that in molten salts. Fig. 30.3 illustrates the strong correlation between viscosity and conductivity for pure, liquid H_3PO_4[20].

Glassy and polymer electrolytes may be considered as liquids[17,21] and Angell[1,22] has defined a *decoupling index* R_τ as the ratio of mechanical (viscosity η) and electrical relaxation time: $R_\tau = \tau(\eta)/\tau(\sigma)$. For some ionic glasses, $R_\tau = 10^{12}$, i.e. the electrical *relaxation time* is much shorter than the mechanical one. The classification can be extended to liquids[1]. 'Strong' liquids show essentially *Arrhenius* behaviour over a wide temperature range, whereas 'fragile' liquids show very pronounced curvature of $\log \sigma = f(10^3/T)$ with the *activation energy* increasing very markedly as

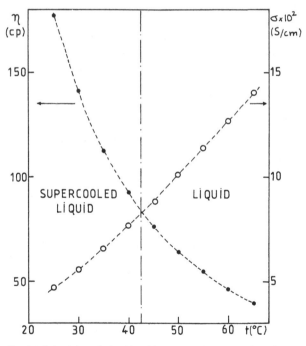

Fig. 30.3. Conductivity (σ) and viscosity (η) versus temperature for H_3PO_4 in the (super-cooled) liquid state[20].

Proton diffusion mechanisms

the temperature falls towards T_g. The behaviour of fragile liquids is associated with changes in free volume whereas the strong one is attributed to the persistence of network structure in the melt. For polymer electrolytes, on the other hand, $R_\tau \leqslant 1$, i.e. local mechanical stress relaxes faster than the electric field. The substantial change in R_τ value must be partly due to low frequencies and large amplitudes associated with polymer segment motions in the elastomeric phase above T_g. In fact, relaxation times are broadly distributed and it is difficult to say precisely which parts are relevant to ion motion. It should be pointed out that the WLF relationship implies strong correlations between these two kinds of relaxations. The microscopic model remains inadequate for the understanding of ionic conduction in highly disordered systems since inertial dynamics and interionic interactions are not taken into account.

30.1.2 Continuous model and hopping model

A review by Dieterich et al.[7] summarizes the distinction between hopping models, in which ions are assumed to be localized on a given set of sites, and dynamical models, in which the ionic motion is described by *Newton's equation* with proper interaction potentials. Molecular dynamics simulations have also been used to solve classical Newton's equations of motion in a potential. Many authors have used site hopping models to discuss ionic motion in solids and a review is given in Murch's book[23]. Hopping models are based on the idea that ions spend most of their time executing relatively low-amplitude vibrations about equilibrium lattice positions and much less time in jumping among various allowed sites (Fig. 30.1). Hopping models such as the *Einstein model* do not take into account inertial effects and solve master equations of the type

$$P_i(t) = \sum_{j \neq i} [P_i \omega_{i \to j}(p) + P_j \omega_{j - i}(p)]$$

where the probability $P_i(t)$ of occupying lattice site i at time t is determined by the probability $\omega_{j \to i}(p)$ for jumping from site j to site i. The interion correlations enter into the dependence of hopping rates on the ionic occupation p.

These models appear adequate for metal hydrides or non- or poorly-conductive halides (oxides). However, the hopping model is of limited value in superionic conductors since hopping and residence times are not so very different. Nevertheless, in the case of protonic conductors, it can be assumed that the proton, because of its specific properties (highly polarizing), keeps a localized behaviour with a very short hopping and long residential time in most solid conductors; the above models ought to be valid at least for the protonic conductors of low conductivity.

Historically, AgI-based (soft framework, a situation similar to that in plastic acid sulphate) materials were the first solid electrolytes to be studied in detail. Their structural disorder is believed to correspond to Ag^+ ion sublattice melting and the transport is governed by the 'immobile', counter ion lattice potential (low density approximation). This quasi-liquid state can be described by generalized *Langevin* dynamics

$$m_i x_i(t) = -e \operatorname{grad}[V_{\text{eff}}(x_i)] - m_i \gamma x_i - m_i \omega_i^2 \int_0^t \mathrm{d}\theta M(t - \theta) x(\theta) + R(t)$$

where m_i is the (effective) mass of the mobile species, e the charge and V_{eff} the effective potential. In this form, m_i is again for an independent particle model. There are the usual terms of *Newton's equation* (acceleration, velocity and displacement) as well as those taking into account the stochastic effect of the thermal bath. Furthermore, consideration of the lattice periodicity allows a link with lattice gas models in their hopping regime.

30.1.2.1 Lattice gas model

Lattice gas models can be applied to numerous systems such as intercalation compounds[24], adsorbed monolayers[11,12] and hydrogen in metals[25,26] (Chapter 6). In these models, particles hop from one site to another by thermal activation. This requires sufficient screening between ions that the host lattice potential becomes dominant, provided two ions do not seek to occupy the same site. Different regimes* are then possible and depend on the values of the following two parameters: (i) temperature versus hopping potential barrier height $(T/\Delta V)$ and (ii) γ-damping to ω_0-vibrational frequency ratio. Fig. 30.4 shows different cases, following the studies of Boughaleb and Gouyet[27].

Hopping models appear well suited at low reduced temperature, $T/\Delta V < 1$. The value of the energy lost to the lattice determines the other types of regime (diffusive, absolute or *Kramers*). Superionic conductors are expected to correspond generally to an underdamped region $(\gamma/\omega_0 < 1)$

---

* In all these descriptions, a separation of slow and fast systems is assumed: on a short time scale, the cation oscillates on a given site, while on a long time scale, it diffuses.

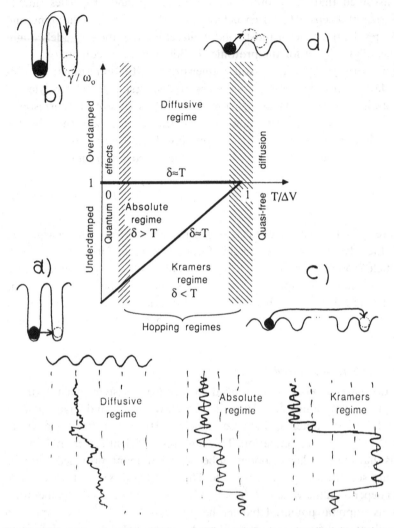

Fig. 30.4. Diagrammatic representation of different behaviour described by lattice gas models as a function of two critical parameters: $T/\Delta V$ (reduced temperature $\equiv$ potential barrier) and γ/ω_0 damping (coupling with lattice). Limiting regimes (a, tunnelling; b, jumping; c, ballistic or Kramer's regime; and d, liquid-like diffusion) are schematized in energy or in direct space. δ is the energy loss during a half oscillation; i.e. the jump of the particle between two consecutive wells. At low temperature quantum effects lead to discrete energy state in each potential well.

as shown by the $S(Q, \omega)$ curves of neutron scattering (Chapter 21) or more indirectly by Raman scattering (Chapter 25). This is manifested by a broad but defined peak near ω_0. In Kramer's model, where the particle diffuses above several wells, damping is small and the hopping model cannot be applied in its usual form. However, a rough notion of velocity can be introduced in such models using the framework of 'correlated random walks' to overcome this limitation[27,28]. A continuous model based on *Langevin's equation* can also be used[29]. The latter can be substituted by a statistical description using *Fokker–Planck* or *Smoluchowski* equations. The physics is the same: a liquid model with a friction term in the velocity in a periodic potential[30]. When the friction is low the particle stays in the bottom of the well and when it jumps it is flying (ballistic particle) inversely proportional to the potential barrier height (Kramer's dynamic correlation). When the friction term is high, the particle 'sticks' to the walls of the well and we have a non-correlated jump model. Most of the theoretical studies have been performed in the high friction limit, but most ionic conductors show a strong oscillating peak indicating that the intermediate or low friction regime appears more appropriate for the description of the dynamics of mobile species.

30.1.3   Ion–ion interaction

The most obvious experimental manifestations of interionic correlation are found in the *Haven ratio* (deduced from diffusion and conductivity measurement, see Chapter 4), in the static structure factor $S(Q)$ (deduced from partial occupation factors measured by X-ray or neutron diffraction) and from the dynamical properties ($S(Q, \omega)$, quasi- and inelastic-neutron scattering, $\sigma(\omega)$ frequency dependent conductivity) and $\varepsilon(\omega)$ dielectric relaxation.

Lattice gas models taking into account interactions between different particles have also been developed. First neighbour interactions are generally considered but also those with Coulomb interactions have been used[31]. The influence of Coulomb forces between mobile ions in combination with the static lattice potential thus plays an important role in non-protonic conductors (e.g. NASICON[32]). The *bounce-back effect* – an ion which performs a jump is expected to return to its initial location more often when repulsive interactions are present – plays an important role in modelling of frequency dependent conductivity (see Chapter 25): this tends to reduce the low frequency conductivity relative to the high frequency values. Conversely, the caterpillar mechanism (a hopping ion tends to push other ions in the same direction) may occur in fast-ion conductors. Transport coefficients and correlation effects have been studied following the 'path probability method'[33,34]. The presence of interactions leads to one or several transitions of the order–disorder type, by analogy with previous work on alloys or spin diffusion. The degeneracy of sites available to ions thus plays an important role[34]. The nature of interactions increases or decreases the conductivity value, in agreement with Monte–Carlo simulations[31,35].

If the interactions are strong – and this is frequently the case for mobile ions at high concentration – a liquid lattice model suits better but is more difficult to develop theoretically. Nevertheless, numerous trials have been attempted[36–38]. These models show the correlation effects between mobile ions to an advantage and in some cases, the models imply an effective potential, the form and barrier height of which depend on the mobile ion concentration*[39,40].

Fig. 30.5 shows the half width of the elastic peak $S(Q, \omega)$ for different models[9,27]. The width at half height (ΔQ) can be measured directly by

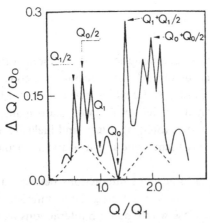

Fig. 30.5. Halfwidth $\Delta Q/\omega_0$ of the elastic peak $S(Q, \omega)$ as a function of the wave vector Q for a model without interaction between mobile species (Chudley–Elliott hopping model, -----)[45] and for a Boughaleb pair model (——)[9,27]. Such an $S(Q, \omega)$ curve can be obtained directly from neutron scattering experiments (see Chapter 21). Peaks on the dashed line are related to the lattice periodicity whereas most of the peaks on the solid line depend on dynamic correlations (with permission[27]).

* Models using the formalism of the dynamical structure factor allow account to be taken of the static (periodicity) and dynamic (coupling) correlations.

neutron scattering (see Chapter 21) but the experiments remain difficult and only a few materials have been studied (protonic, sodium and silver β/β''-aluminas[41,42], AgI[43], $CsHSO_4$[44]). Near the origin (Q, wave vector ~ 0), a hydrodynamic behaviour is observed while the lattice periodicity shows minima due to defined jumping distances according to the continuous hopping model in a periodical lattice but without dynamical interactions established by *Chudley & Elliott* for metal hydrides. This model has been extended by Rowe *et al.* to more complex lattices and used by Dickens *et al.* for analysis of $SrCl_2$[45]. Similar models have also been developed to interpret Raman spectra[46]. According to Boughaleb & Ratner[9], the presence of correlations in the dynamical behaviour of ions modifies strongly the curve by introducing other peaks.

Studies dealing with domain interfaces[47], dynamical percolation[48] and pathway models[49] attempt also to model the conductivity. A cluster bypass model was proposed recently for developing theories of conduction in glass which recognize the essential continuity of glass/melt behaviour. The continuous random network model is modified in order to provide pathways for ion migration which are located in the residual liquid surrounding microdomains (or clusters) within the glass. The existence of this residual liquid at T_g accounts for the high conductivities and the correspondingly high values of Angell's *decoupling index* R_τ[22,50]. Thus, a link is made with a solvent-containing network or with the 'old' weak electrolyte theory of (poorly conducting) glass: only those cations which at any instant are located outside the cluster are 'free' to participate in the conductive and diffusive processes.

30.2   Proton tunnelling

Proton conductors are similar to other ionic conductors, but there are some additional specific features: (i) the proton is localized on an oxygen or nitrogen atom, (ii) the proton, being the lightest nucleus, oscillates with the highest frequency and its energy spectrum cannot be considered as a continuum as is usually the case for heavier particles: i.e. proton energy levels are discrete and their separation is considerable. A 'thermally activated' *proton tunnelling* can thus frequently be expected, at least for non-exotic temperatures[51,52]. A solitonic mechanism has also been considered for materials with hydrogen bonded chains[53].

Conductivity has been calculated for the tunnelling effect, with models related to the previously described hopping models. For example, for a distribution of pair sites for charge carriers, we have

$$\sigma(\omega) = \int_0^\infty \mathbb{N}(R)\alpha(R)\rho(R)\,\frac{\omega^2\tau}{1+\omega^2\tau}\,\mathrm{d}R$$

where $\mathbb{N}(R)$ is the total number of site pairs per unit volume, $\rho(R)\,\mathrm{d}R$ is their spatial distribution, $\alpha(R)$ the polarizability and τ the relaxation time[51]. Conductivity, as a function of structure dimensionality (3D, 2D, 1D), can be calculated either by approximation or by a numerical treatment. These calculations are usually performed using the approximation method[54]. The general law can be represented as $\sigma(\omega) = A\omega^s$, with s taking values between 0 and 1 and more generally with $s \sim 0.5$–1. In this expression, s is either a constant parameter in the case of a tunnelling mechanism or dependent on the energy required to bring a proton from infinity to its site and on temperature for a thermally activated process. In fact, a more correct formula is $\sigma(\omega) = \sigma(0) + A\omega^s$, $\sigma(0)$ being the low frequency limit of conduction; the universal description of Jonscher can be found here[55] (see Chapter 25). If the classical relationship of *Arrhenius*, $\log \sigma T = f(10^3/T)$ is used to plot experimental data, the above relations are characterized by a continuous variation in slope. The best illustration is found in the low temperature phases (< 417 K) of $CsHSO_4$[56,57] (see Chapters 11 & 25). Hainovski *et al.* interpreted the low temperature conductivity in terms of proton tunnelling and their interpretation has been supported by the fact that deuterium substitution lowers the conductivity by two orders of magnitude[58]. In the high temperature superionic phase, on the other hand, there is little difference in the conductivity of H and D crystals and both obey the Arrhenius law with an activation energy of 0.33 and 0.27 eV[56,57], respectively. Here the proton tunnelling mechanism no longer applies. These energy values are similar to energy $\hbar\omega$ for a proton in $H_2(D_2)SO_4$[59,60], i.e. thermal activation is limited by excitation to the first oscillation level and high temperature (superionic) proton transport occurs according to the classical laws of mechanics. The phase transition therefore takes place when the top of the potential barrier is lower than the first thermally excited level of the proton[58].

30.3   Superionic protonic conductivity

Modelling of superionic conduction has received relatively little attention. Experimental studies, concerned mainly with the measurements of conductivity as a function of frequency over a broad range and neutron scattering, are few and/or are too recent to allow a critical analysis (see Chapters 21 & 25). $CsHSO_4$ appears to be the most thoroughly studied and reveals the following features. Raman scattering shows that the superionic phase, unlike the low temperature phase containing infinite chains of hydrogen bonded HSO_4^- ions, consists of open or cyclic $(HSO_4^-)_2$ dimers and that sulphate tetrahedra are undergoing rapid reorientation[44,61]. NMR[62]; elastic neutron scattering[44,61,63], on the other hand, indicates that protons (deuterons) are diffusing with a diffusion coefficient close to 10^{-7} cm^2 s^{-1}. The conductivity mechanism is interpreted in terms of successive steps: first there is an opening of (cyclic) dimers followed by a rotation of HSO_4^- anions and proton jumps to neighbouring anions. It accounts for the experimental spectra and desired jump distances[44]. However, the separation of three distinct motions, proton translation and two different rotations of HSO_4^-, appears too simple (but is consistent with dielectric relaxation[57]) and $S(Q, \omega)$ functions ought to be determined over a large (Q, ω) range in order to observe all the motions. This task is difficult since the available instruments allow only a partial exploration (see Chapter 21). A strong disorder is also evidenced for the Cs ions, both from Raman[61] and NMR[64] and can be related to a local disorder correlating with the reorientations of HSO_4^- tetrahedra.

Comparison of a series of $MDSO_4$ ($MHSeO_4$) compounds reveals that the widest temperature range of a superionic phase is observed for Cs salts[56,65] while the corresponding range in Rb and NH_4^+ salts is much narrower or exists only under pressure[66,67]. Finally, K and Li compounds do not exhibit a superionic phase. It has been concluded that geometrical considerations and alkali lattice disorder play an important role in conduction[44,61]. First, large cations allow the open framework needed for the nearly free rotation of tetrahedra. Second, the large ions are assumed to be moving locally in order to allow a rotational motion of XO_4 tetrahedra and thus proton diffusion*. The superionic phase transition corresponds always to a marked (anisotropic) dilatation[70] and selenates generally exhibit more stable conducting phases. This seems to be corroborated by the study of $Cs_{1-x}M_xHSO_4$ where partial substitition of Cs^+ ions by a smaller cation, M, wipes out completely the phase

---

* Coupled tetrahedra reorientation and ion diffusion are assumed for NASICON[68,69] and the Li_2SO_4 structure[35].

transition and brings together the corresponding electrical properties of the two phases[71]. It has also been shown that, as in the case of electronic conductivity, the presence of static defects decreases the superionic protonic conduction (breaking the dynamic correlation) while in the low temperature phases of lower conductivity, the presence of structural disorder increases the conductivity[61, 72].

Acknowledgements

We would like to thank Dr J. F. Gouyet for fruitful discussions and critical reading of the manuscript.

30.4 References

1. C. A. Angell, *Solid State Ionics* **18/19** (1986) 72–88; C. A. Angell, K. L. Ngai and C. B. Wright (eds), *Relaxation in Complex Systems* (U.S. Government Printing Office, Washington DC (1985)) 205–20.
2. C. A. Vincent, *Progr. Solid State Chem.* **17** (1987) 145–71.
3. M. B. Armand, *Ann. Rev. Mater. Sci.* **16** (1986) 246–71.
4. M. A. Ratner and D. F. Shriver, *Chem. Rev.* **88** (1988) 109–24.
5. H. R. Zeller, P. Bruesch, L. Pietronero and S. Strassler, in *Superionic Conductors*, G. D. Mahan and W. L. Roth (eds) (Plenum Press, New York (1976)) 201–9.
6. Y. Boughaleb and J. F. Gouyet, *Solid State Ionics* **9/10** (1983) 1401–8.
7. W. Dieterich, P. Fulde and I. Peschel, *Adv. Phys.* **29** (1980) 527–40; W. Dieterich, in *High Conductivity Solid Ionic Conductors, Recent Trends and Applications*, T. Takahashi (ed) (World Scientific, Singapore (1989)) 17–44.
8. M. A. Ratner and A. Nitzan, *Solid State Ionics* **28–30** (1988) 3–33; H. Cheradame, in *IUPAC Macromolecules*, H. Benoit and B. Rempp (eds) (Pergamon Press, Oxford (1982)).
9. Y. Boughaleb and M. A. Ratner, in *Solid Electrolytes, Recent Trends*, R. Laskar and S. Chandra (eds) (Academic Press (1989)) 515–51.
10. F. F. Abraham, W. E. Rudge, D. J. Auerbach and S. W. Koch, *Phys. Rev. Lett.* **52** (1984) 445–50.
11. E. Bauer, in *Structure and Dynamics of Surfaces II*, N. Schommers and P. Von Blanchenhagen (eds) *Topics in Current Physics* **43**, (Springer-Verlag, Berlin (1987)) 115–79.
12. M. T. Gillan, *Physica* **13113** (1985) 157–62.
13. P. A. Egelstaff, *Introduction to the Liquid State*, Ch. 11 (Academic Press, New York (1967)); J. C. Lassègues and Ph. Colomban, *Solid State Protonic Conductors II*, J. B. Goodenough, J. Jensen and M. Kleitz (eds) (Odense University Press (1983)) 201–13.
14. Ph. Colomban and A. Novak, *J. Mol. Struct.* **177** (1988) 277–308.

Anhydrous protonic conductors

15. A. A. Volkov, G. V. Kozlov, J. Petzelt and A. S. Rakitin, *Ferroelectrics* **81** (1988) 211–14; A. A. Volkov, G. V. Kozlov and J. Petzelt, *Ferroelectrics* **95** (1989) 23–7.
16. M. L. Williams, R. F. Landel and J. D. Ferry, *J. Am. Chem. Soc.* **77** (1955) 3701–7.
17. S. Smedley, *The Interpretation of Ionic Conductivity in Liquids* (Plenum Press, New York (1980)).
18. H. Tuller, D. P. Button and D. R. Uhlmann, *J. Non-Crystalline Solids* **40** (1980) 93–115.
19. J. N. Mundy and G. L. Jin, *Solid State Ionics* **21** (1986) 305–26.
20. R. A. Munson, *J. Phys. Chem.* **68** (1964) 3374–80; N. N. Greenwood and A. Thompson, *J. Chem. Soc.* (1959) 3485–93; 3864–7; N. N. Greenwood and A. Earnshaw, *Chemistry of the Elements* (Pergamon Press (1976)) 594–9; D.-T. Chin and H. H. Chang, *J. Appl. Electrochem.* **19** (1989) 95–9.
21. D. Ravaine and J. L. Souquet, *Phys. Chem. Glasses* **18** (1977) 27–40.
22. C. A. Angell, *Mat. Chem. Phys.* **23** (1989) 143–69.
23. G. E. Murch, *Atomic Diffusion Theory in Highly Defective Solids* (Trans. Tech., Aedermanshorf (1980)).
24. L. Pietronero and E. Tosatti (eds) *Physics of Intercalation Compounds* (Springer-Verlag, Berlin (1981)).
25. G. Alefeld and J. Völkl (eds) *Hydrogen in Metals I* (Springer-Verlag, Berlin (1978)).
26. R. G. Barnes, F. Borsa, M. Jerosch-Herold, J. W. Han, M. Belhoul, J. Shinar, D. R. Torgeson, D. T. Peterson, C. A. Styles and E. F. W. Seymour, *J. Less-Common Metals* **129** (1987) 279–85.
27. Y. Boughaleb, Thesis, Orsay (1987); Y. Boughaleb and M. A. Ratner, in *Superionic Solids and Solid Electrolytes, Recent Trends*, A. L. Laskar and S. Chandra (eds) (Academic Press) 515–52.
28. T. Gobron and J. F. Gouyet, *Phys. Rev.* **B39** (1989) 12189–99.
29. T. Geisel, *Phys. Rev.* **B20** (1979) 4294–302.
30. P. Bruesch, L. Pietronero, S. Strassler and H. R. Zelker, *Electrochimica Acta* **22** (1977) 717–23.
31. H. Singer and I. Peschel, *Z. Phys.* **B39** (1980) 333–8; A. Bunde, D. K. Chaturvedi and W. Dietrich, *Z. Phys.* **B47** (1984) 209–15.
32. J. P. Boilot, Ph. Colomban and G. Collin, *Solid State Ionics* **28–30** (1988) 403–10; H. Jacobson, H. Salomon, A. Ratner and A. Nitzan, *Phys. Rev.* **B23** (1981) 1580–7.
33. H. Sato and R. Kikuchi, *J. Chem. Phys.* **55** (1971) 677–702; R. Kikuchi and H. Sato, *J. Chem. Phys.* **55** (1971) 702–15; H. Sato in *Solid Electrolytes, Topics in Applied Physics* **21**, S. Geller (ed) (Springer Verlag (1977)) 23–40; T. Ishi, H. Sato and R. Kikuchi, *Solid State Ionics* **28–30** (1988) 108–14.
34. P. M. Richards, in *Fast Ion Transport in Solids*, P. Vashishta, J. N. Mundy and G. K. Shenoy (eds) (Elsevier North-Holland, Amsterdam (1979)), 349–53.
35. G. E. Murch and R. J. Thorn, *Phil. Mag.* **A37** (1978) 85–92; R. W. Impey, M. L. Klein and I. R. MacDonald, *J. Chem. Phys.* **82** (1985) 4690–8.
36. A. R. Bishop, W. Dietrich and I. Peschel, *Z. Physik* **B33** (1979) 187–94.

Proton diffusion mechanisms

37. G. Radons, T. Geisel and J. Keller, *Solid State Ionics* **13** (1984) 75–83.
38. W. Dieterich, *J. Stat. Phys.* **39** (1989).
39. R. O. Rosenberg, Y. Boughaleb, A. Nitzan and M. A. Ratner, *Solid State Ionics* **18/19** (1986) 127–35; 160–8.
40. U. Thomas and W. Dietrich, *Z. Phys.* **B62** (1986) 287–96.
41. J. C. Lassègues, M. Fouassier, N. Baffier, Ph. Colomban and A. J. Dianoux, *J. Physique (Paris)* **49** (1980) 273–80; M. Anne, Thesis, Grenoble (1985).
42. G. Lucazeau, J. R. Gavarri and A. J. Dianoux, *J. Phys. Chem. Solids* **48** (1987) 57–76.
43. K. Funke, *Solid State Ionics* **28/30** (1988) 100–8; K. Funke, in *Superionic Solids and Solid Electrolytes*, A. L. Laskar and S. Chandra (eds) (Academic Press, New York (1989)) 569–630.
44. Ph. Colomban, J. C. Lassègues, A. Novak, M. Pham-Thi and C. Poinsignon, in *Dynamics of Molecular Crystals*, J. Lascombe (ed) (Elsevier, Amsterdam (1987)) 269–74. (See also Chapter 21.)
45. C. T. Chudley and R. J. Elliott, *Proc. Phys. Soc.* **77** (1961) 353–8; J. M. Rowe, K. Sköld, H. E. Flotow and J. J. Rush, *J. Phys. Chem. Solids* **32** (1971) 41–52; M. H. Dickens, W. Hayes, P. Schnabel, M. T. Hutchings, R. E. Lechner and B. Renker, *J. Phys.* **C16** (1983) L1; T. Springer, *Quasi-Elastic Neutron Scattering for the Investigation of Diffusive Motions in Solids and Liquids* (Springer-Verlag, Berlin (1972)).
46. C. H. Perry and A. Feinberg, *Solid State Commun.* **36** (1980) 519–27; R. J. Elliott, W. Hayes, W. G. Kleppmann, A. J. Rushworth and J. F. Ryan, *Proc. R. Soc. Lond.* **A360** (1978) 317–45.
47. W. Van Gool and P. H. Bottleberghs, *J. Solid State Chem.* **7** (1973) 9–17; G. Collin, J. P. Boilot, Ph. Colomban and R. Comes, *Phys. Rev.* **B34** (1986) 5838–49; 5850–60.
48. R. Hilfer and R. Orbach, *Chem. Phys.* **128** (1988) 275–87.
49. F. A. Secco, *Solid State Ionics* **28/30** (1988) 921–3.
50. M. D. Ingram, *Mat. Chem. Phys.* **23** (1989) 51–61.
51. J. C. Giuntini, J. V. Zanchetta and F. Henn, *Solid State Ionics* **28–30** (1988) 142–7.
52. F. Freund, H. Wengeler and R. Martens, *J. Chimie Phys. (France)* **77** (1980) 837–41.
53. A. Gordon, *Solid State Commun.* **68** (1988) 885–7.
54. S. R. Elliott, *Phil. Mag.* **36** (1977) 1291–2; *Solid State Commun.* **27** (1978) 749–54; *Adv. Phys.* **36** (1987) 137–50.
55. A. K. Jonscher, *Dielectric Relaxation in Solids* (Chelsea Dielectrics Press, London (1983)).
56. A. I. Baranov, N. M. Shchagina and L. A. Shuvalov, *J.E.T.P. Lett.* **36** (1982) 459–62.
57. J. C. Badot and Ph. Colomban, *Solid State Ionics* **35** (1989) 143–9.
58. N. G. Hainovsky, Yu. T. Pavlukhin and E. F. Hairetdinov, *Solid State Ionics* **20** (1986) 249–53; N. G. Hainovsky, *Izvest. Sibir. Otd. ANSSSR, Ser. Khim. Nauk.* **5** (1984) 27–9.
59. P. A. Giguère and R. Savoie, *Can. J. Chem.* **38** (1960) 2467–72.

Anhydrous protonic conductors

60. R. Gillespie and E. A. Robinson, *Can. J. Chem.* **40** (1962) 644–57.
61. M. Pham Thi, Ph. Colomban, A. Novak and R. Blinc, *J. Raman Spectrosc.* **18** (1987) 185–94.
62. R. Blinc, J. Dolinsek, G. Lahajnar, I. Zupancic, L. A. Shuvalov and A. Baranov, *Phys. Stat. Sol.* **B123** (1984) K83.
63. A. V. Belushkin, I. Natkaniec, N. M. Plakida and J. Wasicki, *J. Phys. C (Solid State Phys.)* **20** (1987) 671–87.
64. J. Dolinsek, R. Blinc and A. Novak, *Solid State Comm.* **60** (1989) 877–9.
65. N. G. Hainovsky and F. F. Hairetdinov, *Izvest. Sibir. Otd ANSSSR, Ser. Khim. Nauk.* **8** (1985) 33–5.
66. B. V. Merinov, A. I. Baranov, L. A. Shuvalov and B. A. Moksinov, *Kristallografya* **32** (1987) 86–92.
67. A. I. Baranov, E. G. Ponyatowsky, V. V. Sinitsyn, R. M. Fedosyuk and L. A. Shuvalov, *Soc. Phys. Crystallogr.* **30** (1985) 1121–3.
68. Ph. Colomban, *Solid State Ionics* **21** (1986) 97–115.
69. K.-D. Kreuer, *J. Mol. Struct.* **177** (1988) 265–76.
70. Ph. Colomban, M. Pham Thi and A. Novak, *Solid State Ionics* **24** (1987) 193–203.
71. T. Mhiri and Ph. Colomban, *Solid State Ionics* **35** (1989) 99–103.
72. Ph. Colomban, J. C. Badot, M. Pham-Thi and A. Novak, *Phase Transit.* **14** (1989) 55–68.

31 Conduction mechanisms in materials with volatile molecules

KLAUS-DIETER KREUER

Compounds containing loosely bonded molecules form by far the majority of all known solid proton conductors. Most of them are hydrates and their conductivity generally strongly depends on their state of hydration (see Chapter 3). Therefore, conduction mechanisms similar to the ones discussed for aqueous solutions have been suggested to hold also for this family of solids. But this is still a matter of ongoing research and in the following, a recent concept will be put forward using, as a background, the previous models. It provides directly a framework for explaining proton conduction in solid hydrates.

The peculiarity of proton transport in aqueous solutions was recognized early on by the exceptionally high limiting equivalent conductivity of the proton ($350 \, \mathrm{cm}^2 \, \Omega^{-1}$ at 25 °C[1]) compared to that of any other monovalent cation, data for which fall into a very narrow range (39, 50, 74, 78, 77, 74 $\mathrm{cm}^2 \, \Omega^{-1}$ for Li^+, Na^+, K^+, Cs^+, Rb^+, NH_4^+ at 25 °C[1]). The latter similarity is a direct consequence of cation solvation by water molecules, which diffuse at approximately the same rate ($2.26 \times 10^{-5} \, \mathrm{cm}^2 \, \mathrm{s}^{-1}$ at 25 °C).

The earliest ideas about proton conduction in aqueous solutions were stimulated by considerations of the electrolytic decomposition of water in 1806. Grotthuss postulated chains of water dipoles along which electricity may be transported[2]. One basic step which is part of any proton conduction mechanism formulated since the early days of physical chemistry had already been described by several authors at the beginning of this century[3-5]. They recognized that intermolecular proton transfer can lead to charge transport at a rate exceeding that of other species. The first formal theory was attempted by Hückel in 1928[6] at a time when the existence of a discrete H_3O^+ ion had already been suspected[7,8]. Hückel treated this species as a dipole and tried to calculate its reorientational rates into positions favourable for proton transfer to neighbouring water molecules. A first quantum mechanical theory of intermolecular proton transfer was presented by Bernal & Fowler in 1933[9] putting the old concept of Grotthuss chains into a modern dress. The rotation of water molecules, necessary for successive proton transfer, was assumed to be approximately equal to that of a structureless liquid like argon. From their calculation they anticipated a strong isotopic effect for the equivalent conductance of H^+ and D^+ in H_2O and D_2O respectively, which could not be confirmed later[10]. In 1949, Gierer & Wirtz tried to explain the extra conductivity of protons by an elementary process with a single activation enthalpy, but with the temperature dependent, ice-like part of the water structure excluded from this process[11]. In the mid-1950s Conway, Bockris & Linton[12] as well as Eigen and DeMaeyer[13] focused on the question of the rate-limiting step. They all agreed that classical proton transfer is slow, proton tunnelling is much faster and the rotation of hydrogen bonded water molecules near the H_3O^+ ion is the rate-determining reaction. The latter authors describe the extra conductivity as being within the $H_9O_4{}^+$ ion (primary hydration of H_3O^+) which is assumed to be of particular stability. A more recent theoretical contribution by Halle & Karlström[14] should be mentioned in which the non-equilibrium polarization of the solvent (water) is concluded to contribute to charge transport. The role of the solvent is also stressed by the considerations of Hertz[15] who strived for a correct formal description of the *Grotthuss mechanism*. His essential objection is that there must be bodily proton transport, i.e. the displacement of protonic charge (conductance) must be reflected in a corresponding displacement of protonic mass (diffusion) and the trajectories of these displacements must be macroscopically closed from one electrode to the other. This in particular makes it clear that water dipole rotation contributes to the protonic current. Of course this list of contributions is not complete and the reader may find other aspects and comparisons[11-13].

31.1   Proton conduction mechanism in dilute acidic aqueous solutions

All the models presented so far lack quantitative experimental verification. Common to most mechanisms is that the equivalent conductivity of protons (Λ_{H^+}) is thought to be composed of a so-called 'normal' conductivity, i.e. the diffusion of H_3O^+ ions ($\Lambda_{H_3O^+}$) and a so-called '*extra conductivity*' which are thought to be mutually independent.

This is most likely an invalid assumption considering the temperature dependence of the H_2O diffusion coefficient and the diffusion coefficient calculated from the equivalent conductance of protons in dilute aqueous solutions using the *Nernst–Einstein relationship* and the acid molarity (HCl) as the charge carrier concentration (Fig. 31.1). Both show a similar decrease in activation enthalpy with temperature. The temperature dependence of the ratio $D_{cal}(T)/D_{H_2O}(T) = A(T)$ (the *amplification factor*) almost resembles the one of the fraction of broken *hydrogen bonds* in pure water, however with opposite sign (Fig. 31.2). At room temperature, for example, the activation enthalpy for forming a terminal OH is 69 meV compared with -65 meV for the amplification factor. The fraction of the former is thought to be directly related to the *structure of water*[20].

The following model for proton conduction in aqueous solution, therefore, connects molecular diffusion with proton conduction and takes into account the structure of water in aqueous solutions.

In dilute acidic solutions ($m < 0.1$) the structure of water is largely retained. Because of hydrogen bond interactions pure water is frequently thought to be associated forming *clusters* in which the protonic ends of the water dipoles point to the cluster margin[20]. The H_3O^+ carrying the acidic proton finds its most favourable position in the centre of the cluster for simple electrostatic reasons.

There it undergoes normal molecular diffusion, i.e. it changes its

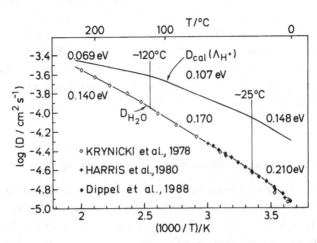

Fig. 31.1. The diffusion coefficient of pure water[16–18] and the diffusion coefficient calculated from the limiting equivalent conductance of protons (Λ_{H^+}) for aqueous solutions of hydrochloric acid.

position with a neighbouring water molecule as shown in the one-dimensional representation of Fig. 31.3. This first step has a correlation time $\tau_D = 5.9 \times 10^{-12}$ s at room temperature with an activation enthalpy of 170 meV (Fig. 31.1). This enthalpy and the corresponding momentum will subsequently dissipate to the aqueous environment. In favourable cases, this will happen in the direction of the molecular jump but anyway to the cluster margin. Via elastic collisions a compression wave propagates from the cluster centre to the cluster margin with longitudinal sound velocity. In this wave the intermolecular distances decrease transiently and the proton transfer probability increased correspondingly. Because of the direction of the proton momentum, this intermolecular proton transfer will turn the orientation of the hydrogen bonded chain from the cluster centre to its margin (second step in Fig. 31.3 $\tau_p = 7.4 \times 10^{-14}$ s). In this way, a new H_3O^+ forms at the cluster margin.

As a third step a new cluster starts to form ($\tau_C = 4.2 \times 10^{-11}$ s) by progressive reorientation of the water molecules surrounding the H_3O^+ before another molecular jump in the centre of the growing cluster initiates the dissolution of its structure.

The above mechanism is 'triggered' by the diffusion of H_3O^+. It is a collective process in which all protons, including the protons of the solvent (H_2O) are involved. Displacements of the individual protons by H_3O^+ diffusion, intermolecular transfer and molecular reorientation are small, as can be seen from Fig. 31.3, but their net effect is a closed trajectory

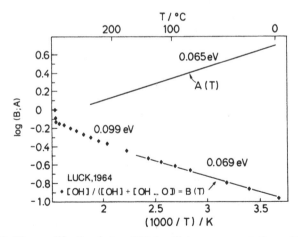

Fig. 31.2. The amplification factor $A(T)$ for dilute aqueous solutions of hydrochloric acid and the portion of broken hydrogen bonds in pure water $B(T)$[19].

describing a displacement of one protonic charge over a distance identical to the radius of the transient cluster. Thus, there are no exceptionally fast individual protons in accordance with many experimental results[21]. One therefore should strictly talk about the equivalent conductivity rather than about the mobility of protons. The quantitative description requires the

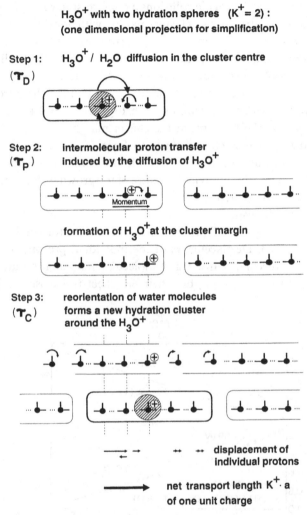

Fig. 31.3. Proton conduction mechanism in acidic aqueous solution: 'structure diffusion triggered by molecular diffusion' (see text). Symbol: a, effective jump length. For 300 K the parameters are: $a \approx 0.208$ nm; $K^+ \approx 2.1$; $A \approx 4.5$; $\tau_D \approx 6 \times 10^{-12}$ s; $\tau_p \approx 7 \times 10^{-14}$ s; $\tau_c \approx 4 \times 10^{-11}$ s.

calculation of the rate of each step. For the initial diffusional step it is simply

$$1/\tau_D = 6 \cdot D_{H_2O}/a^2 \qquad (31.1)$$

where a is the molecular jump distance, (taken as 0.208 nm). $1/\tau_P$ can be estimated from the transient cluster radius and the longitudinal sound velocity in the cluster. For the latter one preferentially takes the extrapolated value for ice[22] rather than that of water because the cluster structure is considered to be similar to that of hexagonal ice[20]. In any case

$$1/\tau_P = v_L/[(A^{1/2} - 1) \cdot a] \text{ (where } v_L \text{ is the longitudinal sound velocity)}$$

$$(31.2)$$

is the highest rate and has little effect on the total proton transport rate. $1/\tau_C$, the rate of the third step, is not directly accessible. The formation of a cluster around H_3O^+ is actually always interrupted by the next jump of the H_3O^+. The final average equilibrium cluster radius in the absence of H_3O^+ diffusion can therefore be observed only for pure water. The spectroscopic results of Luck[19] (Fig. 31.2) allow directly the calculation of this parameter. Assuming a simple spherical shape of the cluster, and with N being the average number of broken hydrogen bonds per water molecule at the cluster margin, the number of hydration shells K for a pure water cluster is calculated as

$$K = 0.98 \cdot N/B - 0.65 \quad \text{(where } B = [OH]/([OH] + [OH \ldots O])) \quad (31.3)$$

Assuming a simple harmonic behaviour for the cluster formation and dissolution the corresponding rate can be expressed readily as

$$\frac{1}{\tau_C} = \frac{1}{\tau_D - \tau_P} \left[\arccos \left(\frac{K - A^{1/2}}{K} \right) \middle/ 2\pi \right] \qquad (31.4)$$

(for $K > 1$ and $\tau_D \ll \tau_C$) where K is the number of hydration spheres of a pure water cluster. All parameters are experimentally accessible and the calculated values for $1/\tau_C$ are presented in Fig. 31.4 together wih $1/\tau_D$, $1/\tau_P$ and results of *dielectric dispersion* measurements on pure water[23,24]. The latter is the fastest absorption process observed in the microwave region, which is frequently attributed to flickering clusters[25]. The reasonable coincidence with the calculated values for $1/\tau_C$ is support for this assignment and the proposed model for proton transport. There is further support from the variation of $A(T)$ with temperature (Fig. 31.2). At about

Proton diffusion mechanisms

300 °C it equals unity, i.e. there is no amplification and proton conductance is simply carried by molecular diffusion. But, there is still significant association of water at that temperature (Fig. 31.2) corresponding to primary hydration of each molecule on average. This configuration does not allow any amplification according to the proposed mechanism (Figs 31.3 and 31.4) in contrast to the model of Eigen & De Maeyer who allot the 'extra conductivity' to the presence of $H_9O_4{}^+$.

It should be noted here that quantum mechanical considerations are not required for calculation of the total proton transport rate. Intermolecular proton transfer which might have contributions from *proton tunnelling*, is triggered by a longitudinal acoustical phonon which can be described by classical mechanics. This is in accordance with a small isotope effect for the equivalent conductance of $^1H^+$ and $^2H^{+}$ [26]. The equivalent conductivity of protons in dilute aqueous solutions can now be expressed in terms of the rates of each step

$$\Lambda_{H^+} = \left[(K)\left(1 - \cos\left(\frac{\tau_D - \tau_P}{\tau_C} \cdot 2\pi \right) \right) \right]^2 \cdot \frac{q^2 a^2 m}{6kT} \cdot \frac{1}{\tau_D} \qquad (31.5)$$

where the first term is the amplification factor A.

The equation demonstrates nicely the ambivalent role of *hydrogen bonds* for proton conduction. On the one hand, they promote the formation of water clusters characterized by the average cluster radius, the cluster

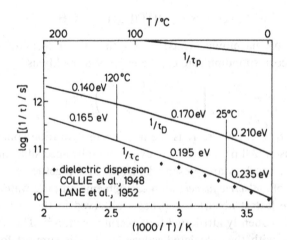

Fig. 31.4. The rates of each step of the proton conduction mechanism presented in Fig. 31.3. $1/\tau_P$ and $1/\tau_D$ are experimentally accessible whereas $1/\tau_C$ is calculated from the model. The latter is compared with the results of dielectric dispersion measurements on pure water[23,24].

formation rate $(1/\tau_C)$ and the amplification factor (A). On the other hand hydrogen bonds reduce the molecular diffusion rate $(1/\tau_D)$ which triggers the entire mechanism. From such considerations we expect an optimum value for the hydrogen bond strength with respect to proton conductivity. Indeed this optimum can be seen in the pressure dependence of the equivalent proton conductivity which, in contrast to that of any other ion, shows a maximum around 0.5 GPa[27].

The interested reader may find more details about the mechanism and the complete data set in reference 28.

31.2 Proton conduction mechanism in concentrated acidic aqueous solutions

Let us now extend the model to proton conductivity in concentrated aqueous solutions, which approaches the situation for solid hydrates. In those, the number of clusters is no longer an intrinsic feature of water, but is determined by the concentration of H_3O^+. The hydration clusters of H_3O^+ cannot extend as deeply into the solvent as is the case for dilute solutions. While the cluster radius decreases with increasing molarity, the cluster formation rate increases and the amplification factor is expected to decrease. Assuming a primary hydration of H_3O^+ by three H_2O molecules, octahedral hydration of the anion and tetrahedral coordination of the solvent in the hydration cluster, the average number of hydration spheres can be estimated. Assuming that $1/\tau_C$ is no longer rate determining the amplification factor then simply reduces to

$$A = (K_{cal}{}^+(m))^2, \; K_{cal}{}^+(m) > 1 \; \text{(primary hydration)} \qquad (31.6)$$

where $K_{cal}{}^+(m)$ is the number of hydration spheres surrounding the H_3O^+ ion. A is now a characteristic number of the average structure and is expected to be independent of temperature, which is indeed observed[29]. Fig 31.5 shows experimental values for A versus the molarity of aqueous HCl solutions at room temperature. They are in reasonable agreement with the curve predicted for high molarities. Also the number of water molecules n per acid proton is indicated.

31.3 Proton conduction mechanism in n solid acidic hydrates

For solid acid hydrates in which the proton is a main constituent of the compound, n is usually smaller than ten according to an amplification factor A not greater than ~ 2 (shaded range in Fig. 31.5). Therefore, if proton conduction in solid hydrates obeys the same model, it is expected to be determined mainly by molecular diffusion.

Whether or not this is really so is discussed next. The most extensively studied compound is probably $H_3OUO_2PO_4 \cdot 3H_2O$(HUP) (see Chapter 17). Its conduction mechanism was deduced from structural/chemical considerations. In analogy to *Grotthuss*-type conductivity contributions claimed for aqueous solutions at that time, transport processes involving intermolecular proton transfer on an infinite hydrogen bonded network and molecular reorientation were assumed[30–35]. The latter step was thought to be rate determining.

Quantitative determination of water reorientational and diffusional modes by NMR and ^{18}O tracer experiments on the analogous HUAs provided no correlation with the data for proton conductivity[36]. Instead a good correlation between molecular diffusion and proton conductivity indicated a mechanism in which H_2O and H_3O^+ are mobile as a whole in the conduction plane without a significant amount of proton transfer between them. The proton conductivity in HUP and HUAs is somewhat higher than that of other monovalent cations in corresponding compounds. This was one of the reasons why molecular diffusion was not

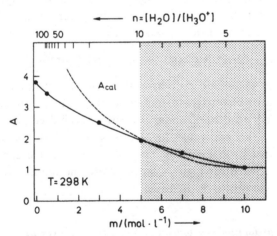

Fig. 31.5. Amplification factor $A(T)$ for aqueous solutions of hydrochloric acid at room temperature as a function of molarity. The dashed line is calculated according to the model (see text) with $1/\tau_D$ being the rate-determining step for proton conductance.

considered at the beginning. The high diffusion coefficient for H_3O^+ compared to that of K^+, for instance, may be due to the higher polarizability and lower symmetry of H_3O^+ and its likeness to H_2O. But compared with the diffusion coefficient in aqueous solution, it is still low. In any case the diffusion of H_3O^+ has to be somewhat cooperative with that of H_2O because of simple geometrical reasons. This frequently gives rise to significant entropy contributions for conduction. As no amplification takes place ($A = 1$), in accordance with the stoichiometry of the conduction plane ($H_9O_4^+$, see Chapter 17), hydrogen bonds are expected to hinder the total conductivity. Because the proton is carried by the water molecule the mechanism is referred to as *vehicle mechanism*[37]. According to this mechanism, protons are transferred by the conductance of H_3O^+ and an equivalent counter diffusion of H_2O. For the practical use of such materials as separators in electrochemical cells it is important to know that the proton is not the only mobile particle and that oxygen, via H_2O diffusion, can also be transferred between redox electrodes[38]. More features characteristic of the vehicle mechanism are collected in reference 36.

Unfortunately HUAs is the only proton-conducting hydrate for which available oxygen diffusion coefficient data allow direct verification of the vehicle mechanism. For other compounds there is just an indication that this may be the mechanism. These are e.g. diffusion bottleneck considerations for ionic conduction in zeolites[38] or the exceptionally high temperature factors for the water oxygen in HUP[39–41] and some heteropoly-acid-hydrates including the high degree of disorder for the water of hydration[42,43]. There is some indication for molecular transport in compounds different from hydrates. Coincidence of the activation enthalpies of ionic conductivity and translational motion in $LiN_2H_5SO_4$ suggests transport of protons by means of N_2H_4 molecules, i.e. $N_2H_5^+$-conductivity[37].

In general, the low thermal stability and high vapour pressure for the anticipated vehicle molecule make plausible a conduction mechanism similar to the one in acid solutions for this family of compounds. If we assume the mechanism presented above, then at least some amplification for proton conduction in the solid hydrate $TSA.28H_2O$ is observed ($A = 1.8$) (Fig. 26.3). This is in perfect agreement with a ratio $[H_3O^+]/[H_2O] = 6$ for this compound (Fig. 31.5). A is indeed independent of temperature and appears to be a structural feature as suggested by the model.

But one has to be careful with generalizations. There is no apparent reason why the structure diffusion of protons should be initiated only by molecular diffusion processes, as is probably the case for aqueous solutions. Why should not local modes, which have a momentum in the direction of a hydrogen bonded chain, be able to trigger reversal of the direction of polarization in the chain? The pressure dependence of the proton conductivity of HUAs, for instance, shows a sudden drop of the activation enthalpy and the pre-exponential factor around 2.5 GPa, suggesting an onset of structure conductivity[44]. Whether the process is released by molecular diffusion has not yet been proven.

Answers can be obtained only from quantitative proton diffusion data and that for the dynamics of the next proton environment. As mentioned above, the self-diffusion coefficient of oxygen in hydrates is of particular interest. In principle, this can be measured by ^{17}O-*PFG-NMR* (see Chapter 26) but the strong quadrupolar relaxation of ^{17}O, i.e. short T_2, restricts the method to a few fast proton conductors with high symmetry, such as heteropolyacidhydrates. For other compounds, e.g. for protonic β''-aluminas, ^{18}O-tracer diffusion measurements should be carried out.

In summary, there are many indications for a proton conduction mechanism in compounds containing volatile molecules (especially hydrates) similar to that in concentrated acid solutions. According to the model presented, the transport of protonic charge occurs via diffusion of H_3O^+ ions which may trigger some 'structure diffusion' leading to an amplification A for more than primary hydration of H_3O^+. With the number of solvent (water) molecules per H_3O^+, n, smaller than 10, as is the case for most known solid proton conductors, A does not exceed two. For $n \leqslant 4$, proton conduction is carried exclusively by molecular diffusion. This limiting case corresponds to the vehicle mechanism of proton transport, which has been confirmed experimentally for proton conduction in HUAs. The answer to the question whether the mechanism presented is of general validity for proton conduction in any compound containing loosely bonded molecules requires more experimental data especially on molecular diffusion (dynamic disorder) and transient structures of molecular ensembles (dynamic order). The interplay of both features is thought to produce macroscopic proton conductivity.

31.4   References

1. G. Kortüm, *Lehrbuch der Elektrochemie* (Verlag Chemie, Weinheim (1972)).
2. C. J. D. von Grotthuss, *Ann. Chim.* **LVIII** (1806) 54.

Materials with volatile molecules

3. Dempwolff, *Physik. Zeitschr.* **5** (1904) 637.
4. S. Tijmstra, *Z. Elektrochem.* **11** (1905) 249.
5. H. Danneel, *Z. Elektrochem.* **11** (1905) 249–52.
6. E. Hückel, *Z. Elektrochem.* **34** (1928) 546–62.
7. H. Goldschmidt and A. Hantzsch, personal communication to E. Hückel.
8. K. Fajans, *Naturwissenschaften* **9** (1921) 729.
9. J. D. Bernal and R. H. Fowler, *J. Chem. Phys.* **1** (1933) 515–48.
10. A. Gierer, *Z. Naturforsch* **5a** (1950) 581–9.
11. A. Gierer and K. Wirtz, *Annalen der Physik* **6** (1949) 257–304.
12. B. E. Conway, J. O'M. Bockris and H. Linton, *J. Chem. Phys.* **24** (1956) 834–50.
13. M. Eigen and L. DeMaeyer, *Proc. R. Soc. Lond.* **A247** (1958) 505–33; M. Eigen, *Angew. Chem.* **75** (1963) 489–588.
14. B. Halle and G. Karlström, *J. Chem. Soc. Faraday Trans. II* **79** (1983) 1047–73.
15. H. G. Hertz, B. M. Braun, K. J. Müller and R. Maurer, *J. Chem. Education* **64** (1987) 777–84.
16. K. Krynicki, Ch. D. Green and D. W. Sawyer, *Faraday Discuss. Chem. Soc.* **66** (1978) 199–206.
17. K. R. Harris and L. A. Woolf, *J. Chem. Soc. Faraday Trans. I* **76** (1980) 377–85.
18. T. Dippel, K.-D. Kreuer and A. Rabenau, *IVth European Conference on Solid State Protonic Conductors*, Exeter (1988).
19. W. A. P. Luck, *Ber. Bunsengesellschaft* **69** (1965) 626–37.
20. W. A. P. Luck, *Fortschr. chem. Forsch.* **4** (1964) 653–781.
21. H. Bertagnolli, P. Chieux and H. G. Hertz, *Z. Phys. Chem. Neue Folge* **135** (1983) 125–40.
22. Landolt-Bernstein, K.-H. Hellwege (ed), Gruppe III, Bd. 1 (Springer–Verlag (1966)).
23. C. H. Collie, J. B. Hasted and D. M. Ritson, *Proc. Phys. Soc.* **60** (1948) 145–60.
24. J. A. Lane and J. A. Saxton, *Proc. R. Soc. Lond.* **A213** (1952) 400–8.
25. H. S. Frank, *Proc. R. Soc. Lond.* **A247** (1958) 481–92.
26. A. Gierer, *Z. Naturforsch.* **5a** (1950) 581–9.
27. E. U. Franck, D. Hartmann and F. Hensel, *Discuss. Faraday Soc.* **39** (1965) 200–6.
28. K.-D. Kreuer, to be published.
29. T. Dippel, K.-D. Kreuer and A. Rabenau, *Solid State Ionics*, **46** (1991) 3–9.
30. A. T. Howe and M. G. Shilton, *J. Solid State Chem.* **34** (1980) 149–55.
31. B. Morosin, *Acta Cryst.* **34** (1978) 3732–4.
32. B. Morosin, *Phys. Lett.* **65A** (1978) 53–4.
33. P. E. Childs, T. K. Halstead, A. T. Howe and M. G. Shilton, *Mat. Res. Bull.* **13** (1978) 609–19.
34. W. A. England, M. G. Cross, A. Hamnett, P. J. Wiseman and J. B. Goodenough, *Solid State Ionics* **1** (1980) 231–49.
35. C. M. Johnson, M. G. Shilton and A. Howe, *J. Solid State Chem.* **37** (1981) 37–43.
36. K.-D. Kreuer, A. Rabenau and R. Messer, *Appl. Phys.* **A32** (1983) 45–53.

37. K.-D. Kreuer, A. Rabenau and W. Weppner, *Angew. Chem.* **94** (1982) 224–5; *Angew. Chem. (Int. Ed. Engl.)* **21** (1982) 208–9.
38. K.-D. Kreuer, W. Weppner and A. Rabenau, *Mat. Res. Bull.* **17** (1982) 501–9.
39. L. Bernard, A. Fitch, A. F. Wright, B. E. Fender and A. T. Howe, *Solid State Ionics* **5** (1981) 459–62.
40. A. N. Fitch, L. Bernard and A. T. Howe, *Acta Cryst.* **C39** (1983) 159–62.
41. A. N. Fitch and B. E. F. Fender, *Acta Cryst.* **C39** (1983) 162–6.
42. M.-R. Spirlet and W. R. Busing, *Acta Cryst.* **B34** (1978) 907–10.
43. C. J. Clark and D. Hall, *Acta Cryst.* **B32** (1976) 1545–7.
44. K.-D. Kreuer, I. Stoll and A. Rabenau, *Solid State Ionics* **9/10** (1983) 1061–4.

A · Energy storage and production

32 Applications of perfluorinated proton conductors (Nafions)

CLAUDE GAVACH AND GERALD POURCELLY

32.1 Introduction

As reported in a previous chapter, under special conditions, perfluorosulphonic ion-exchange polymers behave as proton conducting materials. They possess a set of specific properties which make them suitable for use in various fields.

These *ion-exchange materials* are permeable only to cations – more specifically to protons – and also to water. Some polar organic solvents of low molecular weight can also permeate, to a much lower extent, through this perfluorosulphonic material. The easy water transport through the polymer prevents local drying which could otherwise occur during specific applications. At sufficiently high degrees of *swelling*, the electrical conductivity of *perfluorosulphonic polymers* in the protonated form is high and exceeds that of the material in other cationic forms.

Nafion[R] is the leader of perfluorosulphonic proton conductors. Produced by Dupont de Nemours[1], it is commercially available in different forms: homogeneous or reinforced membranes, powders, tubes and solutions. From solutions of perfluorosulphonic acid or salt in organic solvents, gels and films can also be made. From Nafion[R], many kinds of composite material may also be prepared either by including small, dispersed particles inside the polymeric phase or by using it to coat solid electronic conductors. The adhesive power of this material is important when it is maintained in contact with a wide variety of supports.

Perfluorosulphonic proton conductor is characterized by a high mechanical stability, excellent chemical inertness and very interesting thermal stability. These advantages permit its use in applications involving high temperature ranges and corrosive environments. The main disadvantage of this proton conductor is its high cost. A new perfluorosulphonic ionomer has recently been proposed by Dow Chemicals Company[2-4].

Perfluorosulphonic membranes have also been produced by grafting, under irradiation, fluorinated monomers onto perfluorinated, preformed films[5-7]. A composite perfluorinated proton conductor can also be prepared from porous polyethylene sheets impregnated with a Nafion[R] organic solution[8].

32.2   Solid polymer electrolyte (SPE) technology

Fuel cells, batteries and electrolysers have different uses and compositions. Such electrochemical cells are composed of two electrodes in contact with electrolyte solutions which may be physically separated by a semi-permeable phase called the separator. The general electrochemical cell is described schematically in Fig. 32.1. In the classical configuration in all these electrochemical cells, two electron transfer reactions occur simultaneously between the electrodes and the adjoining solutions which are separated by the semi-permeable sheet.

The separator is an electronic insulator that allows transport of ionic charges. The separators in fuel cells, electrolysers and batteries have a very high ionic conductivity but are not permeable to gas, except for

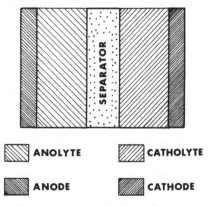

Fig. 32.1. Principle of an electrochemical cell.

sealed batteries. They are homogenous materials to ensure uniform current density distribution. This property prevents dendrite formation on the electrodes of a storage battery. In addition, separators must be highly resistant to degradation by the anolyte, the catholyte and the products of electrochemical reactions.

In all cases of fuel cells, batteries and electrolysers, the distance between the electrodes must be reduced to the minimum possible in order to decrease the internal ohmic drop. The limiting case corresponds to the situation where the electrodes are maintained in physical contact with the separator. In this case, the electrodes are made of a porous electrocatalytic material in order to ensure the triple contact between electrode, separator and the catholyte or anolyte. In practical applications, this configuration can be obtained in two different ways.

In the zero-gap membrane cells, the electrodes and the separator are three distinct components. The electrodes are brought into contact with the two faces of the separator (Fig. 32.2a).

In the *Solid Polymer Electrolyte (SPE)* cell[9, 10], the electrodes do not pre-exist; they are made by deposition onto the separator (Fig. 32.2b).

In all cases, SPE technology involves Nafion[R] membranes and there exist two different ways for obtaining a structurally stable, membrane-electrode assembly.

With *in situ* fabrication, the electrocatalytic material is created directly at the surface of the perfluorinated membrane. By the electrodeless method, platinum deposits are obtained after chemical reduction of chloroplatinate ions dissolved in a solution which is in contact with the membrane. A highly reducing substance, such as $NaBH_4$ or N_2H_4, which is contained in the second aqueous side, diffuses through the membrane and reacts with the chloroplatinate ions. The concentrations of reagents are chosen so that the reduction takes place at the surface of the membrane[11–15]. Aldebert & coworkers[16] obtained a localized precipitation of metal from a cationic platinum species, $Pt(NH_4)_4^{2+}$, this cation being present in the membrane as a counter-ion. The chemical precipitation of reduced, metallic platinum occurs in the close vicinity of the membrane surface and not in the bulk. The reduction of cations which are present in the membrane can also be performed by electrochemical[17–19] and radiochemical methods[20].

An alternative way for forming an SPE assembly consists of preparing the electrocatalyst material before depositing it on the surface of the membrane. Fine powders of electrocatalysts are usually produced by the Adams's method[21]. In order to improve the adhesion of the metallic powder to the membrane surface, pretreatments are necessary which involve binding the metal to another substance (carbon, perfluorinated polymer suspension).

The catalyst is bonded to the membrane surface by vacuum deposition[22] or hot pressing[23]. In the latter case, the temperature range used is between the glass transition temperature of the perfluorosulphonic material and the degradation temperature. The SPE technology gives rise to electrocatalytic deposits of high surface area (e.g. 200 m^2 g^{-1}) because the metal particles are very small. This large active surface area of the electrode is combined with a narrow inter-electrode spacing, which is equal to the membrane thickness. Because the conductivity of Nafion[R] in its protonic form is very high, these properties allow high current densities combined with a moderate ohmic drop. For these reasons, SPE technology involving the perfluorinated proton conductor leads to high

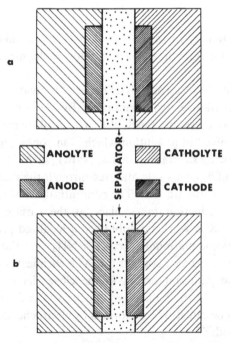

Fig. 32.2. (a) The zero-gap membrane cell. (b) The Solid Polymer Electrolyte (SPE) cell.

efficiency of fuel cells and electrolysers. In addition, the ohmic drop remains low even when the solutions in contact with the electrodes have a high electrical resistance, such as pure water or organic solution. SPE cells can also operate over a large temperature range and under high pressures.

32.3   Fuel cells and electrochemically regenerative cells

The first applications of Nafion^R membranes in SPE technology were in fuel cells for space applications, which have been developed since the end of the 1960s. The perfluorosulphonic polymer is used as a proton conductor which provides to the cathode the protons that have been generated electrochemically at the anode. Perfluorosulphonic SPE is particularly well-suited to this application[24–26]. The principle of the H_2–O_2 fuel cell is shown schematically in Fig. 32.3. At the anode, gaseous hydrogen is reduced following the electrochemical reaction

$$H_2 \rightarrow 2H^+ + 2e^-$$

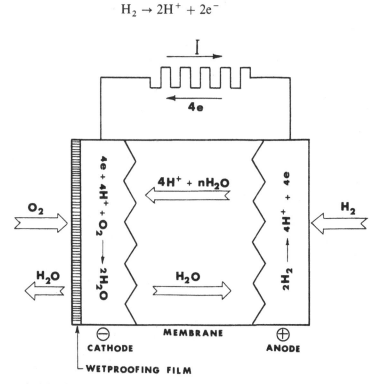

Fig. 32.3. Principle of the H_2–O_2 fuel cell.

The electrons are collected by the metallic electrode while the generated protons enter the membrane. The proton source on the anodic side of the membrane creates a gradient of protons along the membrane thickness. This proton-motive force induces a proton flow between the two electrodes. While crossing the membrane, the proton motion causes a drift of water molecules in the same direction.

At the cathode, the protons are electrochemically combined with gaseous oxygen (air or pure oxygen) and give rise to water formation. The water generated diffuses across the membrane to the anode. So, during operation of the fuel cell, two opposing water fluxes take place simultaneously inside the proton conductor: diffusion of water from cathode to anode and the electro-osmotic flux coupled to proton-migration. The resulting transport number of water – i.e., the number of water molecules transferred when an equivalent of electrical charges is transferred between the two electrodes – ranges between 3.5 and 4.0 at 100 °C[9].

In *methanol fuel cells*[27-29], methanol reacts with hydrogen. The anodic reaction is

$$H_2O + CH_3OH \rightarrow CO_2 + 6H^+ + 6e^-$$

Protons are again generated at the anode and move through the membrane to the cathode. With methanol fuel cells, the diffusion of methanol in the polymeric proton conductor has to be taken into consideration[30].

Storage cell-batteries based on SPE technology have been recently developed[31-33]. In the electrochemically generative hydrogen bromide cell, the overall reaction is the following

$$HBr \underset{\text{discharge}}{\overset{\text{charge}}{\rightleftharpoons}} \tfrac{1}{2}H_2 + \tfrac{1}{2}Br_2$$

These reactions again involve proton transport through perfluoro-sulphonic SPE.

Hydrogen and bromine which are generated by the electrochemical reactions are stored external to the cell. In the discharge cycle, these products are fed back to the cell, which then acts as a fuel cell. The perfluorosulphonic proton conductor has an excellent corrosion resistance in contact with solutions containing bromine.

32.4   *Electrolysers*

The main industrial application of Nafion^R membrane is in *chlor-alkali electrolysis*. The principle is shown schematically in Fig. 32.4. Saturated brine is fed to the anode chamber. Anodic oxidation of chloride ions gives rise to gaseous chlorine production. Under an electrical driving force, sodium ions are transferred through the perfluorinated membrane from the anolyte to the catholyte. In the cathode chamber, sodium ions which have been transferred through the membrane are combined with hydroxyl ions which result from water electrolysis. This latter electrochemical reaction also gives rise to gaseous hydrogen production.

In industrial electrolysers, the cell operates with NaCl anolyte concentration greater than $3 \, \text{mol} \, l^{-1}$ and the sodium hydroxide produced concentration is 20–40%. The current density is about $2000 \, \text{A} \, \text{m}^{-2}$. In this application, the perfluorinated membrane acts as a sodium ion conductor working in a medium of high pH. The counter transport of hydroxyl ion must be minimized for achieving high current efficiency.

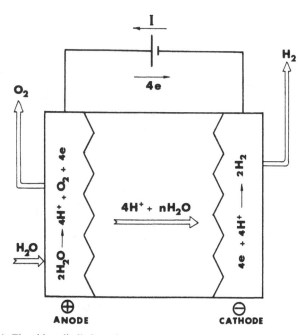

Fig. 32.4. The chlor-alkali electrolyser.

The transmembrane transport of sodium is coupled with co-electro-osmotic transport of water which dilutes the caustic soda formed. Perfluorocarboxylic membranes can be substituted by perfluorosulphonic membranes for reducing the water transport.

Chlorine is also produced by electrolysis of hydrochloric acid solutions. For this purpose, the perfluorosulphonic SPE technology is again used[34] and in this case, the membrane acts as a proton conductor. SPE technology using perfluorosulphonic membranes as proton conductors is well-developed for the electrolysis of water. For this use, the polymer acts as both the separator and the electrolyte. The principle of SPE water electrolysis is shown schematically in Fig. 32.5. Water is consumed at the anode where the electrochemical reaction gives rise to protons and gaseous oxygen. Under the electrical potential, protons are transferred to the cathode where they are electrochemically reduced to produce gaseous hydrogen. Through the membrane, the transport of protons is coupled with an electro-osmotic co-transport of water. In the alternative technique of water electrolysis, concentrated soda solutions are used for reducing the ohmic drop of the aqueous phase. These solutions are corrosive. In

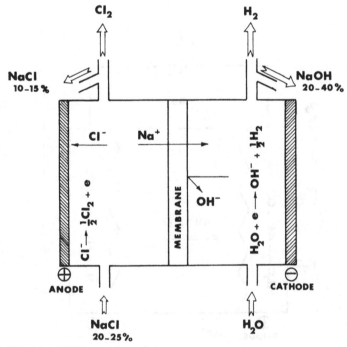

Fig. 32.5. Principle of SPE water electrolysis.

Perfluorinated proton conductors

SPE technology, despite the fact that the conductivity of pure water is low, the working potential difference is minimal due to the low electrical resistance of the proton conducting membrane separating the electrodes.

The cell design is similar to that of fuel cells[35-38]. SPE electrolysers are commercially available for small scale, oxygen generators used in space craft and submarines and also for large scale hydrogen production using off-peak electricity.

32.5 Separation techniques

Ion exchange membranes are widely used in *electrodialysis* and other membrane separation techniques. Perfluorosulphonic membranes are much less used in this area because of their high cost. For separation techniques, less expensive membranes may be used because the physical conditions and the chemical environment are less drastic than in the chlor-alkali electrolysis or the various SPE applications.

The use of Nafion[R] membrane as a proton conductor has been mentioned in a three-compartment electrodialysis set-up for converting sodium salicylate to salicylic acid[39]. Use in Donnan analysis for the recovery of Cu^{2+} ions from dilute solutions has been described[40]. These metal ions are transferred through the membrane. The driving force is the proton-motive force resulting from the proton concentration difference between the two sides of the membrane. The diffusion flux of protons is coupled to a counter-transport of metal ions.

32.6 Catalysis

In homogeneous solutions, fluorosulphonic acid combined with antimony pentafluoride is used as a superacid having some useful catalytic properties in organic chemistry. In its protonated form, perfluorosulphonic Nafion[R] powders are used as the catalyst for a variety of organic syntheses. The catalytic power is higher than that of other solid phase *superacid catalysts*. It enables lower temperatures and pressures to be used and its specificity is higher. Nafion[R] is used as the solid superacid catalyst for the gas phase alkalination of aromatic hydrocarbons[41] and liquid phase esterification or Friedel–Craft reactions[42-44].

32.7   Coated electrodes

The adhesion power of metal electrode to perfluorosulphonic polymer is used to modify the surface of a single electrode for the analysis of electrochemical reactions of special interest. Nafion[R]-coated electrodes have been developed by Rubinstein & Bard[45]. Various electron-conducting materials (glassy carbon, gold, platinum) are used as support for the Teflon[R] layer. With these coated electrodes, the mechanisms of mass and charge transfer in the perfluorosulphonic material have been investigated[45–49] and also the catalytic and photochemical properties of polymer doped with various chemical species[50–58].

Electrodes coated with a perfluorosulphonic acid Nafion[R] membrane can be used to inject protons into non-aqueous media and enable one to exercise additional control on the course of an electro-organic process in which the proton transfer reaction plays an important role[59].

32.8   Conclusions

In the protonic form, perfluorosulphonic membranes are proton conductors which can be used under very drastic chemical conditions. Their chemical inertness makes them suitable for electrochemical applications where they play a role which cannot be obtained with other proton conducting materials. The thermal and chemical stabilities are combined with the ability to bond to electronic conductors or to be doped with dispersed electrocatalytic materials.

The junction between an electronic conductor and a polymeric proton conductor gives a very interesting tool for the study of mass and charge transfer in solids and in solutions. In addition, the perfluorosulphonic proton conductor gives new synthesis routes in organic or electro-organic chemistry. Future developments in this family of proton conductors are to be expected for the preparation of low price perfluorinated materials.

32.9   References

1. W. R. Wolfe, UK Patent no. 1, 184, 321 (1968).
2. B. R. Ezzel, W. P. Carl and W. A. Mod, US Patent, no. 4, 330, 654 (1982).
3. B. R. Ezzel, W. P. Carl and W. A. Mod, US Patent, no. 4, 417, 969 (1983).
4. B. R. Ezzel, W. P. Carl and W. A. Mod, *AIChE Symposium Series* **82** (248) (1983) 45.
5. J. P. Masson, R. Molina, E. Roth, G. Baussens and F. Lemaire, in *Proceed. 3rd WHEC*, Tokyo, **1** (1980) 99.

Perfluorinated proton conductors

6. R. S. Yeo, S. F. Chan and J. Lee, *J. Membrane Sci.* **9** (1981) 273.
7. J. Lee, V. D'Agostino, R. Fried and E. Diebold in *Diaphragms, Separators and Ion-Exchange Membranes*, J. W. Van Zee, R. E. White, K. Kimoshita and H. S. Burney (eds) (The Electrochemical Society Proceedings, Pennington, N.J. **86–13** (1986)) 112–19.
8. R. M. Penner and C. R. Martin, *J. Electrochem. Soc.* **132** (1985) 514–27.
9. A. B. Laconti, A. R. Fragala and J. R. Boyack, in *Proceedings of the Symposium on Electrodes Materials and Processes for Energy Conversion and Storage*, J. D. E. McIntyre, S. Srinivasan, F. G. Will (eds) (The Electrochemical Society, Inc., Princeton, N.J. (1977)) 354.
10. S. Stucki and A. Menth, in *Proceedings of Symposium on Industrial Water Electrolysis*, S. Srinivasan, F. J. Salzano, A. Landgrebe (eds) (The Electrochemical Soc., Princeton, N.J. (1978)) 180.
11. H. Takenaka and E. Torekai, Japanese Patent no. 55, 38934 (1980).
12. H. Takenaka, E. Torekai, Y. Kawami, N. Wakabayashi and T. Sakai, *Denki Kagaku* **53** (1985) 261–4.
13. A. Katayama, Y. Takakuwa, H. Kikuchi, K. Fujikawa and H. Kita, *Electrochim. Acta* **32** (1987) 777–80.
14. H. Kita, F. Fujikawa and H. Nakajima, *Electrochim. Acta* **29** (1984) 1721–4.
15. H. Nakajima, Y. Takakuwa, H. Kikuchi and H. Kita, *Electrochim. Acta* **32** (1987) 791–8.
16. W. T. Grubb and L. W. Niedrach, *J. Electrochem. Soc.* **107** (1960) 131–5.
17. H. Nagel and S. Stucki, US Patent no. 4, 326, 930 (1982).
18. E. Killer, US Patent no. 4, 569, 73 (1986).
19. Commonwealth Scientific and Industrial Research Organization, GB Patent no. 1, 013, 703 (1962).
20. O. Platzer, Thesis, Orsay, Paris Sud (1989).
21. R. Adams and R. L. Shriner, *J. Am. Chem. Soc.* **45** (1923) 2171–5.
22. T. O. Sedgwick and H. Lydtin (eds), in *Proceedings of the Seventh Intern. Conf. on Chemical Vapor Deposition* (The Electrochemical Society, Princeton, N.J. (1979)).
23. R. S. Yeo and S. Srinivasan, *J. Electrochem. Soc.* (*Abstract*) **126** (1979) 378c.
24. L. E. Chapman, in *Proceedings of Seventh Intern. Society Energy Conversion Conference* (American Nuclear Society, La Grange Park, Ill. (1972)) 466.
25. F. J. McElroy, in *National Fuel Cells Seminar Abstracts* (Courtesy Associates, Inc., Washington, D.C. (1978)) 176–9.
26. A. P. Fickett, in *Advances in Ion Exchange Membranes for Rechargeable Fuel Cells*, paper presented at the Colombus Meeting of the Electrochemical Society (1970).
27. *Hydrocarbon Fuel Cell Technology*, B. S. Baker (ed) (Academic Press (1965)).
28. N. A. Hampson and M. J. Williams, *J. Power Source* **4** (1979) 191–205.
29. S. A. Weeks, J. B. Goodenough, A. Hammet, B. J. Kennedy and R. Manoraman, Paper presented at The Electrochemical Society, San Diego, C.A., Oct. 19–24, no. 150 (1986).
30. M. W. Verbrugge, *J. Electrochem. Soc.* **136** (1989) 417–23.

31. D. T. Chin, R. S. Yeo, S. Srinivasan and J. McElroy, *J. Electrochem. Soc.* **126** (1979) 713–21.
32. R. S. Yeo and J. McBreen, *J. Electrochem. Soc.* **126** (1979) 1682–7.
33. R. S. Yeo and D. T. Chin, *J. Electrochem. Soc.* **127** (1980) 549–55.
34. A. B. Laconti, E. N. Balko, T. G. Coker and A. G. Fagala, in *Proceedings of the Symposium on Ion Exchange*, R. S. Yeo and R. P. Buck (eds) (The Electrochemical Society, Pennington, N.J. (1981)) 318.
35. P. W. T. Lu and S. Srinivasan, *J. Appl. Electrochem. Soc.* **9** (1979) 269–73.
36. B. V. Tilak, P. W. T. Lu, J. E. Colman and S. Srinivasan, in *Comprehensive Treatise of Electrochemistry*, J. O'M. Bockris, B. E. Conway, E. B. Yeager and R. E. White (eds) (Plenum Press, New York (1981)) 1.
37. L. J. Nuttal, *Intern. J. Hydrogen Energy* **2** (1977) 395–9.
38. L. J. Nuttal and J. H. Russel, *Intern. J. Hydrogen Energy*, **5** (1980) 75–80.
39. W. G. Grot, in *Proceedings of the 1987 International Congress on Membranes and Membranes Processes*, Tokyo, Japan (1987) 58.
40. J. A. Cox and J. E. Dinunzio, *Anal. Chem.* **49** (1977) 1272–5.
41. G. A. Olah and G. K. S. Prakash, *Sommer. J. Sci.* **206** (1979) 13–19.
42. G. A. Olah, T. Keumi and D. Meidar, *Synth.* **2** (1978) 929–33.
43. P. Beltrame, P. L. Beltrame, P. Carniti and G. Nespoli, *Ind. Eng. Chem. Prod. Res. Dev.* **19** (1980) 205–9.
44. G. A. Olah, D. Meidar, R. Mulhotray and S. C. Narang, *J. Catal.* **61** (1980) 96–100.
45. I. Rubinstein and A. J. Bard, *J. Am. Chem. Soc.* **102** (1980) 6641–4.
46. H. L. Yeager and A. Steck, *J. Electrochem. Soc.* **128** (1981) 1880–4.
47. H. S. White, J. Leddy and A. J. Bard, *J. Am. Chem. Soc.* **104** (1982) 4811–17.
48. C. R. Martin, I. Rubinstein and A. J. Bard, *J. Am. Chem. Soc.* **104** (1982) 4817–24.
49. C. R. Martin and K. A. Doilard, *J. Electroanal. Chem.* **159** (1983) 127–35.
50. D. A. Butry and F. C. Anson, *J. Am. Chem. Soc.* **106** (1984) 59–64.
51. M. Krishnan, X. Zhang and A. J. Bard, *J. Am. Chem. Soc.* **106** (1984) 7371–80.
52. I. Rubinstein, *J. Electroanal. Chem.* **176** (1984) 359–62.
53. F. C. Anson, Y. M. Tsou and J. M. Saveant, *J. Electroanal. Chem.* **178** (1984) 113–27.
54. F. C. Anson, C. L. Ni and J. M. Saveant, *J. Am. Chem. Soc.* **107** (1985) 3442–51.
55. K. Y. Wong and F. C. Anson, *J. Electroanal. Chem.* **237** (1987) 69–79.
56. I. Rubinstein and A. J. Bard, *J. Am. Chem. Soc.* **103** (1981) 5007–14.
57. D. A. Butry and F. C. Anson, *J. Electroanal. Chem.* **130** (1981) 333–8.
58. N. E. Prieto and C. R. Martin, *J. Electrochem. Soc.* **131** (1984) 751–5.
59. M. S. Mubarak and D. G. Petters, *J. Electroanal. Chem.* **273** (1989) 283–92.

33 Synthesis of polycrystalline H_3O^+ and NH_4^+-β''/β-Al_2O_3 and potential applications in steam-electrolysis/fuel-cells

PATRICK S. NICHOLSON

33.1 Introduction

Proton conduction in β- and β''-Al_2O_3 is of interest because of potential use in *steam-electrolysis/H_2/O_2 fuel-cells*. Single crystal $H_3O^+/$-β''-Al_2O_3 was reported by Farrington & Briant[1] to be a fast ion conductor with a proton conductivity of 10^{-4}–10^{-5} (Ω cm)$^{-1}$ at 25 °C. These encouraging results led to the first fabrication, characterization and use of polycrystalline H_3O^+-β''/β-Al_2O_3[2].

A second *electrolyte* with a higher conductivity but a lower thermal stability is NH_4^+-β''-Al_2O_3. This system, NH_4^+/H_3O^+-β''-Al_2O_3, is the best conductor of protons ($\sigma = 7 \times 10^{-4}$ (Ω cm)$^{-1}$ rising to 2×10^{-2} (Ω cm)$^{-1}$ at 200 °C)[3,4]. All previous work has been on single crystals[5,6] (see Chapters 13 & 23) or powders[7]. This chapter describes the synthesis of NH_4^+/H_3O^+-β''-Al_2O_3 polycrystals.

33.2 Ion conducting structure of β''- and β-aluminas

The β-Al_2O_3 family of compounds is constructed from oxygen and aluminium 'spinel blocks' with intervening 'conduction planes'. In β-Al_2O_3 there are two spinel blocks per unit cell and in β''-Al_2O_3 there are three. The upper oxygen layer of the spinel block is mirrored across the conduction plane in β-Al_2O_3 but this symmetry is lost in β''-Al_2O_3. The alkali ion 'Beevers–Ross' site in the β-Al_2O_3 conduction planes is octahedral. The equivalent site in β''-Al_2O_3 is tetrahedral and smaller. Thus K^+ ($r = 0.28$ nm) promotes formation of β-Al_2O_3 and Na^+ ($r = 0.196$ nm), β''- Al_2O_3.

β''-Al_2O_3 has the highest ionic conductivity and so H_3O^+-β''-Al_2O_3 is the desirable precursor. Sintering induces formation of K-β-Al_2O_3, so a mixed-alkali (Na^+/K^+) composition (6N3; 0.6:0.4 Na/K with 3 w/o MgO) was developed with sufficient conductivity to facilitate ion exchange and steam-electrolysis/fuel-cell performance and sufficient *strength* to withstand the H_3O^+-K^+-Na^+ ion exchange steps requisite for synthesis. The 40 v/o β-Al_2O_3 present provided mechanical reinforcement.

Recently, the discovery of unreacted spinel and $NaAl_2O_2$ on the grain-boundaries of the dense but non-equilibrium microstructure, has led to the synthesis of β-Al_2O_3-free material.

The two-dimensional conductivity of the β-aluminas is further restricted by grain boundaries and the random orientation of neighbouring grains. Grain boundary conductivities are 1–2 orders of magnitude lower than that of the grains. This can be reduced by grain–grain alignment (texturing).

These considerations have led to development of the 8N4 composition with a textured microstructure.

33.3 Synthesis of precursor ceramics

Ion exchange of Na for K (c_0 for β''-Al_2O_3 increases from 3.339 to 3.412 nm) induces 0.005 positive strain, resulting in c. 400 MPa stress. There are three possible solutions:

(a) make composite ceramics (β''-Al_2O_3 with dispersed β-Al_2O_3), which reduces stress but maintains satisfactory conductivity;

(b) *ion-exchange at high temperature* to relieve stress; and

(c) texture the microstructure.

Approach (a) has the potential to induce surface flaws during H_3O^+ ion-exchange as the β''-Al_2O_3 of the microstructure shrinks slightly (3.412 nm to 3.40 nm) whereas the β-Al_2O_3 remains unchanged (unexchanged). Residual alkali in the β-Al_2O_3 also causes electrode cohesion problems. Approaches (b) and (c) are now used.

The spray-freeze/freeze-dry process is used[8] and produces high surface area (>10 m² g⁻¹), high-aspect ratio (>5) and low $f(\beta)$ (<0.1), (i.e. $f(\beta'') = 1$), powders. This phase composition required limited sintering times (<2 min) to minimize β-Al_2O_3 development. Fig. 33.1 shows concurrent, density/$f(\beta)$ vs. sintering time. Density is achievable in 2–3 min with $f(\beta) < 0.4$, the critical value. This results in a non-equilibrium microstructure[9] containing residual $MgAl_2O_4$ and $NaAlO_2$

Polycrystalline H_3O^+ and NH_4^+-β''/β-Al_2O_3

from

$$\beta''\text{-}Al_2O_3 + MgO \rightarrow \beta\text{-}Al_2O_3 + MgAl_2O_4 + NaAlO_2$$

The spinel and aluminate cause failure during H_3O^+ ion exchange. Further time reverses this reaction and sintering samples achieve the same $f(\beta)$ value twice.

The discovery of the second low-$f(\beta)$ value led to reproducible polycrystals. The final grain-size of the new ceramic is still fine ($= 3 \, \mu m$), the evident grain-boundary immobility being due to pinning by the slowly assimilating spinel.

The β-Al_2O_3 phase content is further reduced by lowering the K^+ (from 0.4 to 0.2) and increasing MgO (3 w/o to 4 w/o) contents. The resultant ceramic is termed '8N4' and has been textured by uniaxial pressing. An open-ended tube was built by sequentially pressing thin layers to give a through-the-section texture (Fig. 33.2; (006) XRD reflection amplitude). The c_0 axis of the microstructured grains is parallel to the tube axis (Fig. 33.3) and the 25 °C grain-boundary conductivity across the tube section (from left to right of Fig. 33.3) is 2.2×10^{-4} vs. $2.7 \times 10^{-6} \, (\Omega \, cm)^{-1}$ untextured. The texture significantly reduces ion-exchange stresses as the accompanying expansion is parallel to the tube axis.

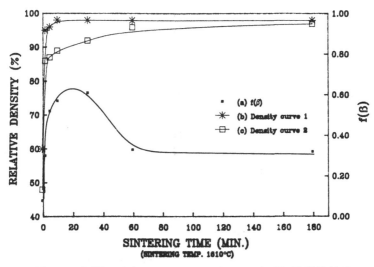

Fig. 33.1. Changes of $f(\beta)$ and density versus sintering time for Na/K-β''/β-Al_2O_3.

33.4   Alkali-ion exchange

The Na^+ in the ceramic is 80% of the alkali content. Stress-development in the β-Al_2O_3 component is markedly different at 850 °C and 1200 °C[10]. At 800 °C, the β-Al_2O_3 phase assimilates K^+ to $X_{K+} \sim 0.3$, then remains constant with increasing mole fraction K^+ in the melt (0.2 to $0.85 X_{K+}$).

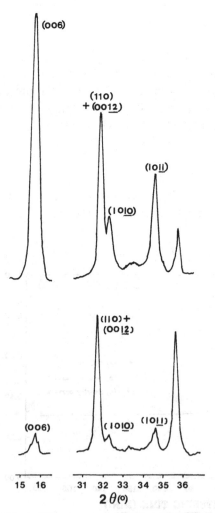

Fig. 33.2. X-ray diffraction traces for textured β''-Al_2O_3 parallel and perpendicular to the pressing axis.

Polycrystalline H_3O^+ and NH_4^+-β''/β-Al_2O_3

The β''-Al_2O_3 expands. The Na^+ retained in the β-Al_2O_3 eventually decohesizes the electrodes of a working electrolyte.

Increasing the temperature to 1200 °C minimizes the stresses. 8N4 is ion-exchanged in KCl vapour at 1200 °C to give 100% $K^+\beta''$-Al_2O_3 textured ceramics.

K^+/Rb^+ ion-exchange is conducted above an RbCl melt at 1200 °C for two days.

33.5   Oxonium and ammonium ion-exchange

Polycrystalline samples are oxonium-exchanged in dilute acetic acid (HAC) ($<10\%$ or 0.1 M) at 80 °C in a d.c. field. The Na^+ level in the acid on ion-exchange completion is higher than the Na^+ content of the β''-Al_2O_3 component of the microstructure. This means some Na^+ must exit the β-Al_2O_3 phase. As H_3O^+ is unable to enter this phase, K^+ fleeing the β''-Al_2O_3 must be replacing this Na^+. This sequence of events was postulated[11] for the 40 v/o β-Al_2O_3 ceramic (6N3) to explain why the mixed-alkali effect that shows the phase (β-Al_2O_3) controls the ion-exchange kinetics, even though it is discontinuous in the microstructure. The β''-Al_2O_3 phase does not exhibit the *mixed-alkali effect*[12].

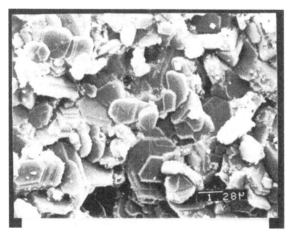

Fig. 33.3. Fracture-surface microstructure parallel to the pressing direction for textured β''-Al_2O_3.

The *field-assisted* current is passed through a sample immersed in 10% HAC at 80 °C. The resultant current–time curves for different applied voltages are shown in Fig. 33.4. Possible sources of sample resistivity are:

(1) development of the critical mixed-alkali-effect composition ahead of the H_3O^+ front;
(2) increased resistivity of the ceramic on K^+ substitution by H_3O^+;
(3) development of a polarization double-layer in the acid adjacent to the electrolyte; and
(4) development of an alkali-exit-interference layer of H_3O^+-β''-Al_2O_3 on the catholyte side.

The low $f(\beta)$ levels (<0.1) of 8N4 reduce the first mechanism to a minor role. Interruption of the power supply resulted in a residual potential of 2 V in the same direction as the applied voltage. This can be generated only by development of a cation concentration gradient through the sample section. As K^+ moves towards the cathode, H_3O^+ enters from the anolyte. If more K^+ exits than slower H_3O^+ enters, the developing polarization will slow down and stop the K^+ exit. As a

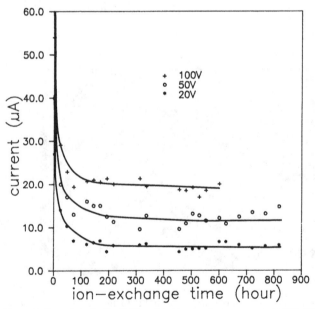

Fig. 33.4. Current–time plots for the field-assisted ion-exchange process at 100, 50 and 20 V.

consequence, anions will also be attracted to the anode by the excess H_3O^+ present to give surface polarization.

Another possible resistivity source is the development of an H_3O^+ layer at the cathode when the ceramic is first introduced into the acid, but replacement of the cathode acid by Hg did not change the current drop.

Field-reversal led to increased current levels (Fig. 33.5). The polarity-reversed sample was completely ion-exchanged after 35 days under 20 V (as opposed to >50 days for single-direction current flow). This result suggests dissipation of anion layers in the liquid at the electrolyte surfaces eliminates fields opposing the applied field.

The mobility of H_3O^+ in the β''-Al_2O_3 increases with temperature and reversing-field-assisted ion-exchange in a high-pressure container at 200 °C in 10% HAC resulted in a 2 day exchanged-layer-thickness equal to that at 80 °C after 34 days.

The Rb^+-β''-Al_2O_3 pellets were immersed in molten NH_4NO_3 and 60% replacement took place after 24 h. Full exchange took 7–10 days. The NH_3 content in the NH_4-β''-alumina pellets was 3.62 ($\pm$0.04) w/o by chemical analysis[13]. DTA showed a large exothermic peak between

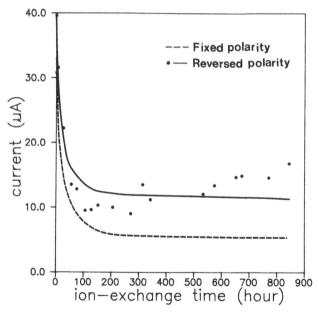

Fig. 33.5. Current–passage plots on reversing the polarity of the field-assisted ion-exchanging ceramic (20 V).

505

300–400 °C, suggesting polycrystals can be used in steam-electrolyser/ fuel-cells up to 300 °C.

33.6   Electrolyte characteristics and preliminary steam-electrolysis/ fuel-cell calculations and performance

The *ionic transference number* and *electronic conductivity* of H_3O^+-β''/β-Al_2O_3 ceramic electrolytes were measured at 293–420 K using electrochemical cells[14]. The ionic transference number is 0.99 (±0.02) and the electronic conductivity is $3.9 \times 10^{-3} \exp(-0.38 \text{ eV}/kT)$. The conductivity and stability of the non-textured H_3O^+-β''-Al_2O_3 electrolyte was studied in high-pressure steam and air between room temperature and 240 °C[15]. The grain-conductivity results are summarized in Fig. 33.6. The conductivity loss suffered in open air (curve 3) is regained on rehydration in a saturated steam atmosphere. The conductivity–temperature behaviour of textured and untextured NH_4^+-β''-Al_2O_3 in air is shown in Fig. 33.7. The marked improvement of the grain-boundary conductivity on texturing is evident.

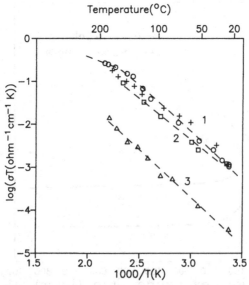

Fig. 33.6. Grain-conductivity as a function of temperature for H_3O^+-β''-Al_2O_3: (1) in saturated steam, (2) in open air, (3) in open air after 8.5 h annealing ('cross' represents the conductivity data measured after rehydration in a saturated steam atmosphere.

Polycrystalline H_3O^+ and NH_4^+-β''/β-Al_2O_3

NH_3-based fuel is attractive for electrolyte stability, its high proportion of hydrogen (75 v/o) and ease of liquefaction. The calculated phase equilibria diagrams[16] show that appreciable decomposition of ammonia to hydrogen and nitrogen starts at *c.* 337 K and 90% ammonia has dissociated at 600 K. The dissociation of ammonia decreases 10–30% between 400–600 K in a 1:3 NH_3/H_2 gas mixture.

The thermodynamic e.m.f. for an (NH_3 + gas/air + 0.031 atm H_2O) fuel-cell was calculated and is compared with H_2 fuel-O_2 in Fig. 33.8. NH_3, NH_3–H_2O and NH_3–H_2 fuels in NH_4^+–H_3O^+-β''-Al_2O_3 electrolyte fuel-cells at > 450 K generate e.m.f.s comparable with those of the H_2–O_2 fuel cell. This result suggests that minor energy loss occurs on formation of nitrogen gas molecules at operating $T > 450$ K.

The H_3O^+-β''/β-Al_2O_3 of the 6N3-type was used in a reversible steam electrolysis/fuel cell unit and the results are shown in Fig. 33.9. The efficiency at the high temperatures (200–300 °C) is substantially below theoretical due to the dehydration of the electrolyte. As this is reversible, a high-pressure steam cell unit was constructed and initial trials showed a substantial improvement in steam-electrolysis/fuel-cell voltage/current-density characteristics.

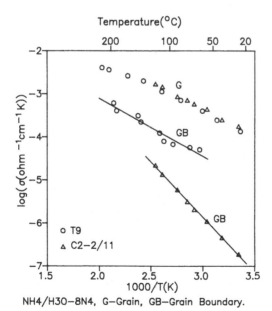

Fig. 33.7. Grain (G) and grain-boundary (GB) conductivity–temperature plots for textured (T9) and non-textured (C2-2/11) NH_4^+-β''-Al_2O_3.

507

NH_4^+/H_3O^+-β''-Al_2O_3 polycrystals have recently been incorporated into 1 atm steam-electrolysis cells.

33.7 Summary

The details of synthesis of H_3O^+- and NH_4^+-β''-Al_2O_3 are discussed and preliminary performance characteristics in steam-electrolysis/fuel-cells outlined. Removal of the β-Al_2O_3 phase is achieved by decreasing the K^+ content and increasing the MgO content of the precursor ceramic, long-time sintering to eliminate non-equilibrium phases (spinel and $NaAlO_2$) on reaction with β-Al_2O_3 to give β''-Al_2O_3 and high-temperature vapour-phase ion-exchange which anneals developing stresses and dispenses with the necessity of the β-Al_2O_3 as microstructural reinforcement.

Ion-exchange rates are increased and associated stresses decreased by texturing the high-aspect-ratio powder obtained from the spray-freeze/freeze-drying process. The texturing of the microstructure also increases the grain-boundary conductivity by 1–2 orders of magnitude.

The increased resistance associated with field-assisted-ion-exchange is minimized by field reversal every 100 s.

The ionic-transference number for H_3O^+-β''-Al_2O_3 is 0.99 (± 0.02) and

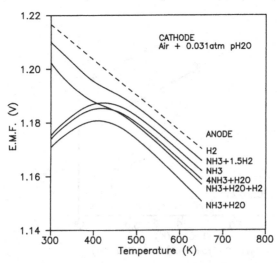

Fig. 33.8. OCV–temperature characteristics for a fuel-cell burning NH_3 fuel in steam (cathode gas: air + 0.031 atm H_2O).

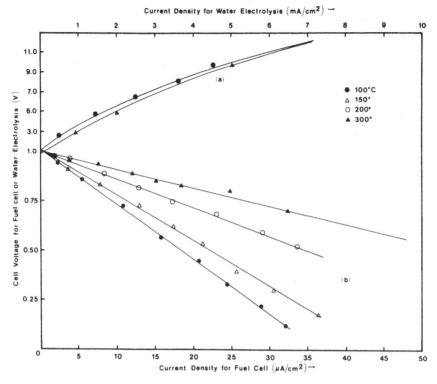

VOLTAGE-CURRENT CHARACTERISTIC OF A FUEL AND STEAM ELECTROLYSIS CELL.

Fig. 33.9. Current density–voltage characteristics for H_3O^+-β''/β-Al_2O_3 electrolyte (6N3 type) in a reversible, 1 atm steam-electrolysis/fuel cell.

its conductivity at 300 °C is re-established in saturated steam at that temperature.

Calculations demonstrate that NH_3 is a feasible fuel for an NH_4^+-β''-Al_2O_3 fuel-cell.

33.8   References

1. G. C. Farrington and J. L. Briant, *Mat. Res. Bull.* **13** (1978) 763.
2. P. S. Nicholson, M. Z. A. Munshi, G. Singh, M. Sayer and M. F. Bell, *Solid State Ionics* **18–19** (1986) 699.
3. G. C. Farrington, K. G. Frase and J. O. Thomas, *Adv. Mat. Sci.* (1984).
4. J. O. Thomas and G. C. Farrington, *Acta Cryst.* B **39** (1983) 227.

5. A. R. Ochadlick Jr, W. C. Bailey, R. L. Stamp, H. S. Story, G. C. Farrington and J. L. Briant, in *Fast Ion Transport in Solids*, Vashita, Mundy and Shenoy (eds.) (Elsevier (1979)) 401–4.
6. A. Hooper and B. C. Tofield, in *Fast Ion Transport in Solids*, Vashita, Mundy and Shenoy (eds) (Elsevier (1979)) 409–12.
7. R. C. T. Slade, P. F. Fridd, T. K. Halstead and P. McGeehin, *J. Solid State Chem.* **32** (1980) 87–95.
8. A. Pekarski and P. S. Nicholson, *Mat. Res. Bull.* **15** (1980) 1517.
9. A. Tan and P. S. Nicholson, *Solid State Ionics* **37** (1989) 51–5.
10. A. Tan and P. S. Nicholson, *Solid State Ionics* **26** (1988) 217–28.
11. P. S. Nicholson, in *Proc. South East Asia Superionics Soc. Conf.* (Singapore 1988), 639–61.
12. T. Tsurimi, G. Singh and P. S. Nicholson, *Solid State Ionics* **22** (1987) 225–30.
13. N. Iyi, A. Grzymek and P. S. Nicholson, *Solid State Ionics* **37** (1989) 11–16.
14. C. K. Kou, A. Tan and P. S. Nicholson (submitted to *Solid State Ionics*).
15. C. K. Kou, A. Tan, P. Sarkar and P. S. Nicholson, *Solid State Ionics* **38** (1990) (in press).
16. C. K. Kou, A. Tan and P. S. Nicholson (submitted to *Solid State Ionics*).

34 Fuel-cells, steam-electrolysis for hydrogen production and hydrogen separation using high temperature protonic conductors

H. IWAHARA

34.1 Introduction

Although there are many protonic conductors which are stable at low temperatures (< 200 °C), they are not suitable for high electrolytic current applications because of their large *electrode polarization*. Therefore, their usage is probably limited to sensors in which only a voltage signal is valuable.

In contrast to this, high temperature protonic conductors are applicable both to sensors and high electrolytic current usage, e.g. electrolysis, galvanic cells for power supply, etc. since, at high temperature, electrode reactions are able to proceed smoothly and, in general, polarization is small. In Table 34.1, possible applications of high temperature proton conducting solid electrolytes are listed, classifying them as having either an electromotive function or a preferential permeation of protons.

In this chapter, the distinctive features of proton conductors as an electrolyte for use in a fuel cell or in a steam electrolyser are discussed in comparison to those of oxide ion conductors. In addition, the possibility of using a proton conductor in a hydrogen gas separator is also described. As examples, the experiments on fuel-cells, steam electrolysers and hydrogen gas separators using proton conducting, perovskite-type oxides are described.

34.2 Fuel-cells

The use of a proton conducting solid electrolyte in a fuel-cell has distinctive features compared with that of oxide ion conductors. As

Table 1. *Possible applications of high temperature proton conducting solid electrolytes*

Function	Phenomena applicable	Application
E.m.f.	Signal	Hydrogen gas sensor
		Hydrogen activity (in fused metal) sensor
		Steam sensor
	Power	Fuel-cell
Electrochemical permeation of hydrogen	Separation	Hydrogen extractor
		Hydrogen pump
		Hydrogen gas controller
	Electrolysis	Steam electrolyser for hydrogen production
		H_2S electrolyser for desulphuration
		HCl electrolyser for Cl_2 recovery
	Reaction	Hydrogenation of organic compounds
		Dehydrogenation of organic compounds

illustrated in Fig. 34.1, when a protonic conductor is used, fuel circulation is unnecessary in a hydrogen fuel-cell because no water molecules are generated at the fuel electrode. Furthermore, in a hydrocarbon fuel-cell, only thermally produced hydrogen is consumed as a fuel and the gaseous products may contain useful reformed products (e.g. ethylene from ethane).

High temperature fuel-cells using proton conducting electrolytes had not been studied until recently for lack of such conductors until the present author and coworkers found some proton-conducting ceramics. As described earlier (Chapter 8), they found that *perovskite*-type oxides based on $SrCeO_3$ exhibit appreciable protonic conduction in a hydrogen-containing atmosphere at high temperatures[1]. $BaCeO_3$-based ceramics also have the same characteristics although they become mixed proton–oxide ion conductors as the temperature is raised[2]. Using these protonic conductors, various kinds of fuel cell can be fabricated[3–10]. In Fig. 34.2 a test cell is illustrated. This uses a thin disc (thickness 0.5 mm, diameter 13 mm) of $SrCeO_3$- or $BaCeO_3$-based ceramic as the solid electrolyte.

Fig. 34.3a shows the typical performance of a hydrogen–air cell based on $SrCe_{0.95}Yb_{0.05}O_{3-\alpha}$ with porous platinum electrodes[11]. The cell worked stably and the relation between terminal voltage and current output is linear. The anode polarization (except the IR drop) is very small over the whole temperature range examined while the polarization of the cathode cannot be neglected at 800 °C. The internal resistance of the cell is attributed mainly to the ohmic resistance of the electrolyte.

When this cell was discharged for a few hours, water vapour was observed to condense at the air exhaust pipe. This means that protons migrate across the solid electrolyte from the hydrogen electrode to the air electrode where they react with oxygen to form water molecules. In fact, the evolution rate of water vapour at the cathode corresponded to the theoretical rate calculated on the basis of Faraday's law[5].

Fig. 34.3b shows the performance of a hydrogen–air fuel-cell using $BaCe_{0.9}Nd_{0.1}O_{3-\alpha}$ solid electrolyte and porous platinum electrodes[9]. Due to the higher conductivity of the solid electrolyte in this cell, the power output is higher than that of a $SrCeO_3$ cell. Water vapour was found to evolve at both electrodes suggesting that the conduction in the

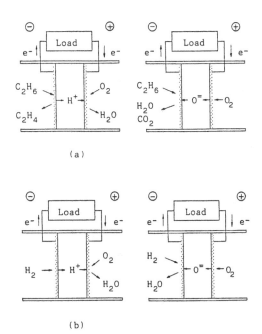

Fig. 34.1. Comparison of proton conductor with oxide ion conductor as the electrolyte in (a) ethane-fuelled and (b) hydrogen-fuelled cells.

513

electrolyte is by both protons and oxide ions. As a replacement for expensive platinum, perovskite-type oxide electronic conductors have been used as the cathode and porous nickel as the anode[5,10], although the internal resistance of the cell was found to be higher than that of the cell with platinum electrodes.

In the case of a hydrocarbon fuel-cell using a protonic conductor, ethane is supplied to the anode compartment and the following pyrolysis reaction occurs exothermically.

$$C_2H_6 \rightarrow C_2H_4 + H_2 \tag{34.1}$$

Hydrogen is then consumed by the cell reaction, but the ethylene is not oxidized to CO_2 (Fig. 34.1a). Therefore, this cell can generate both electric power and ethylene by consuming ethane as a fuel.

An experimental ethane fuel-cell made in the author's laboratory has worked as stably as the hydrogen–air fuel-cell[7]. Exhaust gas from the

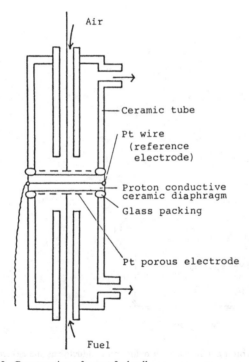

Fig. 34.2. Construction of a test fuel-cell.

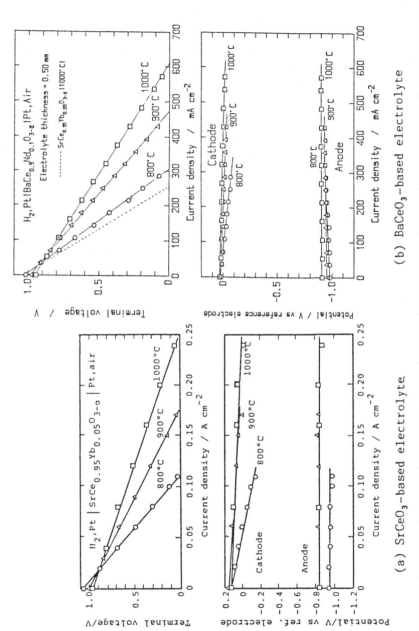

Fig. 34.3. Performances of hydrogen fuel-cells using perovskite-type oxide ceramic as solid electrolyte and porous platinum electrodes. (a) SrCeO₃-based electrolyte; (b) BaCeO₃-based electrolyte.

anode compartment contained H_2, C_2H_4, CH_4 and C_2H_6. Water vapour, the product of cell reaction, was observed to form at the cathode, and carbon dioxide was not formed at the anode. These results also indicate that conduction in the solid electrolyte is protonic. C_1 gas such as methanol or methane can be used as fuel in the perovskite-type oxide electrolyte cells when appreciable amounts of water vapour are added to the fuel[8]. Recently, attempts to use a hydroxyapatite-based proton conductor in high temperature fuel-cells have been made by several investigators[12], but the cell performances were not good because of the low electrolyte conductivity.

34.3   Steam-electrolysis for hydrogen production

The use of a proton conductor has, in principle, the advantage that pure hydrogen can be obtained since only protons migrate across the conductor to the cathode, as shown in Fig. 34.4. Steam-electrolysis using $SrCeO_3$-based ceramic diaphragms as an electrolyte has been tried at high temperature to obtain hydrogen gas[1,13]. Cell construction was the same as that shown in Fig. 34.2. Steam at 1 atm was supplied to the anode compartment and argon gas was passed through the cathode compartment to carry the generated gas to a hydrogen detector.

On passing direct current through the cell at 700–900 °C, evolution of hydrogen gas at the cathode was observed by gas chromatography. Fig. 34.5a shows the relationship between electrolytic current and the hydrogen evolution rate determined from the concentration of hydrogen

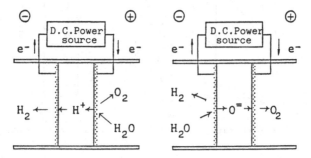

Fig. 34.4. Comparison of proton conductor and oxide ion conductor in a steam electrolyser.

in the argon carrier gas and its flow rate. The current efficiencies for H_2 evolution were about 0.9 in the range 0.1–0.3 A cm^{-2}. Current loss may be ascribed to the electronic (hole) conduction since the proton transport number of the oxides in this condition is not unity, as mentioned elsewhere[1,14].

Fig. 34.5b shows the relation between the electrolytic current and the applied voltage at 800–900 °C. A rather high voltage was necessary to electrolyse the water vapour due to insufficient conductivity of the electrolyte. However, the polarization, except ohmic loss, is rather low, as indicated with dotted lines. This suggests that, if the thickness of the electrolyte is reduced, the electric power required for electrolysis can be reduced to less than that of conventional water electrolysis.

A bench-scale steam electrolyser was fabricated and pure hydrogen gas extracted at the rate of a few litres per hour[15]. Fig. 34.6 represents the cross section of the cell which contains a one-end-closed ceramic tube made of $SeCe_{0.95}Yb_{0.05}O_{3-\alpha}$ (inside diameter 12–14 mm, thickness 1–1.5 mm, length 35–150 mm). As a cathode, porous platinum was baked on the inner side of the tube, and, as an anode, porous palladium was attached to the outer side of the tube. Four cells could be accommodated in a cell furnace and connected electrically either in series or in parallel. Steam

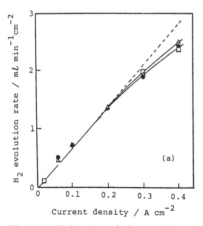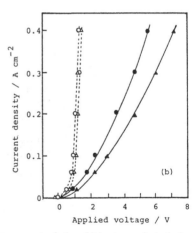

Fig. 34.5. Hydrogen evolution rate (a), and voltage–current relation (b) in steam-electrolysis. (a) $\square$, $SeCe_{0.95}Yb_{0.05}O_{3-\alpha}$, 800 °C; $\bullet$, $SrCe_{0.9}Y_{0.1}O_{3-\alpha}$, 700 °C; $\triangle$, $SrCe_{0.9}Y_{0.1}O_{3-\alpha}$, 800 °C; dotted line shows the theoretical rate. (b) $SrCe_{0.9}Y_{0.1}O_{3-\alpha}$ $\bigcirc$, $\bullet$, 800 °C; $\triangle$, $\blacktriangle$, 900 °C; dotted lines represent the voltage excluding ohmic loss.

was blown against the bottom of the closed end tube and electrolysed to produce hydrogen inside the tube.

Fig. 34.7 shows an example of the performance of the electrolyser which was equipped with three cells in series and operated with 1.8 A per cell at 750 °C. The volume of hydrogen accumulated was proportional to electrolysing time, indicating that performance was very stable. Hydrogen gas evolved was pure and dry (water vapour pressure in the gas less than 1 Torr). Due to limited protonic conductivity of the ceramic tube, a rather high voltage (21 V at 1.8 A for three cells in series) is necessary and, as a result, energy efficiency for conversion is still low. However, if the thickness of the cell tube were reduced, the required voltage would decrease and the energy efficiency be improved.

The electrolysis of hydrogen chloride or hydrogen disulphide to recover chlorine or sulphur, respectively, is, in principle, possible using a high temperature protonic conductor.

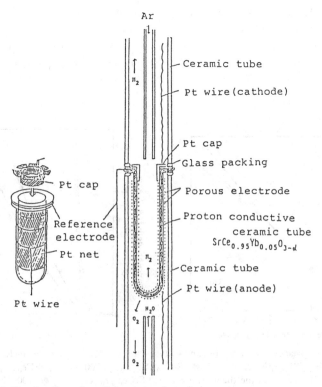

Fig. 34.6. Construction of a cell for steam-electrolysis.

34.4 Hydrogen separation

Electrochemical hydrogen separation from various gas mixtures is possible using a high temperature proton conducting ceramic as an electrolyte diaphragm. The principles are shown schematically in Fig. 34.8 for the separation of hydrogen from a pyrolysed mixture of $CO + H_2O$ and from pyrolysed ethane. To demonstrate this method, a thin disc of $SrCe_{0.95}Yb_{0.05}O_{3-\alpha}$ ceramic diaphragm and porous platinum electrodes were used[15].

Fig. 34.9 shows the results for hydrogen extraction from thermal cracking of ethane at 800 °C. Extraction rates were close to the theoretical calculated from Faraday's law, but a large part of the voltage required resulted from ohmic resistance of the proton conductor. If the ohmic resistance is excluded, however, the voltage required is fairly low as indicated with dotted lines in Fig. 34.9b. This means that, if a very thin and non-porous film of the proton conductor were available, it could be put to practical use as an electrochemical hydrogen separator.

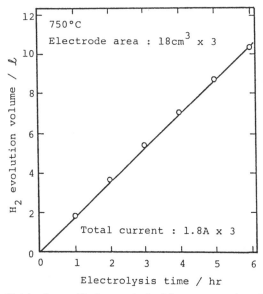

Fig. 34.7. Total volume of hydrogen obtained from a steam electrolyser.

34.5 Other applications

As described elsewhere (Chapter 36), a hydrogen sensor and a steam sensor usable at high temperature can be fabricated using high temperature proton conductors. They are based on the principle of the hydrogen concentration cell. Combining the sensing function with preferential permeation of hydrogen, it is possible to make a regulator of hydrogen partial pressure in the mixed gas[16].

Another possibility is a chemical reactor for *hydrogenation* or *dehydrogenation* of organic compounds. As illustrated in Fig. 34.10, an organic compound can be hydrogenated or dehydrogenated electrochemically using a proton conducting solid. In contrast to conventional heterogeneous catalytic gas reactions, the distinctive features of this method are as follows.

(1) The compounds to be hydrogenated or dehydrogenated are separated from the hydrogen gas to react or to be formed, respectively, by a solid electrolyte diaphragm.

(2) The activity of hydrogen in hydrogenation or dehydrogenation

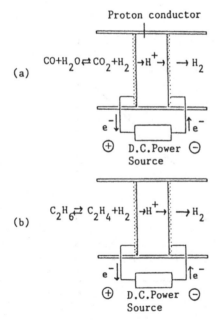

Fig. 34.8. Hydrogen extraction from gas mixture.

reaction sites at the electrode can be controlled by the electrical potential of the electrode.

(3) Reaction rate can be controlled by the electric current through the electrolyte.

No experiments in this field have been reported because the existence of high temperature proton conductors has not widely been known until now. Proton conducting solids are, however, considered to be promising materials for controlling organic reactions if they can be fabricated and used as a thin, dense film.

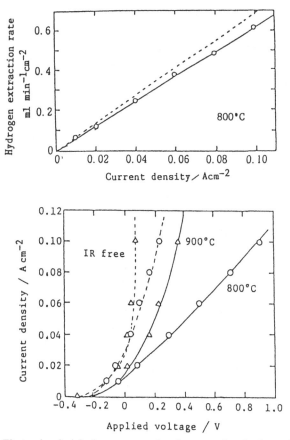

Fig. 34.9. Electrochemical hydrogen extraction from pyrolysed ethane (solid electrolyte; $SrCe_{0.95}Yb_{0.05}O_{3-x}$). (a) Hydrogen extraction rate vs. current density (dotted line shows theoretical rate). (b) Voltage vs. current in extraction cell.

521

Devices

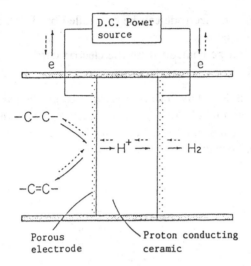

Fig. 34.10. Hydrogenation and dehydrogenation using proton conducting ceramic.

34.6   References

1. H. Iwahara, T. Esaka, H. Uchida and N. Maeda, *Solid State Ionics* **3/4** (1981) 359.
2. H. Iwahara, H. Uchida, K. Ono and K. Ogaki, *J. Electrochem. Soc.* **135** (1988) 529.
3. H. Iwahara, H. Uchida and N. Maeda, *J. Power Sources* **7** (1982) 293.
4. H. Iwahara, H. Uchida and T. Esaka, *Prog. Batter. Solar Cells* **4** (1982) 279.
5. H. Iwahara, H. Uchida and N. Maeda, *Solid State Ionics* **9** (1983) 1021.
6. H. Uchida, S. Tanaka and H. Iwahara, *J. Appl. Electrochem.* **15** (1985) 93.
7. H. Iwahara, H. Uchida and S. Tanaka, *J. Appl. Electrochem.* **16** (1986) 663.
8. H. Iwahara, H. Uchida, K. Morimoto and S. Hosogi, *J. Appl. Electrochem.* **19** (1989) 448.
9. H. Iwahara, H. Uchida and K. Morimoto, *J. Electrochem. Soc.* (submitted).
10. H. Iwahara, H. Uchida and K. Morimoto, *Denki Kagaku* **58** (1990) 178.
11. H. Iwahara, *Proc. of Int. Symp. on SOFC* (Nagoya, 1989) 147.
12. K. Yamashita, H. Owada, T. Umegaki, T. Kanazawa and K. Katayama, *Proc. of Int. Symp. on SOFC* (Nagoya, 1989) 161.
13. H. Iwahara, H. Uchida and I. Yamasaki, *Int. J. Hydrogen Energy* **12** (1987) 73.
14. H. Uchida, N. Maeda and H. Iwahara, *Solid State Ionics* **11** (1983) 117.
15. H. Iwahara, T. Esaka, H. Uchida, T. Yamauchi and K. Ogaki, *Solid State Ionics* **18/19** (1986) 1003.
16. H. Iwahara, H. Uchida and K. Takashima, *Denki Kagaku* **57** (1989) 996.

35 Ice-based devices

I. A. RYZHKIN

35.1 Introduction

To a large extent the electrical properties of ice are similar to those of semiconductors: exponential increase in conductivity with increasing temperature, analogy of positive and negative carriers with electrons and holes, change of conduction upon doping. This similarity makes it possible to consider, in the case of ice, models for numerous devices that have been fabricated using semiconductors. However, it can hardly be expected that ice-based devices will have any advantage (qualitative, technological) against semiconductor analogues. Therefore, consideration of such ice-based devices is of no practical significance.

On the other hand, the electrical properties of ice, principally due to the presence of a large number of carriers and the nature of the carriers, may give rise to *memory effects* which are much more complicated compared with the electrical properties of ordinary semiconductors. Therefore, using ice, one can realize devices which either have no analogues amongst semiconductor devices or which are advantageous compared with the latter. Some such devices are described next.

Before giving detailed descriptions, we shall emphasize two important features. Firstly, we are dealing here with models rather than with operational devices. Secondly, in selecting concrete models, we are basing our consideration on the most interesting physical properties of ice, treated in original research papers.

35.2 Screening effects and capacitance devices

Due to the presence of several types of carriers and memory effects arising during the proton motion, screening of the *blocking electrodes* in ice is rather strongly dependent on the frequency and the ratio of screening length to sample thickness. This dependence may be described in terms of the dielectric permittivity of ice or of a capacitor with ice as the dielectric. It has been shown[1] that the values of low frequency capacitances may exceed, by several orders, the values of high frequency capacitances. In such cases, the relaxation frequencies are determined by concentrations and mobilities of carriers and may be varied by doping or temperature. So, we obtain a whole class of devices, condensors whose capacitance is strongly dependent on frequency.

It has been shown in this work that for sample thickness $L \gg$ the screening length, l, by carriers, the low frequency value of the capacitance is inversely proportional to l. By increasing the concentration of carriers, one can first extend to higher frequency the region where large, low frequency capacitances are observed and, second, decrease l, thus obtaining capacitors of enormous capacity – supercapacitors.

By making the construction somewhat more complicated, one may get a voltage-controlled capacitor. Fig. 35.1 demonstrates a condensor with

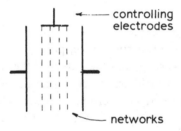

Fig. 35.1. Scheme of capacitor with voltage-controlled capacity.

extra electrodes in the form of networks to which the controlling voltage is supplied. In the absence of the voltage let the condensor behave as the above described supercapacitor of thickness $L \gg l$. The size of the network meshes and the distances between them are sufficient to take equal screening length l; that can easily be done by varying the impurity concentration, i.e. l. Then, upon supplying sufficiently high voltage to the controlling electrodes, the networks block the charge exchange, and we get a capacitor with n-fold reduced capacity, where $n = L/l$. At intermediate voltage values the charge exchange is hindered though still possible. In this case we get a condensor with a small capacitance but with a controlled leakage current. Note that the role of the network may be played by proton-conducting membranes, whose permittivities depend on some controlling signal.

35.3   Devices based on memory effects

Due to polarization of hydrogen bonds, a system may have memory associated with the passing of particular defects[2]. Thus, having passed a certain current of H_3O^+-defects, an ice sample becomes polarized. It is then capable of passing the same defects only in the opposite direction. It seems attractive to employ this effect in memory devices or logic schemes. One has to take into account that there is always a final concentration of depolarizing defects, and we may be dealing only with dynamic memory devices.

35.4   Field transistors

This is the sole example of a device that has been experimentally demonstrated[3]. The device has two current electrodes frozen into the near-surface layer of ice and a gate across the dielectric layer, located above the region between the current electrodes. By supplying a voltage of different polarities across the *controlling electrode*, the carrier concentration in the near-surface region may be varied. By studying the frequency dispersion of the near-surface conduction, one may determine the mobilities of all the four types of carriers in ice: H_3O^+, OH^-, D and L defects. Naturally, a field transistor may be used not only for research purposes but, also, as a power amplifier or memory device.

35.5   Sensors

A number of phenomena may be employed for constructing sensors. We shall give some examples. The pseudopiezoeffect of ice[4] is associated with the occurrence of charges at the boundaries of a crystal subjected to pressure. Employing this effect, one may construct *piezosensors*.

A dramatic change in the electrical properties of ice upon doping with particular materials may be used in devices for tracing these materials. For instance, doping with HF, which is easily dissolved in ice, was experimentally accomplished by applying a piece of cotton wool soaked with HF to an ice sample. In this case, the electrical properties of the sample changed over a period of some minutes. The sensitivity of devices using this effect may be improved considerably by measuring the surface rather than bulk conduction.

The electrical properties of ice depend appreciably on the presence of dislocations, cracks and polycrystalline texture. Therefore, the electrical properties, being most easily measurable, may be used for the determination and prediction of mechanical properties, namely, strength and plasticity. Here, one may be dealing not only with ice, but, also, with ice-containing materials.

35.6   *References*

1. V. F. Petrenko and I. A. Ryzhkin, *Phys. Stat. Sol.* (*b*) **121** (1984) 421.
2. C. Jaccard, *Phys. Kondens. Materie* **3** (1964) 99.
3. V. F. Petrenko and N. Maeno, *J. Physique* **48** (1987) C1–11.
4. A. A. Evtushenko, V. F. Petrenko and I. A. Ryzhkin, *Phys. Stat. Sol.* (*a*) **86** (1984) K31.

36 Solid-state gas sensors operating at room temperature

NORIO MIURA and NOBORU YAMAZOE

36.1   Introduction

Some proton conductors have relatively high conductivities at room temperature. Introduction of these materials into electrochemical cells brings about attractive chemical sensors workable at room temperature. Potentiometric or amperometric detection of chemical components at room temperature would create new fields of application for sensors especially in bioprocess control and medical diagnosis. With an all-solid-state structure, the sensors would be compatible with micro-fabrication and mass production, and small power consumption associated with their ambient-temperature operation would be intrinsically suited for cordless or portable sensors.

As listed in Table 36.1, a good deal of research has been carried out so far on proton conductor-based *gas sensors* workable at ambient temperature. Various inorganic and organic ion exchangers, such as *hydrogen uranyl phosphate* (HUP), *zirconium phosphate* (ZrP), *antimonic acid* (AA), and *NAFION* membrane, have been utilized in the form of a disc, thick- or thin-film. The ionic conductivities of these proton conductors, in the range 10^{-4}–10^{-2} S cm^{-1}, are modest but seem to be still sufficient for chemical sensing devices.

A problem with these conductors is that humidity in the ambient atmosphere affects the conductivities and hence sensing characteristics when an amperometric mode of operation is adopted, as described later. The sensing electrodes have been provided mostly by Pt, and the counter electrodes by Ag, PdH$_x$, TiH$_x$, H$_{0.35}$MoO$_3$, H$_x$WO$_3$, etc. The gases to be measured extended from H$_2$ in inert gases to H$_2$ in air in 1982, and now seem to also include CO, NH$_3$ and NO, all diluted in air, as well as O$_2$ in the gas phase or condensed phase. Besides the examples given in Table 35.1, proton conductors have been used in humidity sensors, e.g. NAFION membrane in room temperature operation type[28], antimonic

Devices

Table 36.1. *Solid-state gas sensors using proton conductors operative at ambient temperature*

Sensor structure (remarks)	Objective gas	Conc. range	Temp.	Year	Ref.
Pt, $H_2/\beta,\beta''$-alumina/Pt (disc type)	H_2 in N_2 O_2 in N_2	5–80% 20–100%	23 °C	1974	1
Pd ref./β,β''-alumina/Pt (thin-film type)	H_2 in N_2	0.01–10%	RT	1981	2
Pt/NAFION/Pt (applied voltage)	CO in air NO in air	$\sim$1500 p.p.m. $\sim$260 p.p.m.	25 °C	1981	3
Pt, $H_2/HUO_2PO_4 \cdot 4H_2O(HUP)/Pt$ (disc type)	H_2 in N_2	1–10%	20 °C	1982	4
Pd ref./HUP thick-film/Pt	H_2 in N_2	0.01–10%	RT	1982	5
H_xWO_3/HUP disc/Pd	H_2 in Ar	30–6000 p.p.m.	25 °C	1983	6
PdH_x/PVA with H_3PO_4/Pt	H_2 in N_2	0.02–100%	RT	1985	7
$H_{0.35}MoO_3$/HY zeolite/Pd	H_2 from steel	–	RT	1986	8
PbO_2/HUP disc/Pt black	H_2 in N_2	10^{-12}–1 atm	17 °C	1988	9
$TiH_x/Zr(HPO_4)_2 \cdot H_2O(ZrP)/Pt$ (thick-film type)	H_2 in N_2 H_2 in air	10 p.p.m.–10% 10 p.p.m. $\sim$1%	25 °C 25–200 °C	1989	10
Pt/NAFION/Pt (applied voltage)	O_2 in N_2	$\sim$30%	15–50 °C	1989	11
Pt, air/NAFION/Pt black	H_2 in air	0.02–2%	RT	1982	12
Ag ref./ZrP disc/Pt black	CO in air	20–1000 p.p.m.	RT	1983	13, 14

528

Table 36.1. (*cont.*)

Sensor structure (remarks)	Objective gas	Conc. range	Temp.	Year	Ref.
Pt ref./Sb$_2$O$_5$. 4H$_2$O(AA)/Pt (amperometric type)	H$_2$ in air	~1.5%	20 °C	1984 1986	15 16
Pt, air/NAFION/Pt Ag ref./AA, ZrP/Pt	O$_2$ in N$_2$ O$_2$ in N$_2$	0.2–20% 1–100%	25 °C 80–120 °C	1984	17
Pt, air/AA, ZrP/Pt	Dissolved O$_2$ Dissolved H$_2$O$_2$	0.3–50 p.p.m. 10^{-9}–10^{-4} M	10–70 °C 25 °C	1987	18
Pt, air/NAFION/Pt (applied voltage)	O$_2$ in N$_2$	~80%	25 °C	1988	19
Pt ref./AA/Pt	NH$_3$ in air	50–5000 p.p.m.	RT	1987, 1988	20, 21
Pt ref./AA/Pt (four-probe type)	H$_2$ in air	0.05–1.3%	25 °C	1987 1989	22 23
SnO$_2$(+Pt)/AA/WO$_3$ (+Pt) (amperometric type)	CO in air	100–1000 p.p.m.	RT	1989	24
Pt/AA thick film/Pt (applied voltage)	O$_2$ in N$_2$ Dissolved O$_2$	~100% ~20 p.p.m.	30 °C 30 °C	1989	25
Au or Pt/AA thick film/Pt	H$_2$ in air CO in air	200 p.p.m.–1% 20–1000 p.p.m.	30 °C 30 °C	1988, 1990	26, 27

acid[29] and antimony phosphate[30] in intermediate temperature type (100–250 °C), and SrCe$_{0.95}$Yb$_{0.05}$O$_{3-\delta}$ in high temperature type (300–1000 °C)[31,32]. The last sensor is also applicable to the detection of H$_2$ in inert gases[33] and molten metals[34] at higher temperatures.

In this section, development of proton conductor-based chemical sensors is briefly described by focusing on *room temperature operation type H$_2$ sensors*.

36.2   *Principle of* **potentiometric sensors**

An electrochemical cell is constructed when two compartments are separated by a proton conducting disc or membrane attached with electrodes (such as Pt black). If both compartments contain H_2 (in an inert gas), the cell becomes a hydrogen concentration cell

$$H_2(p'), Pt \mid \text{proton conductor} \mid Pt, H_2(p'') \qquad (36.1)$$

Both electrodes undergo the same electrode reaction (36.2), generating an e.m.f. between them according to the *Nernst equation* (36.3)

$$H_2(\text{gas}) \rightleftarrows 2H^+(\text{proton conductor}) + 2e^-(Pt) \qquad (36.2)$$

$$E(\text{e.m.f.}) = (RT/2F) \ln(p'/p'') \qquad (36.3)$$

Here p' and p'' are the partial pressures of hydrogen, F is the Faraday constant, R the gas constant, and T absolute temperature. At $T = 298$ K, E is linear with $\log(p'/p'')$ giving a Nernst slope of 30 mV decade^{-1}. When p' is known, p'' can be evaluated from E, and this is the basis for concentration cell-type H_2 sensors, which were developed in earlier stages[1,2,4].

The present authors have shown that the same cell structure is also applicable to the detection of dilute H_2 in air[12,13]. In this case, the cell is expressed as follows:

$$\text{air}, Pt \mid \text{proton conductor} \mid Pt, H_2(p) \text{ in air} \qquad (36.4)$$

As shown by the insert in Fig. 36.1, the cell responds very sharply to 0.2% H_2 in air at room temperature with a 90% response time of c. 20 s. With change in H_2 concentration, the e.m.f. apparently follows Nernst-type behaviour as shown in Fig. 36.1, but its slope, c. 140 mV decade^{-1} instead of 30 mV decade^{-1} for the above concentration cell (36.1), suggests a different operation mechanism, which is mentioned below[13]. Anyway, a sharp response to dilute H_2 in air assures the potential of this type of sensor for new applications.

In the electrochemical cell (36.4), the sensing electrode (Pt) is exposed to two electrode-active components, i.e. H_2 and O_2, which undergo the following reactions

$$H_2 \rightarrow 2H^+ + 2e^- \qquad (36.5)$$

$$\tfrac{1}{2}O_2 + 2H^+ + 2e^- \rightarrow H_2O \qquad (36.6)$$

while only reaction (36.6) takes place on the counter electrode.

Solid-state gas sensors

The two reactions proceed eventually at an equal rate to determine a *mixed potential* of the electrode, which is given as an intersection of the polarization curves for H_2 and O_2 as shown in Fig. 36.2. It is noted that reaction (36.5) enters into the diffusion-limited region under low H_2 concentrations, and this fact provides a basis for the H_2-concentration dependent e.m.f. of the present sensor. The observed 'Nernst slope' of c. 140 mV decade^{-1} is understood as originating from the polarization curve for reaction (36.6).

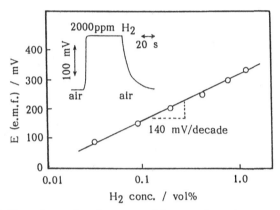

Fig. 36.1. Dependence of e.m.f. response of the potentiometric proton conductor sensor on H_2 concentration in air and response transient of the sensor to 2000 p.p.m. H_2 in air at room temperature[13] (with permission, Kodansha Ltd).

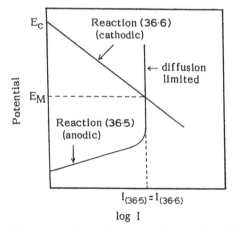

Fig. 36.2. Schematic polarization curves for reactions (36.5) and (36.6)[13]. E_M: mixed potential (with permission, Kodansha Ltd).

531

36.3 New sensing modes

The above potentiometric mode generates an e.m.f. in proportion to the logarithm of H_2 concentration, which is well suited for detecting broadly varying gas concentrations. For narrow concentration changes, however, an amperometric mode is known to give more precise data because the signal output (current) is proportional to the gas concentration. Usually, such amperometric sensing has been achieved under an external applied voltage as can be seen in the cases of CO^3 and $O_2^{11,19}$ sensors. Quite notably, however, our electrochemical cell (36.4) has been found to change into an *amperometric mode* (Table 36.2) when the two electrodes are simply connected electronically[15,16]. The short-circuit current is proportional to the concentration of H_2 in air under a constant relative humidity (R.H.) as shown in Fig. 36.3a. The 90% response time to 0.2% H_2 is about 10 s at room temperature, being a little shorter than that for the potentiometric mode.

The origin of the short-circuit current is the imbalance of electrode reactions (36.5) and (36.6). The protons produced by (36.5) cannot be consumed completely by (36.6) under the present conditions. Excess protons migrate through the proton conductor (AA) towards the counter electrode to be consumed there, accompanied by an equivalent electronic

Table 36.2. *Various sensing modes of proton conductor H_2 sensor*

Sensing mode	Sensor signal	Sensing characteristics	Remarks	Ref.
(1) potentiometric	e.m.f. (ΔE)	$\Delta E \propto \log C_{H2}$	Gas concentration cell mixed potential	1, 2, 4–7, 9, 10, 12, 13, 17, 18, 21, 27
(2) amperometric	Short-circuit current (I_s)	$I_s \propto C_{H2}$	Humidity dependent	15, 16, 24, 27
(3) amperometric	Electrolytic current (I_e)	$I_e \propto C_{H2}$	Needs applied voltage and gas-diffusion layer	3, 11, 19, 25
(4) four-probe type	Inner potential difference (ΔE_i)	$\Delta E_i \propto C_{H2}$	Insensitive to humidity	22, 23, 26

current passing through the external circuit. The magnitude of the short-circuit current (I) depends strongly on the electrical resistance (R) of the proton conductor used, and this is the reason for the R.H.-dependent output shown in Fig. 36.3a. It has been shown that the product, $I \times R$, is constant irrespective of R.H.[22] This means that the driving force for the short-circuit current is the same, but the flux is determined by the resistance.

R.H.-independent signal output has been achieved in the *four-probe type sensor* shown in Fig. 36.4, where two additional Ag probes are inserted in the proton conductor bulk (AA) beneath the Pt electrodes[23]. One of the Pt electrodes is covered by a layer of AA sheet, which acts as a sort of gas diffusion layer. The short-circuit current flowing between the two Pt electrodes is proportional to H_2 concentration but dependent on R.H., just as in the previous amperometric sensor. On the other hand, the difference in potential between the two Ag probes (inner potential difference, ΔE_{Ag}) with the outer Pt electrodes short-circuited is shown to be not only proportional to H_2 concentration but also independent of R.H. as shown in Fig. 36.3b and Table 36.2. This mode of sensing has no precedence, and is noted as a new method to overcome the greatest difficulty in using proton conductor-based devices, i.e. their R.H. dependence.

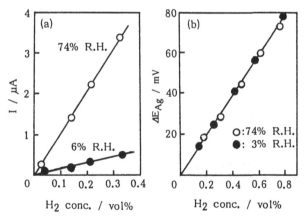

Fig. 36.3. (a) Dependence of short-circuit current (I) of the amperometric proton-conductor sensor on H_2 concentration, and (b) dependence of ΔE_{Ag} (inner potential difference) of the four-probe type sensor on H_2 concentration in air at different relative humidity (25 °C)[16,23] (reprinted by permission of The Electrochemical Society, Inc.).

Devices

36.4 Simplification of sensor elements

The prototype sensor element in the gas concentration cell needs a reference gas (air or H_2) and is not necessarily suited for simplification and microminiaturization. Elimination of the reference gas flow to change it into a probe type element is highly desirable. Several methods have been proposed for this purpose as follows.

   (i) Replacement of the Pt counter (reference) electrode by other H_2-inert metal electrodes such as Ag and Au[13,17,27].

   (ii) Use of hydrides having constant hydrogen activities such as PdH_x and TiH_x for the counter electrode[2,5–8,10].

   (iii) Covering the Pt counter electrode with a gas diffusion limiting layer[16,21,23,25,27].

In the potentiometric sensing of H_2, (ii) is useful for H_2 in inert gases while (i) is especially suited for H_2 in air. On the other hand, only (iii) is applicable to the amperometric sensor and the four-probe type sensor.

The probe-type elements (without reference gas) thus obtained are convenient for microminiaturization and integration. In fact, several *micro sensors* have been fabricated by combining proton conductor films and hydride counter electrodes (PdH_x or TiH_x) and their good sensing characteristics have been reported[2,5,10]. For detection of H_2 in air, we have reported a potentiometric element (a) and an amperometric element

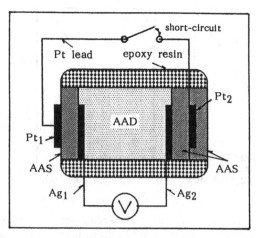

Fig. 36.4. Structure of the four-probe type sensor. AAD, antimonic acid disc; AAS, antimonic acid sheet[23] (reprinted by permission of The Electrochemical Society, Inc.).

534

Solid-state gas sensors

(b) using a proton conductor film about 10–20 μm thick, as shown in Fig. 36.5[26, 27]. The film was spin-coated on a porous alumina substrate from a paste suspending AA in a polyvinyl alcohol (PVA) solution. Element (a) has a planar structure with a sputtered Pt sensing electrode and an evaporated Au reference electrode; the latter electrode was selected according to (i) above. This element exhibited almost the same sensing characteristic to dilute H_2 in air as the previous prototype element, i.e. quick response and e.m.f. change in proportion to the logarithm of H_2 concentration. Element (b) has a laminated structure in which the proton conductor film is sandwiched between sputtered Pt electrodes. The inner Pt electrode is in the neighbourhood of a porous substrate which acts as a gas diffusion limiting layer. Thus this probe-type element is based on

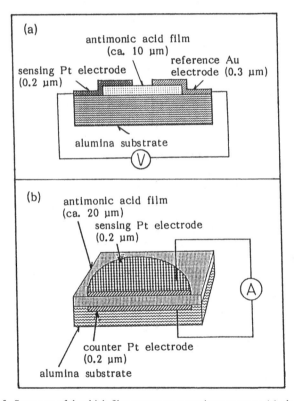

Fig. 36.5. Structure of the thick-film type proton conductor sensor, (a) planar-type potentiometric sensor, (b) laminated-type amperometric sensor[27] (reprinted by permission of Elsevier Sequoia S.A.).

535

(iii) above. The short-circuit current as an output signal of this element has been shown to be linear with H_2 concentration, as in the prototype amperometric sensor. Introduction of such thick- or thin-film technology in the proton conductor-based devices seems to be promising for further development of micro, integrated, and intelligent sensors.

36.5   Extension of proton conductor sensors

Most proton conductor sensors have aimed at sensing of H_2 in inert gases or air, as mentioned above. However, other chemical components are also sensitive whenever they produce or consume protons through electrode reactions. Typical of such examples are CO[3,13], NH_3[20,21] and O_2[17,18,25]. In the cases of CO and NH_3, the following electrode reactions are involved, respectively, and compete with the proton-consuming reaction (36.6).

$$CO + H_2O \rightarrow CO_2 + 2H^+ + 2e^- \qquad (36.7)$$

$$\tfrac{2}{3}NH_3 \rightarrow \tfrac{1}{3}N_2 + 2H^+ + 2e^- \qquad (36.8)$$

In the presence of water vapour, CO can be sensitive with devices usually having Pt electrodes. However, the sensitivity (short-circuit current) is about one tenth of that of H_2 at the same concentration[16]. The choice of other electrodes may be varied to increase CO sensitivity. For example, the use of a pair of oxide electrodes, Pt-loaded WO_3 and Pt-loaded SnO_2, in an amperometric element has been shown to increase CO sensitivity to about seven times as high as the H_2 sensitivity[24].

For sensing of NH_3, Pt electrodes show sufficiently high sensitivity. The four-probe type element (Fig. 36.4) can reportedly detect 50–5000 p.p.m. NH_3 in air at room temperature[21].

It was reported in 1974 that gaseous O_2 is detectable with proton conductor sensors[1]. As O_2 consumes protons through reaction (36.6), a change in O_2 partial pressure under a fixed water vapour pressure generates a change in e.m.f. Recently amperometric sensors have been also reported for gaseous O_2, dissolved O_2 and H_2O_2[18], which operate under the external applied voltages. Thick-film proton conductor elements are also available for such purposes[25]. O_2 and H_2O_2 are the most popular components consumed or generated in enzymatic reactions. With the capabilities of proton conductor sensors described so far, it is not so unpractical to expect that new all-solid-state *biosensors* will be developed

Solid-state gas sensors

by using proton conductors, as in the case of a new type of glucose sensor using a fluoride ion conductor (LaF_3)[35].

36.6 References

1. J. S. Lundsgaard and R. J. Brook, *J. Mat. Sci.* **9** (1974) 2061–2.
2. Ph. Schnell, G. Velasco and Ph. Colomban, *Solid State Ionics* **5** (1981) 291–4.
3. A. B. Laconti, M. E. Nolan, J. A. Kosek and J. M. Sedlak, *Chemical Hazards in the Workplace* (ACS (1981)) 551–73.
4. J. S. Lundsgaard, J. Malling and M. L. S. Birchall, *Solid State Ionics* **7** (1982) 53–6.
5. G. Velasco, J. Ph. Schnell and M. Croset, *Sensors & Actuators* **2** (1982) 371–84.
6. S. B. Lyon and D. J. Fray, *Solid State Ionics* **9&10** (1983) 1295–8.
7. A. J. Polak, A. J. Beuhler and S. Petty-Weeks, *Digest of Tech. Papers of Transducers '85*, Philadelphia (1985) 85–8.
8. J. Schoonman and J. L. 'de Roo, *Proc. 2nd Int. Meet. on Chemical Sensors*, Bordeaux (1986) 319–22.
9. R. V. Kumar and D. J. Fray, *Sensors & Actuators* **15** (1988) 185–91.
10. G. Alberti and R. Palombari, *Solid State Ionics* **35** (1989) 153–6.
11. Y. He-Qing and L. Jun-Tao, *Sensors & Actuators* **19** (1989) 33–40.
12. N. Miura, H. Kato, N. Yamazoe and T. Seiyama, *Denki Kagaku* **50** (1982) 858–9.
13. N. Miura, H. Kato, N. Yamazoe and T. Seiyama, *Proc. 1st Int. Meet. on Chemical Sensors*, Fukuoka (1983) 233–8.
14. N. Miura, H. Kato, N. Yamazoe and T. Seiyama, *Chem. Lett.* **1983** (1983) 1573–6.
15. N. Miura, H. Kato, Y. Ozawa, N. Yamazoe and T. Seiyama, *Chem. Lett.* **1984** (1984) 1905–8.
16. N. Miura, H. Kato, N. Yamazoe and T. Seiyama, *Fund. & Appl. Chemical Sensors* (ACS (1986)) 203–14.
17. N. Miura, H. Kato, N. Yamazoe and T. Seiyama, *Denki Kagaku* **52** (1984) 376–7.
18. S. Kuwata, N. Miura and N. Yamazoe, *Nippon Kagaku Kaishi* **1987** (1987) 1518–23.
19. S. Kuwata, N. Miura and N. Yamazoe, *Chem. Lett.* **1988** (1988) 1197–200.
20. N. Miura and W. L. Worrell, *Chem. Lett.* **1987** (1987) 319–22.
21. N. Miura and W. L. Worrell, *Solid State Ionics* **27** (1988) 175–9.
22. N. Miura, H. Kaneko and N. Yamazoe, *J. Electrochem. Soc.* **134** (1987) 1875–6.
23. N. Miura, T. Harada and N. Yamazoe, *J. Electrochem. Soc.* **136** (1989) 1215–19.
24. N. Miura, K. Kanamaru, Y. Shimizu and N. Yamazoe, *Chem. Lett.* **1989** (1989) 1103–6.
25. N. Miura, N. Yoshida, N. Matayoshi, Y. Shimizu, S. Kuwata and N. Yamazoe, *J. Ceramic Soc. Jpn* **97** (1989) 1300–3.

Devices

26. N. Miura and N. Yamazoe, *Chemical Sensor Tech.*, Vol. 1 (1988) 123–39.
27. N. Miura, T. Harada, Y. Shimizu and N. Yamazoe, *Sensors & Actuators* **B1** (1990) 125 9.
28. A. Yamanaka, T. Kodera, K. Fujikawa and H. Kita, *Denki Kagaku* **56** (1988) 200–1.
29. N. Miura, I. Yashima and N. Yamazoe, *Nippon Kagaku Kaishi* **1985** (1985) 1644–9.
30. N. Miura, H. Mizuno and N. Yamazoe, *Jpn J. Appl. Phys.* **27** (1988) L931–3.
31. H. Iwahara, H. Uchida and J. Kondo, *J. Appl. Electrochem.* **13** (1983) 365–70.
32. K. Nagata, M. Nishino and K. S. Goto, *J. Electrochem. Soc.* **134** (1987) 1850–4.
33. H. Uchida, K. Ogaki and H. Iwahara, in *Proc. Symp. on Chemical Sensors*, Honolulu (1987) 172–9.
34. H. Iwahara, H. Uchida, T. Nagano and K. Koide, *Denki Kagaku* **57** (1989) 992–5.
35. N. Miura, N. Matayoshi and N. Yamazoe, *Jpn J. Appl. Phys.* **28** (1989) L1480–2.

37 All solid-state protonic batteries

J. GUITTON, C. POINSIGNON and J. Y. SANCHEZ

37.1 A solid-state battery with a proton conducting electrolyte

A solid protonic electrolyte battery presents the general advantages of all solid-state systems such as absence of liquid leakage and ease of handling. It must have two electrodes to react with the protons, as in aqueous liquid electrolyte batteries. The main aim for using solid protonic conductors resides in the possibility of using similar electrodes to those in current commercial batteries.

37.2 Advantages and problems of batteries with a liquid proton conducting electrolyte

The main advantages of batteries with a *liquid electrolyte* come from the high conductivity exhibited by the usual concentrated solutions, sulphuric acid and potash, and from the good interface between electrolyte and electrode material, which allows high current densities to be achieved. Electrode materials of these batteries have been in continuous improvement throughout this century; they now have a good faradic efficiency and a reasonable cost.

The main reaction involved with all these electrode materials is a proton insertion/deinsertion which supplies protons to the electrolyte when *electrodes* are anodically polarized and catches protons from the electrolyte when they are cathodically polarized. This protonic cathode–anode exchange can be pointed out in the following reactions.

In Leclanché primary cells

cathode (+) $2MnO_2 + 2H^+ + 2e^- \rightarrow 2MnOOH$

anode (−) $Zn \rightarrow 2e^- + Zn^{2+} + H_2O \rightarrow Zn(OH)_2 + 2H^+$

In lead–acid secondary cells

$(+)$ PbO_2 + H_2SO_4 + $2e^-$ + $2H^+$ $\underset{\leftarrow \text{ anode}}{\overset{\text{cathode} \rightarrow}{}}$ $PbSO_4$ + $2H_2O$

$(-)$ Pb + H_2SO_4 $\underset{\leftarrow \text{ cathode}}{\overset{\text{anode} \rightarrow}{}}$ $PbSO_4$ + $2e^-$ + $2H^+$

In cadmium–nickel oxide secondary cells

$(+)$ $2NiOOH$ + $2e^-$ + $2H^+$ $\underset{\leftarrow \text{ anode}}{\overset{\text{cathode} \rightarrow}{}}$ $2Ni(OH)_2$

$(-)$ Cd + $2H_2O$ $\underset{\leftarrow \text{ cathode}}{\overset{\text{anode} \rightarrow}{}}$ $Cd(OH)_2$ + $2e^-$ + $2H^+$

However, these electrode materials are to varying degrees unstable in liquid aqueous electrolytes.

Some of them (for example, zinc) are corroded, with hydrogen evolution leading to a local desiccation of the electrode which makes the cell swell, if it is sealed. An interfacial pH increase can also occur leading to a local passivation of the electrode.

Others are corroded by oxygen dissolved in the electrolyte.

Others are acidically or basically dissolved (like Ag_2O), assisted by migration and/or diffusion into the electrolyte; the dissolved ionic species may reach the counter-electrode where they react, making necessary the use of a special separator to prevent this parasitic reaction.

In some cases, the electrode reaches some potential value at which an electrochemical dissolution of its components into the electrolyte occurs: for example, the positive grid of lead–acid batteries contains antimony; at the high potential of the grid, Sb can be oxidized to soluble SbO^+ which can be reduced at the negative lead electrode. An antimony deposit then occurs leading to an acceleration of the self-discharge corrosion reaction.

For such reasons, some materials cannot be used over certain pH ranges although this would give a more advantageous potential: this is illustrated by γ-MnO_2[1]. Its potential is 1.01 V versus NHE in an acidic medium, and only 0.71 in neutral medium. However, this oxide should dissolve as Mn^{2+} in an acidic aqueous electrolyte during cathodic discharge and the electrode material should consequently disappear.

All these reactions produce or consume ionic species different from protons (i.e. metallic or non-metallic cations and anions): they can occur in the presence of liquid aqueous electrolyte because a liquid medium allows the species to migrate and diffuse.

The main problems of batteries using a liquid aqueous electrolyte result from the smaller than unity value of the proton transport number in this electrolyte.

37.3 Characteristics of an ideal all solid-state battery

In an ideal *solid-state battery* involving protonic reactions, the solid electrolyte must be only a protonic conductor: the proton transference number should be equal to unity ($t_{(H^+)} = 1$). Then all the previous reactions cannot occur and it is theoretically possible to use electrode materials which were unstable in the presence of a liquid aqueous electrolyte.

For example, the positive electrode material γ-MnO_2 can be associated with a solid protonic conductor containing H^+ at a high concentration and can then be used to give the benefit of its high acidic potential. If $t_{(H^+)}$ is equal to unity in the electrolyte, and if the negative electrode delivers protons into this electrolyte when anodically polarized, the only cathodic reaction will be the insertion of protons into MnO_2, and the formation of Mn^{2+} is not possible.

Another interesting case is given by PbO_2 into which protons cannot be cathodically inserted in the presence of sulphuric acid because of the lead sulphate interfacial precipitation. This insertion is possible with a solid protonic conductor (SPC) as electrolyte.

Let us consider an ideal battery of this type, Fig. 37.1. During discharge, the negative electrode must supply protons to the solid electrolyte in which they are carried[2] to the other electrode by a translocation/vehicular mechanism. Then they must penetrate into the positive electrode material.

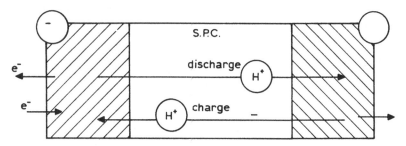

Fig. 37.1. Principle of an 'all solid' battery with a protonic conductor as electrolyte and two mixed protonic–electronic conductors as electrodes.

Both electrodes must present a mixed conductivity – electronic and protonic – with an internal RedOx couple giving a potential high enough (versus a hydrogen electrode) for the positive electrode and sufficiently low for the negative one.

The proton conduction paths must be the same in the three materials so that proton transfer reactions at both interfaces should be rapid, and overpotentials at both electrodes negligible.

37.4   The first attempts

Such an ideal system is almost obtained for the positive electrode and the electrolyte in some complete protonic batteries whose characteristics have been published[3,4].

> Two or three solid hydrated acids are stable and good protonic conductors at room temperature. This is the case, for example, with hydrogen uranyl phosphate (HUP) which exhibits a protonic conductivity of about $10^{-3}\,\Omega\,cm^{-1}$ at 25 °C, allowing good power densities, even with a 1 mm thickness of electrolyte.
>
> Several metallic oxides present a good reversibility for proton electrochemical insertion when associated with these electrolytes. Composite positive electrodes are obtained by mixing a powder of one of these materials with electrolyte and acetylene black. Kinetic tests of these electrodes were carried out in three-electrode pellet-shaped cells, with reference and counter electrodes made of a platinum layer in the presence of hydrogen gas. Their electrochemical capacity and their power are interesting.

However, a problem remains at the level of negative electrodes. Their characteristics presented in the early studies were not as good as those of the positive electrodes. Except for gaseous electrodes (hydrogen), the only negative materials used were hydrides such as PtH_x, PdH_x, NiH_x or $TiNiH_x$, associated with HUP as electrolyte[5].

The proton conduction paths are obviously different in the hydride and in HUP: the couple HUP/hydride leads to higher overpotentials than with HUP–metallic oxide interface. For that reason, current densities are smaller.

The major inconvenience of using hydrides as negative electrodes comes from the insufficiently negative value of their potential (about $-0.1\,V$ versus SHE), and, consequently, the e.m.f. of the battery is not very high.

All solid-state protonic batteries

For example, with the $TiNiH_{1.5}/HUP/MnO_2$ cell, the e.m.f. is only 1.23 V and the battery voltage during discharge is no higher than 1.15 V for $i = 1 \, \text{mA cm}^{-2}$.

In order to increase the value of the e.m.f., some authors[6-8] have described all solid systems using a metal layer (Zn, Pb ...) pressed on the solid electrolyte (Fig. 37.2) as negative electrode.

Even with sufficient e.m.f., the practical capacity, power and energy densities of these cells are generally not given or have uncertain values. That is not surprising since the metallic cations produced during discharge cannot easily migrate in the protonic conductor electrolyte, and the metallic electrodes cannot supply to the solid electrolyte the protons that are consumed at the positive electrode. Moreover, the absence of metallic cations in the electrolyte leads consequently to a non-rechargeable cell.

37.5   Recent improvements

These concern essentially the negative electrode and the electrolyte.

The negative electrode can present good characteristics if two essential conditions are fulfilled: a reversible exchange of protons with the electrolyte, and a potential as negative as that of zinc. This is obtained by adding, to the *composite electrode* (metal and electrolyte grains), a third component (Fig. 37.3) able to exchange metallic cations M^{n+} with the metal, and protons with the electrolyte.

When using a solid hydrated electrolyte, such an electrode component will be advantageously hydrated too: for instance, a neutral or weakly acidic compound (XY, nH_2O) in which M^{n+} can be anodically inserted while nH^+ ions are extracted towards the electrolyte. The overall anodic

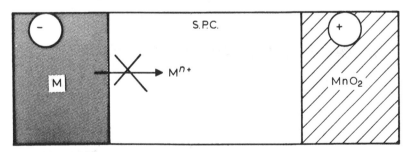

Fig. 37.2. Unfeasibility of using a metal as the negative electrode with a solid protonic conductor as electrolyte.

oxidation occurring at the negative electrode during discharge of the battery can be as depicted in Scheme 37.1.

It is clear that the electrochemical capacity of this composite electrode depends on its water content. When using zinc as metal and HUP as electrolyte, the hydrated sodium phosphate $Na_3PO_4 . 12H_2O$ is very well adapted for the insertion of zinc cations, but the results are better if hydrated zinc sulphate is added to this mixture. The potential of this electrode is -0.75 V versus H_2.

In order to obtain a complete battery, three layers (negative mixture powder, electrolyte powder, positive mixture powder) are pressed together

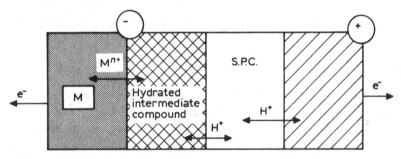

Fig. 37.3. Principle of an 'all solid' battery with a protonic conductor as electrolyte and negative metallic composite electrode exchanging protons with this electrolyte.

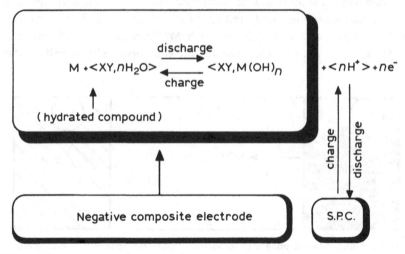

Scheme 37.1.

All solid-state protonic batteries

at 700 MPa for half an hour. A solid pellet (diameter 13 mm; height 3 mm) is obtained.
The electrochemical behaviour of the cell (1.77 g)

$$(-)Zn-ZnSO_4 . 7H_2O-Na_3PO_4 . 12H_2O/HUP/HUP-(\alpha + \beta)PbO_2-\text{acetylene black}(+)$$
$$(500 \text{ mg} + 200 \text{ mg} \quad + 250 \text{ mg} \quad /300 \text{ mg}/ \; 300 \text{ mg} \quad + 200 \text{ mg} \quad + 20 \text{ mg})$$

during discharge, at various constant currents down to 2 V, is given in Fig. 37.4. It can be noticed that, even for this non-optimized battery (for these first tests the thickness of the electrolyte layer and consequently the ohmic drop are also very important) the voltage is higher than 2 V for 6 h at 0.5 mA cm^{-2} and higher than 2 V for 10 s at 10 mA cm^{-2} which makes it suitable as a back-up power source for protection against momentary mains power shut-down[9].

It should be also emphasized that these materials and cells are entirely prepared, pressed, used and stored at ambient atmosphere and room temperature.

37.5.1 *Protonic batteries with polymer electrolyte*

HUP is one of the best solid superprotonic conductors, easily prepared but exhibiting the major inconvenience of having uranium in its chemical formula, which makes it difficult to handle. Furthermore, its layered structure induces a stacking anisotropy which increases the internal resistance of a device, when the layers are perpendicular to the flow lines. Another possibility for making solid-state batteries with solid electrolytes involves the use of hydrated proton conducting polymers, new organic–inorganic polymer electrolytes developed in LIESG[10].

This polymer electrolyte is an organic–inorganic polymer of the *Ormolyte* (ORganically MOdified silicate electroLYTE) type prepared by a sol–gel process by hydrolysis and condensation of modified trialkoxy silanes. The organic phase yields plasticity and chemical reactivity, the inorganic backbone an amorphous state and a good mechanical strength. The choice of inorganic chains is determined by the type of applications planned. In this particular case, protonic conductivity is conferred by sulphonic groups grafted on the silica backbone. Conductivity mechanism is of the vehicular type, as for HUP, and makes the presence of water necessary to ensure good protonic conductivity, typically 5.8×10^{-3} Ω cm^{-1}, at room temperature and ambient humidity.

The sol–gel route goes through a liquid–solid transition and then allows

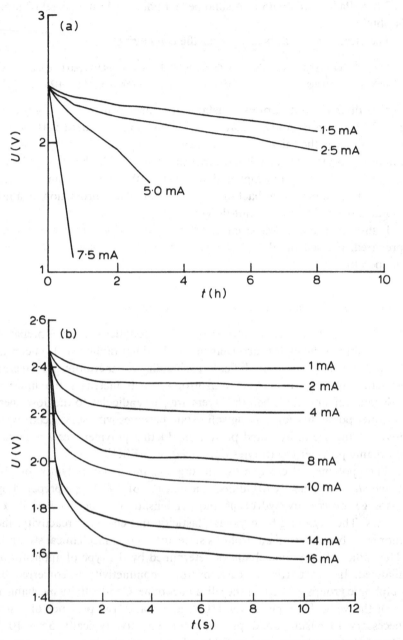

Fig. 37.4. Galvanostatic discharge curves of the $Zn/Na_3PO_4 \cdot 12H_2O/HUP/PbO_2$–HUP battery.

the fabrication of good composite electrodes by inducing polymerization of the polymer during preparation of the composite electrode: Ormolyte in the liquid phase wets the pores of the electrode material, and then ensures the realization of a good electrolyte/electrode interface leading to a low internal cell resistance able to support high current density.

The electrode Ormolyte/γ-MnO$_2$ was studied by slow cathodic voltammetry[11]: a unique peak is observed (Fig. 37.5a) at 1.03 V/PdH$_x$. Compared with the voltammogram obtained in KOH media for the same γ-MnO$_2$, some differences are noticeable (Fig. 37.5b). In KOH electrolyte, the first peak on the voltammogram is noticed around 1.2 V/Hg/HgO and is attributed to the perfectly reversible insertion on surface sites, related to the surface charge density, and due to the amphoteric dissociation of surface-OH groups[12]. Indeed, γ-MnO$_2$ is known to be a weak acid with a Zero Point Charge (ZPC) at pH 4.85. For pH values higher than 4.85, the surface OH groups are dissociated and give MnO$^-$ sites, saturated by cations present in majority in the bulk; they must be neutralized in basic conditions. On the contrary, for pH values less than ZPC, the surface OH groups attract protons, giving rise to Mn-OH^{2+}, which induces a surface state favourable to proton insertion. The disappearance, in acid pH, of the first insertion peak comes from the charge surface modification with pH conditions and is associated with the increase in e.m.f. observed for cells using acid protonic solid electrolytes.

This electrolyte was checked in the complete battery.

$$\text{PdH}_{0.5}/\text{Ormolyte}/\gamma\text{-MnO}_2-\text{Ormolyte–acetylene black}$$
$$(188 \text{ mg}/ \quad 250 \text{ mg} \quad / 500 \text{ mg} \quad + \quad 100 \text{ mg})$$

A slow external voltage sweep from OCV (1.23 V) down to 0.8 V makes a peak (of intensity 1.28 mA) appear at 1.18 V (e_p) (Fig. 37.6a). Compared with the one in Fig. 37.6b related to the behaviour of the cell

$$\text{Zn}/\text{HUP}/\gamma\text{-MnO}_2-\text{HUP–acetylene black}$$

and obtained with similar slow voltage sweep conditions, it can be noticed that both voltammograms exhibit the same shape: the first insertion peak observed in basic pH (Fig. 37.5b) has disappeared. This confirms that γ-MnO$_2$ is, in both cases, present in the same acidic surrounding.

These experiments single out the advantages of the polymer electrolyte versus the crystalline one: the intensity is five times higher, and the voltage difference between OCV and e_p is three times smaller. These results are good enough for a future active development of this new technique in our laboratory.

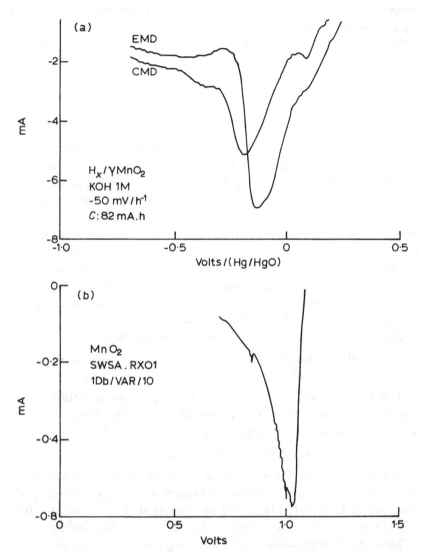

Fig. 37.5. (a) Voltammograms recorded during proton insertion from KOH 7N in CMD (Chemically Manganese Dioxide (Sedema WSA)) and EMD (Electrochemically Manganese Dioxide (Tekkosha)). (b) Voltamperogram recorded during proton insertion from a protonic polymeric conductor of the Ormolyte type, in MnO_2 (Sedema WSA) with $PdH_{0.5}$ as reference electrode and PdH_x as negative electrode.

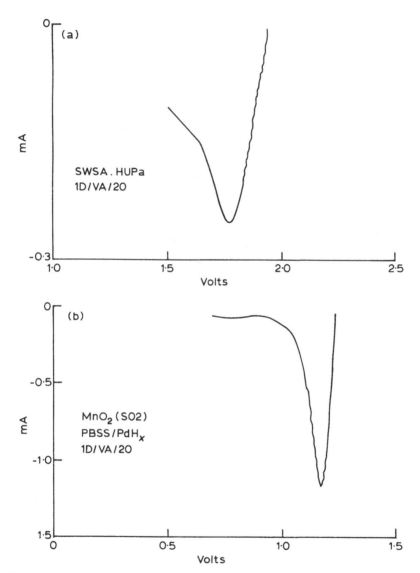

Fig. 37.6. (a) Voltamperogram recorded during proton insertion from protonic crystallized conductor (HUP: $H_3O^+UO_2PO_4$, $3H_2O$) in MnO_2 (Sedema WSA) with zinc as negative and comparison (or 'reference') electrode. (b) Voltamperogram recorded during proton insertion from protonic polymeric conductor of ormolyte type, in MnO_2 (Sedema WSA) with PdH_x as negative and comparison (or 'reference') electrode.

37.6   References

1. M. Pourbaix, *Atlas d'équilibres électrochimiques* (Gauthier-Villars, Paris (1963)) 290.
2. C. Poinsignon, A. Fitch and B. E. F. Fender, *Solid State Ionics-83*, M. Kleitz, B. Sapoval and Y. Chabre (eds) (North-Holland (1983)) 1049–54.
3. H. Kahil, E. J. L. Schouler, M. Forestier and J. Guitton, *Solid State Ionics-85*, Lawrence Berkeley Laboratory (ed) (North-Holland, 18–19 (1986)) 892–6.
4. H. Kahil, M. Forestier and J. Guitton, *Solid State Protonic Conductors III-84*, J. Goodenough, J. Jensen and A. Potier (eds) (Odense University Press (1985)) 84–99.
5. P. de Lamberterie, M. Forestier, J. Guitton, A. Rouault, R. Fruchart and D. Fruchart, *C.R. Acad. Sci. Paris* **300** (1985) sér. II, no 14, 663–6.
6. G. W. Mellors, European Patent no. 013120A2.
7. T. Takahashi, S. Tanase and O. Yamamoto, *J. Appl. Electrochem.* **10** (1980) 415–41.
8. K.-D. Kreuer, W. Weppner and A. Rabeneau, European Patent no. 070020A2.
9. J. Guitton, B. Dongui, R. Mosdale and M. Forestier, *Solid State Ionics-87*, W. Weppner and H. Schulz (eds) (North-Holland (1988)) 847–52.
10. C. Poinsignon, *Mat. Scien. Eng.* B **3** (12) (1989) 31–7.
11. C. Poinsignon, Y. Chabre, J. Pannetier, M. Ripert, A. Denoyelle and J. Y. Sanchez, *Journées Européennes d'Etudes de la SEE:* 'Accumulateurs électrochimiques, évolution et techniques récentes' 21–22 Novembre 1989, Gif-sur-Yvette, France.
12. H. Tamura, T. Oda, M. Nagayama and R. Furuichi, *J. Electrochem. Soc.* **136** (1989) 2782–6.

38 Applications of proton conductors in electrochomic devices (Ecds)

ODILE BOHNKE

38.1 Introduction

Some mixed conductors, i.e. electronic and ionic, into which ions can be rapidly and reversibly inserted can undergo a colour change. This is, for example, the case of the hydrogen 'bronzes', mentioned earlier in this book. The colour change can be either from transparent to coloured or from one colour to another. This phenomenon, which can be produced electrochemically, is called electrochromism. It is broadly defined as the production of an absorption band in a display material caused by an applied electric field or current. Such a property is currently under intensive study because of its potential use for passive information display[1-4], glare-free rearview mirrors for automotives[5-6], solar control windows or 'smart' windows[7-10], thermal sensors[11] and projection systems if matrixable.

The electrochromic devices (ECDs) exhibit many attractive features and offer some superior display qualities compared with liquid crystal displays (LCDs) and light emitting diodes (LEDs). The use of ECDs might improve some of the disadvantageous properties of LCDs and LEDs. They typically show a good aesthetic appearance and a good colour contrast especially under high levels of ambient light where emissive displays lose contrast. They have a wide angle of view which is an advantage over most liquid crystal displays. They may exhibit a continuously variable intensity of coloration and possess memory in either the bleached or coloured state without power consumption. They need low driving voltage to operate. Moreover, large area devices can be fabricated as well as all solid-state displays. In this latter configuration and with some solid electrolytes, the lack of a threshold voltage and the presence of a back electromotive force (e.m.f.) are no longer undesirable properties for matrix-addressed displays. Despite all these attractive properties, we should not forget that ECDs have a slower response time ($\simeq 100$ ms) then LCDs ($\simeq 1$ ms) and need higher power consumption to operate.

38.2   Structure of electrochromic devices

The basic device configuration typically consists of: (i) an electrochromic layer, or mixed conductor, deposited onto a transparent electronic conductor; this layer can be coloured or bleached by double injection or ejection of electrons and ions; (ii) an electrolyte, or purely ionic conductor, in contact with the above described layer; (iii) a counter electrode for storing the ions involved in the electrochromic reaction. Depending on the application, the electrolyte, the counter electrode and the whole cell may possess certain specific characteristics.

For display devices, a pigment has to be mixed into the electrolyte to provide a diffuse background[12] (Fig. 38.1a). Moreover, very fast switching properties and a lifetime of 10^7 cycles are needed. The counter electrode needs only to be a material which is electrochemically reversible to the ion involved in the electrochromic reaction without having any particular optical properties.

For a window application, a transmissive device is necessary. Then both the electrolyte and the counter electrode have to be transparent. This latter requirement severely restricts the number of suitable materials. Moreover, since an application is directed to improve the energy efficiency of buildings and vehicles, it is highly desirable to have spectrally selective reflection modulation[7] as shown in Fig. 38.1b: reflection of infrared radiation (IR) in both coloured and bleached states to save energy, transmission of the high energy solar radiation (HES) and the near infrared radiation (NIR) in the bleached state and reflection of part of these HES and NIR radiations in the coloured state to provide comfort and reduce lighting energy loads. For this application, very fast switching is not necessary but intermediate coloration between transparent and opaque states is important as well as the operating temperature compared with display applications. It has to be noted that most of the actual electrochromic materials undergo absorption during coloration leading to production of heat. However, Goldner *et al.*[13,8] have demonstrated that polycrystalline films of WO_3 present a reflection modulation of NIR energy.

For a mirror application, a totally reflective device is necessary. The reflector may be either behind the counter electrode or included into the

Electrochromic devices

display as shown in Fig. 38.1c. In the former case, the system is a transmissive cell plus a reflector and optical transparency of both the electrolyte and the counter electrode is then required. In the latter case, the counter electrode has to be optically reflective. Fast switching is

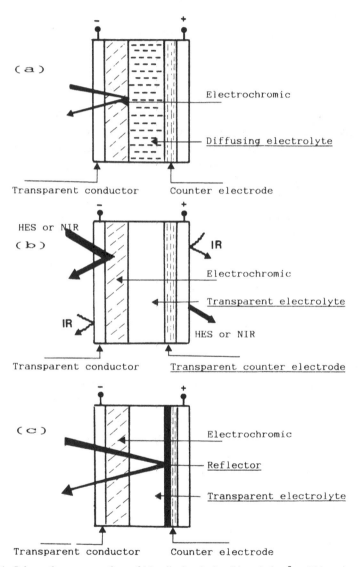

Fig. 38.1. Schematic representations of (a) a display device, (b) a window[7] and (c) a mirror.

not necessary. As for window applications, intermediate coloration is needed.

Finally, for thermal sensors, we need an electrolyte whose conductivity is strongly dependent on temperature[11].

This present study covers only electrochromic devices using proton conductors. Indeed, since the proton is the smallest ion, proton conductors have been extensively studied and used in electrochromic cells either as the solid electrolyte or as the electrochromic material or counter electrode.

38.3 *Mixed conductors as electrochromic materials*

Organic and inorganic materials, exhibiting both good ionic and electronic conductivity, may rapidly and reversibly undergo colour change during ion insertion. Their optical properties depend directly on the number of inserted ions. Moreover, they are characterized by an internal electromotive force (e.m.f.) which is related to the chemical potential of the inserted species. This e.m.f. has a major role in electrochromic processes. These materials, which are generally deposited in the form of thin films, may then be used as electrochromic layers. In this paper we consider only mixed conductors that have been used in hydrogen ion insertion cells, with emphasis on methods of preparation and electrical properties (electronic conductivity and proton diffusion coefficient).

38.3.1 *Organic materials*

The presence of a multicolour electrochromic effect in rare-earth diphthalocyanines was first reported in 1970 by Moskalev *et al.*[14]. Lutetium diphthalocyanine, $LuH(Pc)_2$, has been studied extensively at Rockwell International Corporation by Nicholson and co-workers[15–18]. This material may display different colours when polarized either anodically (red, orange) or cathodically (blue, violet) from its initial rest potential. The initially green complex film is obtained by vacuum evaporation. The anodic reaction occurs by the insertion of anions into the film and extraction of electrons rather than by loss of protons[15]. The cathodic reaction, leading to blue and violet products, occurs by the insertion of cations. When protons are present in the electrolyte, the reaction is the following[17].

$$LuH(Pc)_2(green) + nH^+ + ne^- \rightarrow LuH_{n+1}(Pc)_2(violet)$$

Electrochromic devices

The dark violet product appears to be a protonic conductor. It has a hydrogen-ion mobility of $8 \times 10^{-7} \, \text{cm}^2 \, \text{V}^{-1} \, \text{s}^{-1}$ and a bulk resistivity of $\simeq 1800 \, (\Omega \, \text{cm})$ at room temperature.

38.3.2 *Inorganic materials*

Some inorganic materials, generally transition metal oxides, exhibit electrochromism through a reversible proton insertion mechanism. Some of them, such as W, Mo, V, Nb and Ti oxides, are coloured when reduced (or when protons and electrons are inserted) while a few others, such as Ir, Ni and Co oxides, are coloured when oxidized (or when protons and electrons are extracted). Of these materials, tungsten trioxide (WO_3) and iridium oxide (Ir_2O_3) have attracted much attention for application with protonic electrolytes. Indeed, since acidic aqueous solutions have been mostly used as electrolytes, Mo and V oxides have been excluded because of their high solubility in these electrolytes[19-21]. Moreover, for Nb_2O_5[22] and TiO_2[23], it has been observed that coloration overlaps the onset of hydrogen evolution and for Ni and Co anodic oxides the coloration overlaps the onset of oxygen evolution[3]. These effects may then give rise to long-term lifetime problems.

The mechanism of coloration and bleaching in thin films of cathodic materials may be written as follows

$$MO_y(\text{transparent}) + xH^+ + xe^- \leftrightarrows H_xMO_y(\text{blue})$$

with $0 < x < 0.33$.

Although it is now well established that coloration and bleaching occur, in either crystalline or amorphous WO_3 thin films, by double injection of cations and electrons, it is not so obvious whether a single mechanism is applicable to the large variations in this oxide produced by different technologies. Indeed, depending on the method of preparation, the crystallinity of the oxide may vary from very fine grained (or amorphous) material to a completely ordered oxide, as shown in Table 38.1. Based on the optical absorption observed during insertion, several mechanisms have been proposed. For amorphous films, Faughnan *et al.*[24] proposed an intervalence charge transfer mechanism. The electron injected is trapped at a W^{6+} site to give a W^{5+} ion. The optical absorption is then due to a resonance transfer of the localized electron amongst the equivalent W atoms. The injected cation remains ionized in an interstitial site. Another model proposed by Schirmer *et al.*[25] suggests the formation of a small

Devices

Table 38.1. *Conductivity and proton diffusion coefficient into* WO_3 *and* MoO_3 *films prepared by different techniques*

Material	Technique	$\sigma\ (\Omega\,cm)^{-1}$ at 25 °C	$D_{H^+}\ (cm^2 . s^{-1})$ at 25 °C	Ref.
Amorphous WO_3	Thermal evaporation	10^{-6}	10^{-11}–10^{-9}	2, 39, 40–42
	Sputtering (d.c.)	10^{-3}		43
	Sputtering (rf)	10^{-7}–10^{-11}		49
	Anodic oxidation of W		5×10^{-8}	44–46
Crystalline WO_3	Polycrystalline evaporated layers	10^{-3}	6×10^{-12}–2×10^{-11}	40, 47, 50
	CVD	10^{-2}		30
Amorphous MoO_3	Sputtering (rf)	10^{-10}–10^{-12}		48
Crystalline MoO_3	CVD	10^{-2}		30

polaron. The electrons remain localized on the tungsten ions. However, a local structural distortion is brought about by the electron trapping and the W^{5+} formation. The optical transfer of the polaron also occurs from a W^{5+} site to a neighbouring W^{6+} site. This transfer is no longer a resonance process. On the other hand, for crystalline films, a free electron like behaviour has been observed[8, 25–27]. The injected electrons would become delocalized. This model, based on Drude theory, would explain the change of optical properties upon crystallization of the layers.

As well as optical properties, the method of thin film preparation also influences the electrical properties. As-deposited WO_3 films are n-type semiconductors. The room temperature conductivity usually lies in the range 10^{-2}–10^{-6} $(\Omega . cm)^{-1}$, as shown in Table 38.1. Moreover, it influences the degree of hydration and the porosity of the films[2]. These last parameters may strongly enhance the diffusion coefficient of protons within the host matrix (Table 38.1) leading to higher coloration and bleaching kinetics. The electrochemical coloration of WO_3 is accompanied by a large increase in electronic conductivity. Crandall *et al.*[28] reported for $H_{0.32}WO_3$ a room temperature conductivity of 6 $(\Omega.cm)^{-1}$. They suggested a variable-range hopping model for $x < 0.32$. At this composition, an insulator to metal transition is observed.

Electrochromism has also been observed in either amorphous[29] or crystalline[30] films of MoO_3. The mechanism of coloration is thought to be similar to that of WO_3 films. To improve electrochromism, mixed oxides WO_3/MoO_3 have been studied. The absorption band of these materials extends to higher energies, i.e. to a region of higher display efficiency[31,32].

As mentioned above, electrochromism may also occur during oxidation. For example, hydrated iridium oxide shows a transparent to blue-black transition. The proposed reaction for coloration is based on one of two mechanisms: (i) proton extraction from the film or (ii) hydroxide ion insertion. These films can be prepared by various methods: reactive sputtering of iridium metal to lead (SIROF)[33], cyclic anodization of iridium metal (AIROF)[34], and more recently, periodic reverse electrolysis of sulphatoiridate complex solution (PRIROF)[35]. McIntyre *et al.*[36] and Yoshino *et al.*[37] gave some evidence that, in aqueous solutions of low pH, coloration results from the double extraction of electrons and protons. However, coloration may also occur by injection of OH^-, F^- or CN^-[38]. In terms of switching time, coloration efficiency and temperature response, iridium oxide is comparable to or better than tungsten oxide[3]. Moreover, the association of WO_3 and Ir_2O_3 in an electrochromic cell seems to be very attractive, especially for transmissive systems. However, despite the number of studies which exist on these materials, a lot of problems remain before we can clearly explain the structural and electronic changes induced by the electron and ion injections into these oxides.

38.4   Proton conductors as electrolytes in ECD devices

In an ECD, the electrolyte is used as the reservoir of cations needed for injection into the electrochromic films for either coloration or bleaching. Such a device needs fast response, chemical and electrochemical stability, reversibility and memory properties. To achieve fast coloration kinetics with WO_3, the use of protonic electrolytes is very attractive since protons have the highest mobility in the electrochromic films compared with other cations. Their use would then lead to the fastest ECD performances. However, the presence of protons is not sufficient and a conductivity of the order of 10^{-2} $(\Omega.\text{cm})^{-1}$ would be required to obtain a reasonably good electrochromic response. Such a conductivity is not generally obtained with most protonic electrolytes, since at room temperature, it generally lies in the range $10^{-8}-10^{-9}$ $(\Omega.\text{cm})^{-1}$, too low for electrochromic applications. Such a requirement restricts considerably the choice of the electrolyte. Moreover, to obtain a memory effect of the coloration, it is necessary to use an electronic insulator; this is the case with ionic conductors which show very low electronic conductivity.

Most of the earlier studies on proton insertion in ECDs, have been devoted to acidic aqueous electrolytes. However, since most of the electrochromic materials are chemically unstable in acidic medium, commercially viable ECDs must use other electrolytes. Recently, the use of solid anhydrous proton electrolytes seems to be very promising. We present some of the materials which have been used in ECDs.

38.4.1 Hydrated dielectric layers

To avoid the presence of acids in the reservoir of protons, hydrated dielectric films have been used in ECDs. Materials such as MgF_2[51], Ta_2O_5[52], Cr_2O_3[53], LiF[54, 55], SiO_x[56], ZrO_2[57] have been reported. The water, contained in the dielectric layer, is dissociated by electrochemical reactions and provides the protons for coloration. Because of the high electrical resistance of these materials, high applied voltages are necessary. Moreover, the decomposition of water molecules leads to the formation of H_2 bubbles (at the cathode) and O_2 bubbles (at the anode), shortening considerably the life of the ECDs, although good electrochromic response times may be achieved. Ionic conductors are then more suitable for this application.

38.4.2 Hydrated crystalline electrolytes

Table 38.2 gives the conductivity of crystalline materials, all containing some water molecules, and used in ECDs. Phosphotungstic acid (PWA) has been used either as an electrochromic electrolyte able to undergo a colour change during oxidation[58] or as an electrolyte[59] or associated with WO_3[60]. Zirconium phosphate showed a good conductivity but a fast dissolution of WO_3 was observed similar to that found with WO_3 in pure water[60]. However, both compounds are susceptible to variations of the hydrated water content and react with WO_3 films. Compared to the above

Table 38.2. *Conductivity of some protonic electrolytes for ECDs*

	$\sigma\ (\Omega.cm)^{-1}$ 25 °C	Ref.
Hydrated crystalline electrolytes		
$H_3PO_4(WO_3)_{12}.29H_2O$ (PWA)	6×10^{-2}	60
$H_3PO_4(MoO_3)_{12}.29H_2O$	1.8×10^{-1}	61
$ZrO(H_2PO_4)_2.1.6H_2O$	5×10^{-4}	60
$ZrO(H_2PO_4)_2.7H_2O$	2×10^{-2}	60
$HUO_2PO_4.4H_2O$ (HUP)	4×10^{-3}	62
$Sn(HPO_4)_2.H_2O$	$\simeq 10^{-5}$	63
$Zr(HPO_4)_2.H_2O$	$8 \times 10^{-3}\text{--}3 \times 10^{-5}$	64
Hydrated polymer electrolytes		
NAFION ($6H_2O$/equiv)	2×10^{-3}	65, 74
PSSA ($6H_2O$/equiv)	10^{-2}	65
PESA ($6H_2O$/equiv)	10^{-1}	65
poly-AMPS ($6H_2O$/equiv)	2×10^{-2}	65, 73
p-TsOH (60% humidity)	$10^{-2}\text{--}10^{-3}$	66
Anhydrous polymer electrolytes		
$P(EO)_4 - NH_4HSO_4$	2×10^{-4}	67
$P(AA)_4 - NH_4HSO_4$	2.4×10^{-5}	67
$PEO - 0.5H_3PO_4$	3×10^{-5}	68
$PVP - 1.5H_3PO_4$	3×10^{-5}	68
$PVP - 1.75H_2SO_4$	10^{-4}	68
$PEI - 0.2H_2SO_4$	4×10^{-5}	69
$PEI - 0.2H_3PO_4$	5×10^{-6}	69
$PVA - H_3PO_4$	10^{-5}	70
$BPEI - 0.5H_2SO_4$	10^{-4}	71
$Paam - 1.2H_2SO_4$	10^{-2}	71

proton conductors, hydrogen uranyl phosphate (HUP) was insoluble[62]. Finally, tin phosphate and zirconium phosphate were also used but present a lower conductivity[63, 64].

38.4.3   Polymer electrolytes

The first attempt to use ion-containing polymers in ECDs was made by Randin in 1982[65]. He investigated polystyrene sulphonic acid (PSSA), polyethylene sulphonic acid (PESA), poly-2-acrylamide-2-methylpropane sulphonic acid (poly-AMPS) and polyperfluorosulphonic acid (Nafion). All these polymers contain sulphonic acid groups and water molecules. Their ionic conductivity generally increases with increasing water content. Since WO_3 was stable in contact with ion-containing polymers containing less than $6H_2O$ per equivalent, a trade-off had then to be found between WO_3 stability and polymer conductivity. Another solid proton conductor, composed of *p*-toluenesulphonic acid (*p*-TsOH) and urea has been used by Shizukmishi and co-workers[66]. The major problem with these electrolytes was the stability of tungsten oxide in contact with an acidic electrolyte containing water molecules. However, Randin showed that WO_3 was extremely stable with these electrolytes provided the water content was in the bound water region[65]. Up to six molecules of water per acid equivalent, these water molecules were bound. Out of these electrolytes, poly-AMPS gave the best results with regard to chemical and electrochemical stability. Moreover good electrical contact could be achieved because poly-AMPS is a water soluble polymer.

Since it seems clear that it is the combination of acid functions and free water in the electrolytes that leads to the chemical instability of WO_3, the use of anhydrous protonic polymers may be attractive for this kind of application. Recently, anhydrous polymers have been synthesized and studied. However, only a few studies are reported on the performances of such complete ECDs. Table 38.2 presents the values of conductivity, at room temperature, of some of these polymers. These materials are promising for an electrochromic cell with WO_3 but their conductivity has to be improved to be of the order of 10^{-2} $(\Omega.cm)^{-1}$. Either ammonium salts[67] or acids[68–72] have been added to polymers such as poly(ethylene oxide) (PEO), polyvinylpirolydone (PVP), poly(ethylene imine) (PEI), poly(vinylalcohol) (PVA), poly(acrylic acid) (PAA), branched poly(ethylene imine) (BPEI), poly(acrylamide) (Paam) to obtain anhydrous protonic conductors.

38.5   ECD performances

In this chapter, we present the best results obtained with these different materials in order to give an ECD system. Table 38.3 presents the performances of these displays and we will discuss the limitations observed.

The best results, with a dielectric layer, Ta_2O_5 as 'electrolyte', were

Electrochromic devices

Table 38.3.

Material	Response time	Efficiency $(cm^2\,C^{-1})$	Cycle	Remarks	Ref.
	ITO/WO$_3$/Proton cond./C.E.				
Ta$_2$O$_5$	Colour = 800 ms Bleach = 130 ms	360	10^6 2 Hz	Reflection mode at $\lambda = 633$ nm for $\Delta(OD) = 0.5$	52
PWA electrolyte – EC	25 ms	100–150	10^5 2 Hz	Reflection mode at $\lambda = 600$ nm for $\Delta V = 1$ V $\Delta(OD) = 0.3$	58
HUP	300 ms		5 × 10^5		62
Sn(HPO$_4$)$_2$.H$_2$O	1 s		10^6	Reflection mode $\Delta(OD) = 0.6$ $\Delta V = \pm 3$ V	63
poly(AMPS)	60–90 s		10^4	Transm. mode at $\lambda = 550$ nm $\Delta(OD) = 0.5$ $\Delta V = -0.5$ V $+0.6$ V	75
p-TsOH	400 ms	100	10^5	Reflection mode $\lambda = 633$ nm $\Delta(OD) = 0.4$ $\Delta V = 2.5$ V	66
PVP – H$_3$PO$_4$	50 s at 30 °C			Reflection mode $\lambda = 633$ nm $\Delta(OD) = 0.1$ $V = -0.5$ V/NHE	76
	ITO/AIROF/NAFION/AIROF				
Nafion	2 s	34	10^6	Reflection mode $\lambda = 633$ nm $\Delta(OD) = 0.33$ $\Delta V = 1.5$ V	74

obtained with the following configuration[52].

$$ITO/WO_3/Ta_2O_5/Ir_2O_3/ITO$$

Generally, in this kind of system with a dielectric layer, it was found that the electrochemical reaction at the interface between electrolyte layer and electrode causes gas to be produced, leading to rapid degradation of the device.

In the case of cells with crystalline hydrated electrolytes, HUP was the most studied. The structure was the following[62].

$$ITO/H_xWO_3/HUP/H_yWO_3/ITO$$

Good performances have been obtained but in order to provide a source of hydrogen, a voltage in excess of 2 V was required for electrolysis, leading to gas evolution.

PWA was used as an electrochromic electrolyte. Pellets were pressed between two electronic conductors[58].

$$SnO_2/PWA/Graphite$$

The shelf-life of these systems was unsatisfactory and a short memory effect due to residual oxygen was observed.

It was also used as an electrolyte in Mohapatra's system[60].

$$ITO/H_xWO_3/PWA/WO_3/ITO$$

Generally, in all these hydrated materials, where water content is fixed, there is a reaction with the WO_3 film which leads to the degradation of the cell.

ECDs with hydrated polymer electrolytes were investigated. In these materials water content may be changed to lead to chemical stability towards WO_3. Good conductivity has been obtained; poly(AMPS), Nafion and p-TsOH have been used. The following cells have been studied[66, 73].

$$ITO/WO_3/poly(AMPS)/poly(N-benzylaniline)$$

This cell contains an anodic electrochromic material associated with WO_3.

$$ITO/WO_3/p\text{-}TsOH/metal$$

Both of these systems show a degradation of WO_3 due to the residual water in poly(AMPS), as reported by Dao et al.[73]. The ambient humidity generally affects the performances of cells as well as the cell resistance[66]. Since the anodic iridium oxide film (AIROF) does not degrade in the presence of water the following cell has been studied[74].

AIROF/Nafion/AIROF

It seems clear that, when the source of protons is made of water, contained either in a dielectric layer or in a crystalline electrolyte, hydrolysis is necessary to lead to proton formation and it is then difficult to avoid gas evolution at the interfaces of the cell. When protons are provided through acid functions, WO_3 is no longer stable if the electrolyte contains free water. Then, the use of acid anhydrous electrolytes seems to be very attractive. We report in Table 38.3 some results obtained in ECDs, but very little information is yet available on the characteristics of these cells. The structure of the cell is the following

ITO/WO_3/polymer/ITO or metal

The availability of an electrolyte which is plastic seems to offer very interesting possibilities for thin ECD devices, although the room temperature conductivity of most of these polymers has to be improved to be used in ECDs.

38.6   Conclusions

The use of proton conductors in ECDs is very attractive since the proton is the smallest ion and can rapidly diffuse into WO_3. However, the attempts made with hydrated materials lead either to gas evolution at the electrochromic/electrolyte interface or to the degradation of the electrochromic layer leading to a poor lifetime of the cell. The discovery of polymeric materials, free from water, has opened a new area of research into ECDs with proton conductors. It has now been ascertained that a variety of polymer electrolytes may be obtained. However, the polymer complexes have generally a conductivity in the range 10^{-6}–10^{-5} $(\Omega.cm)^{-1}$, at room temperature. Some studies, reported in the literature, show that higher conductivity can be achieved (i.e. 10^{-2} for Paam–$1.2H_2SO_4$). Information on ECDs using such polymers should shortly become available.

Devices

38.7 References

1. B. W. Faughnan and R. S. Crandall, *Display Devices, Topics in Applied Physics*, Vol. 40, J. J. Pankove (ed) (1980) 181–211.
2. W. C. Dautremont-Smith, *Displays* 4 (1982) 3–22; 67–80.
3. S. A. Agnihotry, K. K. Saini and S. Chandra, *Indian J. Pure Appl. Phys.* 24 (1986) 19–40.
4. B. Scrosati, *Phil. Mag.* B 59 (1989) 151–60.
5. F. G. K. Baucke, *Schott Information* 1 (1983) 11–17.
6. F. G. K. Baucke, K. Bange and T. Gambke, *Displays* (1988) 179–87.
7. C. M. Lampert, *Solar Energy Materials* 11 (1984) 1–27.
8. R. B. Goldner, T. E. Haas, G. Seward, K. K. Wong, P. Norton, G. Foley, G. Berera, G. Wei, S. Schulz and R. Chapman, *Solid State Ionics* 28/30 (1988) 1715–21.
9. A. M. Anderson, A. Talledo and C. G. Granqvist, in *Proc. 176th Meeting of Electrochem. Soc.* 90–2, 261–73 (1989) Hollywood.
10. P. Baudry and D. Deroo, in *Proc. 176th Meeting of Electrochem. Soc.* 90–2, 274–87 (1989) Hollywood.
11. O. Bohnke and C. Bohnke, in *Proc. 176th Meeting of Electrochem. Soc.* 90–2, 312–21 (1989) Hollywood.
12. E. Ando, K. Kawakami, K. Matsuhiro and Y. Masuda, *Displays* (1985) 3–10.
13. R. B. Goldner, R. L. Chapman, G. Foley, E. L. Goldner, T. E. Haas, P. Norton, G. Seward and K. K. Wong, *Solar Energy Mater.* 14 (1986) 195–203.
14. P. N. Moskalev and I. S. Kirin, *Opt. i Spektrosk.* 29 (1970) 414–19.
15. M. M. Nicholson and F. A. Pizarello, *J. Electrochem. Soc.* 126 (1979) 1490–5.
16. M. M. Nicholson and F. A. Pizarello, *J. Electrochem. Soc.* 127 (1980) 821–7.
17. M. M. Nicholson and F. A. Pizarello, *J. Electrochem. Soc.* 128 (1981) 1740–3.
18. A. F. Samuels and N. U. Pujare, *J. Electrochem. Soc.* 133 (1986) 1065–6.
19. S. K. Deb and J. A. Chopoorian, *J. Appl. Phys.* 37 (1966) 4818–21.
20. T. C. Arnoldussen, *J. Electrochem. Soc.* 123 (1976) 527–31.
21. R. J. Colton, Ph.D Thesis (1976) U. Pittsburgh.
22. B. Reichman and A. J. Bard, *J. Electrochem. Soc.* 128 (1981) 344–6.
23. C. K. Dyer and J. S. Leach, *J. Electrochem. Soc.* 125 (1978) 23–9.
24. B. W. Faughnan, R. S. Crandall and P. M. Heyman, *RCA Rev.* 36 (1975) 177–97.
25. O. F. Schirmer, V. Wittmer, G. Baur and G. Brandt, *J. Electrochem. Soc.* 124 (1977) 749–53.
26. S. F. Cogan, T. D. Plante, M. A. Parker and R. D. Rauh, *Solar Energy Mater.* 14 (1986) 185–9.
27. J. S. E. M. Svensson and C. G. Granqvist, *Appl. Phys. Lett.* 45 (1984) 828–30.

28. R. S. Crandall and B. W. Faughnan, *Phys. Rev. Lett.* **39** (1977) 232–5.
29. R. J. Colton, A. M. Guzman and J. W. Rabalais, *J. Appl. Phys.* **49** (1978) 409–15.
30. A. Donnadieu, in *Large-area Chromogenics: Materials and Devices for Transmittance Control*, C. M. Lampert and C. G. Granqvist (eds) SPIE Inst. Series, vol. **154** (1990).
31. B. W. Faughnan and R. S. Crandall, *Appl. Phys. Lett.* **31** (1977) 834–7.
32. M. Kitao, S. Yamada, Y. Hiruta, N. Suzuki and K. Urabe, *Appl. Surf. Sci.* **33/34** (1988) 812–17.
33. K. S. Kang and J. L. Shay, *J. Electrochem. Soc.* **130** (1983) 766–9.
34. S. Gottesfeld and J. D. E. McIntyre, *J. Electrochem. Soc.* **126** (1979) 742–6.
35. T. Yoshino, N. Baba and K. Arai, *Jpn J. Appl. Phys.* **26** (1987) 1547–9.
36. J. D. E. McIntyre, W. F. Peck and S. Nakahara, *J. Electrochem. Soc.* **127** (1980) 1264–8.
37. T. Yoshino, N. Baba, H. Masuda and K. Arai, in *Proc. 176th Meeting of Electrochem. Soc.* **90**–2, 89–98 (1989) Hollywood.
38. G. Beni, C. E. Rice and J. L. Shay, *J. Electrochem. Soc.* **127** (1980) 1342–8.
39. B. W. Faughnan, R. S. Crandall and M. A. Lampert, *Appl. Phys. Lett.* **27** (1975) 275–80.
40. J. P. Randin and R. Viennet, *J. Electrochem. Soc.* **129** (1982) 2349–54.
41. C. Bohnke, Thesis Doctorat d'Etat (1986) Besançon.
42. O. Bohnke, Thesis Doctorat d'Etat (1984) Besançon.
43. W. C. Dautremont-Smith, M. Green and K. S. Kang, *Electrochim. Acta* **22** (1977) 751–9.
44. P. Falaras, Thesis U. Paris VI (1986) Paris.
45. M. Rezrazi, Thesis U. Franche-Comté (1987) Besançon.
46. B. Reichman, A. J. Bard and D. Laser, *J. Electrochem. Soc.* **127** (1980) 647–54.
47. M. L. Hitchman, *Thin Solid Films* **61** (1979) 341–3.
48. N. Miyata and S. Akiyoshi, *J. Appl. Phys.* **58** (1985) 1651–5.
49. K. Miyake, H. Kaneko, N. Suedomi and S. Nishimoto, *J. Appl. Phys.* **54** (1983) 5256–61.
50. K. Miyake, K. Kaneko, M. Sano and N. Suedomi, *J. Appl. Phys.* **55** (1984) 2747–53.
51. A. Deneuville, P. Gérard and R. Billat, *Thin Solid Films* **70** (1980) 203–23.
52. M. Watanabe, Y. Koike, T. Yoshimura and K. Kiyota, *Proc. Japan Display* (1983) P2.19, 372–5.
53. E. Inoue, K. Kawaziri and A. Iwaza, *Jpn J. Appl. Phys.* **16** (1977) 2065–6.
54. P. V. Ashrit, F. E. Girouard, V. V. Truong and G. Bader, *Proc. SPIE* **562** (1985) 53–60.
55. N. Egashira and H. Kokado, *Jpn J. Appl. Phys.* **25** (1986) L462–4.
56. R. Hurditch, A. E. Hughes and P. Lloyd, in *19th Electronic Materials Conf.* (1977) Cornell.
57. Y. Sato and I. Shimizu, *Proc. Japan Display* (1983) P2.20, 376–8.
58. B. Tell, *J. Electrochem. Soc.* **127** (1980) 2451–4.
59. S. P. Maheswari and M. A. Habib, *Solar Energy Mater.* **18** (1988) 75–82.

60. S. K. Mohapatra, G. D. Boyd, F. G. Storz, S. Wagner and F. Wuld, *J. Electrochem. Soc.* **126** (1979) 805–8.
61. O. Nakamura, T. Kodama, I. Ogino and Y. Miyake, *Chem. Lett.* (1979) 17–18.
62. A. T. Howe, S. H. Sheffield, P. E. Childs and M. G. Schilton, *Thin Solid Films* **67** (1980) 365–70.
63. K. Kuwabara, S. Ichikawa and K. Sugiyama, *J. Electrochem. Soc.* **135** (1988) 2432–6.
64. G. Alberti, M. Casciola, U. Costantino, G. Levi and G. Ricciardi, *J. Inorg. Nucl. Chem.* **40** (1978) 533–7.
65. J. P. Randin, *J. Electrochem. Soc.* **129** (1982) 1215–20.
66. M. Shizukmishi, E. Kaga, I. Shimizu, H. Kokado and E. Inoue, *Jpn J. Appl. Phys.* **20** (1981) 581–6.
67. M. F. Daniel, B. Desbat and J. C. Lassègues, *Solid State Ionics* **28/30** (1988) 632–6.
68. F. Defendini, Thesis (1987) U. Grenoble.
69. M. F. Daniel, B. Desbat, F. Cruege, O. Trinquet and J. C. Lassègues, *Solid State Ionics* **28/30** (1988) 637–41.
70. A. Polak, S. Petty-Weeks and A. J. Beuhler, *Chem. Eng. News* (1985) 28–32.
71. J. C. Lassègues, B. Desbat, O. Trinquet, F. Cruège and C. Poinsignon, *Proc. Exeter (1988) Solid State Ionics* **35** (1989) 17–25.
72. *Polymer Electrolyte Reviews – I*, J. R. McCallum and C. A. Vincent (eds) (Elsevier Appl. Sci. (1987)).
73. My T. Nguyen and L. H. Dao, *J. Electrochem. Soc.* **136**(7) (1989) 2131–2.
74. W. C. Dautremont-Smith, G. Beni, L. M. Schiavone and J. L. Shay, *Appl. Phys. Lett.* **35**(7) (1979) 565–7.
75. C. S. Harris and C. B. Greenberg, *SAE Technical Paper Series* (1989) Wichita, 1–4.
76. D. Pedone, Thesis (1987) Grenoble.

39 Supercapacitors and interfacial charge accumulation devices

PHILIPPE COLOMBAN AND MAI PHAM-THI

39.1 Introduction

The energy consumption of electronic devices has decreased gradually with progress in integrated circuit technology. The stable working voltage required for power supplies is today considerably lower than in the first transistor devices. Furthermore, the development of amorphous silicon films makes possible a new design for many portable systems. The demand for miniaturized high energy density and low leakage current capacitors as, for example, a stand-by power source for RAM devices, for actuators or sensors as well the substitution of batteries by long-life power supplies offers new opportunities (pace-makers, watches where the supercapacitor is associated with a balance-wheel...) for capacitors using the electric double layer developed at a highly polarizable, blocking electrode/electrolyte interface.

The first attempts to achieve such a device were made in the 1970s. The cell had a non-symmetric $Ag/RbAg_4I_5/C$ structure with a double-layer capacitance at the carbon electrode in the range of $10–40\ \mu F\ cm^{-1}$ interface area[1]. However, due to the redox reaction, the working voltage was too low (<0.7 V) for the electronic devices of the time. Furthermore, the reversibility of the Ag electrode was poor and it was difficult to use fully the surface area of the ultrafine carbon. More recently, double-layer capacitors using acidic solution (H_2SO_4) as liquid electrolyte were developed by NEC (Nippon Electric Co. Ltd)[2]. A liquid electrolyte allows most of the surface area of the carbon electrode to be used. The electrical characteristics of the devices can be classified in relation to the properties of each material as follows.

> The working voltage is fixed by the voltage gap between anodic and cathodic reactions (typically 0.5–3 V), and is thus also a function of the device structure (symmetric or non-symmetric).

Parallel connection of each elementary cell allows the resulting voltage to increase, but its resistance is also increased.

The autodischarge of the device is related to the electrical properties of the electrolyte. Solid electrolytes exhibit only one type of ionic carrier and have a very low electronic conductivity. They thus give a lower leakage current and have a larger voltage safeguard than liquid electrolytes in which other charge carriers are also present.

The series resistance is directly related to the conductivity of the electrolyte and the electrodes used.

The capacitance is related to the surface area of the blocking electrode – carbon black – and thus indirectly to the possibility of covering this surface with the electrolyte.

Protonic conductors have the possibility to be fabricated easily in water and thus to allow a good coverage of the high surface area carbon with a material of very low electronic conductivity. Furthermore, all-solid systems exhibit a specific advantage: the slow rate of electrochemical reaction in the solid state prevents degradation due to transient over-voltages, and the products of reaction remain in place at the electrode/ electrolyte interface which favours reversible reaction (self-recovery).

39.2 Device fabrication

The device studied by the Laboratoire Centre de Recherche THOMSON-CSF uses $H_3OUO_2PO_4 . 3H_2O$ (HUP) as solid electrolyte and various activated carbons (NORIT RBX or CORAX L6)[3,4]. Before use, the carbon powders are thermally treated at 400 °C in vacuum in order to remove impurities that are absorbed on the surface or arise from incomplete carbonation. These defects lower the electronic conductivity and complicate the redox reactions.

Fig. 39.1a shows a schematic representation of the HUP/C super-capacitor: a pure solid electrolyte membrane separates two composite electrodes composed of a mixture of highly polarizable electrode particles and electrolyte crystallites. Current collectors and plastic encapsulation complete the component. The manufacturing process involves (i) synthesis of electrolyte powder for the pure electrolyte membrane; (ii) synthesis of the intimate mixture of composite electrode parts; (iii) pressure performing of the membrane; (iv) assembly of the two composite electrodes and the inner electrolyte separator and pressure sintering; and (v) mounting of current collectors and plastic encapsulation.

The HUP powder is synthesized in the form of microcrystals (platelets of up to 10 μm side, 0.1–1 μm thick): a 2.3 M H_3PO_4 aqueous solution is mixed with a 2.3 M uranyl nitrate solution (at room temperature). After complete precipitation, the material is quickly washed with water, decanted, dried, annealed with a defined H_2O partial pressure ($\sim 60\%$) and carefully ground. In order to achieve a dense and translucent membrane, the powder must be quickly sintered. Composite electrode powder is synthesized in the same way, the carbon black being introduced into the uranyl nitrate solution. Depending on precipitation conditions and the nature of carbon blacks (porosity, particle size), different situations occur: micrographs show that the carbon black particles can encircle the HUP crystallites for one type of carbon powder (Corax L6) or alternatively that the HUP platelets are precipitated on and in the voids of the large Norit 'sponge'. Depending on the size and porosity of these carbon particles, the relative proportions of the powders are optimized to produce the lowest resistance, the highest interfacial area and thus the highest

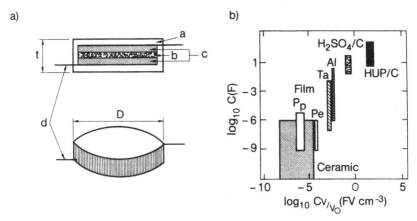

Fig. 39.1. (a) Schematic representation of a microsupercapacitor; copper current collector (a), HUP pure ionic membrane (b), composite HUP/C electrodes (c), and plastic encapsulation (d). Diameter (D) is 10 mm and thickness (t) is 2 mm. The capacity is 1 F. (b) Comparison of specific energy storage capability (capacitance x voltage per unit volume $-CV/V_0$) vs. capacitance (C). Pp, polypropylene film capacitors; Pe, polyethylene film capacitors; Ta, tantalum electrolytic capacitors; Al, aluminium electrolyte capacitors; H_2SO_4/C, liquid–solid supercapacitors; HUP/C, all-solid supercapacitor[3] (with permission, American Institute of Physics).

capacitance. The optimum corresponds to the double percolation of both electronic conduction and ionic conduction along the electronic conductor electrode and the electrolyte, respectively. The threshold values vary from 5 to 10% of carbon by weight in HUP, the smaller value occurring for the finest powder coated by HUP[4]. Electrodes and the electrolyte membrane are pressure sintered together at 30–40 °C and 2t cm^{-2} for times of 10–30 min in order to achieve a densification ratio higher than 90%. HUP solid electrolyte can be optimized by potassium doping which smooths the electrical characteristics in the −50 to 80 °C temperature range[5]. Finally, an 'electrical formation step', similar to that usually performed in (Al or Ta) electrolytic capacitor manufacture, is carried out for 48 h under 1 V a.c. voltage at 50 °C.

39.3 Electrical properties

Table 39.1 compares the d.c. performance of a HUP/C supercapacitor with that of other supercapacitors (all-solid and liquid–solid devices). With a 1 V working a.c. voltage, no degradation was found after 10^6 cycles. The leakage current of HUP/C supercapacitor was much lower than in other devices and high energy density was obtained (Fig. 39.1b).

This new device appears to combine the advantages of electrolytic capacitors (very long life time) and of lithium microbatteries (high energy densities). Furthermore, the component can be miniaturized. Use of an asymmetric device (one electrode is not polarizable, e.g. a metal layer) allows the capacity to increase, but the internal resistance is also increased and the life time is strongly reduced.

39.4 Giant accumulation layer at the Si–HUP interface

It was recently shown that giant electron accumulation layers can be created at the n-Si:Acetonitrile electrolyte interface[6]. Surface densities as high as 8×10^{13} electrons cm^{-2} have been obtained. For such densities only the lowest sub-band is significantly occupied, so that this system provides the unique attainment of a two-dimensional electron gas at room-temperature. However, the scattering of electrons in such layers is not satisfactorily understood. Such a device allows the manufacture of electrochemical field induced transistors[7]. The difficulty of varying the temperature over a wide range for a liquid electrolyte has led Benisty *et al.*[8] to attempt the fabrication of a similar system with non-liquid

Table 39.1. Comparison of d.c. characteristics of various double-layer capacitors (with permission, American Institute of Physics[3])

Cell	C/RbAg$_4$I$_5$/C	C/H$_2$SO$_4$/C	C/Propylene/C carbonate	C/Rb$_2$Cu$_8$I$_3$ Cl$_7$/Cu$_2$S	C, HUP/HUP/ C, HUP	C/PEO- LiClO$_4$/C
Working voltage (V)	0.66	1.66	1.6	1.6	1.6	5[a]
Capacitance (F)	5	1	10	1	1	1
Energy density (J cm^{-3})	1.3	0.5	20	1.3	5	3
Leakage current (A)	6.6	120	100	5	3	10
Operating temperature range (°C)	25–70	−25–70	−25–70	−10–100	−50–100	20

[a]Five unit cells are connected together.

electrolytes, that might be cooled down after the accumulation layer has been obtained. Furthermore, only all-solid systems can really be used. Thus Si–HUP interfaces have been prepared. Analysis of the capacitance of the interface at 300 K measured at 1000 Hz with a 20 mV modulation shows three domains: in the anodic region, the capacitance is limited by the n-Si depletion layer (it obeys the classical Schottky–Mott law); in the middle region, the capacitance peak originates from surface states. In the cathodic region, free carriers accumulate near the Si crystal surface. As the potential becomes more negative, the capacitance of this accumulation layer smoothly increases until too much continuous faradaic current flows through the interface; it is very likely that this current arises from reduction of the protonic species (at about 2 V vs Ag/Ag$^+$). The charge and hence the surface density are deduced by integrating the area under the capacitance curve, up to the potential where proton reduction takes place. The maximum observed density was 1.2×10^{13} electrons cm^{-2}. This illustrates the possibilities to design new devices using solid proton conductors.

39.5   References

1. J. E. Oxley, in *Proceedings of the 24th Power Sources Symposium*, edited by the P.S. Publication Committee (1970) pp. 20–3; and Gould Ionics Inc. Catalog, p. 48 Nov. 1971.
2. K. Sanada and M. Hosokawan, *NEC-Nippon Electric Co. Res. Dev.* **55** (1979) 21–30.
3. M. Pham-Thi, Ph. Adet, G. Velasco and Ph. Colomban, *Appl. Phys. Lett.* **48** (1986) 1348–50.
4. M. Pham-Thi, G. Velasco and Ph. Colomban, *J. Mater. Sci. Lett.* **5** (1986) 415–17.
5. G. Velasco, Ph. Adet, Ph. Colomban and M. Pham-Thi, French Patent 8509653, European Patent EP 208589-14 Janv. 1987.
6. A. Tardella and J. N. Chazalviel, *Phys. Rev.* B **32** (1985) 2439–48.
7. H. Benisty and J. N. Chazalviel, to be published.
8. H. Benisty, Ph. Colomban and J. N. Chazalviel, *Appl. Phys. Lett.* **87** (1987) 1121–3.
9. X. Andrieu and R. Vic, *Proc. of the Ministère de la Recherche et de la Technologie (MRT) Passif Component Meeting*, MRT-Contract no. 84-E-1415, Paris, 6–7 December 1988.

Index

Index

Index

Index

Index

Index

Index